Biological Membranes

A Molecular Perspective
from Computation and Experiment

Biological Membranes

A Molecular Perspective
from Computation and Experiment

Kenneth M. Merz, Jr. and Benoît Roux
Editors

Birkhäuser
Boston • Basel • Berlin

Kenneth M. Merz, Jr.
Department of Chemistry
Pennsylvania State University
152 Davey Laboratory
University Park, PA 16802
USA

Benoît Roux
Groupe de Recherche en Transport
  Membranaire (GRTM)
Départements de physique et de chimie
Université de Montréal
C.P. 6128, succ. Centre-Ville
Montréal, Québec
Canada H3C 3J7

Library of Congress Cataloging-In-Publication Data

Biological membranes : a molecular perspective from computation and
  experiment / Kenneth M. Merz, Jr., and Benoît Roux, editors.
    p. cm.
   Includes bibliographical references and index.
    ISBN 0-8176-3827-X (hardcover : acid free). -- ISBN 3-7643-3827-X
(hardcover : acid free)
    1. Membranes (Biology) 2. Lipid membranes. 3. Membrane lipids.
4. Membrane proteins. I. Merz, Kenneth M., 1959– . II. Roux,
Benoît, 1958– .
QH601.B485   1996
574.87′5--dc20

95-51668
CIP

Printed on acid-free paper
© 1996 Birkhäuser Boston

Birkhäuser

ISBN 0-8176-3827-X
ISBN 3-7643-3827-X
Cover design by David Gardner, Dorchester, MA.
Typeset by University Graphics, York, PA.
Printed and bound by Braun-Brumfield, Ann Arbor, MI.
Printed in the U.S.A.

9 8 7 6 5 4 3 2 1

Contents

Preface vii

List of Contributors xi

PART I: COMPUTATIONAL ISSUES REGARDING BIOMEMBRANE SIMULATION

1. Time Scales of Lipid Dynamics and Molecular Dynamics
   Richard W. Pastor and Scott E. Feller 3

2. An Empirical Potential Energy Function for Phospholipids:
   Criteria for Parameter Optimization and Applications
   Michael Schlenkrich, Jürgen Brickmann,
   Alexander D. MacKerell, Jr., and Martin Karplus 31

3. Statistical Mechanics and Monte Carlo Studies of
   Lipid Membranes
   H. Larry Scott 83

4. Strategic Issues in Molecular Dynamics Simulations of
   Membranes
   Eric Jakobsson, Shankar Subramaniam, and H. Larry Scott 105

PART II: EXPERIMENTAL PROBES OF BIOMEMBRANE STRUCTURE AND DYNAMICS

5. The Liquid-Crystallographic Structure of Fluid Lipid
   Bilayer Membranes
   Stephen H. White and Michael C. Wiener 127

6. Infrared Spectroscopic Determination of Conformational Disorder
   and Microphase Separation in Phospholipid Acyl Chains
   Richard Mendelsohn and Robert G. Snyder 145

7. Membrane Structure and Dynamics Studied with
   NMR Spectroscopy
      Michael F. Brown 175

PART III: SMALL MOLECULES AND PEPTIDES IN BIOMEMBRANES

8. Movement of Small Molecules in Lipid Bilayers:
   Molecular Dynamics Simulation Studies
      Terry R. Stouch and Donna Bassolino 255

9. Structural Basis and Energetics of Peptide Membrane Interactions
      Huey W. Huang 281

10. Computational Refinement Through Solid State NMR and
    Energy Constraints of a Membrane Bound Polypeptide
       Randal R. Ketchem, Benoît Roux, and Timothy A. Cross 299

11. Bilayer-Peptide Interactions
       K. V. Damodaran and Kenneth M. Merz, Jr. 323

PART IV: MEMBRANE PROTEINS

12. Peripheral Membrane Proteins
       Barbara A. Seaton and Mary F. Roberts 355

13. Thermodynamics of the Interaction of Proteins with
    Lipid Membranes
       Thomas Heimburg and Derek Marsh 405

14. Role of Lipid Organization and Dynamics for Membrane
    Functionality
       Ole G. Mouritsen and Paavo K. J. Kinnunen 463

15. Prediction of the Structure of an Integral Membrane Protein:
    The Light-Harvesting Complex II of *Rhodospirillum molischianum*
       Xiche Hu, Dong Xu, Kenneth Hamer, Klaus Schulten,
       Juergen Koepke, and Hartmut Michel 503

16. Monte Carlo Models of Spontaneous Insertion of Peptides into
    Lipid Membranes
       Jeffrey Skolnick and Mariusz Milik 535

17. Molecular Dynamics of Pf1 Coat Protein in a
    Phospholipid Bilayer
       Benoît Roux and Thomas B. Woolf 555

Index 589

Preface

The interface between a living cell and the surrounding world plays a critical role in numerous complex biological processes. Sperm/egg fusion, virus/cell fusion, exocytosis, endocytosis, and ion permeation are a few examples of processes involving membranes. In recent years, powerful tools such as X-ray crystallography, electron microscopy, nuclear magnetic resonance, and infra-red and Raman spectroscopy have been developed to characterize the structure and dynamics of biomembranes. Despite this progress, many of the factors responsible for the function of biomembranes are still not well understood. The membrane is a very complicated supramolecular liquid-crystalline structure that is largely composed of lipids, forming a bilayer, to which proteins and other biomolecules are anchored. Often, the lipid bilayer environment is pictured as a hydrophobic structureless slab providing a thermodynamic driving force to partition the amino acids of a membrane protein according to their solubility. However, much of the molecular complexity of the phospholipid bilayer environment is ignored in such a simplified view. It is likely that the atomic details of the polar headgroup region and the transition from the bulk water to the hydrophobic core of the membrane are important. An understanding of the factors responsible for the function of biomembranes thus requires a better characterization at the molecular level of how proteins interact with lipid molecules, of how lipids affect protein structure and of how lipid molecules might regulate protein function. Computer simulations of detailed atomic models based on realistic microscopic interactions represent a powerful approach to gain insight into the structure and dynamics of complex macromolecular systems such as a biomembrane. At the present time, even qualitative information gained from such computer simulations is valuable. Nevertheless, extension of current computational methodologies to simulate biomembrane systems still represents a major challenge. However, this field is just in its infancy, and it is likely that both experimental and theoretical tools will be needed to solve these problems. It is the goal of the present volume to provide a concise overview of computational and experimental advances in the understanding of lipid bilayers and protein/lipid interactions at the molecular level.

It can be reasonably expected that molecular simulations will play an increasingly important role in the future. While most trajectories to date are confined to the 1 ns time regime, this clearly will not be the case in the coming years. Just what kinds of time scales will we be able to simulate in the next five years? As an illustration, let us consider the following analysis. A typical simulation for a biological system now consists of ~15,000 atoms, which on modern parallel and vector supercomputers requires approximately 1 hour to generate 1 ps of a trajectory. Thus, one individual running MD simulations continuously can generate at most ~9 ns of trajectories in one year utilizing modern vector and parallel hardware resources available at many universities and supercomputer centers. Furthermore, it can be expected that algorithmic capabilities will be enhanced by a factor of 2 to 10-fold in the next year. This suggests that in one year we will go from a 9 ns capability to a 18 ns–90 ns capability just by improving our computational algorithms. In addition, it is generally agreed that there is a twofold speed increase every 18 months in computer technology. Thus, by the end of five years we can estimate a tenfold increase in computer power alone. Hence, the 9 ns/year we estimate now will increase to 90 ns based only on an increase in computer performance. Including an algorithmic improvement of 2 to 10-fold on top of this leads to an estimate of a 180 ns to 0.9 micros/year capability for one individual only. Obviously, this analysis is just an estimate and neglects many factors (e.g., increased system sizes, increased potential function complexity, etc.), but we think it is clear that molecular simulations on phospholipid bilayers will reach near microsecond capabilities in the next five years.

It is clear that theoretical methods will evolve to the point where we can address very long time scale issues. How about experimental techniques? Clearly, new approaches to solve experimental problems involving biomembranes will be developed in the coming years. Furthermore, we can also expect significant improvements in the techniques used to study biomembranes. For example, NMR has enjoyed tremendous growth over the years and this will continue in the coming years. The field strength of magnets have continued to grow, which provides higher resolution information that can be used to analyze biomembrane structure and dynamics. Moreover, new Solid State NMR techniques for oriented samples continue to be developed that will improve the ability to analyze biological membrane systems. Similarly, other experimental techniques like neutron scattering, X-ray, IR, CD, etc., will also continue to be improved upon in the coming years. Hence, the combination of improved computational and experimental techniques indicate that there is a bright future for the continued investigation of the structure, function, and dynamics of biological membranes at the molecular level.

This volume is separated into four sections. In section I, the basic theoretical and computational issues regarding biomembrane structure and dynamics are addressed. These issues range from basic statistical mechanics to force field development and evaluation. Thus, this section contains the necessary information required for anyone interested in attempting to model biomembranes using molecular dynamics or Monte Carlo methods. Section II then moves onto a series of chapters describing experimental probes that can be used to assess biomembrane structure. These include X-ray, IR and NMR techniques, and all are capable of providing microscopic or macroscopic insights that can be used to enhance our understanding of biomembrane structure and dynamics. Moreover, these experimental techniques generate information that can be used to assess and verify theoretical studies. Section III gives both a theoretical and experimental perspective on the interaction of peptides with biomembranes. Many peptides are membrane active and deserve study in their own right, but these systems can also serve as powerful models of protein/lipid interactions. Hence, by understanding these smaller (and hopefully less complicated) systems we will increase our understanding of the larger integral membrane class of proteins. Finally, Section IV gives a broad theoretical and experimental perspective of protein/lipid interactions. In this section the chapters give insights into the thermodynamics of protein lipid interactions as well as provide structural details of these systems.

The editors would like to thank the staff at Birkhäuser and the contributors for helping us produce an outstanding volume on recent advances towards understanding the structure, function and dynamics of biomembranes and lipid/protein interactions.

List of Contributors

Donna Bassolino, Bristol-Myers Squibb Pharmaceutical Research Institute, P.O. Box 4000, Princeton, NJ 08543-4000, USA

Jürgen Brickmann, Institut für Physikalische Chemie, Technische Hochschule Darmstadt, Petersenstraße 20, D-6100 Darmstadt, Federal Republic of Germany

Michael F. Brown, Department of Chemistry, University of Arizona, Tucson, AZ 85721, USA

Timothy A. Cross, Center for Interdisciplinary Magnetic Resonance at the National High Magnetic Field Laboratory, Institute of Molecular Biophysics and Department of Chemistry, Florida State University, 1800 E. Paul Dirac Dr., Tallahassee, Fl 32306-3016, USA

K.V. Damodaran, Department of Chemistry, Pennsylvania State University, 152 Davey Laboratory, University Park, PA 16802, USA

Scott E. Feller, Biophysics Laboratory, Center for Biologics Evaluation and Research, Food and Drug Administration, 1401 Rockville Pike, Rockville, MD, 20852-1448, USA

Kenneth Hamer, Theoretical Biophysics, Beckman Institute, University of Illinois at Urbana-Champaign, Urbana, IL 61801, USA

Thomas Heimburg, Abteilung Spektroskopie, Max-Planck-Institute für Biophysikalische Chemie, Postfach 2841, D-37018 Göttingen, Germany

Xiche Hu, Theoretical Biophysics, Beckman Institute, University of Illinois at Urbana-Champaign, Urbana, IL 61801, USA

Huey W. Huang, Department of Physics, Rice University, 6100 Main Street, Houston, TX 77005-1892, USA

Eric Jakobsson, National Center for Supercomputing Applications, 4039 Beckman Institute, University of Illinois, 405 North Mathews Ave, Urbana, IL 61801, USA

Martin Karplus, Department of Chemistry, Harvard University, 12 Oxford St, Cambridge, MA 02138, USA; and Laboratoire de Chimie Biophysique, Institut le Bel, Université Louis Pasteur, 67000 Strasbourg, France

Randal R. Ketchem, Center for Interdisciplinary Magnetic Resonance at the National High Magnetic Field Laboratory, Florida State University, 1800 E. Paul Dirac Dr., Tallahassee, Fl 32306-3016, USA

Paavo K.J. Kinnunen, Department of Medical Chemistry, Institute of Biomedicine, POB 8, FIN-00014, University of Helsinki, Finland

Juergen Koepke, Max-Plank-Institut für Biochemie, Abteilung Molekulare Membranbiologie, 6000 Frankfurt, Germany

Alexander D. MacKerell, Jr., Department of Pharmaceutical Sciences, School of Pharmacy, University of Maryland at Baltimore, 20 North Pine Street, Baltimore, MD 21201-1180, USA; and Department of Chemistry, Harvard University, 12 Oxford Street, Cambridge, MA 02138, USA

Derek Marsh, Abteilung Spektroskopie, Max-Planck-Institute für Biophysikalische Chemie, Postfach 2841, D-37018 Göttingen, Germany

Richard Mendelsohn, Department of Chemistry, Newark College, Rutgers University, 73 Warren Street, Newark, NJ 07102, USA

Kenneth M. Merz, Jr., Department of Chemistry, Pennsylvania State University, 152 Davey Laboratory, University Park, PA 16802, USA

Mariusz Milik, Department of Molecular Biology, Scripps Research Institute, 10666 N. Torrey Pines Road, MB1, La Jolla, CA 92037, USA

Hartmut Michel, Max-Plank-Institut für Biochemie, Abteilung Molekulare Membranbiologie, 6000 Frankfurt, Germany

Ole G. Mouritsen, Department of Physical Chemistry, Technical University of Denmark, Bldg 206, DK-2800 Lyngby, Denmark

Richard W. Pastor, Biophysics Laboratory, Center for Biologics Evaluation and Research, Food and Drug Administration, 1401 Rockville Pike, Rockville, MD, 20852-1448, USA

Mary F. Roberts, Merkert Chemistry Center, Boston College, Chestnut Hill, MA 02167, USA

Benoît Roux, Groupe de Recherche en Transport Membranaire (GRTM), Départements de physique et de chimie, Université de Montréal, C.P. 6128, succ. Centre-Ville, Montréal, Québec, Canada H3C 3J7

Michael Schlenkrich, Institut für Physikalische Chemie, Technische Hochschule Darmstadt, Petersenstraße 20, D-6100 Darmstadt, Federal Republic of Germany; and Silicon Graphics, Erlenstraesschen 65, CH-4125 Riehen, Switzerland; and Department of Chemistry, Harvard University, 12 Oxford St, Cambridge, MA 02138, USA

Klaus Schulten, Theoretical Biophysics, Beckman Institute, University of Illinois at Urbana-Champaign, Urbana, IL 61801, USA

H. Larry Scott, Department of Physics, Oklahoma State University, Stillwater, OK 74078-3072, USA

Barbara A. Seaton, Structural Biology Group, Department of Physiology, Boston University School of Medicine, 80 E. Concord St., Boston, MA 02118, USA

Jeffrey Skolnick, Department of Molecular Biology, Scripps Research Institute, 10666 N. Torrey Pines Road, MB1, La Jolla, CA 92037, USA

Robert G. Snyder, Department of Chemistry, University of California, Berkeley, CA 94720-1460, USA

Terry R. Stouch, Bristol-Myers Squibb Pharmaceutical Research Institute, Room H3812, P.O. Box 4000, Princeton, NJ 08543-4000, USA

Shankar Subramaniam, National Center for Supercomputing Applications, 4041 Beckman Institute, University of Illinois, 405 North Mathews Ave, Urbana, IL 61801, USA

Stephen H. White, Department of Physiology and Biophysics, School of Medicine, University of California at Irvine, Irvine, CA 92717-4560, USA

Michael C. Wiener, Department of Biochemistry and Biophysics, University of California at San Francisco, San Francisco, CA 94143-0448, USA

Thomas B. Woolf, Department of Physiology, Johns Hopkins University School of Medicine, 725 N. Wolfe Street, Baltimore, MD 21205, USA

Dong Xu, Theoretical Biophysics, Beckman Institute, University of Illinois at Urbana-Champaign, Urbana, IL 61801, USA

Part I

Computational Issues Regarding Biomembrane Simulation

This section describes the basic background material for molecular simulations as they relate to biomembrane containing systems. Biomembrane modeling, while in many respects very similar to modeling other biomolecules (e.g., proteins, DNA, etc.), has its own set of technical vagaries that must be considered prior to beginning a simulation. MD trajectories based on atomic models are typically limited to a few ns, while many membrane phenomena take place over much longer time scales. In Chapter 1, Rich Pastor and Scott Feller describe the time scales of lipid bilayer motions and how they affect the outcome of molecular dynamics (MD) simulations. This first chapter provides a critical discussion of the limitations of current computational models. In Chapter 2, Michael Schlenkrich, Jurgen Brickmann, Alex Mackerell, and Martin Karplus describe the parametrization of the all-atom CHARMM PARAM 22 molecular mechanical force field for phospholipid molecules. They stress the factors that need to be considered and indicate the limitations of this type of approach. Larry Scott, in Chapter 3, gives an overview of the basic statistical mechanics of lipid bilayers and also presents details regarding how Monte Carlo (MC) methods can be used to advantage when studying the configurations of lipid assemblies. MD and MC methods have strengths and weaknesses that can be used to advantage when trying to understand the dynamics of lipid containing phases. In the present volume, calculations based on MD are described in Chapters 1, 2, 3, 8, 11, 15, and 17; calculations based on MC are described in Chapters 2, 3, 10, 13, 14, and 16. Further methodological issues, such as application of current simulation methods to model a lipid bilayer under constant pressure are also addressed by Eric Jakobsson, Shankar Subramanian, and Larry Scott in Chapter 4. Their chapter also illustrates how a molecular simulation provides insight into the origin of the membrane surface potential.

1

Time Scales of Lipid Dynamics and Molecular Dynamics

RICHARD W. PASTOR AND SCOTT E. FELLER

Introduction

It is finally possible to carry out a molecular dynamics (MD) computer simulation of a protein or peptide in a lipid bilayer. Simulation programs with reasonable potential energy parameters are readily available, computer workstations are affordable, and plausible initial conditions can be constructed by combining the polypeptide with lipid configurations taken from simulations of pure lipid bilayers. Clearly, there are many questions to ask. Does the protein somehow order the nearby lipids or perturb the water structure at the headgroup/solution interface? If the membrane contains a mixture of lipids, do some selectively condense around the protein? What are the lateral diffusion constants and isomerization rates for the lipids and protein, and are they perturbed from the pure state? These sorts of effects might be important to the protein's function, or they might modulate the rate that substrates pass through the bilayer. They could change the interfacial tension, making it easier for the membrane to bend or even fuse with another. A peptide with potential drug applications might disrupt the bilayer, aggregate to form channels, or bind to a membrane protein.

As this chapter shows, only some of these questions can be answered at present with a conventional MD simulation, which, for a bilayer/protein system, can produce a trajectory of about 100 picoseconds (ps) to one nanosecond (ns). The following section provides a brief overview of the molecular dynamics method and some specifics pertaining to simulations of lipid bilayers, including constant pressure algorithms. Lipid motions are then discussed in order of their accessibility on the MD time scale: isomerization (reasonably good); rotational

Biological Membranes
K. Merz, Jr. and B. Roux, Editors
© Birkhäuser Boston 1996

relaxation (borderline); and lateral diffusion (just out of reach). We examine these motions using a combination of analytic theory and Brownian dynamics simulations of simple model systems, and molecular dynamics simulations of a dipalmitoylphosphatidylcholine (DPPC) lipid bilayer. The analyses, though restricted to pure lipid systems, should provide a sense of the motions on the ps to ns time scale, and assist readers in assessing simulations and other modeling of more complex membranes.

The Molecular Dynamics Method

Overview

For a system with constant particle number, volume, and energy (*NVE*, or micro-canonical ensemble), the molecular dynamics method simply involves numeri-cally solving Newton's equations for each particle (Allen and Tildesley, 1987). First the initial conditions (positions and velocities) and interparticle forces must be specified. The form of the potential function and the type of algorithm then dictate the time step (too large a time step will lead to an unacceptable drift in the total energy). Finally, the nature of the problem and the available computer resources determine the length of the simulation. As such, an MD simulation is similar to a numerical simulation of orbiting planets and moons. Aside from the interparticle forces, there are several important differences:

(1) The initial velocities of the atomic system are typically obtained from a Maxwell-Boltzmann distribution, and are scaled (or rerandomized) until kinetic and potential energies are in equipartition and the target temperature is reached. As a result, the statistical nature of the system is introduced early on.

(2) The assignment of initial positions for simple systems (such as atomic fluids) can proceed from a relatively ordered configuration at the appropriate density, while complex systems like membranes often require artistry. In any case, the system should be well equilibrated before the production phase of the simulation begins.

(3) Periodic boundary conditions are usually imposed in order to eliminate so-called wall effects and thereby better model a bulk fluid. Hence, although the terms *box* or *cell* are commonly used, the walls are completely porous: a particle leaving the cell through one side reenters through the opposite face. This technique can lead to difficulties when the number of particles is small (e.g., the particle interacts strongly with its own image); it also sets an upper limit on the wavelengths of undulations or collective modes that can be studied.

It is not obvious that molecular dynamics simulations should work: one could imagine that it would require a nearly macroscopic number of particles and very accurate potential energy functions to produce anything comparable with experiment. (To verify the assertion that fast computers and complicated theories do not guarantee good results, check the weather report or business section of today's newspaper). Nevertheless, by the 1960's MD simulations (and a cousin, Monte Carlo) were able to reproduce structural and dynamic properties of fluids using simple potentials and only several hundred particles. These early successes motivated the application to more complex systems, and computer simulation rapidly became an important complement to formal statistical mechanical theory. Simulations of biopolymers (Brooks et al, 1988; van Gunsteren et al, 1993) and lipid bilayers (Brasseur, 1990; Pastor, 1994b) followed, and, despite occasional grumbling by nonpracioners, molecular dynamics has become a central technique in biophysics. The remainder of this section describes some important technical details of bilayer simulations.

Dynamics at Constant Pressure

Unfortunately, it is difficult to simulate membranes using only constant volume algorithms. Both the height normal to the interface and the surface area must be specified correctly because the properties of surfactants are sensitive to both the normal pressure and the surface area per molecule (Cevc and Marsh, 1987; Small, 1986). While quantities such as interlamellar spacing and molecular areas have been determined for some lipids (Nagle, 1993; Rand and Parsegian, 1989), experiments give indirect guidance at best for assigning the appropriate simulation cell dimensions for bilayers made up of mixtures of lipids, or for ones containing peptides and proteins. Consequently, some allowance for volume and/or shape adjustments of the simulation cell is necessary in most cases.

Fluctuations in cell dimensions are most naturally accomplished by simulating in ensembles other than the microcononical. Because the appropriate variables are not always obvious, it is useful to start with the thermodynamics. We assume that the bilayer/water interface is planar with surface area A, and normal to the z direction. From the condition of hydrostatic stability, the pressure normal to the interface, P_n, equals the bulk pressure, P. Then, from the First Law of Thermodynamics:

$$dE = TdS - P_n dV + \gamma dA + \sum_{i=1}^{2} \mu_i dN_i, \qquad (1)$$

where E is the internal energy, T the temperature, S the entropy, V the volume, γ the interfacial tension, N_i the number of particles of liquid i, and μ_i its chemical potential. We see that there are four pairs of variables (μ, N), (T, S), (P_n, V) and (γ, A); when one is fixed, the other fluctuates. If the system is isolated (i.e., constant particle number and no heat exchange with the surroundings), then $\mu_1 dN_1 = \mu_2 dN_2 = TdS = 0$, and Equation (1) becomes

$$dE = -P_n dV + \gamma dA \qquad (2)$$

The consistency of the simulations at NVE is clear from Equation (2): the extensive variables volume and surface area are constant, while their conjugate intensive variables, normal pressure and surface tension, respectively, can be evaluated by averaging over the trajectory. Surface tensions of liquid/vapor interfaces, including monolayers, are reasonably calculated from simulations in the *NVE* ensemble. Ensembles describing interfaces in which various intensive thermodynamic variables are constant have recently been described (Zhang et al, 1995). Two particulary useful ones are:

(1) *NPAH*, where $H = E + P_n V$. The surface area is fixed, but because the applied normal pressure is constant, the box height (and hence the volume) fluctuates. This ensemble is useful for calculating surface tensions of liquid/liquid systems, including the "microscopic" surface tension of lipid bilayers.

(2) *NPγH*, where $H = E + P_n V - \gamma A$. Both the surface area and volume fluctuate, while the surface tension and normal pressure are constant. This ensemble is useful for expanding or contracting bilayers, or allowing the bilayer to relax under a constant applied surface tension (Feller et al, 1995b).

Developing equations of motions for particles outside the *NVE* ensemble is not particularily straightforward. A method for simulating isotropic systems at constant pressure was introduced by Andersen (1980) and then generalized for solids (where a pressure tensor is required) (Nose and Klein, 1983; Parrinello and Rahman, 1981). The Andersen, or extended system, approach (others are possible, as reviewed in Allen and Tildesley, 1987) is based on incorporating into the Lagrangian an additional degree of freedom, corresponding to the volume but commonly called a piston; the force on the piston is proportional to the difference of the instantaneous and reference pressures. The resulting equations of motion produce trajectories in which the system volume adjusts to and fluctuates about a value consistent with the reference pressure. The extended system equations for the *NPγH* ensemble under an applied normal pressure, P_{n0}, and surface tension, γ_0, are (Zhang et al, 1995):

$$\dot{x}_i = \frac{p_{xi}}{m_i} + \frac{\dot{h}_x}{h_x} x_i, \qquad \dot{y}_i = \frac{p_{yi}}{m_i} + \frac{\dot{h}_y}{h_y} y_i, \qquad \dot{z}_i = \frac{p_{zi}}{m_i} + \frac{\dot{h}_z}{h_z} z_i,$$

$$\dot{p}_{xi} = f_{xi} - \frac{\dot{h}_x}{h_x} p_{xi}, \qquad \dot{p}_{yi} = f_{yi} - \frac{\dot{h}_y}{h_y} p_{yi}, \qquad \dot{p}_{zi} = f_{zi} - \frac{\dot{h}_x}{h_x} p_{zi}, \qquad (3)$$

$$M_x \ddot{h}_x = h_y (\gamma_0 - \tilde{\gamma}_{xx})$$

$$M_y \ddot{h}_y = h_x (\gamma_0 - \tilde{\gamma}_{yy})$$

$$M_z \ddot{h}_z = h_x h_y (P_{zz} - P_{n0})$$

where x_i, p_{xi}, f_{xi} are the position, momentum and force in x for the ith particle, respectively; h_x is the length in x of the simulation cell, and M_x is the mass of this extended degree of freedom. Variables in y and z are defined similarily, and the symbols dot and double dot have their usual meanings as time derivatives. Finally,

$$\tilde{\gamma}_{xx} = h_z (P_{n0} - P_{xx})$$

$$\tilde{\gamma}_{yy} = h_z (P_{n0} - P_{yy}) \qquad (4)$$

where P_{xx}, P_{yy} and P_{zz} are the components of the pressure tensor along x, y, and z. Note that Newton's equations are recovered when the box lengths are constant. The equations of motion for the $NPAH$ ensemble are obtained by setting the area constant (i.e., $\dot{h}_x = \dot{h}_y = 0$), and the Nose-Klein (1983) equations are recovered when $\gamma_0 = 0$.

Dynamics at Constant Temperature

Temperature drifts of several degrees are a common problem in constant energy (or enthalpy) simulations of complex systems. For example, in the 170 ps bilayer simulation of Venable et al (1993), the temperature increased from 323 to 326 K as the potential dropped and the kinetic energy increased (to maintain a constant total energy). Temperature changes also occur when surface areas of interfaces are adjusted (Feller et al, 1995b). Ad hoc occasional rescaling of the particle velocities is the simplest way to correct temperature drift, but, because discontinuities in the trajectory are introduced, this technique is generally restricted to the equilibration phase of the simulation. A variety of isothermal methods based on rigorous statistical mechanical arguments are available (Allen and Tildesley, 1987), including an extended system version (Nose, 1984); constant temperature and pressure extended system algorithms have been combined in several different ways (Hoover, 1985; Martyna et al, 1994). Isothermal-isobaric conditions can also be imposed using the Langevin Piston (LP) method (Feller et al, 1995a). In this algorithm, random force and dissipative forces acting on the piston damp volume and pressure fluctuations and maintain the temperature. These features are useful for equilibrating small systems. The generally perceived drawback of stochastic methods is their basis in phenominological descriptions (e.g., the Langevin equation for the LP method).

Initial Conditions for Lipid Bilayers

Lipid bilayers take on essentially three phases which, in order of increasing transition temperature, are denoted: crystal or L_c; gel or L_β; and liquid crystalline, fluid, or L_α. The acyl chains in the L_c phase are well ordered and essentially all-trans. Crystal structures of a number of lipid bilayers are available (Pascher et al, 1992; Small, 1986), and are adequate initial conditions for simulations of this phase; such simulations are typically carried out to test potential energy functions (Williams and Stouch, 1993; see also the chapter by Schlenkrick et al in this volume). Gel phases are slightly more disordered in the hydrocarbon region than L_c phases, and can be characterized by the direction of chain tilt (which is an extremely sensitive function of hydration) (Smith et al, 1990; Katsaras, 1995). Even though bilayer dimensions and densities are available for some lipids (Tristran-Nagle et al, 1993), simulation of gel phases currently requires nontrivial model building for initial conditions. The results of gel phase simulations should provide useful tests of methodology, and will be important for understanding phase separation in biological membranes (Glaser, 1993). Most bilayer simulations have been concerned with the biologically active L_α phase. Here the chains are only weakly oriented with respect to the bilayer normal and, in DPPC for example, contain 3–4 gauche dihedral angles on average (Yellin and Levin, 1977). The bilayer surface tension is also a strong function of surface area (Feller et al, 1995b), which should be kept in mind when attempting to simulate in constant surface tension ensembles. Hence, developing initial conditions for the L_α is not as straightforward as for the L_c phase. There have been two basic approaches to the problem of initial conditions for MD simulations:

(1) Begin with ordered lipids (as might be modeled from a crystal structure of a related lipid and therefore easy to pack onto a lattice). The fluid phase is then developed during equilibration. Unless special techniques are employed (e.g., simulated annealing), this approach can require very long equilibration times (Stouch, 1993; Pastor and Venable, 1993).

(2) Assemble the bilayer with conformationally averaged lipids. For the DPPC simulation soon to be described, for example, individual lipid configurations were first obtained by sampling from a mean field (Hardy and Pastor, 1993), and then packed into a bilayer. Woolf and Roux (1994) used a similar approach in generating an initial condition for their simulations of a membrane containing gramicidin A (see the chapter by Roux and Woolf in this volume). Equilibration times appear to be shorter with this approach; its drawback is that arguments for the validity of mean field descriptions are still indirect.

Many groups, ours included, freely distribute coordinate sets of equilibrated bilayer systems. Although modifications and further equilibration might be necessary (e.g., if different lipids or parameter sets are used), beginning with a previously simulated bilayer may be faster than starting from scratch.

Simulating the Bilayer

A bilayer MD simulation is still a too big for a Mac or PC. The DPPC bilayer in excess water described in the next sections consists of 72 lipids and 2511 waters, for a total of almost 17,000 atoms (hydrogens were included on all molecules). Each picosecond of simulation took approximately 8 hours of processing time on a Hewlett Packard 9000/735 workstation or 15 hours on an IBM 3090 (time steps of 0.001 ps were used throughout). Hence, some patience and a long term perspective is still required. In order to investigate longer times described in this chapter, we took the liberty of combining three different simulations that were run sequentially, but under slightly different conditions. The first 170 ps segment was carried out at *NVE* (Venable et al, 1993) and provided initial conditions for a 75 ps simulation at *NPAH* (Feller et al, 1995b); both of these simulations were based on the CHARMM potential energy parameter set, PARM22b2, a developmental version of PARM22b4b (see the chapter by Schlenkrich et al in this volume). Upon switching to PARM22b4b, the trajectory was continued for an additional 175 ps at *NPAT* (Feller et al, 1995b.) Hence, the combined trajectory is 420 ps (including the approximately 75 ps of reequilibration following the change of parameter sets). The discontinuities in methods do

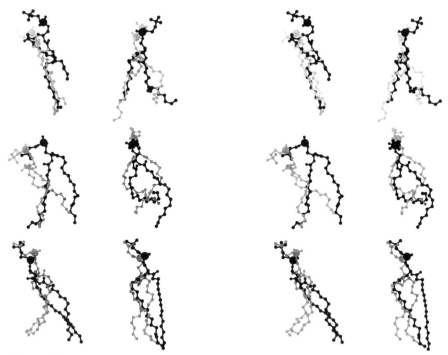

Figure 1. Stereo view of 6 lipids from the molecular dynamics simulation of a DPPC bilayer described in the text. Lipids at $t = 0$ and 420 ps are in grey and black, respectively.

not affect the qualitative analysis that is our present concern. To give a sense of the dynamics that lipids undergo on this time scale, Figure 1 overlays the 0 and 420 ps configurations for 6 lipids. We begin our analysis of the motion with the most rapid of the three primary components.

Isomerization

A Comparison of Heads and Tails

As will be described more fully in the chapter by Schlenkrick et al, the dihedral (or torsion) angles of hydrocarbon chains have three minima: trans (or t) at 180° (in IUPAC convention); gauche plus (g_+) at approximately −60°; and gauche minus (g_-) at approximately 60°. The trans minimum is 0.5–0.9 kcal/mol lower in energy than the gauche minima (Snyder, 1992), and the barrier between trans and gauche is approximately 3 kcal/mol.

Aside from the libration of the chains in their torsional minima, transitions between the t and g states are among the fastest molecular level motions in lipid bilayers, with jump rates of 10–20 ns^{-1} (Venable et al, 1993); transitions in the headgroup region, however, are decidedly slower (Stouch, 1993). Figure 2

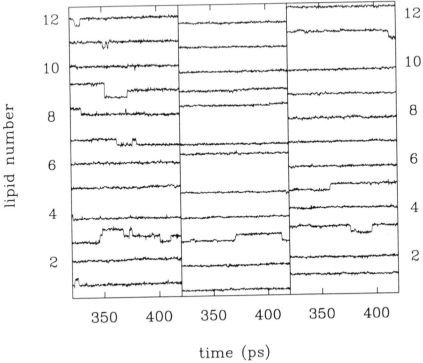

Figure 2. Time series of representative dihedral angles of 12 lipids from the 320–420 ps segment of the MD simulation: ϕ_7 of chain 1 (*left panel*); τ_1 of the glycerol group (*middle*); α_5 of the choline, (*right*). Denoting the carbonyl carbon of chain 1 as C1, ϕ_7 is the torsion angle of carbons C7-C8-C9-C10. Headgroup nomenclature follows the convention of Sundaralingam (1972): τ_1 is defined by the three glycerol carbons and the oxygen bonded to the phosphorus; α_5 is centered between the two carbon atoms on the choline group.

illustrates dihedral behavior for twelve lipids from the final 100 ps the $NPAT$ bilayer simulation of Feller et al (1995b). The left, middle and right columns show representative angles from the chain (ϕ_7), glycerol (τ_1) and choline (α_5) (nomenclature is described in the figure caption). Because the isomerizations are between well defined minima and there does not appear to be substantial recoiling or spinning of angles, we may proceed with an analysis based on simply counting the number of isomerizations as described in the caption of the Table; otherwise, a correlation function approach would be necessary (Chandler, 1977). Ignoring the distinction between $t \rightarrow g$ and $g \rightarrow t$, a total of 23 transitions can be counted for ϕ_7, although some angles had no transitions while one had 7. As can be calculated from the distribution for all 72 chains over the same 100 ps interval (Table 1), the average number of transitions per chain is 2.0. From this we can define a rate constant for ϕ_7 as the number of transitions divided by the total time, or $k_{\phi 7} = 20$ ns^{-1}. Following the same procedure, $k_{\tau 1} = 1.2$ ns^{-1} and

Table 1. Distribution of Transitions of Representative Dihedral Angles from Chain 1 (ϕ_7), the Glycerol (τ_1) and Choline (α_5) Groups[a]

m	ϕ_7	τ_1	α_5
0	22	64	53
1	11	7	11
2	15	1	7
3	8	0	1
4	9	0	0
5	2	0	0
6	3	0	0
7	1	0	0
8	0	0	0
9	0	0	0
10	1	0	0

[a] Occurrences in which dihedral angle passed over a barrier, entered the product minima and, without oscillating, returned to the original minima were not counted as transitions. Limits and guidelines for counting methods are discussed in Zhang and Pastor (1994).

$k_{\alpha 5} = 3.9$ ns^{-1}. It seems reasonable to surmise from this data that the chains isomerize much more rapidly than the heads.

Poisson Processes

To put some teeth in the above assertion, it is useful to analyze these data as Poisson jump processes (Hoel et al, 1972). The probability of m events (i.e., transitions) for a process with rate constant k in a trajectory of length T_{run} is then given by the Poisson distribution:

$$P(X = m) = \frac{\lambda^m}{m!} e^{-\lambda} \tag{5}$$

where $\lambda = kT_{run}$ is the estimated mean. So, for example, if $\lambda = 1$ for a particular torsion angle, Equation (5) predicts that there will be an equal number of chains with 1 and 0 transitions. Additionally, because the mean and variance of a Poisson distribution both equal λ, the standard deviation, $\sigma = \sqrt{\lambda} = 1$. Readers may find it interesting to try fitting the data in the Table to Equation (5), keeping in mind that the sample sizes are fairly small.

There are two more results from statistics that are important for our discussion. First, the sum of Poisson distributed random variables is also Poisson distributed. This implies that the *total* number of observed transitions should be also described by Equation (5), but with λ multiplied by the total number of chains. Hence, if we ran many more 100 ps trajectories with slightly different initial conditions (or broke up a very long single trajectory), we would expect to find Poisson distributions with average values of approximately 150, 10 and 30 for ϕ_7, τ_1 and α_5, respectively. The second result is that when λ is large, the Poisson distribution can be well approximated by a normal distribution with mean λ and variance λ; e.g., if there are 100 transitions observed for each angle, the standard deviation of the expected distribution should be around 10.

With these tools, we can compare the transition data in various ways. The first, which is usually a good thing to do anyway, is a quick estimate. If we assume that the distribution is normal, and mark out two standard deviations from the average of each to approximate the 95% confidence intervals (de Groot, 1975), we obtain the ranges 120–170 (ϕ_7), 3–15 (τ_1) and 17–39 (α_5). The normal approximation is a little drastic when $\lambda = 9$, so we can look up or calculate the same confidence intervals explicitly for the Poisson distribution with these values of λ (Crow and Gardner, 1959). They are: 122–170 (ϕ_7), 5–17 (τ_1) and 19–40 (α_5). Because there is no overlap in preceding ranges, we can state (with 95% confidence) that the difference in numbers of transitions is not an artifact of incomplete sampling. Conversely, if the ranges had overlapped one should be reluctant to conclude the rates are different. This first pass approach can be confirmed with formal hypothesis testing (de Groot, 1975). A discussion of this topic from the simulator's point of view and an application to dihedral transitions is contained in Brown et al (1995).

A Comparison of Different Parameter Sets

As noted at the end of the previous section, our simulations of DPPC were carried out with two different parameter sets. The primary revision involved increasing the potential energy difference between the trans and gauche wells for the hydrocarbon torsions. The effect of the change is seen clearly in Figure 3, in which the average number of gauche conformations per chain decreases from 4.2 to 3.2 over the course of 75 ps. Given that the number of transitions observed over the last 100 ps of the simulations was 110 with the original set, can we deduce that $k_{\phi 7}$ has been increased by switching parameters? Assuming Poisson statistics, the 95% confidence interval for $\lambda = 110$ is 89–132 (almost identical to the range obtained from assuming a normal distribution), and does overlap the new range (122–170). For this example, one would be required to look at additional data (e.g., the transition rates of ϕ_7 in chain 2 or the rates of neighboring dihedrals) before making strong conclusions. In a more detailed analysis it would also be appropriate to distinguish $t \rightarrow g$ and $g \rightarrow t$ rates, given that the conformational equilibrium between trans and gauche has been altered.

In closing this section, we stress that while simply counting the number of transitions is useful for giving a sense of the time scale of isomerization, transition rates in alkanes depend on the precise configuration of the chain (Skolnick and Helfand, 1980); i.e., there are many different rate constants in the system. To the extent that these are independent of each other, the overall transition behavior can be treated as the the sum of many individual Poisson processes, and

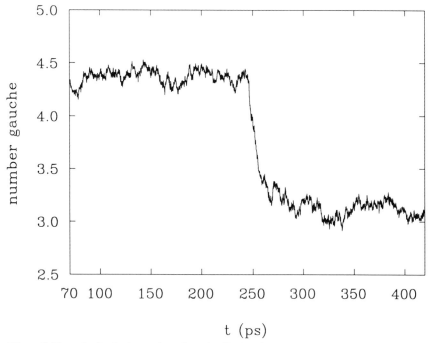

Figure 3. The reduction in the number of gauche dihedrals in the lipid chains following the change of parameter sets at $t = 245$ ps.

therefore still fit a model of a Possion process with a single parameter. However, some isomerization pathways in lipid chains have been proven to involve concerted mechanisms, implying nonindependence of neighboring dihedrals (Brown et al, 1995). This can lead to deviations from a simple Poisson model.

Rotation

Identifying It

The model of a lipid molecule "wobbling in a cone" is a familiar one in the membrane literature (Lipari and Szabo, 1980; Petersen and Chan, 1977). In this model the internal motions (primarily isomerization) rapidly average, and the lipid molecule, as an effective cylinder with one end fixed at the membrane/water interface, undergoes diffusive rigid body rotation in a pendulum-like fashion. However, a phospholipid is not rigid (like benzene) and does not even have a rigid core (like cholesterol) but, rather, is made up of flexible chains. Hence, we consider both the time scale and the physical picture for overall rotation of lipids.

When discussing rotational motion of molecules it is important to appreciate that our entry point is through the angular reorientation of particular molecular vectors (e.g., a CH bond or dipole moment) (Berne and Pecora, 1976). Because of its connection with numerous experiments, we work with the correlation function

$$C_2(t) \equiv \langle P_2(\hat{\mu}(0) \cdot \hat{\mu}(t)) \rangle = \left\langle \frac{3}{2}(\hat{\mu}(0) \cdot \hat{\mu}(t))^2 - \frac{1}{2} \right\rangle \tag{6}$$

where $\hat{\mu}(t)$ and $\hat{\mu}(0)$ are unit vectors separated by time t, the brackets signify an average over all initial times, and $P_2(x)$ is the second Legendre polynomial:

$$P_2(x) = \frac{3}{2}x^2 - \frac{1}{2} \tag{7}$$

Figure 4 plots $C_2(t)$ for the CH vectors of carbons 9 and 15 of both chains for the entire 420 ps spliced MD trajectory. A decay in the 10–50 ps range is clearly evident and, from results of the previous section, can be attributed to the isomerization of the chain dihedrals.

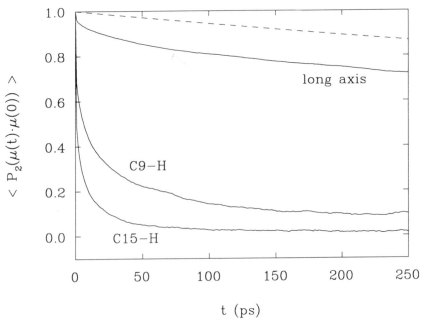

Figure 4. Reorientational correlation functions evaluated from the entire 420 ps MD trajectory for the CH bond vectors of carbons 9 and 15 of the chains, and the lipid long axis vector (defined by the eigenvector corresponding to the smallest eigenvalue of the instantaneous moment of inertia tensor). Dashed line is the approximate analytic result for diffusive wobbling of a rod in an ordering potential (Equation 10), with parameters determined by a fit to experimental relaxation data (see text).

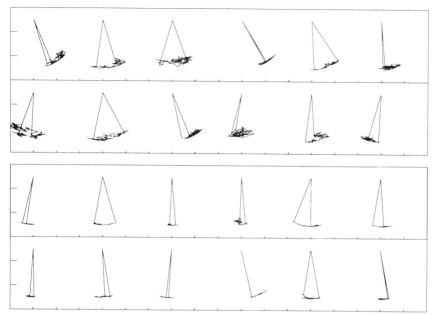

Figure 5. Representative 420 ps trajectories for the long axis vectors whose correlation functions are shown in Fig. 4. The top two panels show the projections onto the xz and yz planes, respectively, from the 420 ps molecular dynamics simulation (the same 6 lipids are shown in Fig. 1); the bottom panels are from Brownian dynamics simulations, in which the initial positions were obtained by randomly choosing points from a long trajectory. For each trajectory a line connects the origin to the initial and final positions of the vector.

Is there evidence of overall rotation? To answer this question we calculated the moment of inertia tensor with respect to carbon 2 of the glycerol group at 5 ps intervals along the trajectory. This is a reasonable way, though not the only way, to define a rotational frame; the eigenvector corresponding to the smallest eigenvalue then defines the long, or wobble, axis of the lipid. The top two panels of Figure 5 plot the motion of this eigenvector in the xz and yz planes for the 6 lipids shown in Figure 1. Over their 420 ps trajectories there is what might loosely be described as "half a swing". Fortunately, this statement can be quantified by calculating the reorientational correlation function. As shown in Figure 4, the dominant feature in $C_2(t)$ for the long axis vector is a slow decay with a relaxation time, $\tau_\perp = 1.3$ ns (obtained from a fit to the curve over 100–200 ps).

As a technical aside, it is generally dangerous to extract a relaxation time that is longer than the simulation time. This was accomplished in the present case by: (1) including all 72 lipids in the average; and (2) assuming that there is only one slow process. The statistical error in $\tau_\perp$ can then be estimated from the formula of Zwanzig and Ailawadi (1969) as $\sqrt{2\tau_\perp/T_{run}}$, or approximately 30%.

Modeling the Reorientational Correlation Function for Wobble

Now consider a very simple model of a lipid: a rigid rod, tethered at one end and diffusing with a rotational diffusion constant $D_\perp$ in the Maier-Saupe potential (de Gennes, 1974)

$$U(\theta) = -\phi P_2(\cos\theta) \tag{8}$$

where θ is the angle made by the rod and the z axis, and ϕ is the field strength. $U(\theta)$ has the effect of aligning the rod along z; it is continuous and for simulations is preferable to the cone model (where $U(\theta)$ is either 0 or ∞). We assume here that the rod does not flip. The order parameter, for the rod axis, $\langle P_2\rangle$, is then obtained by averaging over the potential in the usual way:

$$\langle P_2\rangle = \frac{\int_0^{\pi/2} P_2(\cos\theta)\exp(-U(\theta)/k_B T)\sin\theta\,d\theta}{\int_0^{\pi/2}\exp(-U(\theta)/k_B T)\sin\theta\,d\theta} \tag{9}$$

where k_B is Boltzmann's constant and T is the temperature.

There is no closed form solution for the reorientational correlation functions for rotational diffusion in a potential (Wang and Pecora, 1980). However, Szabo (1984) has developed approximate analytic expressions that compare very well with results obtained from simulations (Pastor and Venable, 1993). For the tethered rod model, $C_2(t)$ for the long axis reorientation is conveniently written in terms of averages over powers of $z = \cos\theta$:

$$
\begin{aligned}
C_2(t) = {} & \left[\frac{3}{2}\langle z^2\rangle - \frac{1}{2}\right]^2 \\
& + \frac{9}{4}\left[\langle z^4\rangle - \langle z^2\rangle^2\right]\times\exp\left\{-4D_\perp t\left[\frac{\langle z^2\rangle - \langle z^4\rangle}{\langle z^4\rangle - \langle z^2\rangle^2}\right]\right\} \\
& + 3\left[\langle z^2\rangle - \langle z^4\rangle\right]\times\exp\left\{-D_\perp t\left[\frac{1 - 3\langle z^2\rangle + 4\langle z^4\rangle}{\langle z^2\rangle - \langle z^4\rangle}\right]\right\} \\
& + \frac{3}{4}\left[1 - 2\langle z^2\rangle + \langle z^4\rangle\right]\times\exp\left\{-4D_\perp t\left[\frac{1 - \langle z^4\rangle}{1 - 2\langle z^2\rangle + \langle z^4\rangle}\right]\right\}
\end{aligned}
\tag{10}
$$

The correlation function is the sum of three exponentials and a plateau value (equal to $\langle P_2\rangle^2$). When $\phi = 0$ (no ordering), $\langle P_2\rangle = 0$, $\langle z^2\rangle = \frac{1}{3}$, $\langle z^4\rangle = \frac{1}{5}$, and we recover the result for isotropic reorientation (Berne and Pecora, 1976):

$$C_2(t) = \exp(-6D_\perp t) \tag{11}$$

To model DPPC, we set $D_\perp = 10^8$ s^{-1} and $\phi = 3k_B T$, leading to $\langle P_2\rangle = 0.605$. These values of $D_\perp$ and $\langle P_2\rangle$ are in the range obtained by Pastor et al (1988b) from a combined fit of data from NMR relaxation (Brown et al, 1983) and Brownian dynamics simulation (Pastor et al, 1988a). As readers can verify, with these parameters $C_2(t)$ is a slowly decaying function with a relaxation time of 1.1 ns. When $\langle P_2\rangle = 0$ (Equation 11), $\tau_\perp = 1.7$ ns; i.e., the presence of a potential decreases the relaxation time. As shown in Figure 4, the decay of the rigid rod and moment of inertia eigenvector extracted from the MD simulations are qualitatively very similar between 100 and 250 ps.

Brownian Dynamics Simulation of Wobble

The good agreement of the correlation functions from the diffusive rod model and MD simulation leads to the question, "What does the rod motion look like over 420 ps?" This is very easily answered using Brownian dynamics (BD) simulations (Ermak, 1975; Pastor, 1994a). The equations of motion for BD are obtained in the high friction limit of the Langevin equation, and hence BD is also referred to as diffusive dynamics.

For simplicity, we model the tethered rod as a single point particle with a translational diffusion constant, $D = 10^{-6}$ cm^2/s, attached to an immobile point by a spring of equilibrium length $r_0 = 10$ Å and frequency $\omega = 100$ ps^{-1}. Since the spring is stiff and hydrodynamic interaction is neglected, the translational friction constant of the particle, $\zeta = k_B T/D$, and the friction constant for rotation of the bond vector, $\zeta_\perp = k_B T/D_\perp$ are related by

$$\zeta_\perp = \zeta r_0^2 \qquad (12)$$

or $D_\perp = D/r_0^2 = 10^8$ s^{-1}. By reducing the complex system of 72 lipids in water to a single tethered particle diffusing in a field, a 420 ps trajectory can be generated in moments. For further discussion of Langevin and Brownian dynamics methods and the code for the present application, see Pastor (1994a).

The lower two panels of Figure 5 show six trajectories of the tethered particle (each started with a different random seed, and from a different initial position). While the long vector obtained from the MD simulation clearly moves more rapidly than the diffusive rod between 5 ps steps (as was already evident from the correlation functions shown in Figure 4), the features of the trajectories on the 420 ps timescale are remarkably similiar.

The reader should now take another look at Figure 1, with stereo glasses if possible. The simple model appears to have captured what are obviously very complicated motions. Whether the agreement of the correlation times is fortuitous is a question probably best resolved by longer MD simulations or more detailed stochastic simulations (De Loof et al, 1991). Nevertheless, let us press the model a little bit more and derive an order of magnitude estimate for the effective viscosity of the membrane. Assume first that the diffusion constant of the point particle is that of a Stokes' Law sphere with radius R in a medium of viscosity η,

$$D = \frac{k_B T}{6\pi \eta R} \qquad (13)$$

A plausible value of R can be estimated from the volume of the cylinder occupied by the lipid chains. If the surface area and length are now assumed to be 70 Å^2 and 15 Å, respectively (we neglect the headgroup), $R = 6.3$ Å and, consequently, $\eta = 3.8$ centipoise (cP). For reference, the viscosity of neat hexadecane at 323 K is 1.9 cP (Small, 1986). Hence, in agreement with our earlier estimates of the membrane viscosity based on fast motions (Venable et al, 1993), we conclude that the interior of the membrane is qualitatively very similar to a liquid alkane on the ns time scale.

Rotation on Longer Time Scales

The wobbling just described is implicitly a single molecule motion. Collective modes (e.g., bending and undulations of the entire bilayer) also rotationally relax the lipids in a membrane (Brown et al, 1983). However, from the functional form of the frequency dependent NMR measurments (Marqusee et al, 1984; Rommel et al, 1988), it appears that these primarily take place on a much longer time scale (up to seconds) than presently accessible by MD simulation. Nevertheless, the distinction between diffusive wobbling and concerted rotations of nearest neighbors (or small groups of lipids) is a fine one: molecules must get out of each other's way, yet don't necessarily do it in the same manner each time. Hence, determining from MD that a motion is described better as single molecule or as (locally) collective requires careful analysis.

Reorientation and the Deuterium Order Parameter

An important measure of molecular orientation in membranes is the deuterium order parameter, $S_{CD} = \langle P_2(\cos\theta)\rangle$, where θ is the angle between the CD vector and the bilayer normal (Seelig and Macdonald, 1987; Seelig and Seelig, 1980). It also has an important relationship to $C_2(t)$: because lipid reorientation is axially symmetric about the bilayer normal, $C_2(\infty)$ (or the plateau value) for the CH vector equals S_{CD}^2 (under the reasonable assumption that CD and CH bond vectors behave similarly). In a sufficiently long simulation, the value obtained for S_{CD} from a direct trajectory average and from the plateau region of the correlation function should be consistent. Lastly, assuming independence of motions, S_{CD} equals the product of the order parameters associated with internal motion, wobbling, and collective motions averaged within 10^{-5} s. The validity of this common approximation should eventually be determined from MD simulation. Regardless of this assumption, the presence of a 1 ns relaxation associated with wobble puts a longer time scale on the evaluation of S_{CD} from simulation than would be surmised from considering only gauche/trans isomerizations (Pastor and Venable, 1993).

Translation

The Trajectory

Translational motion is defined by center of mass (CM) displacement, and therefore requires no preprocessing to visualize. Because the motion of interest is in the xy plane (i.e., along the bilayer surface), we define $L(t)$, the lateral displacement at time, t, as

$$L(t) = [(x(t) - x(0))^2 + (y(t) - y(0))^2]^{\frac{1}{2}} \tag{14}$$

and the root mean squared lateral displacement, $L_{rms} = \langle L^2 \rangle^{\frac{1}{2}}$. Figure 6 plots the CM trajectories for all 72 lipids in the 420 ps simulation. The lipids have not gone very far on this time scale: at 420 ps, $L_{rms} = 2.7$ Å, although 4 of them traversed between 5 and 6 Å from their initial positions.

The lateral displacement correlation function $C_L(t) = \langle L^2(t) \rangle$ is just the mean squared displacement averaged over all possible time offsets. Figure 7 shows $C_L(t)$ averaged over the lipids.

Free Diffusion

For most practical purposes, molecular translation at long times is well modeled as Brownian, or diffusive, motion on a flat potential surface. In one dimension, this means that if many particles were placed at the origin at time $t = 0$, the distribution of the CM positions along the x axis at time t would be normal with a variance $2Dt$, where D is the diffusion constant. The graph of $\langle x^2 \rangle$ versus t (or displacement correlation function) would be linear with slope $2D$. For isotropic diffusion the CM distributions for each dimension are independent, and thus, for two dimensions

$$\langle L^2(t) \rangle = 4Dt \tag{15}$$

Self diffusion constants of lipids are on the order of 10^{-8} cm^2/s when measured with long time techniques such as photobleaching recovery (Cherry, 1979); to avoid notational clutter, units of cm^2/s for translational diffusion constants are henceforth assumed in the text.

To provide a picture of free diffusion and to compare with the MD results, we carried out diffusive dynamics simulations on the xy plane for 36 noninteracting particles; each particle was initially placed on a lattice point of a square grid with 8 Å sides (i.e., a surface area per particle of 64 Å^2), and simulated for 420 ps. This is an even simpler model than was described in the previous section because the particles are not tethered to the surface and therefore diffuse freely. Figure 8 plots the trajectories obtained for diffusion constants of 10^{-7} (top) and 10^{-6} (bottom). Even with $D = 10^{-7}$ (approximately ten times larger than experiment) the diffusive trajectories are considerably more compact than those from the MD simulation (Figure 6). (These results, of course, could have been anticipated from Equation (15): $L_{rms} = 1.3$ Å at 420 ps when $D = 10^{-7}$.) The diffusive trajectories run with $D = 10^{-6}$ are somewhat similar to those in Figure 6, though a little more spread out. Nevertheless, as Figure 7 clearly shows, the displacement correlation function calculated for $D = 10^{-6}$ (dot-dashed line) only agrees with the simulation at very short times. Hence, 100 ps is still short time for translation of lipids, and the free diffusion model isn't directly applicable.

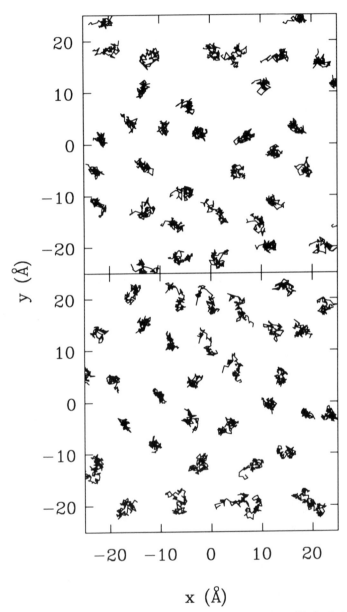

Figure 6. Center of mass trajectories showing lateral motion over the 420 ps MD simulation for the top and bottom halves of the bilayer.

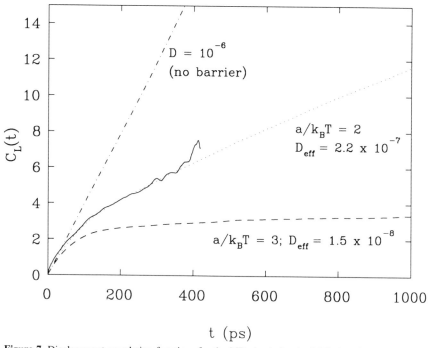

Figure 7. Displacement correlation functions for the MD simulation (*solid line*) and the BD simulations on the potential surface defined by Equation (18).

Diffusive motion, of course, is idealized. Simulations of small molecules and atomic fluids have shown that on the subpicosecond time scale inertial motions (banging and recoiling in a cage of neighbors) lead to deviations from linearity in the displacement correlation function (Allen and Tildeley, 1987; Rahman, 1964). For flexible molecules, isomerization and rotation couple with translation to produce fast center of mass motion. This is the origin of the so-called microscopic diffusion constant of 10^{-6} to 10^{-7} for lipids in bilayers obtained from experiments sensitive to fluctuations on the 100 ps time scale (Vaz and Almeida, 1991). As seen in Figure 7, the present MD simulation yields $D_{micro} \approx 10^{-6}$ when a linear fit is are carried out between 1–10 ps, and 2×10^{-7} for the range 200–1000 ps.

Back to the Poisson Process

An alternative to a continuous diffusion model is a jump model. In the limit of long times (many jumps) the distributions from the jump and diffusion models become equivalent (Feller, 1950; Wax, 1954). At very short times, however, the behavior is like that seen in Figure 6: the particles have not made any jumps. At some later time, substantial jumps (e.g., to a different site on the lattice) will be observed, and at very long times all of the particles will have made many

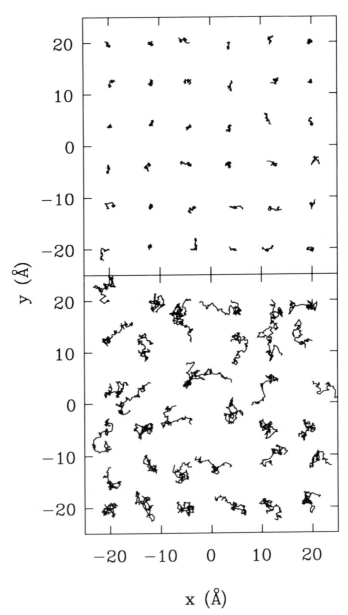

Figure 8. Brownian dynamics trajectories, each 420 ps, for 36 noninteracting particles undergoing free diffusion with $D = 10^{-7}$ and 10^{-6} cm^2/s (*top* and *bottom* panels, respectively).

transitions. Our interest here is to estimate how long we must simulate in order to be reasonably certain of seeing some.

There are two parameters in a simple jump model: the frequency, v, and the length, b. The mean squared displacement in any dimension is given by nb^2, where n is the number of jumps. Thus, in two dimensions:

$$\langle L^2 \rangle = nb^2 = 4Dt \tag{16}$$

Since $v = n/t$,

$$v = \frac{4D}{b^2} \tag{17}$$

Although b is a free parameter, for the model to be convincing its value is expected to approximately equal a molecular diameter. Since the surface area of a lipid is $60 - 70$ Å^2 in a fluid phase bilayer, a jump length of 8 Å is typically assumed (Galla et al, 1979). With the preceding value for b and $D = 10^{-8}$, we obtain $v = 6 \times 10^6$ s^{-1}.

We can now calculate the probability that out of a total of N lipids, *none* have jumped by time t. This is just the value of the Poisson distribution (Equation 5) for $m = 0$ and Poisson parameter $\lambda = Nvt$. For $D = 10^{-8}$, $N = 72$ and $t = 420$ ps, $P(X = 0) = 0.83$ and $P(X = 1) = 0.15$; i.e., we would expect to find no jumps in 83 out of 100 simulations of length 420 ps, and only one jump in 15 of 100 simulations. The results shown in Figure 6 are not surprising in this light. How long must we run so that, on average, every lipid makes a jump? That is just the value of t when $\lambda = N$; i.e., v^{-1}, or about 170 ns.

Diffusion with Barriers

Even if we can't at present carry out a 170 ns MD trajectory of a bilayer, simulation studies can still help in understanding the long time translational dynamics of lipids. To begin, we return to the diffusion model and add a potential. This is a very traditional thing to do. To take an example from biophysics, a rigorous treatment of sedimentation involves adding a chemical potential term to Fick's law of diffusion (Tanford, 1962); in chemical physics, one dimensional diffusion over a barrier is a popular model of the kinetics of activated processes (Hanggi et al, 1990; Schulten et al, 1981). Consider the following potential containing mimima on a square lattice of sides b, separated by barriers of height $2a$:

$$U(x, y) = a \cos \frac{2\pi x}{b} + a \cos \frac{2\pi y}{b} \tag{18}$$

Using BD simulations we can determine if this simple potential function can reproduce the short time features found in the MD simulation and the long time translational diffusion constant. To maintain a connection with the modeling done so far, we fix $D = 10^{-6}$ (cf. Equation 12) and $b = 8$ Å (Equation 16); units of $k_b T$ will be assumed for a.

Figure 7 plots the displacement correlation function evaluated from 100 ns simulations with two different barrier heights. The effective diffusion constants,

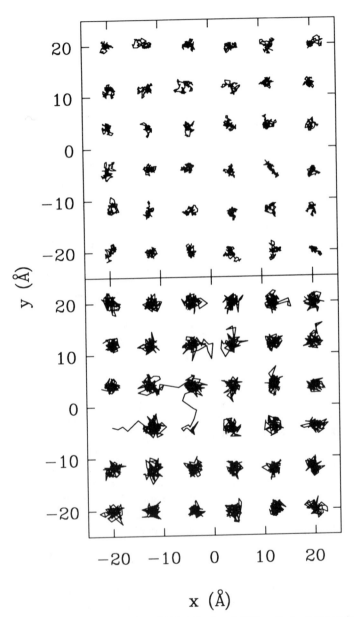

Figure 9. Brownian dynamics trajectories of 420 ps (*top*) and 4200 ps (*bottom*) for particles on a potential surface given by Equation (18) with $a = 3k_BT$, $b = 8$ Å and $D = 10^{-6}$ cm^2/s.

D_{eff}, calculated from the long time slope of the correlation function, are 2.2×10^{-7} and 1.5×10^{-8} for $a = 2$ and 3, respectively. Interestingly, $a = 3$ does a good job at reproducing the experimental diffusion constant, while $a = 2$ nicely models the 100–400 ps time scale found in the MD simulation.

Figure 9 shows the behavior of the set of 36 Brownian particles for simulations with $a = 3$ and length 420 ps (top, no jumps) and 4200 ps (bottom, three jumps). In both simulations the particles oscillate in their minima, producing a rattle-like motion, although the oscillations are more isotropic and, on the 420 ps time scale, smaller in amplitude than observed in the MD simulation. The observed number of transitions is consistent with jump model.

Towards a More Complete Model

The preceding model, in which the lipid is represented as a point particle on a two dimensional potential energy surface, cannot reproduce both time scales found in the MD *and* the experimental diffusion constant (D_{eff}). However, it is clear that we can introduce another time scale by allowing rotational diffusion. The simplest version entails removing the constraint from the tethering point in the rod model presented in the previous section: the lipid could then rattle, wobble, and jump. Another degree of freedom could be added in a natural way by replacing the rod by a cylinder with an explicit value of $D_{\parallel}$ to allow for rotation of the long axis.

Summary

Primarily by using a spliced 420 ps simulation of a DPPC bilayer, we have illustrated the range of lipid dynamics captured on the time scale of a modern molecular dynamics simulation. Isomerization of the chain dihedral angles is relatively fast, occurring at the rate of about 20 ns^{-1}. Because there were 72 lipids in the simulation, good sampling was obtained. The rates of representative headgroup torsions were significantly slower (approximately 1 and 4 ns^{-1} for the glycerol and choline groups). To a first approximation, the statistics of dihedral angle transitions can be described as a simple Poisson process. Consequently, when the number of transitions, N, becomes large (50 or more), the distribution is approximately normal with variance, N, and the 95% confidence interval $\approx N \pm 2\sqrt{N}$.

Rotation of the lipid molecule is much slower than isomerization. The lipid long axis (modeled as the eigenvector corresponding to the smallest eigenvalue of the instantaneous inertia tensor) decayed with relaxation time $\tau_{\perp} \approx 1$ ns. While this value agrees nicely with other analyses of lipid wobble, a close look at 6 configurations (Figure 1) reveals that a model of diffusive rigid body reorientation should be carefully scrutinized as longer (i.e., ns) MD simulations become available. We did not analyze collective motions involving many lipids, because both the size and time scale of our simulation were too small. However, determining the extent to which collective (or concerted) motions of small groups of lipids play a role in the 1 ns relaxation will be important for understanding membrane dynamics.

Lipid translation is very difficult to study with MD. Based on the experimental lateral diffusion constant and an analysis based on a Poisson jump model, we estimated that it would take 170 ns for all 72 lipids in our simulation to (on average) jump once. Interchanging lipid positions using Monte Carlo techniques is probably the most reasonable way of dealing with the problem of translational averaging in the near term. Nevertheless, it will be interesting when the first of the translational jumps appear in MD simulations.

In addition to the MD simulation, we presented the results of single particle Brownian dynamics simulations of rotation (a tethered rod) and translation (a free particle and a particle on a two dimensional potential surface). Such simulations are trivial to run and therefore play the role of "back of the envelope calculations". They are particularly useful for comparing simple models and detailed trajectories. Finally, we proposed that a model of an untethered rod in an orientation (Equation 8) and position (Equation 18) dependent potential can reproduce, with reasonable parameters, the lipid center of mass displacement over the ps to μs time scales. The effective membrane viscosity that is consistent with this model is about that of a neat alkane and the barrier height modulationing translational diffusion is only $6k_BT$ (or about 3–4 kcal/mol). The results of careful MD simulations should aid in the refinement of this and other simple models, and thereby provide insight into the longer time scales of membrane dynamics.

REFERENCES

Allen MP, Tildesley DJ (1987): *Computer Simulation of Liquids.* Oxford: Clarendon

Andersen HC (1980): Molecular dynamics simulations at constant pressure and/or temperature. *J Chem Phys* 72:2384–2393

Berne BJ, Pecora R (1976): *Dynamic Light Scattering.* New York: Wiley-Interscience

Brasseur R, ed. (1990): *Molecular Description of Biological Membrane Components by Computer Aided Conformational Analysis, Vol. I.* Boca Raton: CRC Press

Brooks CL, Pettitt BM, Karplus M (1988): *Proteins: A Theoretical Perspective of Dynamics Structure, and Thermodynamics.* New York: Wiley-Interscience

Brown MF, Ribeiro AA, Williams GD (1983): New view of lipid bilayer dynamics from ^{2}H and ^{13}C relaxation time measurements. *Proc Natl Acad Sci USA* 80:4325–4329

Brown ML, Venable RM, Pastor RW (1995): Method for characterizing transition concertedness from polymer dynamics computer simulations. *Biopolymers* 35:31–46

Cevc G, Marsh D (1987): *Phospholipid Bilayers.* New York:Wiley-Interscience

Chandler D (1977): Statistical mechanics of isomerization dynamics in liquids and the transition state approximation. *J Chem Phys* 68:2959–2970

Cherry RJ (1979): Rotational and lateral diffusion of membrane proteins. *Biochem Biophys Acta* 559:289–327

Crow EL, Gardner RS (1959): Confidence intervals for the expectation of a Poisson variable. *Biometrika* 46:441–453 de Gennes PG (1974): *The Physics and Chemistry of Liquid Crystals*. Oxford: Clarendon

DeGroot MH (1975): *Probability and Statistics*. Reading, MA: Addison-Wesley

De Loof H, Segrest JP, Harvey S, Pastor RW (1991): Mean field stochastic boundary molecular dynamics simulation of a phospholipid in a membrane. *Biochemistry* 30:2099–2113

Ermak DL (1975): A computer simulation of charged particles in solution. I. Technique and equilibrium properties. *J Chem Phys* 62:4189–4196

Feller W (1950): *An Introduction to Probability Theory and its Applications*. New York: John Wiley and Sons

Feller SE, Zhang Y, Pastor RW, Brooks BR (1995a): Constant pressure molecular dynamics simulation: the Langevin piston method. *J Chem Phys* 103:4613–4621

Feller SE, Zhang Y, Pastor RW (1995b): Computer simulations of liquid/liquid interfaces. II. Surface tension-area dependence of a bilayer and monolayer. *J Chem Phys* 103:10267–10276

Galla H-J, Hartmann W, Theilen U, Sackmann E (1979): On two-dimensional passive ranndom walks in lipid bilayers and fluid pathways in biomembranes. *J Membrane Biol* 48:215–236

Glaser M (1993): Lipid domains in biological membranes. *Current Opinion in Structural Biology* 3:475–481

Hanggi P, Talkner P, Borkovec M (1990): Reaction-rate theory: fifty years after Kramers. *Rev Mod Phys.* 62:251–341

Hardy BH, Pastor RW (1994): Conformational sampling of hydrocarbon and lipid chains in an ordering potential. *J Comput Chem* 15:208–226

Hoel PG, Port SC, Stone CJ (1972): *Introduction to Stochastic Process*. Boston: Houghton Mifflin

Hoover WG (1985): Canonical dynamics: equilibrium phase-space distributions. *Phys Rev A* 31:1695–1697

Katsaras J (1995): Structure of the subgel ($L_{c'}$) and gel ($L_{\beta'}$) phases of oriented dipalmitoylphosphatidylcholine multilayers. *J Phys Chem* 99:4141–4147

Lipari G, Szabo A (1980): Effect of librational motion on fluorescence depolarization and nuclear magnetic resonance relaxation in macromolecules and membranes. *Biophys J* 30:489–506

Marqusee JA, Warner M, Dill KA (1984): Frequency dependence of NMR spin lattice relaxation in bilayer membranes. *J Chem Phys* 81:6404–6405

Martyna GJ, Tobias DL, Klein ML (1994): Constant pressure molecular dynamics algorithms. *J Phys Chem.* 101:4177–4189

Nagle JF (1993): Area/lipid of bilayers from NMR. *Biophys J* 64:1476–1481

Nose S (1984): A molecular dynamics method for simulations in the canonical ensemble. *Mol Phys* 52:255–268

Nose S, Klein ML (1983): Constant pressure molecular dynamics for molecular systems. *Mol Phys* 50:1055–1076

Parrinello M, Rahman A (1981): Polymorphic transitions in single crystals: a new molecular dynamics method. *J Appl Phys* 14:7182–7190

Pascher I, Lundmark M, Nyholm, P-G, Sundell S (1992): Crystal structures of membrane lipids. *Biochem et Biophys Acta* 1113:329–373

Pastor RW (1994a): Techniques and applications of Langevin dynamics simulations. In *The Molecular Dynamics of Liquid Crystals* Luckhurst GR and Veracini CA, eds. Dordrecht: Kluwer Academic Publishers

Pastor RW (1994b): Molecular dynamics and Monte Carlo simulations of lipid bilayers. *Curr Opin Struct Biol* 4:486–492

Pastor RW, Venable RM (1993): Molecular and stochastic dynamics simulation of lipid membranes. In: *Computer Simulation of Biomolecular Systems: Theoretical and Experimental Applications*, van Gunsteren WF, Weiner PK, Wilkinson AK, eds. Leiden: ESCOM Science Publishers

Pastor RW, Venable RM, Karplus M (1988a): Brownian dynamics simulation of a lipid chain in a membrane bilayer. *J Chem Phys* 89:1112–1227

Pastor RW, Venable RM, Karplus M, Szabo A (1988b): A simulation based model of NMR T_1 Relaxation in lipid bilayer vesicles. *J Chem Phys* 89:1128–1140

Petersen NO, Chan SI (1977): More on the motional state of lipid bilayer membranes: interpretation of order parameters obtained from nuclear magnetic resonance experiments. *Biochemistry* 16:2657–2667

Rahman A (1964): Correlations in the motion of atoms in liquid argon. *Phys Rev* 136A:405–411

Rand RP, Parsegian VA (1989): Hydration forces between phospholipid bilayers. *Biochem Biophys Acta* 998:351–376

Rommel E, Noack F, Meier P, Kothe G (1988): Proton spin relaxation dispersion studies of phospholipid membranes. *J Phys Chem* 92:2981–2987

Schulten K, Schulten Z, Szabo A (1981): Dynamics of reactions involving diffusive barrier crossing. *J Chem Phys* 74:4426–4432

Seelig J, Mcdonald PM (1987): Phospholipids and proteins in biological membranes. ^{2}H NMR as a method to study structure, dynamics and interactions. *Acc Chem Res* 20:221–228

Seelig J, Seelig A (1980): Lipid conformation in model membranes and biological membranes. *Quart Rev of Biophys* 13:19–61

Skolnick J, Helfand E (1980): Kinetics of conformational transitions in chain molecules. *J Chem Phys* 72:5489–5500

Small DM (1986): *The Physical Chemistry of Lipids*. New York: Plenum

Smith GS, Sirota EB, Safinya CR, Plano RJ, Clark NA (1990): X-ray structural studies of freely suspended ordered hydrated DMPC miltimembrane films. *J Chem Phys* 92:4519–4529

Snyder RG (1992): Chain conformation for the direct calculation of the Raman spectra of the liquid alkanes C_{12}-C_{20}. *Faraday Trans* 13:1823–1833

Stouch TR (1993): Lipid membrane structure and dynamics studied by all-atom molecular dynamics simulations of hydrated phospholipid bilayers. *Mol Sim* 10:335–362

Sundaralingam M (1972): Molecular structures and conformations of the phospholipids and sphingomyelins. *Ann N Acad Sci* 195:324–355

Szabo A (1984): Theory of fluorescence depolarization in macromolecules and membranes. *J Chem Phys* 81:150–167

Tanford C (1961) *Physical Chemistry of Macromolecules*. New York: John Wiley and Sons

Tristram-Nagle S, Zhang R, Suter RM, Worthington CR, Sun WJ, Nagle JF (1993): Measurement of chain tilt angle in fully hydrated bilayers of gel phase lecithns. *Biophys J* 64:1097–1109

van Gunsteren WF, Weiner PK, Wilkinson AK, eds (1993): *Computer Simulation of Biomolecular Systems: Theoretical and Experimental Applications*. Leiden: ESCOM Science Publishers

Vaz WLC, Almeida PF (1991): Miscoscopic versus macroscopic diffusion in one-component fluid phase bilayer membranes. *Biophys J* 60:1553–1554

Venable RM, Zhang Y, Hardy BJ, Pastor RW (1993): Molecular dynamics simulations of a lipid bilayer and of hexadecane: an investigation of membrane fluidity. *Science* 262:223–226

Wang CC, Pecora R (1980): Time correlation functions for restricted rotational diffusion. *J Chem Phys* 72:5333–5340

Wax N (1954): *Noise and Stochastic Processes*. New York: Dover

Williams DE, Stouch TR (1993): Characterization of force fields for lipid molecules: applications to crystal structures. *J Comp Chem* 14:1066–1076

Woolf TB, Roux B (1994): Molecular dynamics simulation of the gramicidin channel in a phospholipid bilayer. *Proc Natl Acad Sci (USA)* 91:11631–11635

Yellin N, Levin I (1977): Hydrocarbon chain trans-gauche isomerization in phospholipid bilayer gel assemblies. *Biochemistry* 16:642–647

Zhang Y, Pastor RW (1994): A comparison of methods for computing transition rates from molecular dynamics simulation. *Mol. Sim.* 13:25–38

Zhang Y, Feller SE, Brooks BR, Pastor RW (1995): Computer simulations of liquid/liquid interfaces. I. Theory and application to octane/water. *J Chem Phys* 103:10252–10266

Zwanzig R, Ailawadi NK (1969): Statistical error due to finite time averaging in computer experiments. *Phys Rev* 182:280–283

2

An Empirical Potential Energy Function for Phospholipids: Criteria for Parameter Optimization and Applications

MICHAEL SCHLENKRICH, JÜRGEN BRICKMANN,
ALEXANDER D. MACKERELL JR., AND MARTIN KARPLUS

Introduction

Lipid membranes are an essential component of all living cells. A molecular description of the structure and dynamics of such membranes from either experimental or theoretical approaches is still lacking. This is due in part to the two-dimensional fluid character of membranes (Singer and Nicolson, 1972), which makes difficult a detailed analysis by X-ray diffraction, neutron diffraction, or nuclear magnetic resonance. Detailed structural data of lipid molecules based on X-ray crystallography are available only for the nearly anhydrous crystalline state (Pascher et al, 1992; Small, 1986).

Theoretical studies, including molecular dynamic simulations, can provide information concerning the molecular structure of and motions in biological membranes. A prerequisite for such theoretical studies is a reliable empirical potential energy function for phospholipids. Several potential energy functions are available and have been used in simulations of lipid bilayers (Charifson et al, 1990; Damodaram et al, 1992; Marrink and Berendsen, 1994; Stouch et al, 1991). In this paper we describe the derivation of an all-atom potential energy function for phospholipids. Even though the parameters have not been published before now, several studies have been performed with the potential energy function (Heller et al, 1993; Venable et al, 1993; Woolf and Roux, 1994a, Woolf and Roux, 1994b) and on a sodium dodecyl sulfate micelle (MacKerell, 1995; see also the chapter by Feller and Pastor in this volume). All of these indicate that the properties of phospholipids are well described by the potential function. In the final section of this chapter, we discuss briefly some of the results obtained in the simulations and indicate how they support the criteria used in the parameter development.

Biological Membranes
K. Merz, Jr. and B. Roux, Editors
© Birkhäuser Boston 1996

The chapter focuses on two of the major components of biological membranes, phosphatidylcholine (PC) and phosphatidylethanolamine (PE). Since membranes involve an aggregation of many lipid molecules in an approximately planar double layer, special attention was given to the potential governing the interactions between the molecules. The parameters were derived to be consistent with the all-atom CHARMM 22 potential for peptides and for proteins (MacKerell and Karplus, 1996a; MacKerell et al, 1992) and nucleic acids (MacKerell et al, 1995). This makes it possible to simulate complex membrane systems that contain peptides or proteins.

For the parameter optimization process the lipids were subdivided into functional groups for which model compounds were defined. Ab initio calculations were performed on the model compounds which, together with experimental data, were used as the basis for the determination of the parameters. Some of the parameters (i.e., the aliphatic and phosphate parameters) were taken from the protein or nucleic acid sets (MacKerell et al, 1995). We concentrated here on the trimethylammonium group and the ethanolamine group, which are required for the PC head group and the PE head group, respectively, as well as on the ester link to the aliphatic chains.

Tests of the derived empirical energy function were performed via crystal simulations. Three phospholipid crystals were used. They have known structures with molecular arrangements that are similar to that of a membrane bilayer. The first crystal is 3-lauroylpropanediol-1-phosphorylcholine (LPPC) (Hauser et al, 1980). The lipid has only one aliphatic chain, which results in a fully interdigitated organization of the chains. The second crystal is 2,3-dimyristoyl-D-glycero-1-phosphorylcholine (DMPC) (Pearson and Pascher, 1979); it has the same headgroup as LPPC but has two fatty acid chains. The molecules form a double layer structure with a zigzag arrangment of the headgroups. Finally, the first crystallized structure of a phospholipid, 2,3-dilauroyl-DL-glycero-1-phosphorylethanolamine (DLPE) (Elder et al, 1977; Hitchcock et al, 1974) was studied. The arrangement of the aliphatic chains is also a double layer structure, allowing for an additional test of the overall reliability of the energy function.

The next section of the chapter presents the CHARMM potential energy function and the strategy applied in the optimization of the parameters. Some details of the calculations are given. The results and a discussion of the optimization procedure used for the individual models compounds are presented subsequently. The following part of the chapter is concerned with the minimization and simulation studies of the three lipid crystals. Finally, we outline the conclusions and include an overview of published simulation results that makes use of the potential. The parameters are listed in the Appendix.

Potential Energy Function and Parametrization Strategy

The lipid parameters are designed to be used with already existing CHARMM22 protein and nucleic acid potential energy function; a set of sugar parameters (Guyan and Brady, 1996) will complete the all-hydrogen set for biomolecules.

Consequently, many of the parameters were chosen to be equal to those determined for corresponding groups found in proteins or nucleic acids (MacKerell et al, 1995; MacKerell et al, 1992). Further to obtain a consistent set of parameters, the fitting procedure used for the lipid parameter determination is the same as that employed in the protein and nucleic acid parametrization. The parameters were optimized for use with the CHARMM modified TIP3P water model (Jorgensen et al, 1983; Reiher, 1985; Neria et al, 1996).

Calculations were performed with version 22 of the CHARMM program (Brooks et al, 1983). The empirical CHARMM energy function has the form:

$$E_{tot} = \sum_{bonds} k_b(r - r_0)^2 + \sum_{angle} k_\alpha(\alpha - \alpha_0)^2 + \sum_{UB} k_{1-3}(r^{1-3} - r_0^{1-3})^2$$

$$+ \sum_{improper} k_\gamma(\gamma - \gamma_0)^2 + \sum_{dihedrals} V[\cos(n\tau - \tau_0) + 1] \tag{1}$$

$$+ \sum_{nonbonded} \left\{ \frac{q_i q_j}{4\pi\varepsilon_0 r_{ij}} + \varepsilon \left[\left(\frac{\sigma}{r_{ij}}\right)^{12} - \left(\frac{\sigma}{r_{ij}}\right)^6 \right] \right\}$$

In Eq. (1), nonstandard definitions of the force constants (e.g., k_b) and ε are used to eliminate multiplications. The first four sums in the potential function are harmonic functions describing the internal coordinates: bond length, valence angle, Urey-Bradley, and out-of-plane displacements. The out-of-plane coordinate is modeled through an improper torsion (Brooks et al, 1983). The fifth sum accounts for the torsion potential through a cosine function of the dihedral angle which may be expanded in a Fourier series. Except for the out-of-plane and Urey-Bradley terms all possible internal coordinates contribute to the intramolecular potential function. The out-of-plane and Urey-Bradley coordinates are used only where necessary (see below). The Urey-Bradley term is introduced to achieve a satisfactory description of the vibrational modes. The final summation includes the Coulomb and Lennard-Jones interactions between all pairs of atoms not bonded together nor having common valence-angle coordinates. The 1–4 interactions between the terminal atoms of a dihedral angle are not scaled as they are in the polar hydrogen parameter set, but in a few cases modified 1–4 Lennard Jones parameters are used. The Lennard Jones parameters between different atom types are derived by use of the Lorentz-Berthelodt combination rule; i.e., for the ε values the geometric mean is used and for the σ values the arithmetic mean. To assign partial atomic charges, the molecules were subdivided into groups carrying an integral $(-2, -1, 0, +1, +2)$ electron charge. This allows use of the group option in the CHARMM program, which is required for the extended electrostatic treatment of the long-range Coulomb interactions (Stote et al, 1991). This is particularly important for membrane systems because they have a high charge density. The charge partitioning also simplifies the modular construction of other molecules.

The parametrization was divided into three steps associated with different parts of the empirical force field. The force constants and equilibrium distances of the harmonic terms for bonds and angles were fitted to reproduce the experimental equilibrium geometries and vibrational frequencies of the model compounds presented below. The dihedral parameters were next fitted to reproduce the relative energies of different conformers and barriers between conformers of model compounds. Urey-Bradley and improper torsion parameters were introduced only where the bond, angle, and dihedral terms alone were insufficent to reproduce the data. Finally, the interaction parameters (partial atomic charges and Lennard-Jones parameters) were optimized based on ab initio results for water/model compound interactions and from macroscopic thermodynamics properties calculated via Monte Carlo simulations (see below). These three steps were repeated in an iterative fashion until the parameters converged to a self-consistent set of intramolecular and intermolecular values.

Parameters for the intramolecular portion of the force field were primarily derived from experimental data, supplemented with ab initio results as required. Equilibrium bond lengths and angles were optimized to reproduce gas phase geometries. Force constants were optimized to reproduce the experimental and ab initio frequencies. To insure that the frequency assignments were correct, ab initio vibrational spectra were calculated. Using Raman and IR activity considerations the ab initio modes were assigned to the experimental frequencies. The normal mode eigenvectors from the empirical force field and the ab initio results were used to determine the respective potential energy distributions (PED). Through this analysis, parameters were adjusted to fit the frequencies and the normal modes resulting from the force field. This was made possible by use of the MOLVIB program (Kuczera et al, 1993), incorporated in CHARMM.

The dihedral parameters were also adjusted to reproduce the energy differences between conformers and transition states. Ab initio results for different conformers of selected model compound were used as the basis of the optimization. The effects of different basis sets and the inclusion of electron correlation, through second order Möller-Plesset perturbation theory (MP2) (Möller and Plesset, 1934), were studied to verify the reliability of the results.

Interaction energy parameters were optimized to reproduce ab initio determined water-model compound complexes, dipole moments and macroscopic pure solvent, and aqueous solvation properties. Typically, initial Lennard-Jones parameters were selected from the protein and nucleic acid parameter sets. The partial atomic charges and Lennard-Jones parameters were then optimized to reproduce the interaction energies and geometries of ab initio water-model-compound complexes. Model compound, water geometies were selected to test the interactions at hydrogen bonding sites. Ab initio calculations were performed at the HF/6-31G(d) level of theory, and the resulting interaction energies were scaled, using a strategy developed by MacKerell and Karplus (1991). The partial atomic charges were then adjusted to reproduce the ab initio goal data with respect to both the minimum interaction energy and geometry using the OPLS35 program (Gao and Jorgensen, 1992) with the same Z-matrix as in the ab initio calculations (see below). For the charge distribution, each molecule is subdivided into groups which are neutral or carry an integral charge. This group approach allows for the modular construction of larger molecules without complicated readjustment of the charges. If two group are linked together, the charges of the atoms that are deleted are added to the charge of the next bonded atom within the group. This preserves the total charge of the group and also results in an integral charge for the resulting molecule. Application of this approach to nucleic acid parameter determination has yielded satisfactory results (MacKerell et al, 1995).

Further adjustment of the interaction parameters was performed via pure solvent simulations and simulations in water. This allowed a comparison of heats of vaporization or heats of solvation and molecular volumes from the empirical force field with experimental results. The pure solvent simulations were especially important for the optimization of the Lennard-Jones terms. These calculations were performed via Monte-Carlo simulations using the BOSS program (Jorgensen, 1983) in a periodic box of 128 molecules using a cutoff of 9.5 Å; a cubic spline function was used to truncate interaction over the outermost 1 Å. Runs were equilibrated for one-million (1M) configurations with production sampling over 2M configurations. Aqueous solvation simulations were performed in a periodic system of 264 TIP3P water molecules with the solute molecule placed in the center of the box. All water molecules that have interaction energies with the solute greater than 0.5×10^5 kcal/mole were removed. The systems were equilibrated for 1M configurations and sampled for 4.5M configurations. In all simulations the move size was adjusted to obtain an approximately 40% acceptance rate, and the internal geometry of the molecule was rigid except for torsional rotations. Solvation energies were determined by taking the difference in total energies between the solute-solvent system and a pure solvent simulation that contained an identical number of water molecules; the latter are in good agreement with experiment. If any adjustment of either the partial atomic charges or Lennard-Jones parameters was required to obtain agreement for the solvation energy, the water-model complex interactions were reanalyzed and additional optimization performed. The iterative procedure involving the interactions between model compounds and water molecules and the condensed phase simulations was continued until convergence of the interaction parameter optimization. This insured the proper balance among the solvent-solvent, solvent-solute and solute-solute aspects of the interaction parameters. Such a balance was essential to obtain satisfactory simulation results in aqueous solution.

Ab initio calculations were performed with the Gaussian88 and Gaussian90 (Frisch et al, 1990) programs. All geometries were optimized at the HF/6-31G(d) (Hariharan and Pople, 1972) level of theory, unless noted. Frequency calculations were performed analytically. MP2 energies were calculated in the frozen core approximation in which only the valence electrons were included in the perturbation calculation. The notation MP2/6-31G(2d,p)//HF/6-31G(d) indicates the MP2/6-31G(2d,p) energy was calculated for the HF/6-31G(d) optimized structure. Water-model compound complex calculations were performed by first optimizing model compounds at the HF/6-31G(d) level. The water-model compound interactions were then optimized with respect to the distance and, in some cases, a single angle, while maintaining the model compound monomer HF/6-31G(d) geometry and the experimental water geometry (Cook et al, 1974).

Crystal minimizations were initiated from the experimentally determined crystal non-hydrogen atom positions. Atoms in the asymmetric unit were moved in the minimization, as were the image particles generated via transformations of the asymmetric unit (Brooks et al, 1983). In the first step, hydrogen atoms were added in standard geometries and optimized using a cutoff of 15 Å with all heavy atoms fixed in position. These structures were the starting configurations for the minimizations with different cutoffs (15 Å and 30 Å). Truncations were performed using the shift, atom-based scheme (Brooks et al, 1983). The minimizations were carried out in two steps. First the unit cell parameters were fixed at the crystal values, and the molecules in the asymmetric unit were minimized. Then, the unit cell parameters (the a, b, and c axis lengths and the β angle of the monoclinic cells) were included in the optimization in addition to the atom positions. For all minimizations we used the adopted basis Newton-Raphson (ABNR) method and the CRYSTAL facility implemented in CHARMM22. Minimizations were continued until the RMS gradient was less then 5.0×10^{-6} kcal/mole/Å.

Parametrization: Application to Specific Fragments and Results

A schematic structure of the lipids studied in the present work is shown in Figure 1. The structure can be subdivided into groups as shown in the figure.

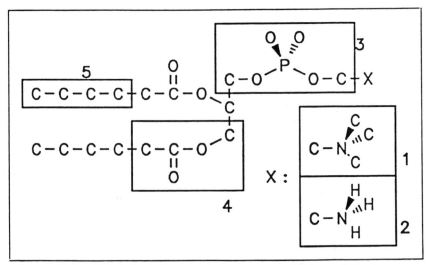

Figure 1. Schematic representation of a phospholipid molecule. The decomposition into different groups is shown by the five boxes. Box 1 encloses the positive trimethylammonium group of the phosphotidylcholine and box 2 the NH3$^+$ group of phosphotidylethanolamine. Box 3 contains the negatively charged phosphate group. Box 4 contains the ester link between the head group and the aliphatic chains. Box 5 represents the aliphatic chains.

These groups are used to determine the model compounds for the parameterization, including the calculations of the partial atomic charge distribution. The first group (1) is the positively charged trimethylammonium moiety of the phosphatidylcholine head group. No comparable group exists in either nucleic acids or proteins which requires that all parameters associated with this group be derived in the present study. Parameter optimization was based on the model compounds tetramethylammonium (TMA), ethyltrimethylammonium (ETMA), and choline. The corresponding group in phosphatidylethanolamine is $-NH_3^+$ (2 in Figure 1); parameters for this moiety were taken from the amino acid lysine. Ethanolamine was also examined using ab initio data to evaluate the torsion parameters. Parameters of the phosphate group (3 in Figure 1) are identical to those in the nucleic acid parameter set (MacKerell et al, 1995), so that no reinvestigation was performed. The ester link (4 in Figure 1) is another portion of lipids which is not represented in proteins or nucleic acids. As model compounds methyl acetate, methyl propionate, and ethyl acetate were used. Parameters for the aliphatic chains (5 in Figure 1) were taken directly from those for the protein aliphatic side chains (MacKerell et al, 1996).

Trimethylammonium Fragment

INTRAMOLECULAR PARAMETERS OF TETRAMETHYLAMMONIUM (TMA)

TMA experimental gas phase data are lacking due to the charged nature of the compound. However, several crystal structures with counter ions (Cl^-, Br^-, I^-) are available (Vegard and Sollesnes, 1927; Wyckof, 1928). The presence of a counterion destroys the T_d-symmetry of the molecule and the N–C bond length varies between 1.48 Å and 1.52 Å. An optimized ab initio geometry for isolated TMA was obtained at the RHF/6-31G(d) level (Table 1). The ab

Table 1. Geometry of Optimized Tetramethylammonium

	Ab initio		CHARMM	
	Minimum	TS[a]	Minimum	TS[a]
r_{N-Ct}	1.496	1.515	1.497	1.497
r_{N-C}		1.497		1.497
r_{Ct-H}	1.079	1.078	1.084	1.084
r_{C-Ht}		1.079		1.085
r_{C-H}		1.079		1.085
α_{N-Ct-H}	109.04	109.47	111.21	111.24
α_{N-C-Ht}		109.06		111.15
α_{N-C-H}		108.92		111.17
α_{Ct-N-C}	109.47	110.00	109.47	109.51

[a] For the transition state geometry: Ct and Ht indicate the methyl group atoms in eclipsed positions. Distances in angstroms; angles in degrees.

initio determined N–C bond length of 1.496 Å is in agreement with the TMA crystal values of 1.48–1.52 Å. The transition state (TS) geometry for rotation of a methyl group was also determined; it corresponds to an eclipsed position. To obtain good force field results for the geometry and vibrational spectra of TMA, a new atom type (NT) for the nitrogen atom was defined because no appropriate nitrogen is present in the protein or nucleic acid parameter sets. As an initial model from the CHARMM22 protein parameters, we chose the sp^3 nitrogen (NH3). Optimization of the bond and angle equilbrium values, r_0 and α_0, respectively (Equation 1), was based on the ab initio determined geometries. Agreement between the empirical and ab initio geometries was seen to be satisfactory (Table 1), although the difference between the minimum and TS geometries was significantly smaller in the empirical model than in the ab initio calculations.

TMA has 45 vibrational degrees of freedom, but only 19 different frequencies due to the T_d symmetry of the molecule. Of these frequencies, 14 are Raman active, while only seven are IR active; four frequencies are not accessible by either experimental technique (see Table 2). All IR measurements were performed on crystalline TMA containing counter ions. This meant the frequencies were shifted from the gas phase values and the degenerate modes were split (Berg,

Table 2. Frequencies of Tetramethylammonium $(cm^{-1})^a$

Nr.	Typ	Exp 1. Zeolite	Exp 2. aq	Ab initio*0.9	CHARMM	Assignment
1	A2	206^b		181	250	torsion
2	F1	293^b		284	286	torsion
5	E	369	370	349	380	CN4 asym.def.
7	F2	493	455	440	453	CN4 asym.def.
10	A1	761	752	713	768	CN4 sym.str.
11	F2	955	950	926	957	CN4 asym.str.
14	F1			1061	1039	CH3 rock
17	E	1172	1171	1165	1082	CH3 rock
19	F2	1290	1290	1297	1289	CH3 rock
22	F2	1419	1420	1431	1495	CH3 sym.def.
25	F1			1457	1419	CH3 asym.def.
28	E	1452	1453	1468	1437	CH3 asym.def.
30	A1			1490	1535	CH3 sym.def.
31	F2			1497	1453	CH3 asym.def.
34	F2	3039	3040	2935	2925	CH3 sym.str
37	A1	2986	2988	2946	3057	CH3 sym.str.
38	F1			3026	2918	CH3 asym.str.
41	E			3027	2921	CH3 asym.str.
43	F2	2928	2929	3033	3043	CH3 asym.str.

^{a} Absolute rms dev. (ab initio -minus) of modes 5–31: ±5.3 cm^{-1}.
^{a} Relative rms dev. (ab initio -minus) of modes 5–31: 0.5%.
^{b} Inelastic neutron scattering data from Brun et al, 1987.

1977; Bottger and Geddes, 1965; Mahendra et al, 1984; Mahendra et al, 1982). Bottger et al showed the frequencies to shift in a linear fashion with respect to $1/R_{N...X}$, where $R_{N...X}$ is the distance between the nitrogen of TMA and the counter ion. This relationshiop may be used to estimate the frequency of a free TMA cation by extrapolating $1/R_{N...X} \to 0$. The frequencies of free TMA in solution or in zeolite cages (Dutta et al, 1986) have been observed by Raman measurements (see Table 2). The frequency differences between the two environments are within 10 cm^{-1}, except for mode 7, whose frequency in water is 455 cm^{-1}, while it is 493 cm^{-1} in the zeolite cage. We chose the frequencies in water for our fit since they are more similar to the the extrapolated IR frequencies (not shown). For the low frequency region (below 300 cm^{-1}) only inelastic neutron scattering results for TMA in an omega zeolite are available (Brun et al, 1987). To access the nonactive frequencies, we performed an ab initio HF/6-31G(d) frequency calculation. The resulting frequencies were scaled by a factor of 0.9 (Florián and Johnson, 1994), which yielded good agreement with the Raman active modes (see Table 2). A normal mode analysis of the ab initio force constant matrix was made to determine the potential energy distribution. This information was used in the fitting procedure to allow for correct assignments of the normal modes. Initially, Urey-Bradley terms were not used because they have little influence on the general frequency distribution. However, the correct splitting of the methyl group stretching and deformation modes could only be reproduced by adding Urey-Bradley terms (see Appendix). The Urey-Bradley force constants are of the same magnitude as the corresponding bend force constants. All low frequencies (below 2000 cm^{-1}), except four modes, are within 20 cm^{-1} of the corresponding Raman or scaled ab initio frequency. One CH$_3$ rock (mode 17) at 1172 cm^{-1} is 90 cm^{-1} too low, while a symmetric CH$_3$ deformation at 1420 cm^{-1} (mode 22) is about 60 cm^{-1} too high.

Optimization of the HL-CT3L-NTL-CT3L torsion parameters was performed based on the rotational barrier of the methyl group. The barrier to rotation was obtained from the energy difference of the ab initio fully optimized transition state and the minimum energy structures. The values of the internal coordinates are given in Table 1 in which the subscript t denotes the atoms of the methyl group that have an eclipsed transition state conformation. The total energies and their differences are given in Table 3 for two basis sets and HF and MP2 levels of theory for the HF/6-31G(d) geometries. It can be seen that the barrier height varies from 4.47 kcal/mol to the 'best' value of 4.29 kcal/mol upon increasing the basis set and including electron correlation via MP2. The torsion parameter was adjusted to reproduce the latter value, yielding an empirical rotation barrier of 4.29 kcal/mole. The two low torsion frequencies of TMA (see Table 2) have been measured by inelastic neutron scattering and shown to be split by about 90 cm^{-1} (Brun et al, 1987). The present potential function, which contains only one free parameter (HL-CT3L-NTL-CT3L) associated with the torsional mode, produced a smaller splitting of 40 cm^{-1}. The higher torsion frequency was within 10 cm^{-1} of the experimental value, while the lower frequency was 44 cm^{-1} higher than the corresponding experimental frequency. As discussed above, the

Table 3. Ab Initio Energetic Results for the Choline Fragments[a]

		6-31G(d)		6-311G(d,p)	
Conformation		HF	MP2	HF	MP2
Tetramethylammonium					
minimum	Td	−212.685141	−213.395182	−212.741964	−213.537194
saddle	C3V	−212.678010	−213.389658	−212.735001	−213.530354
ΔE		4.47	4.34	4.37	4.29
Ethyl-trimethyammonium					
trans	CS	−251.712105	−252.542125	−251.786821	−252.735288
cis	CS	−251.720761	−252.533342	−251.778153	−252.726520
ΔE		5.43	5.51	5.44	5.50
Choline					
gauche	C1	−326.570001	−327.566950	−326.662950	−327.804416
trans	CS	−326.562894	−327.557782	−326.656353	−327.795973
ΔE		4.46	5.75	4.14	5.30
cis	CS	−326.561720	−327.559246	−326.654701	−327.795878
ΔE		5.20	5.46	5.18	5.36
TS	C1	−326.560056	−327.555469	−326.653235	−327.793445
ΔE		6.24	7.20	6.10	6.88
Ethanolamine					
trans	CS	−209.458190	−210.055586	−209.529326	−210.211128
gauche	C1	−209.471864	−210.071115	−209.542247	−210.225456
ΔE		8.58	9.74	8.11	8.99

[a] Absolute energies in hartree; distances in Å and energy differences in kcal/mole.
[a] The geometries were optimized at HF/6-31G(d) level of theory.

change of internal coordinates upon going from the minimum to the TS was smaller in the empirical model compared to the ab initio result (Table 1). This indicates that TMA is "stiffer" in the barrier region with the empirical force field, which may be responsible for the lower splitting of the low-frequency torsional modes. The larger ab initio structural changes in the transition state may also reflect aspects of the molecular structure which are not modeled by the harmonic formulation of the intramolecular potential function.

NONBONDED INTERACTION PARAMETERS OF TETRAMETHYLAMMONIUM (TMA)

Partial atomic charges and Lennard-Jones parameters were derived based on ab initio water-TMA supermolecule calculations (see above). From these calculations interaction energies and geometries were obtained, which were used as input for the fitting of the intermolecular parameters. Three water-TMA arrangements were investigated, as shown in Figure 2. Only one variable, the nonbonded distance marked by the dashed line, is varied in the HF/6-31G(d) optimization. The resulting minimum interaction energies and intermolecular distances are given in Table 4. It can be seen that complex B has the most favorable inter-

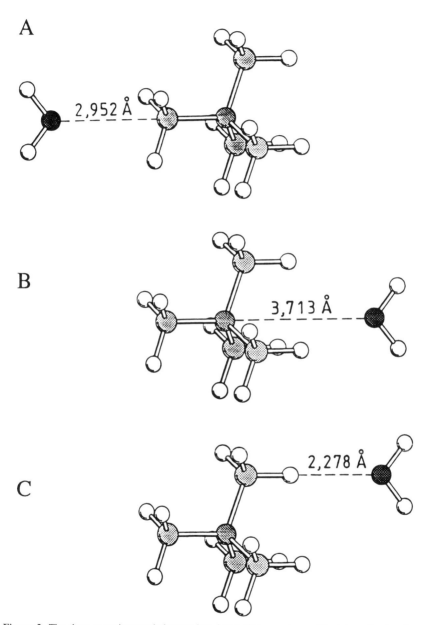

Figure 2. The three water/tetramethylammonium interaction complexes. The intermolecular distances obtained at the HF/6-31G(d) level of theory are given.

Table 4. Minimum Interaction Energies and Geometries of the TMA/Water Complexes[a]

Interaction	Ab initio			CHARMM		
	ΔE_{HF}	r_{N-O}	r_{C-O}	ΔE	r_{N-O}	r_{C-O}
A	−8.02	4.45	2.95	−7.55	4.60	3.10
B	−10.80	3.71	3.51	−10.46	3.54	3.35
C	−8.70	4.09	3.36	−8.18	3.93	3.17

[a] Distances in Å and energies in kcal/mole. Interaction geometries are shown in Figure 2.

action energy. In this orientation the water molecule is closest to the central nitrogen atom, which dominates the electrostatic potential of TMA. Complex C, in which the water approaches along the C–H bond, has the second most favorable interaction energy. In Complex A the water molecule approaches the methyl group along the N–C axis. The favorable energy of C is due to the polar character of the methyl hydrogen atoms in TMA (see below), which allows the formation of hydrogen-bond-like complexes.

The partial atomic charges and Lennard-Jones parameters of TMA were chosen to reproduce the ab initio interaction energies and geometries. Initial charges were obtained from a HF/6-31G(d) Mulliken population analysis, and the initial Lennard-Jones parameters were obtained from the CHARMM22 protein force field. Manual, iterative adjustment of both the charges and Lennard-Jones parameters led to a set of parameters that adequately reproduced the ab initio data, as shown in Table 4. During the optimization it was found that the standard nitrogen and carbon Lennard-Jones parameters were sufficent. However, to reproduce the energetics of the hydrogen-bond-like complex in orientation C, Lennard-Jones parameters for the hydrogen atom were selected with a radius intermediate between the aliphatic and polar hydrogen parameters (see Appendix). The partial charge distribution from the Mulliken population analysis gave good agreement with the ab initio data and therefore was modified only slightly. The partial polar character of the hydrogen atoms of TMA was not included in other empirical parameter sets (Charifson et al, 1990; Rao and Singh, 1989; Stouch et al, 1991). Such an omission resulted in a reversal of the energetic ordering of the interaction energies between orientations A and C. Polar aliphatic hydrogens had been introduced previously (e.g., for histidine, MacKerell and Karplus, 1996b).

In previous parameter development, scaling factors for both ab initio energies and distances had been introduced in the fitting procedure (Jorgensen, 1986; Jorgensen and Swenson, 1985; MacKerell and Karplus, 1991; Reiher, 1985). In accord with that procedure, the empirical optimized interaction distances were approximately 0.2 Å shorter, compared with the ab initio results. This shortening is necessary as the intermolecular Hartree-Fock distances are systematically too long due to the neglect of electron correlation, which is responsible for the attractive dispersion interactions. In addition, many-body polarization effects expected in liquid water were not included in the ab initio calculations. A systematic study of the influence of basis set and electron correlation on hydrogen bonds showed that a decrease of approximately 0.2 Å occurs when electron correlation is included in ab initio calculations (Reiher, 1985). Test calculations with more than one water molecule simultaneously interacting with TMA showed that the interaction energies are not significantly changed, indicating many-body polarization effects to be minimal in TMA (Schlenkrich, 1992).

Final testing of the TMA interaction parameters was done by calculating the heat of solvation and the molecular volume in aqueous solution, for which experimental values are available. The calculated heat of solvation was −52.7 kcal/mole, which compares well with the experimental values of −41 and −53 kcal/mole (Aue et al, 1976; Boyd, 1969). The calculated molecular volume was 74 Å^3, compared to the experimental value of 54 Å^3 (Millero, 1971). Estimated errors in the calculated values were approximately 3 kcal/mole for the heat of solvation and 20 Å^3 for the molecular volume, both of which were determined from the difference between two large, fluctuating numbers. Thus, the microscopically determined interaction parameters yielded reasonable agreement with these experimentally determined macroscopic properties.

Ethyl Trimethylammonium

To parameterize the linkage between the tetramethylammonium moiety and the hydroxyl group of choline, ethyl trimethylammonium was selected as the model compound to optimize the C–C–N angle and C–C–N–C torsion parameters (see Figure 3). Ab initio optimized structures were obtained at the HF/6-31G(d) level with the C–C–N–C dihedral in the trans orientation, corresponding to the mini-

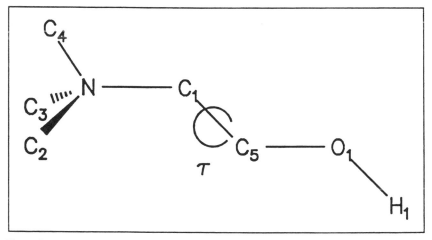

Figure 3. Schematic representation of choline indicating the dihedral τ that defines the conformation of the molecule.

Table 5. Values of Internal Coordinates of the Optimized Ethyl
Trimethylammonium Structures[a,b]

	Ab initio		CHARMM	
	trans	cis	trans	cis
r_{N-C1}	1.518	1.543	1.515	1.518
r_{N-C4}	1.497	1.497	1.503	1.502
$r_{N-C(2,3)}$	1.495	1.497	1.503	1.503
r_{C1-C5}	1.522	1.522	1.539	1.539
$\alpha_{C1-N-C4}$	107.8	113.8	110.4	111.4
$\alpha_{C1-N-C(2,3)}$	111.1	109.1	110.2	110.0
$\alpha_{N-C1-C5}$	116.0	116.9	114.6	115.8
$\tau_{C5-C1-N-C(2,3)}$	61.1	120.7	59.9	120.1

[a] The atom numbering is: C5-C1-N-C(2,3,4). Trans and cis are defined
through the dihedral C4-N-C1-C5.
[b] Bond lengths in Å and angles in degrees.

mum energy structure, and with the dihedral in the cis orientation, corresponding
to the transition state conformation. The energy results are presented in Table 3,
while the geometries are included along with the empirical optimized values in
Table 5. For the C–C–N angle, only the equilibrium angle was fit, while the
force constant was taken directly from the CT2-CT2-CT2 aliphatic parameter.
An equilibrium angle of 115° was found to reproduce the ab initio result for the
minimum structure. This value is similar to the average angle found in crystal
structures containing choline (Gelin and Karplus, 1975). The C–C–N–C torsion
parameter was adjusted to reproduce the energy difference between the trans
minimum energy and the cis transition state structures (see Table 3). The fitted
empirical force field yielded a barrier energy of 5.5 kcal/mole above the trans
minimum, identical to the MP2/6-311G(d,p)//HF/6-31G(d) value. Analysis of
changes in the internal geometry (Table 5) in going from the trans to cis orientation showed that the direction of the ab initio trends is reproduced by the
CHARMM22 parameters, but that the magnitude of the differences is smaller
in the force field calculations.

Choline

Final testing of the parameters for the choline head group was performed using the choline fragment itself. Conformations of choline can be defined by the
orientation of the hydroxyl group relative to the trimethylammonium group, as
shown in Figure 3. Two stable conformers were obtained at the HF/6-31G(d)
level of theory; namely the gauche conformer with a N–C–C–O dihedral angle
of 56° and the trans conformer with a value of 180° (see Table 6). These conformers are shown in Figure 4a and 4b, respectively, together with two transition
state structures of choline. The cis ($\tau_{N-C-C-O} = 0°$) conformer (Figure 4c) is

Table 6a. Values of Internal Coordinates of the Optimized Choline Structure[a]

	Ab initio		CHARMM	
	trans	gauche	trans	gauche
r_{N-C}	1.510	1.516	1.522	1.531
r_{C-C}	1.528	1.521	1.529	1.556
r_{C-O}	1.391	1.395	1.417	1.425
α_{N-C-C}	116.8	116.6	117.6	119.1
α_{C-C-O}	102.5	109.6	104.4	111.6
$\tau_{N-C-C-O}$	180.0	56.0	180.0	59.9

[a] Bond lengths in Å and angles in degrees.

Table 6b. NCC and CCO Angles as a Function of Conformation in Choline and Ethanolamine[a]

	Theory	cis	gauche	TS	trans
Choline					
NCC	HF/6-31G(d)	119.1	116.6	116.2	116.8
	CHARMM	121.4	119.1	117.7	117.6
CCO	HF/6-31G(d)	112.1	109.6	105.2	102.5
	CHARMM	115.4	111.6	104.7	104.4
Ethanolamine					
NCC	HF/6-31G(d)	109.6	107.9	111.3	111.0
	CHARMM	112.7	111.2	112.9	112.9
CCO	HF/6-31G(d)	106.5	104.9	105.6	103.6
	CHARMM	112.6	110.0	108.6	107.7

[a] Angles in degrees.

the transition state structure between the two symmetry equivalent gauche conformers and the conformer named TS ($\tau_{N-C-C-O} = 130.5°$, Figure 4d) is the transition state structure between trans and gauche.

In Table 3 the total energies and the energy differences relative to the minimum energy gauche conformer are given for two different basis sets at the HF and MP2 levels of theory. The energy of the trans structure (Figure 4b) is 4.46 kcal/mole above the minimum (Figure 4a) at the HF/6-31G(d) level and 5.30 kcal/mol at the MP2/6-311G(d,p)//HF/6-31G(d) level. The strong preference for the gauche conformation arises from the nonbonded intramolecular interaction between the hydroxyl group and the trimethylammonium moiety. As shown in Figure 4a the nonbonded O to H distance is 2.24 Å in the HF/6-31G(d) gauche structure. This interaction is stabilized by the partial polar character of the TMA hydrogens. The gauche conformer as the minimum energy structure is in agreement with crystal structures that contain the choline fragment for which a gauche conformation is found in the majority of cases (Hauser et al, 1981).

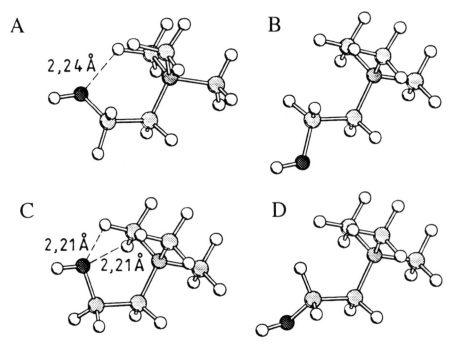

Figure 4. HF/6-31G(d) optimized structures of choline. A) Minimum energy gauche, B) saddle-point trans, C) cis transition state and D) TS transition structures (see text). Selected nonbonded intramolecular distances are given.

The adiabatic energy profile from the empirical force field as a function of dihedral angle is given in Figure 5. In determining these curves, the central N–C–C–O dihedral was constrained while all other degrees of freedom were allowed to relax. The energy of the cis structure is underestimated while the energies of the trans minimum and the TS structure are in good agreement with the ab initio data. Direct application of the aliphatic X-CT2L-CT2L-X parameter to the N–C–C–O dihedral angle leads to the same energy for the gauche and the trans minima. To separate them, NTL-CT2L-CT2L-OHL torsion parameters were introduced. A combination of 1-fold and 3-fold torsion parameters for the N–C–C–O dihedral were included and optimized to obtain the correct energy difference (see Appendix). A negative force constant for the 3-fold term was required to increase the barrier height at 0° and lower the barrier height at 120°.

Figure 5 also shows the contributions from the individual energy terms to the potential energy function on the adiabatic surface. The dihedral term dominates the surface, but all terms make some contributions. In the region of the TS ($\approx$120°), the electrostatic and VDW terms make favorable contributions that partly counterbalance the unfavorable dihedral term. Thus, the gauche structure is not stabilized by polar terms in the empirical force field; it is the dihedral term that determines the position of the minima. The gauche minimum from the

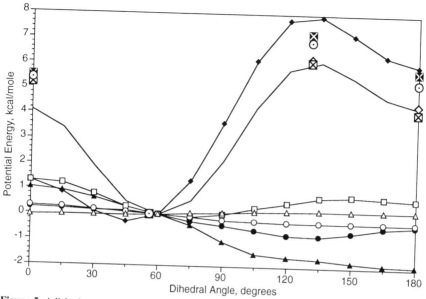

Figure 5. Adiabatic potential energy surface for the N–C–C–O dihedral angle of choline. Shown are the total (line), electrostatic (●), VDW (▲), bond (○), angle (□), Urey (△) and dihedral (◆) energy contributions to the surface. Ab initio data is represented by ◇ (HF/6-31G(d)), ▨ (MP2/6-31G(d)//HF/6-31G(d)), ⊠ (HF/6-31G(d,p)//HF/6-31G(d)) and ⊙ (MP/6-31G(d,p)//HF/6-31G(d)).

empirical energy calculation has a O to H distance equal to 2.13 Å, in satisfactory agreement with the 2.24 Å distance in the HF/6-31G(d) gauche structure. The electrostatic interactions occurring between the hydroxyl and trimethylammonium moieties are complex. There are repulsive contributions between the hydroxyl oxygen and the TMA carbon and nitrogens and the hydroxyl hydrogen and the TMA hydrogens. Attractive interactions exist between the hydroxyl oxygen and the TMA hydrogens, as well as between the hydroxyl hydrogen and the TMA carbon and nitrogen. The balance between these interactions yields a relatively small total electrostatic contribution so that an explicit dihedral term is required. Since the charges have to be chosen to obtain correct intermolecular interactions, the intramolecular properties must be adjusted by bonding terms. The relation of the force field results to the physical origin of effects in the ab initio calculation is difficult to access.

Changes of the internal geometry of ethyl trimethylammonium as a function of conformation provide an indirect check of the validitiy of the method used to reproduce the ab initio data for the N–C–C–O dihedal surface. Such changes reflect the influence of the electrostatic and Lennard-Jones forces on the intramolecular geometry, particularly on the angles, since there are no explicit coupling terms; e.g., there is no term that depends directly on a product of the bond angles and dihedral angles, as in some force fields. Table 6 shows that for the gauche and trans structures the empirical force field results corresponds to the ab initio results with the exception of the r_{C-C} and α_{N-C-C} terms. The increase of both terms in the empirical model appears to be due to Lennard-Jones repulsion between the hydroxyl and trimethylammonium moieties of choline. A more detailed comparison of the ab initio and empirical values of the NCC and CCO angles as a function of the conformation is shown in Table 6b. The NCC angle is largest for both the ab initio and empirical models in the cis conformation. In the empirical force field the angle decreases from the gauche to the TS to the trans conformer, whereas in the ab initio calculations the NCC angle is smallest in the TS conformer and then opens slightly in going to the trans minimum. The monotonic change of the NCC angles in the empirical model is due to the continuous decrease in the van der Waals energy upon going from the cis to the trans structure (see Figure 5).

Similar trends are observed for the CCO angle. A large decrease occurs upon going from the cis to the gauche conformer and upon going to the TS conformer; the differences are larger in the empirical than the ab initio structures. In contrast to the NCC angle, the large decrease continues upon going to the trans conformer in the ab initio calculations, while a smaller change occurs in the empirical model. The decreased changes in the CCO angle in the empirical model correspond to the small electrostatic contribution to the surface in this region. The discrepancies suggest that additional optimization of the van der Waals interactions could be done. However, because the differences are unimportant for most of the phospholipid properties of interest, no adjustments were made.

Ethanolamine

Most of the parameters for the hydroxyl and amine moieties of ethanolamine were extracted from the protein parameters. However, parameters related to the angle and dihedral terms involved in the aliphatic connection between the two moieties do not exist in the protein set. To evaluate these parameters ab initio calculations were performed on the N–C–C–O dihedal surface of ethanolamine; the naming convention used for choline was used for ethanolamine (see Figure 3). The global minimum gauche conformer and the trans saddle point structure are shown in Figure 6a and 6b, respectively, and the cis (g+ to g−) and TS (gauche to trans) transition structures are shown in Figures 6c and 6d, respectively. The energies and their relative differences with respect to the gauche structure are given in Table 3. The energy difference between the gauche and the trans conformer is even larger than that for choline; at the MP2/6-311G(d,p)//HF/6-31G(d) level, the difference is 8.99 kcal/mol. As was found for choline, there is an intramolecular hydrogen bond type interaction between the oxygen of the hydroxyl group and the ammonium moiety ($r_{O...H} = 2.13$ Å). The N–C–C–O torsion parameter was adjusted to reproduce the ab initio calculated energy difference. In contrast to choline, only a onefold term was required to reproduce the ab initio result. The adiabatic potential energy curve is shown in Figure 7, which includes the contributions of the individual terms in the potential energy function.

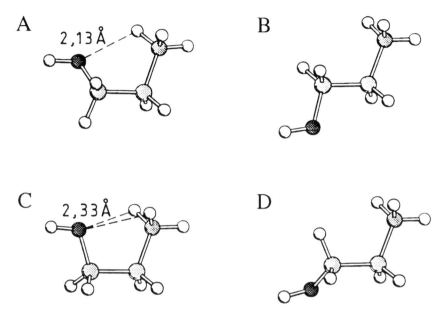

Figure 6. HF/6-31G(d) optimized structures of ethanolamine. A) Minimum energy gauche, B) saddlepoint trans, C) cis transition state and D) TS transition structures are shown. Selected nonbonded intramolecular distances are given.

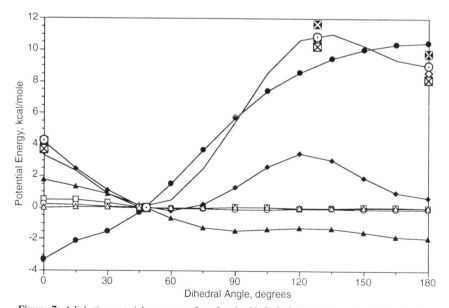

Figure 7. Adiabatic potential energy surface for the N–C–C–O dihedral angle of ethanolamine. Shown are the total (line), electrostatic (●), VDW (▲), bond (○), angle (□), Urey (△) and dihedral (◆) energy contributions to the surface. Ab initio data is represented by ◇ (HF/6-31G(d)), ⊠ (MP2/6-31G(d)//HF/6-31G(d)), ⊠(HF/6-31G(d,p)//HF/6-31G(d)) and ⊙ (MP/6-31G(d,p)//HF/6-31G(d)).

In ethanolamine the electrostatic interactions dominates the surface; significant contributions are also made by the Lennard-Jones and dihedral terms.

The changes in the NCC and CCO angles as a function of conformation for ethanolamine are presented in Table 6b. For the NCC angle the empirical force field reproduces the ab initio trend, although, the magnitude of the changes among conformers differs somewhat. For the CCO angle the trends are similar to those observed in the NCC angle; however, the changes are less pronounced and the relative values differ. As with choline, the empirical force field satisfactorily reproduces the direction of the ab initio changes in the NCC and CCO angles as a function of conformation, although some differences in magnitude are present. Such agreement provides an indication of the proper balance between the intra and intermolecular portions of the force field.

Ester Link

Parameters for the ester link between the glycerol backbone and the fatty acid chains were obtained using methyl acetate as the model compound. Interaction parameters were determined making use of ab initio calculations for the methyl acetate, water system, and the intramolecular parameters were fit to the experimental geometry and frequencies available for the molecule. Torsion parameters were adjusted based on the energy differences between the conformers of methyl acetate. Methyl propionate and ethyl acetate were also included as model compounds. Methyl propionate provides additional information on parameters associated with the ester link to the aliphatic chain, and ethyl acetate yields additional information on the ester link and the glycerol backbone.

INTRAMOLECULAR PARAMETERS OF METHYL ACETATE

For methyl acetate most of the parameters were available from the protein and nucleic acid parameter sets. Only parameters related to the ester oxygen (OS) and the carbonyl carbon had to be optimized. As a starting point, parameters for the ester oxygen were obtained from the methyl ester oxygen of dimethylphosphate. The equilibrium bond lengths and angles were set to the experimental values, except for the H–C–C=O dihedral which was set to the geometry of the ab initio energy minimum (Wiberg and Laidig, 1988; 1987), which is the Z-conformer. The Z-conformer is defined by a O=C–O–C dihedral angle of 0. The E-conformer is a local minimum in which the ester methyl group is trans to the carbonyl oxygen. Geometric information from microwave experiments (Williams et al, 1971), HF/6-31G(d), ab initio calculations (Wiberg and Laidig, 1988; 1987) and the CHARMM force field are presented in Table 7. Since the Z-conformation is found in lipid molecules, we concentrated on the geometry of Z-methyl acetate in the fitting procedure. As seen from Table 7 the ab initio and experimental structures are in good agreement. Calculated internal coordinates for the Z-conformer have bond lengths within 0.02 Å and bond angles with 1° of the experimental stucture. Upon going from the Z to the E-conformer,

Table 7. Geometries of Methyl Acetate[a]

	Exper Z^b	Ab initio		CHARMM	
		Z	E	Z	E
r_{C-C}	1.52	1.504	1.512	1.512	1.517
$r_{C=O}$	1.20	1.188	1.183	1.216	1.217
r_{C-O}	1.33	1.326	1.335	1.334	1.343
r_{O-C}	1.43	1.416	1.407	1.439	1.435
α_{C-C-O}	109	111.4	118.1	109.6	114.0
$\alpha_{O-C=O}$	125.87	123.4	119.3	125.9	124.1
α_{C-O-C}	114.78	116.9	122.8	114.3	121.6

[a] Bond lengths in Å and angles in degrees.
[b] Williams et al, 1971.

the empirical force field reproduces the direction of the change obtained in the ab initio calculations, with the exceptions of the C=O bond length, where the differences are 0.005 Å or less in both the empirical and ab initio models.

Optimization of the bond and angle force constants was based on both experimental and ab initio vibrational data. Almost all frequencies above 250 cm^{-1} were availible from gas-phase IR measurements (Hollenstein and Günhthard, 1980). To aid in the assignment of these frequencies the ab initio HF/6-31G* level frequency spectrum was used. Table 8 shows the experimental, scaled ab initio, and empirical optimized frequencies. The empirical frequencies can be compared with the 17 experimentally observed frequencies below 2000 cm^{-1}. The differences are generally 25 cm^{-1} or less. Two modes, 13 and 20, which are associated with the H–C–H and H–C–O bends of the methyl group in the ester bridge have deviations of approximately 60 cm^{-1} and 140 cm^{-1} from the ab initio and experimental results, respectively. The discrepancy of mode 20 is related to the 1–4 electrostatic interactions between the positively charged carbonyl carbon and the hydrogen atoms. No attempt was made to reduce this discrepancy because of the good quality of the fit for the remaining low frequency modes. A Urey-Bradley term was included for the OBL-CL-OSL angle term to obtain the large splitting between modes 8 and 14 (see Appendix, Figure 1e).

Acetic acid was used to verify the use of the OBL-CL-OSL Urey-Bradley term in methyl acetate. In addition, acetic acid is present in the crystal of DLPE which is analyzed in a later section. The standard O–C–O angle parameters from acetic acid, containing no Urey-Bradley term, were replaced with the values determined for methyl acetate. As presented in Table 9, addition of the new parameters leads to changes of approximately 70 cm^{-1} and 130 cm^{-1} for modes 4 and 6, respectively, of the CHARMM results. Only minor changes occur for the remaining modes. Both modes 4 and 6 shift towards the experimental values (Hollenstein and Günhthard, 1980); the results are now within 30 cm^{-1} and 40 cm^{-1} of experiment, instead of the original deviations of 50 and 170 cm^{-1}. Thus, a Urey-Bradley term or other off-diagonal term is essential to reproduce

Table 8. Frequencies of Methyl Acetate $(cm^{-1})^a$

Nr.	Type	Exper.[b]	Ab initio*0.9	CHARMM
1	A''	(73)	84	90
2	A''	(136)	148	107
3	A''	(185)	172	165
4	A'	295	273	288
5	A'	438	408	435
6	A''	603	601	625
7	A'	636	627	634
8	A'	840	856	834
9	A'	976	976	992
10	A''	(1053)	1064	1050
11	A'	1058	1084	1034
12	A'	1160	1206	1155
13	A''	1194	1169	1131
14	A'	1246	1292	1220
15	A'	1372	1406	1376
16	A'	1437	1452	1419
17	A'	1440	1472	1428
18	A''	1447	1458	1430
19	A''	1462	1479	1454
20	A'	1469	1485	1606
21	A'	1769	1817	1753
22	A'	2950	2908	2895
23	A'	2964	2926	2917
24	A''	3003	2965	2977
25	A''	3003	2999	2917
26	A'	3028	3006	2914
27	A'	3028	3014	2976

[a] Absolute rms. deviation (ab initio -emp) of frequencies 4 to 21: ±9.

[a] Relative rms deviation (ab initio -emp) of frequencies 4 to 21: 0.7%.

[b] From Hollenstein et al (1980); calculated values are in parentheses.

the experimental frequencies for this case. This is supported by a valence force field (VFF) previously developed for acetic acid (Hollenstein and Günhthard, 1980). The VFF contains an off-diagonal $k_{r(OS-OB)}$, r_{OS-OB} force constant, which is one magnitude larger than all other off-diagonal terms. Given the diagonal $k_{r(C-OB)}$, r_{C-OB} and $k_{r(C-OS)}$, r_{C-OS} force constants, a simple estimation indicates that the large off-diagonal term corresponds to the high Urey-Bradley force constants in the present force field. A readjustment of the other force constant yields even better agreement between the empirical and experimental frequencies. Also, the PED values from HF/6-31G(d) ab initio calculations are in good agreement with the empirical results, as shown in Table 9. Details of the acetic acid parameterization will be published elsewhere (Kuczera et al, 1996).

Final adjustment of the methyl acetate intramolecular force field dealt with the torsion terms. Previous ab initio studies on methyl acetate (Wiberg and Laidig, 1988, 1987) had shown that there are two stable conformers, the Z- and

Table 9. Frequencies of Acetic Acid (cm^{-1})

Nr.	Type	Exper[a]	Ab initio*0.9	CHARMM Old	CHARMM Old + UB	CHARMM New
1	A″	(92)	92	163	164	73
2	A′	428	406	440	433	413
3	A″	535	528	499	496	498
4	A′	581	574	529	600	583
5	A″	639	653	667	663	663
6	A′	847	844	672	806	835
7	A′	987	1003	978	977	971
8	A″	1044	1066	1050	1049	1043
9	A′	1181	1219	1174	1219	1193
10	A′	1280	1335	1306	1281	1257
11	A′	1380	1417	1377	1377	1372
12	A″	1434	1458	1438	1438	1424
13	A′	1439	1452	1431	1431	1418
14	A′	1779	1838	1765	1767	1762
15	A′	2944	2909	2895	2895	2916
16	A″	2996	2966	2955	2955	2976
17	A′	3051	3010	2955	2955	2976
18	A′	3566	3650	3691	3690	3690
Absolute rms deviation of modes 2–14				±15	±6	±4.5
Relative rms deviation of modes 2–14				1.9%	0.8%	0.7%

[a] Calculated values from Hollenstein et al, 1980.

E-ester. The highest level of theory used in that study, MP3/6-311G(d,p)//HF/6-31G(d) yielded an energy difference between Z- and E-conformers of 8.66 kcal/mol and a barrier height of 13.19 kcal/mole. Experimental studies have yielded a Z/E energy difference of 8.5 ± 1 kcal/mol in the gas phase and an estimate of the barrier height of 10–15 kcal/mol (Blom and Günthard, 1981). Additional ab initio calculations performed at the HF and MP2 6-31G(d) and 6-311G(d,p) levels of theory in the present study (see Table 10), are in good agreement with the previously published values. The empirical dihedral parameters were adjusted based on the results from the MP2/6-31G(2d,p)//HF/6-31G(d) calculations. Figure 8 shows the empirical O-C-O-C adiabatic surface, including the contributions of the individual potential energy terms and the ab initio data. Agreement between the empirical and ab initio surfaces is good for both the relative Z/E energies and the energy of the transition state in the vicinity of 90°. As with choline (Figure 5), the surface contains significant contributions from the majority of the terms in the potential energy function, with the dihedral term dominating the surface. To obtain the desired behavior, a sum of onefold and twofold torsion terms (OBL-CL-OSL-CTL3, see Appendix) were used of the O-C-O-C dihedral. The position of the trans to gauche barrier was 98° for the CHARMM surface, which compares well with a value of 103° from the ab initio HF/6-31G(d) surface (Wiberg and Laidig, 1987).

Table 10. Ab Initio Results for Methyl Acetate, Ethyl Acetate, and Methyl Propionate[a]

| | | 6-31G(d) | | 6-31G(2d,p) | |
	Conformation	HF	MP2	HF	MP2
Methyl acetate					
Z	CS	−266.836830	−267.589907	−266.856383	−267.691362
TS(Z−E)	C1	−266.815558	−267.567291	−266.835519	−267.669366
ΔE		13.35	14.19	13.09	13.80
E	CS	−266.821845	−267.575865	−266.842193	−267.678606
ΔE		9.40	8.81	8.90	8.00
Methylpropionate					
trans	180	−305.872599	−306.735933	−305.896356	−306.882650
TS	105.2	−305.870653	—	−305.894376	−306.881252
ΔE		1.22	—	1.24	0.88
gauche	64.9	−305.871065	−306.734695	−305.894830	−306.881573
ΔE		0.96	0.78	0.96	0.68
cis	0	−305.869972	−306.733797	−305.893641	−306.880567
ΔE		1.65	1.34	1.70	1.31
Ethylacetate					
gauche	83.8	−305.875891	−306.740308	−305.899807	−306.886568
trans	180	−305.876629	−306.740189	−305.900523	−306.886284
ΔE		−0.46	0.07	−0.45	0.18
TS	124.2	−305.874601	−306.738754	−305.898554	−306.884839
ΔE		0.96	0.98	0.79	1.09
cis	0	−305.864315	—	−305.888218	−306.874819
ΔE		7.26	—	7.27	7.37

[a] Angles in degrees and energies in hartrees except the energy differences, which are in kcal/mole.
[a] The geometries were optimized at HF/6-31G(d) level of theory.

INTERACTION PARAMETERS OF METHYL ACETATE

Supermolecule water/methyl acetate ab initio HF/6-31G(d) results were obtained from Briggs et al (Briggs et al, 1991; see Figure 9). Their approach was the same as that used here. The results are listed in Table 11. To account for correlation and polarization effects in polar neutral compounds, all interaction energies were scaled by a factor of 1.16, as discussed in MacKerell and Karplus (1991) and Reiher (1985). The intermolecular distances were decreased by 0.2 Å as required to reproduce the proper pure solvent density (Jorgensen, 1986; Jorgensen and Swenson, 1985). For the Lennard-Jones parameters, standard values from the protein and nucleic acid parameter sets were used. The partial charges were adjusted to fit the scaled ab initio results, with the HF/6-31G(d) Mulliken charge distribution used as an initial guess. As may be seen in Table 11, both the interaction energies and geometries of the acetate, water minima are reproduced by the empirical force field. The largest discrepancy occurs in orientation D. Attempts to improve the agreement lead to larger disagreement for the remaining interactions. Since orientation D represents the least favorable interaction, which

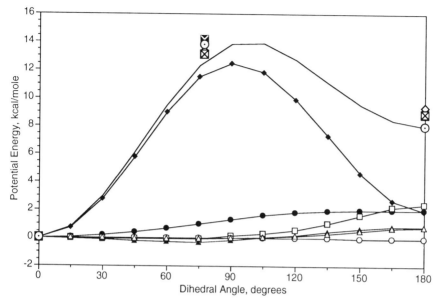

Figure 8. Adiabatic potential energy surface for the O–C–O–C dihedral angle of methylacetate. Shown are the total (line), electrostatic (●), VDW (▲), bond (○), angle (□), Urey (△) and dihedral (▲) energy contributions to the surface. Ab initio data is represented by ◇ (HF/6-31G(d)), ⊠(MP2/6-31G(d)//HF/6-31G(d)), ⊠(HF/6-31G(d,p)//HF/6-31G(d)) and ⊙ (MP/6-31G(d,p)//HF/6-31G(d)).

is, therefore, the least important, the charges were kept as is. The dipole moment from the empirical charge distribution of methyl acetate is 2.45 D. This is larger than the experimental gas-phase value of 1.72 D, corresponding to the effective mean polarization of methyl acetate in the condensed phase. This is analogous to the overestimation of the gas phase dipole moment of water by the TIP3P water model, 2.35 Debye, as compared to the experimental gas phase value of 1.855 Debye (Reiher, 1985).

Pure liquid simulations of methyl acetate were performed to test the interaction parameters. A MC simulation was performed for a box of 128 methyl acetate molecules. From these simulations values of 9.04 ± 0.02 kcal/mole and 133.9 ± 0.3 Å^3 were determined for the heat of vaporization and molecular volume, respectively. Comparison with the experimental values of 7.76 kcal/mole (Briggs et al, 1991) and 132.7 Å^3 (Timmermans, 1950) showed that the empirical force field overestimates the heat of vaporization while giving good agreement for the molecular volume. Attempts to improve agreement with experiment were undertaken (not shown); the van der Waals parameters of the carbonyl carbon and ester oxygen were varied along with small adjustments of the charges. Simultaneous fit of both liquid parameters could not be obtained without destroying the water interaction results. This suggests that the polarization of methyl acetate in water is larger than it is in the pure liquid.

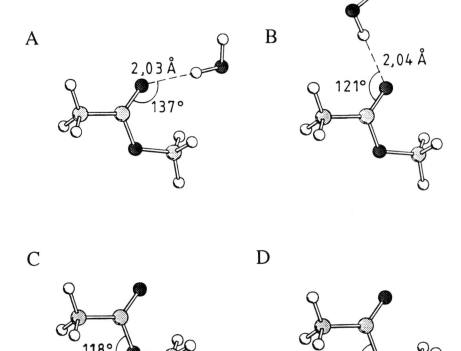

Figure 9. The four water/methyl acetate interaction complexes from Briggs et al. The HF/6-31G* optimized intermolecular distances and angles are given.

Table 11. Minimum Interaction Energies and Geometries between Water and Methyl Acetate

Interaction	Ab initio[a]			CHARMM		
	ΔE^{b}	r_{O-Ow}	a_{C-O-Ow}	ΔE	r_{O-Ow}	a_{C-O-Ow}
A	−6.77	2.03	137	−6.48	1.77	143
B	−6.43	2.04	121	−6.36	1.76	126
C	−3.58	2.15	118	−3.36	1.95	118
D	−3.20	2.15	134	−2.18	2.03	148

[a] The water-methyl acetate ab initio data was obtained from Briggs et al, 1991.
[b] Interaction energies have been scaled by 1.16 in accord with MacKerell and Karplus, 1991. Distances in Å, angles in degrees and energies in kcal/mole. Interaction geometries are shown in Figure 9.

Ethyl Acetate and Methyl Propionate

To investigate the conformation energies of additional terminal methyl groups, we performed ab initio calculations on ethyl acetate and methyl propionate in the Z-conformation (see Table 10). Of interest were the relative energies in the trans and gauche conformations associated with the additional methyl groups and the barrier heights between the stable conformers (cis for gauche → antigauche and TS for gauche → trans). The position of the methyl group was defined by the dihedral angle $\tau_{C-C-C-Oe}$ (methyl propionate) and $\tau_{C-C-Oe-C}$ (ethyl acetate), where Oe is the ester oxygen. These four structures were optimized at HF/6-31G(d) level of theory. The resulting structures are shown in Figures 10 and 11 for methylpropionate and ethylacetate, respectively. The dihedral angles from the HF/6-31G(d) optimized structures are given in Table 9 along with the relative energies obtained from single point calculations at the MP2/6-31G(2d,p)//HF/6-31G(d) level of theory.

In Figure 12 the potential energy of the methyl propionate Z-conformer is plotted as a function of the orientation of the terminal methyl group. The ab initio energy difference between gauche and trans is 0.68 kcal/mol (MP2/6-31G(2d,p)//HF/6-31G(d)). This theoretical value is somewhat lower than the experimentally estimate of 1.1±0.3 kcal/mol (Moravie and Coret, 1974). Predicted barrier heights at the MP2/6-31G(2d,p)//HF/6-31G(d) level are 0.88 kcal/mol (TS; gauche → trans) and 1.31 kcal/mol (cis; gauche → antigauche). To repro-

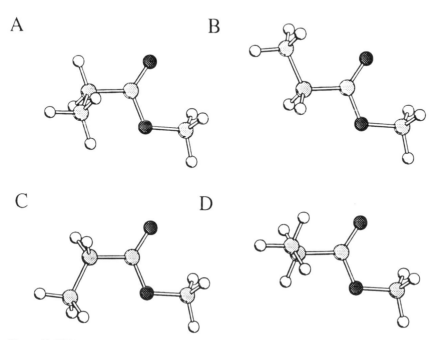

Figure 10. HF/6-31G(d) optimized structures of methylpropionate. A) Minimum energy gauche, B) saddlepoint trans, C) cis transition state and D) TS transition structures are shown.

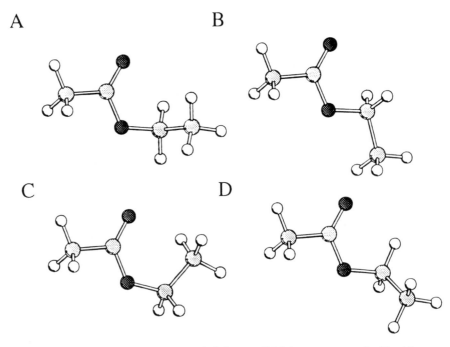

Figure 11. HF/6-31G(d) optimized structures of ethylacetate. A) Minimum energy gauche, B) saddle-point trans, C) cis transition state and D) TS transition structures are shown.

duce this relatively flat energy profile the sum of onefold and twofold torsional terms was used (see Appendix). The onefold term was included to optimize both the gauche/trans minima relative energies and the cis barrier height. Contributions from the individual terms in the potential energy function show significant contributions from the angle, electrostatic and van der Waals terms, in addition to the dihedral term.

Adjustment of the torsion parameters associated with rotation of the ethyl group in ethyl acetate was made to reproduce ab initio data. Table 10 presents the ab initio results, including the C–C–O–C dihedral angles of the various minima and barriers. The energy difference between the trans and gauche structures is small, varying from −0.5 to 0.2 kcal/mole at the MP2/6-31G(2d,p)//HF/6-31G(d) level. A similar result has prevously been reported by Mannig et al (1986) with HF/4-21G calculations. The TS and cis barriers equal 0.7 and 7.4 kcal/mol, respectively, at the MP2/6-31G(2d,p)//HF/6-31G(d) level. Figure 13 presents the potential energy of ethyl acetate as a function of the orientation of the added methyl group. The agreement between the empirical and the ab initio MP2/6-31G(2d,p)//HF/6-31G(d) results is good. In the empirical calculation significant contributions are made by the angle, van der Waals, and Urey-Bradley terms. The dihedral contribution is relatively small; only a onefold

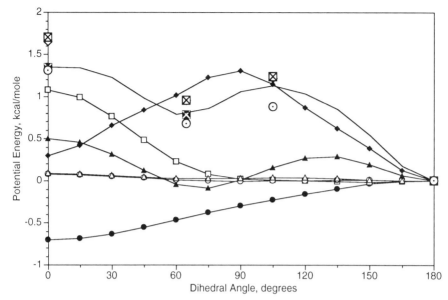

Figure 12. Adiabatic potential energy surface for the C–C–C–O dihedral angle of methylpropionate. Shown are the total (line), electrostatic (●), VDW (▲), bond (○), angle (□), Urey (△) and dihedral (◆) energy contributions to the surface. Ab initio data is represented by ◇ (HF/6-31G(d)), ⊠(MP2/6-31G(d)//HF/6-31G(d)), ⊠(HF/6-31G(d,p)//HF/6-31G(d)) and ⊙ (MP/6-31G(d,p)//HF/6-31G(d)).

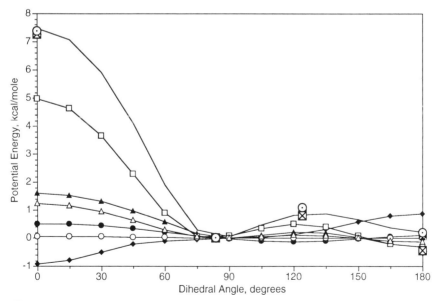

Figure 13. Adiabatic potential energy surface for the C–O–C–C dihedral angle of ethylacetate. Shown are the total (line), electrostatic (●), VDW (▲), bond (○), angle (□), Urey (△) and dihedral (◆) energy contributions to the surface. Ab initio data is represented by ◇ (HF/6-31G(d)), ⊠(MP2/6-31G(d)//HF/6-31G(d)), ⊠(HF/6-31G(d,p)//HF/6-31G(d)) and ⊙ (MP/6-31G(d,p)//HF/6-31G(d)).

torsion term with a force constant of 0.7 kcal/mole/deg. was required to obtain the final surface (see Appendix).

Application of the Force Field to Lipid Crystals

Simulations of crystals provide an important test of the empirical force field in the condensed phase because they permit detailed comparisons of the theoretical results with experimental data. Only a small number of lipid crystal structures are known (Hauser et al, 1981). The arrangement of the lipid molecules in the crystals that are analyzed correspond to multilaminal bilayers, comparable to those ocurring in membrane bilayers. Three lipid crystals were studied (see Figure 14):

(1) 3-lauroylpropanediol-1-phosphorylcholine monohydrate (LPPC, Hauser et al, 1980). The crystal is monoclinic with the space group $P2_{1/c}$ (see Figure 14a and Table 12), and the asymmetric unit contains one LPPC molecule and one water molecule. The structure was refined to an R-factor of 25%. Each LPPC molecule has one aliphatic chain in a fully interdigitated arrangement and a tilt angle of 41° with respect to the layer plane. From the cell parameters, a layer thickness of 24.5 Å can be derived; the polar region of the molecule spans 7.3 Å of the total 24.5 Å. The surface area of one LPPC molecule in the crystal is 52.1 Å^2.

(2) 2,3-Dimyristoyl-D-glycero-1-phosphorylcholine   dihydrate   (DMPC) (Pearson and Pascher, 1979). DMPC crystallizes in the monoclinic space group $P2_1$ (see Figure 14b and Table 13), there are two lipids and four water molecules in the asymmetric unit, and the R-factor is 17.4%. The DMPC molecules form a double layer structure in the crystal with a thickness of 54.9 Å. Headgroups of the independent DMPC molecules exhibit almost mirror image symmetry, while their position and orientation differ. The individual headgroups are referred to as DMPC1 and DMPC2. A 17° inclination occurs between the dipole arising from the negative phosphate group and the positive trimethylammonium group and the bilayer plane for DMPC1, and a 27° angle occurs for DMPC2. In addition both monomers are displaced in the direction of the layer normal by 2.5 Å, giving a zigzag, or sawtooth arrangement (Hussin and Scott, 1987). There are two aliphatic chains per molecule, which have a minor tilt angle of 12°. This tilt accomodates the slightly larger head group area of 54.9 Å^2, as compared to LPPC.

(3) 2,3-dilauroyl-DL-glycero-1-phosphorylethanolamine acetic acid (DLPE) (Hitchcock et al, 1974). The crystal is monoclinic with space group $P2_{1/c}$ (see Figure 14c and Table 14). The asymmetric unit contains one lipid molecule and one acetic acid; the R-factor of 28%, suggesting that the structure is not very accurate. The DLPE is arranged in the crystal as a double layer struc-ture with a thickness of 47.7 Å. The phosphorylethanolamine headgroup is almost parallel to the bilayer plane and the aliphatic chains are perpendicular to the plane.

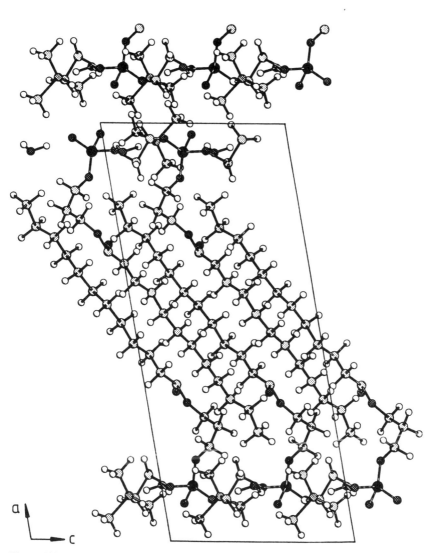

Figure 14(A). Experimental structures of the three lipid crystals as obtained from X-ray difraction. A) LPPC; B) DMPC (only the heavy atoms are shown) and C) DLPE. For each crystal the orientation was selected to optimize viewing of the aliphatic tails. All figures contain two of the unit cell dimensions as lines. The dimensions shown are indicated in the figure. See text for original references.

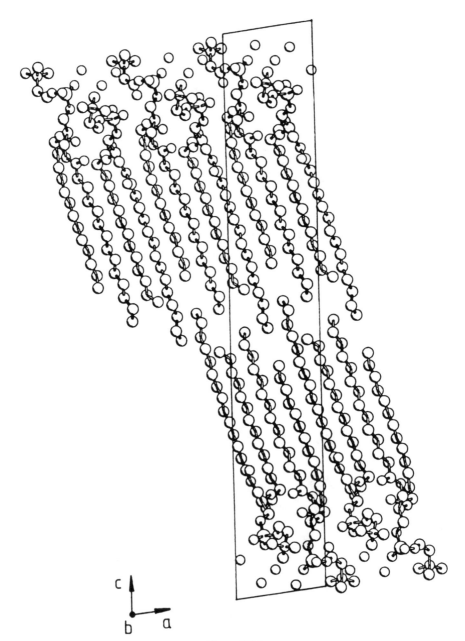

Figure 14(B).

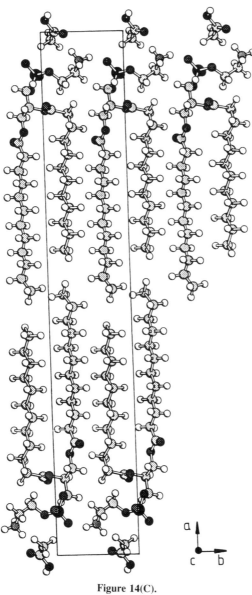

Figure 14(C).

Table 12. Experimental and Optimization Data for the LPPC Crystal[a]

		No lattice opt.		Lattice opt.	
	Crystal	15 Å	30 Å	15 Å	30 Å
Total Energy		−135.6	−139.2	−139.9	−143.6
a	24.82			24.22	24.46
b	9.53			9.07	8.98
c	10.94			10.86	10.79
β	99.7			100.8	101.0
$V_{unitcell}$	2552.0			2343.4	2324.9
RMS_{asym}		0.33	0.33	0.66	0.58
RMS_{LPPC}		0.28	0.28	0.24	0.25

[a] Energies in kcal/mole, distances in Å and volumes in $Å^3$.

Table 13. Experimental and Optimization Data for the DMPC Crystal[a]

		No lattice opt.		Lattice opt.	
	Crystal	15 Å	30 Å	15 Å	30 Å
Total Energy		−322.8	−331.0	−332.4	−339.0
a	8.72			8.49	8.49
b	8.92			8.74	8.71
c	55.4			53.45	53.71
β	97.4			91.6	91.6
$V_{unitcell}$	4273.3			3963.6	3971.7
RMS_{asym}		0.68	0.62	1.31	1.18
RMS_{DMPC1}		0.35	0.34	0.40	0.41
RMS_{DMPC2}		0.34	0.33	0.40	0.41

[a] Energies in kcal/mole, distances in Å and volumes in $Å^3$.

Table 14. Experimental and Optimization Data for the DLPE Crystal[a]

		No lattice opt.		Lattice opt.	
	Crystal	15 Å	30 Å	15 Å	30 Å
E		−223.8	−232.8	−225.3	−235.0
a	47.730			47.989	47.983
b	7.773			7.746	7.685
c	9.953			9.469	9.478
β	92.02			90.92	90.45
$V_{unitcell}$	3690.3			3519.3	3494.8
RMS_{asym}		0.42	0.51	0.71	0.67
RMS_{DLPE}		0.16	0.18	0.18	0.18

[a] Energies in kcal/mole, distances in Å and volumes in $Å^3$.

Crystal Minimizations

The minimizations were done in two stages, as described above. In the minimizations two different cutoffs were used to analyze the effect of truncation of the long-range Coulomb interactions. Those interactions between the large dipole moments of the lipid headgroups can lead to long-range effects. Minimizations where the unit cell parameters are varied along with the atomic positions allow for the lipid molecules to rearrange themselves and obtain their desired molecular volume. In Tables 12, 13, and 14 the unit cell parameters and volumes for the experimental and minimized structures for each crystal are listed. Also, the root mean square (RMS) displacements of the molecules between the minimized and X-ray structures are given. The energy listed in the tables include intra- and intermolecular interactions of one asymmetric unit with itself and with all image particles. The RMS of the asymmetric unit (RMS_{asym}) is calculated for all heavy atoms, including any solvent molecules. The RMS_{lipid} of an individual lipid molecule in the asymmetric unit was also calculated. Here the displacements are obtained after optimum superposition of the minimized coordinates with the X-ray coordinates of the molecule. Thus, RMS_{asym} accounts for the rearrangement of both the internal structure of the lipid and solvent molecules and of their position in the crystal, while RMS_{lipid} only contains contributions from changes in the internal geometries.

The results in Tables 12, 13, and 14 show that the energies of the all three crystals become more favorable upon increasing the cutoff distance. This is consistent with the role of long-range electrostatic interactions in a crystal. Based on the changes observed in the other crystal properties, it appears that truncation effects in the range of 15 Å to 30 Å have only a very small influence on these properties. All three systems showed a contraction in the minimization where the unit cell parameters were optimized (LPPC: −9%; DMPC: −7%; DLPE: −5%). This result is consistent with the fact that the minimizations yield structures corresponding to 0 K, whereas the experimental structures were determined at finite temperatures. For example, NPT crystal simulations of histidine showed a small increase in the unit cell volume as compared to crystal minimizations (MacKerell and Karplus, 1996b). With the exception of lattice parameter a in the DMPC crystal, all the unit cell parameters contracted. The uniformity of the contraction indicates that the overall balance of the parameters is satisfactory. The increase of lattice parameter A in the DMPC crystal was 0.5%, which is the smallest change of all the unit cell parameters. As expected, the RMS differences in the atomic positions are larger when the unit cell parameters are optimized; the largest changes occured in DMPC where RMS_{asym} deviations up to found 1.3 Å were calculated. The larger differences in DMPC system may be related to the presence of two lipid molecules in the asymmetric unit.

A detailed analysis of changes in the conformation of the individual lipids was performed by comparing the experimental and minimized dihedral angles (see Table 15 and Figure 15). The maximum difference in the dihedral angles between the 15 Å and 30 Å cutoff minimization results (not shown) was ±2°, indicating that the 15 Å cutoff results were essentially converged. Further anal-

Table 15. Structural Results from Crystal Minimization of LPPC, DMPC, and DLPE[a]

	LPPC		DMPC1		DMPC2		DLPE	
	Cryst.	30 Å	Cryst.	30 Å	Cryst.	30 Å	Cryst.	30 Å
$\alpha 1$	162	165	163	168	177	160	−154	−142
$\alpha 2$	87	63	62	59	−74	−72	58	60
$\alpha 3$	44	69	68	60	−47	−43	66	61
$\alpha 4$	130	111	143	139	−150	−146	106	107
$\alpha 5$	84	74	−64	−81	54	71	67	64
$\alpha 6$	166	180	179	180	176	172		
$\theta 1$	28	63	58	62	168	168	−52	−62
$\theta 2$			177	−178	−82	−72	65	61
$\theta 3$	78	72	−178	179	166	−175	−172	−168
$\theta 4$			63	58	51	65	69	67
$\beta 1$			82	78	120	144	97	104
$\beta 2$			172	180	179	174	179	172
$\beta 3$			−81	−87	−134	−145	−119	−123
$\beta 4$			45	73	67	59	65	70
$\beta 5$			171	170	180	−175	−178	180
$\gamma 1$	156	166	−177	156	102	136	−178	−176
$\gamma 2$	176	177	168	172	176	177	173	179
$\gamma 3$	−175	−172	−173	163	180	−172	179	−176
$\gamma 4$	178	176	178	174	180	−178	−171	−179
$\gamma 5$	179	177	180	177	−170	−178	−173	−176

[a] The dihedral are given in degrees and defined in Figure 15.

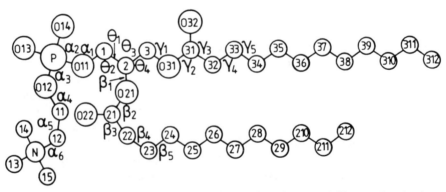

Figure 15. Definition of the lipid dihedral angles taken from Hauser et al. The α angles give the head group orientation and the θ angles the conformation of the glycerine back bone. The linkage with the aliphatic chains are given by the β and γ angles. The dihedral angles θ_1, θ_2, θ_3, and θ_4 are defined as O11-C1-C2-C3, O11-C1-C2-O21, C1-C2-C3-O31 and O21-C2-C3-O31, respectively.

ysis was performed on the 30 Å cutoff structures. In the head group region (α dihedrals) the largest discrepancies occurred in the LPPC crystal. Dihedrals defining the P–O–C–C–N backbone changed between 10° to 25° relative to the X-ray value. These differences had almost no effect on the essential P to N distance, which defines the magnitude of the head group dipole moment. The experimental distance was 4.54 Å and it increased to 4.60 Å in the minimized structure. For the glycerol backbone (θ dihedrals), DMPC2 had the largest discreancies (10° to 19°). Dihedrals involved in the linkages to the aliphatic chains (β and γ dihedrals) were generally close to the experimental structure in the DMPC crystal, although some changes as large as 34° occurred. These deviations did not lead to significant changes in the overall structure. The DMPC lipid crystals were relatively poorly defined as evidenced by the R-factor of 17.4%, so it is possible that there are alternative minima, such as those found in the minimizations, that are only little higher in energy than the experimental structure.

Crystal Dynamics Simulations

As a more rigorous test of the parameters a number of crystal dynamics simulations were made. The molecular dynamics simulations were performed with the full unit cell for 100 ps using a time step 1fs and a cutoff of 12 Å. The simulations were in the NVE ensemble with the experimental lattice parameters. The simulations were started by gradually heating to 300° K. The average temperatures from the simulations were 310° for LPPC, 305° for DMPC, and 294° for DLPE. Coordinates were stored every 20 fs for analysis.

   In all three systems no significant changes in the conformations of the lipids were observed. The average displacements from the equilibrium positions were ≈0.1 to 0.4 Å for the non-hydrogen atoms. The only exceptions were the hydrophobic tails of DMPC and DLPE. Displacement of both chains in DMPC increased from 0.2 to 1.1 Å upon going from the minimizations to the simulations. The DLPE chains showed an asymmetry and the terminal carbon of the gamma chain had an average deviation of 1.3 Å from the experimental position. The terminal carbon of the DLPE beta chain showed an average deviation of 2.5 Å from the initial position. The larger deviations in the DMPC and DLPE crystals were consistent with the large values of the observed temperature (Debye-Waller) factors. In both DLPE and DMPC a significant increase in the Debye-Waller factors occurred at the termini of the aliphatic chains. Such an increase was not found in LPPC, possibly due to the interdigitated structure of the aliphatic tails.

Discussion of Applications to Membranes

The present empirical energy function has been used successfully in a number of calculations on lipid membranes. Since such systems are notorious for being very sensitive to the force field, they are a good test of the parameter set. In addition, they are the system of primary interest for the lipid force field. We comment on certain aspects of the force field that play an important role in the stability of the lipid membrane simulations and in the structural and dynamic results. Systems discussed include 1-palmitoyl-2-oleoyl-*sn*-glycero-3-phosphatidylcholine (POPC) lipid bilayers (Heller et al, 1993), with the head group partial atomic charges from the present study, a dipalmitoyl phosphatidylcholine (DPPC) bilayer (Venable et al, 1993), o-phosphorylcholine and o-phosphorylethanolamine in solution (Woolf and Roux, 1994a) and a sodium dodecyl sulfate micelle (MacKerell, 1995). In addition, calculations have been published for a simulation of gramicidin in a lipid bilayer (Woolf and Roux, 1994b) that test the consistencies of the protein and lipid force fields. All of these simulations included explicit solvent. The agreement between theory and experiment obtained in these studies demonstrates the validity of the parameters for simulations of lipid bilayers.

Simulation of the DPPC bilayer (Venable et al, 1993) included 72 lipid molecules and 2533 water molecules with periodic boundary conditions. Results from the simulations showed good agreement with experiment for the fast reorientation correlation times of the aliphatic chains and the microscopic lateral diffusion of the lipids. These reorientation times depended on the quality of both the intra- and intermolecular aliphatic parameters. To obtain the proper torsional surface in the aliphatic chains, the dihedral parameters were optimized such that the torsional surface of butane agreed with a range of experimental and ab initio data. Both the van der Waals and electrostatic terms made a significant contributions to the butane surface, emphasizing the importance of the balance between the intra- and intermolecular parameters obtained via the iterative optimization approach used in the parameterization. Lateral diffusion of the lipids appeared to be dominated by the molecular interactions at the bilayer surface. Since lipid-lipid, lipid-solvent, and solvent-solvent interactions all contributed to these interactions, the good agreement with experiments validates the balance of the intramolecular parameters. It should be emphasized that optimization of the present force field is based on the use of the TIP3 water model; other water models may not be consistent with the solute parameters.

The balance of the interaction parameters is also supported by the recent study on the conformational flexibility of o-phosphorylcholine (PC) and o-phosphorylethanolamine (PE) (Woolf and Roux, 1994a). In that study potential of mean force calculations were performed in aqueous solution with respect to the N–C–C–O dihedral angles. Stabilization of the trans conformer was observed for both PC and PE; the extent of stabilization was in qualitative agreement with experimental data obtained from NMR studies. Since the parameter optimization of the torsional surfaces of choline and ethanolamine was for the gas phase (see above), the agreement for the solution conformation was particularly important. Analysis of the gas phase energy surfaces (Figures 5 and 7 for choline and ethanolamine, respectively) showed that the van der Waals and electrostatic terms made significant contributions to the surfaces, especially in the case of ethanolamine. The magnitude of this contribution was due to the fact that intermolecular interactions between atoms in a 1,4 configuration (connected through three bonds) are not scaled, unlike some other force fields (e.g., the CHARMM 19 polar hydrogen model and Weiner et al, 1986). The good agreement of the empirical gas phase surfaces with ab initio data in Figures 5 and 7 and the agreement of the conformational flexibility of PC and PE in solution with experiment support the absence of 1,4 scaling. The agreement of the solution results also supports the methodology used to combine the model compounds employed in the parameterization into larger species and the compatibility of the lipid and nucleic acid parameter sets. As noted above, the phosphate parameters in the present force field were transferred directly from the nucleic acid parameter set. No tests were performed on PC or PE to validate the quality of the transfer prior to their use by Woolf and Roux (1994a).

Simulation of a sodium dodecyl sulfate micelle in solution tests the use of ions with present parameters. The sulfate (MacKerell, 1994) and sodium (Roux, 1990) parameters were optimized with the same parameterization philosophy as in the present study. The simulation involved 60 lipid molecules, 60 sodiums, and 4398 water molecules with periodic boundary conditions. All ionic species were fully charged. Sodium to sulfate interactions were observed to occur primarily via the second hydration shell, in agreement with experiment. Direct ion contact pairs were observed, but they were transitory in nature. These results again emphasize the importance of the balance in the interaction potential; e.g., if the solute-solute interactions were too favorable, stable ion contact pairs would be dominant. Good agreement with experiment was also found for the overall size of the micelle and the distribution of the terminal aliphatic tail methyl groups with respect to the micelle center of mass. This agreement depends on the quality of the aliphatic parameters, which provide for the correct packing in the micelle interior.

Protein-lipid interactions were studied in a simulation of gramicidin in a lipid bilayer by Woolf and Roux (1994b). The system included 16 dimyristoyl phosphatidylcholine molecules (DMPC), one gramicidin channel and 649 water molecules, and applied periodic boundary conditions; the system was set up to mimic that used in several NMR solid-state studies. Concerning the DMPC bilayer, a local minimum in the alkane chain density at the center of the bilayer was obtained from the calculations, in agreement with neutron scattering experiments. This indicated that intermolecular parameters for the aliphatic and polar head groups of the lipids and those for the solvent were well balanced. Good agreement was also obtained between calculation and experiment for ^{15}N chemical shifts, ^{15}N–^{1}H dipolar coupling of the tryptophans and ^{2}H quadrupolar splitting of the gramicidin Ca^{+2} sites, and tryptophans. This demonstrates that the overall structural properties of the gramicidin channel were maintained; the RMS difference from the experimental structure was 1.2 Å. Such behavior requires that the protein and lipid parameters are well balanced. Further support for this is provided by the agreement of the conformation of the Trp side chains with experiment. In the simulation the Trp side chains were at the surface of the membrane, oriented such that the N–H group of the side chains were facing the polar head groups and solvent. A decrease in the percent of gauche conformers of the lipid aliphatic chains was calculated due to the presence of the gramicidin channel, again in agreement with NMR and FTIR experimental data.

Overall, the quality of agreement with experiment in these simulations demonstrated that the same parametrization method, when applied to different types of systems, can insure the quality of the parameters for the individual systems as well as for interactions between them.

Conclusion

A new all-atom force field for empirical energy calculations on phospholipid molecules has been presented. It is compatible with the corresponding CHARMM22 force field for proteins and nucleic acids. In all cases, a major focus in the development was the balance between the intramolecular and intermolecular portions of the force field and the balance among the solvent-solvent, solvent-solute, and solute-solute terms in the intermolecular terms. Proper treatment of these aspects of the force field is essential for accurate thermodynamic and structural simulations of condensed phase systems. Parameter optimization was based on a wide range of experimental and ab initio data that included microscopic and macroscopic properties. Of note is the choice of the Lennard-Jones terms for the methyl hydrogen atoms in TMA. They were treated as intermediate between aliphatic and polar hydrogens, due to their partial polar character. This allowed for the proper treatment of hydrogen bonding interactions. Urey-Bradley terms were introduced to fit the vibrational frequencies and assignments in specific instances. Similarly, a two-term series was used for certain angles to reproduce ab initio dihedral energy surfaces.

Optimization of the partial atomic charges is based primarily on the reproduction of ab initio determined interaction energies and geometries between model compounds and water molecules interacting with polar sites on the model compounds. In the ab initio calculations the water-model compound interactions are calculated individually such that the induced polarization at a particular site on the molecule due to the interaction with the water molecule is included in the interaction energy and geometry. These effects are neglected in partial charge optimization procedures that use single molecule calculations, such as those based on quantum mechanical electrostatic potentials (Bayly et al, 1993; Chirlan and Francl, 1987; Singh and Kollman, 1984).

Adiabatic dihedral potential energy surfaces for choline, ethanolamine, methylacetate, methylpropionate, and ethylacetate have significant contributions from both bonding and nonbonding terms. This emphasizes the importance of obtaining the proper balance between the intra- and intermolecular portions of the empirical force field; i.e., alteration of one requires readjustment of the other. The balance also plays a role in the reproduction of ab initio results on changes in bond angles as a function of conformation.

Force field calculations on three lipid crystals and on lipid membrane systems indicate that the force field is valid for condensed phase simulations. This is confirmed by its successful use in several solvated liquid bilayer and solution simulations, including one for gramicidin in a lipid bilayer.

ACKNOWLEDGMENTS

This work was supported in part by grants from the National Institutes of Health and from the National Science Foundation. We thank Richard Pastor and Benoit Roux for helpful discussions. ADM Jr. acknowledges a NIH Postdoctoral fellowship at Harvard University, and M.S. thanks the Deutsche Forschungs Gemeinschaft and Deutscher Akademischer Austausch Dienst, which made possible the work at Harvard University. M.S. thanks P. Bopp for helpful discussions and introduction to the field of molecular dynamic simulations. We thank Horst Vollhardt for helping in generating the molecule plots and for preliminary results concerning the crystal molecular dynamic simulations.

REFERENCES

Aue DH, Webb HM, Bowers MT (1976): A thermodynamic analysis of solvation effects on the basicities of alkylamines. An electrostatic analysis of substituent effects. *J Am Chem Soc* 98:318–329

Bayly CI, Cieplak P, Cornell WD, Kollman PA (1993): A well-behaved electrostatic potential based method using charge restraints for deriving atomic charges: The RESP model. *J Phys Chem* 97:10269–10280

Berg RW (1977): The vibrational spectrum of the normal and perdeuterated tetramethylammonium ions. *Spectrochim Acta* 34A:655–659

Blom CE, Günthard HH (1981): Rotational isomerism in methylformate and methylacetate: A low-temperature matrix infrared study using thermal molecular beams. *Chem Phys Lett* 84:267–271

Bottger GL, Geddes AL (1965): The infrared spectra of the crystalline tetramethylammonium halides. *Spectrochim Acta* 21:1701–1708

Boyd RH (1969): Lattice energies and hydration thermodynamics of tetra-alkylammonium halides. *J Chem Phys* 51:1470–1474

Briggs JM, Nguyan TB, Jorgensen WL (1991): Monte Carlo simulations of liquid acetic acid and methyl acetate with the OPLS potential functions. *J Phys Chem* 95:3315–3322

Brooks BR, Bruccoleri RE, Olafson BD, States DJ, Swaminathan S, Karplus M (1983): CHARMM: A program for macromolecular energy, minimization, and dynamics calculations. *J Comput Chem* 4:187–217

Brun TO, Curtiss LA, Iton LE, Kleb R, Newsam JM, Beyerlein RA, Vaughn DEW (1987): Inelastic neutron scattering from tetramethylammonium cations occluded with zeolites. *J Am Chem Soc* 109:4118–4119

Charifson PS, Hiskey RG, Pedersen LG (1990): Construction and molecular modeling of phospholipid surfaces. *J Comp Chem* 11:1181–1186

Chirlan LE, Francl MM (1987): Atomic charges derived from electrostatic potentials: A detailed study. *J Comp Chem* 8:894–905

Cook RL, DeLucia FC, Helminger P (1974): Molecular force field and structure of water: Recent microwave results. *J Mol Spect* 53:62–76

Damodaram KV, Merz KM Jr, Gaber BP (1992): Structure and dynamics of the dilauroylphosphatidylethanolamine lipid bilayer. *Biochemistry* 31:7656–7664

Dutta PK Del Barco B, Shieh DC (1986): Raman spectroscopic studies of the tetramethylammonium ion in zeolite cages. *Chem Phys Lett* 127:200–204

Elder M, Hitchcock P, Mason R, Shipley GG (1977): A refinement analysis of the crystallography of the phospholipid, 1,2-dilauroyl-DL-phosphatidylethanolamine, and some remarks on lipid-lipid and lipid-protein interactions. *Proc R Soc Lond A* 354:157–170

Florián J, Johnson BG (1994): Comparison and scaling of hartree-fock and density functional harmonic force fields. 1. Formamide monomer. *J Phys Chem* 98:3681–3687

Frisch FMJ, Head-Gordon M, Trucks GW, Foresman JB, Schlegel HB, Raghavachari K, Robb M, Binkley JS, Gonzalez C, Defrees DJ, Fox DJ, Whiteside RA, Seeger R, Melius CF, Baker J, Martin RL, Kahn LR, Stewart JJP, Topiol S, Pople JA (1990): Gaussian 90 (computer program). Revision. Pittsburgh, PA: Gaussian, Inc.

Gao J, Jorgensen WL (1992): personal communication

Gelin BR, Karplus M (1975): Role of structural flexibility in conformational calculations. Application to acetylcholine and β-methylacetylcholine. *J Am Chem Soc* 97:6996–7006

Guyan L, Brady J (1996): All-hydrogen empirical force field parameters for carbohydrates. (Manuscript in preparation)

Hariharan PC, Pople JA (1972): The effect of d-functions on molecular orbital energies for hydrocarbons. *Chem Phys Lett* 66:217–219

Hauser H, Pacher I, Sundell S (1980): Conformation of phospholipids: Crystal structure of a lysophosphatidylcholine analogue. *J Mol Biol* 137:249–264

Hauser H, Pascher I, Pearson RH, Sundell S (1981): Preferred conformation and molecular packing ofphosphatidylethanolamine and phosphatidylcholine. *Biochim Biophys Acta* 650:21–51

Heller H, Schaefer M, Schulten K (1993): Molecular dynamics simulation of a bilayer of 200 lipids in the gel and in the liquid-crystal phases. *J Phys Chem* 97:8343–8360

Hitchcock PB, Mason R, Thomas KM, Shipley GG (1974): Structural chemistry of 1,2 dilauroyl-DL-phosphatidylethanolmaine: molecular conformation and intermolecular packing of phospholipids. *Proc Natl Acad Sci USA* 71:30363040

Hollenstein H, Günhthard HH (1980): A transferable valence field for polyatomic molecules. *J Mol Spec* 84:457–477

Hussin A, Scott HL (1987): Density and bonding profiles of interbilayer water as a function of bilayer separation: A Monte Carlo study. *Biochim Biophys Acta* 897:423–430

Jorgensen WL (1986): Optimized intermolecular potential functions for liquid alcohols. *J Phys Chem* 90:1276–1284

Jorgensen WL (1983): Theoretical studies of medium effects on conformational equilibria. *J Phys Chem* 87:5304–5312

Jorgensen WL, Swenson CJ (1985): Optimized intermolecular potential functions for amides and peptides. Structure and properties of liquid amides. *J Am Chem Soc* 107:569–578

Jorgensen WL, Chandrasekhar J, Madura JD, Impey RW, Klein ML (1983): Comparison of simple potential functions for simulating liquid water. *J Chem Phys* 79:926–935

Kuczera K, Wiorkiewicz-Kuczera J, Karplus M (1993): MOLVIB: Program for Vibrational Spectroscopy, Program Charmm (computer program). Version 22

Kuczera K, Gao J, MacKerell AD Jr, Karplus M (1996): Empirical parameters for the ionic species of amino acids and protein termini. (Manuscript in preparation)

MacKerell AD Jr (1994): Empirical force field parameters for sulfate and methylsulfate. (unpublished)

MacKerell AD Jr (1995): Molecular dynamics simulation analysis of a sodium dodecyl sulfate micelle in aqueous solution: Decreased fluidity of the micelle hydrocarbon interior. *J Phys Chem* 99:1846–1855

MacKerell AD Jr, Karplus M (1991): Importance of attractive van der Waals contributions in empirical energy function models for the heat of vaporization of polar liquids. *J Phys Chem* 95:10559–10560

MacKerell AD Jr, Karplus M (1996a): All-atom empirical energy function for simulations of peptides and proteins. (Manuscript in preparation)

MacKerell AD Jr, Karplus M (1996b): Parameterization of histidine for molecular modeling and molecular dynamics simulations. (Manuscript in preparation)

MacKerell AD Jr, Bashford D, Bellott M, Dunbrack RL Jr, Field MJ, Fischer S, Gao J, Guo H, Ha S, Joseph D, Kuchnir L, Kuczera K, Lau FTK, Mattos C, Michnick S, Ngo T, Nguyen DT, Prodhom B, Roux B, Schlenkrich M, Smith J, Stote R, Straub J, Wiorkiewicz-Kuczera J, Karplus M (1992): Self-consistent parameterization of biomolecules for molecular modeling and condensed phase simulations. *Biophys J* 6:A143

MacKerell AD Jr, Wiorkiewicz-Kuczera J, Karplus M (1995): An all-atom empirical energy function for the simulation of nucleic acids. *J Amer Chem Soc*: 117:11946–11975

MacKerell AD Jr, Field MJ, Fischer S, Watanabe M, Karplus M (1996): All-hydrogen alkane potential for use in aliphatic groups of macromolecules. (Manuscript in preparation)

Mahendra P, Agarwal A, Khandelwal DP, Bist HD (1984): Dynamic disorder of $(CH_3)_4N^+$ in $(CH_3)_4NX$ (X = Cl, Br and I) as studied by Raman spectroscopy. *J Mol Struct* 112:309–316

Mahendra P, Raghuvanshi GS, Bist HD (1982): Vibrational studies and phase transitions in tetramethylammonium chloride. *Chem Phys Lett* 92:85–92

Mannig J, Klimkowski VJ, Siam K, Ewbank JD, Schäfer L (1986): Ab initio structural investigation of methyl and ethyl carbamate and carbamy choline. *J Mol Struct (THEOCHEM)* 139:305–314

Marrink S-J, Berendsen HJC (1994): Simulation of water transport through a lipid membrane. *J Phys Chem* 98:4155–4168

Millero FJ (1971): The molal volumes of electolytes. *Chem Rev* 71:147–176

Möller P, Plesset MS (1934): Note an an approximation treatment for many-electron systems. *Phys Rev* 46:618–622

Moravie RM, Coret J, (1974): Conformational behaviour and vibrational spectra of methyl propionate. *Chem Phys Lett* 26:210–214

Neria E, Fischer S, Karplus M (1996): Simulation of activation free energies in molecular systems. *J Chem Phys* (Submitted)

Pascher I, Lundmark M, Nyholm P-G, Sundell S (1992): Crystal structures of membrane lipids. *Biochem Biophys Acta* 1113:329–373

Pearson RH, Pascher I (1979): The molecular structure of lecithin dihydrate. *Nature* 281:499–501

Rao BG, Singh UC (1989): Hydrophobic hydration: A free energy perturbation study. *J Am Chem Soc* 111:3125

Reiher WE III (1985): Theoretical studies of hydrogen bonding (dissertation). Cambridge, MA: Harvard University

Roux B (1990): Theoretical study of ion transport in the gramicidin a channel (dissertation). Cambridge, MA: Harvard University

Schlenkrich M (1992): Entwicklung und anwendung eines kraftfeldes zur simulation von phospholipidmembransystemen (dissertation). Darmstadt, Germany: Technischen Hochschule

Singer SJ, Nicolson GL (1972): The fluid mosaic model of the structure of cell membranes. *Science* 175:720–731

Singh UC, Kollman PA (1984): An approach to computing electrostatic charges for molecules. *J Comp Chem* 5:129–145

Small DM (1986): The physical chemistry of lipids. New York: Plenum

Stote RH, States DJ, Karplus M (1991): On the treatment of electrostatic interactions in biomolecular simulation. *Chimie Physique* 88:2419–2433

Stouch TR, Ward KB, Altieri A, Hagler AT (1991): Simulations of lipid crystals: Characterization of potential energy functions and parameters for lecithin molecules. *J Comp Chem* 12:1033–1046

Timmermans J (1950): *Physico-Chemical Constants of Pure Organic Compounds*. Elsevier: Amsterdam

Venable RM, Zhang Y, Hardy BJ, Pastor RW (1993): Molecular dynamics simulations of a lipid bilayer and of hexadecane: An investigation of membrane fluidity. *Science* 262:223–226

Vegard L, Sollesnes K (1927): Structure of isomorphic substances $N(CH_3)_4X$. *Phil Mag* 4:985–1001

Weiner SJ, Kollman PA, Nguyen DT, Case DA (1986): An all atom force field for simulations of proteins and nucleic acids. *J Comp Chem* 7:230–252

Wiberg KB, Laidig KE (1987): Barriers to rotation adjacent to double bonds. 3. The C−O barrier in formic acid, methyl formate, acetic acid, and methyl acetate. The origin of ester and amide "Resonance". *J Am Chem Soc* 109:5935–5943

Wiberg KB, Laidig KE (1988): Acidity of (Z)- and (E)-methyl acetates: Relationship of meldrum's acid. *J Am Chem Soc* 110:1872–1874

Williams G, Owen NL, Sheridan J (1971): Spectroscopic studies of some substituted methyl formates. *Trans Faraday Soc* 67:922–949

Woolf TB, Roux B (1994a): Conformational flexibility of o-phosphorylcholine and o-phosphorylethanolamine: A molecular dynamics study of solvation effects. *J Am Chem Soc* 116:5916–5926

Woolf TB, Roux B (1994b): Molecular dynamics simulation of the gramicidin channel in a phospholipid bilayer. *Proc Natl Acad Sci USA* 91:11631–11635

Wyckof RWG (1928): The crystal structure of tetramethylammonium halides. *Z Kristallogr* 67:91–105

Appendix

Table I. Bond Parameters[a]

Atom types		K_b	b_0	Source
CTL3	CL	200.0	1.522	methyl acetate
CTL2	CL	200.0	1.522	methyl acetate
CTL1	CL	200.0	1.522	methyl acetate
OBL	CL	750.0	1.220	methyl acetate
OSL	CL	150.0	1.334	methyl acetate
OHL	CL	230.0	1.40	methyl acetate
HOL	OHL	545.0	0.960	acetic acid
CTL3	HAL	322.00	1.111	alkanes (protein)
CTL2	HAL	309.00	1.111	alkanes (protein)
CTL1	HAL	309.00	1.111	alkanes (protein)
CTL3	OSL	340.0	1.43	phosphate (nucleic acids)
CTL2	OSL	340.0	1.43	phosphate (nucleic acids)
CTL1	OSL	340.0	1.43	phosphate (nucleic acids)
OSL	PL	270.0	1.60	phosphate (nucleic acids)
O2L	PL	580.0	1.48	phosphate (nucleic acids)
OHL	PL	237.0	1.59	phosphate (nucleic acids)
NH3L	HCL	410.0	1.04	ethanolamine
NH3L	CTL2	261.0	1.51	ethanolamine
NTL	CTL2	215.00	1.51	tetramethylammonium
NTL	CTL3	215.00	1.51	tetramethylammonium
CTL3	HL	300.00	1.08	tetramethylammonium
CTL2	HL	300.00	1.08	tetramethylammonium
CTL1	CTL1	222.500	1.500	alkanes (protein)
CTL1	CTL2	222.500	1.538	alkanes (protein)
CTL1	CTL3	222.500	1.538	alkanes (protein)
CTL2	CTL2	222.500	1.530	alkanes (protein)
CTL2	CTL3	222.500	1.528	alkanes (protein)
CTL3	CTL3	222.500	1.530	alkanes (protein)
OHL	CTL1	428.0	1.420	glycerol
OHL	CTL2	428.0	1.420	glycerol
OHL	CTL3	428.0	1.420	glycerol
SL	O2L	540.0	1.448	methylsulfate
SL	OSL	250.0	1.575	methylsulfate

[a] K_b in kcal/mole/$Å^2$ and b_0 in $Å$.

Table II. Bond Angle Parameters[a]

Atom types			K_θ	θ_0	K_{UB}	S_0	Source
OBL	CL	CTL3	70.0	125.0	20.0	2.442	methyl acetate
OBL	CL	CTL2	70.0	125.0	20.0	2.442	methyl acetate
OBL	CL	CTL1	70.0	125.0	20.0	2.442	methyl acetate
OSL	CL	OBL	90.0	125.9	160.0	2.2576	acetic acid
CL	OSL	CTL1	40.0	109.6	30.0	2.2651	methyl acetate
CL	OSL	CTL2	40.0	109.6	30.0	2.2651	methyl acetate
CL	OSL	CTL3	40.0	109.6	30.0	2.2651	methyl acetate
HAL	CTL2	CL	33.0	109.5	30.00	2.163	methyl acetate
HAL	CTL3	CL	33.0	109.5	30.00	2.163	methyl acetate
CTL2	CTL2	CL	52.0	108.0			alkanes (protein)
CTL3	CTL2	CL	52.0	108.0			alkanes (protein)
OSL	CL	CTL3	55.0	109.0	20.00	2.3260	methyl acetate
OSL	CL	CTL2	55.0	109.0	20.00	2.3260	methyl acetate
OHL	CL	OBL	50.0	123.0	210.0	2.2620	acetic acid
OHL	CL	CTL3	55.0	110.5			acetic acid
OHL	CL	CTL2	55.0	110.5			acetic acid
HOL	OHL	CL	55.0	115.0			acetic acid
OSL	CTL1	CTL2	75.7	110.10			acetic acid
OSL	CTL1	CTL3	75.7	110.10			acetic acid
OSL	CTL2	CTL1	75.7	110.10			acetic acid
OSL	CTL2	CTL2	75.7	110.10			acetic acid
OSL	CTL2	CTL3	75.7	110.10			acetic acid
HAL	CTL2	HAL	35.5	109.00	5.40	1.802	alkanes (protein)
HAL	CTL3	HAL	35.5	108.40	5.40	1.802	alkanes (protein)
HAL	CTL1	OSL	60.0	109.5			phosphate (nucleic acids)
HAL	CTL2	OSL	60.0	109.5			phosphate (nucleic acids)
HAL	CTL3	OSL	60.0	109.5			phosphate (nucleic acids)
CTL2	OSL	PL	20.0	120.0	35.0	2.33	phosphate (nucleic acids)
CTL3	OSL	PL	20.0	120.0	35.0	2.33	phosphate (nucleic acids)
HOL	OHL	PL	30.0	115.0	40.0	2.30	phosphate (nucleic acids)
OSL	PL	OSL	80.0	104.3			phosphate (nucleic acids)
OSL	PL	O2L	98.9	111.6			phosphate (nucleic acids)
OSL	PL	OHL	48.1	108.0			phosphate (nucleic acids)
O2L	PL	O2L	120.0	120.0			phosphate (nucleic acids)
O2L	PL	OHL	98.9	108.23			phosphate (nucleic acids)
NTL	CTL2	HL	40.0	109.5	27.	2.13	tetramethylammonium
NTL	CTL3	HL	40.0	109.5	27.	2.13	tetramethylammonium
HL	CTL2	HL	24.0	109.50	28.	1.767	tetramethylammonium
HL	CTL3	HL	24.0	109.50	28.	1.767	tetramethylammonium
CTL3	NTL	CTL2	60.0	109.5	26.	2.466	tetramethylammonium
CTL3	NTL	CTL3	60.0	109.5	26.	2.466	tetramethylammonium
HL	CTL2	CTL2	33.43	110.10	22.53	2.179	alkanes (protein)
HL	CTL2	CTL3	33.43	110.10	22.53	2.179	alkanes (protein)
HAL	CTL1	CTL1	34.5	110.10	22.53	2.179	alkanes (protein)
HAL	CTL1	CTL2	34.5	110.10	22.53	2.179	alkanes (protein)
HAL	CTL1	CTL3	34.5	110.10	22.53	2.179	alkanes (protein)
HAL	CTL2	CTL1	26.5	110.10	22.53	2.179	alkanes (protein)

(continued)

Table II. (Continued)

Atom types			K_θ	θ_0	K_{UB}	S_0	Source
HAL	CTL2	CTL2	26.5	110.10	22.53	2.179	alkanes (protein)
HAL	CTL2	CTL3	34.6	110.10	22.53	2.179	alkanes (protein)
HAL	CTL3	CTL1	33.43	110.10	22.53	2.179	alkanes (protein)
HAL	CTL3	CTL2	34.60	110.10	22.53	2.179	alkanes (protein)
HAL	CTL3	CTL3	37.50	110.10	22.53	2.179	alkanes (protein)
NTL	CTL2	CTL2	67.7	115.00			tetramethylammonium
NTL	CTL2	CTL3	67.7	115.00			tetramethylammonium
HCL	NH3L	CTL2	33.0	109.50	4.00	2.056	ethanolamine
HCL	NH3L	HCL	41.0	109.50			ethanolamine
NH3L	CTL2	CTL2	67.7	110.00			ethanolamine
NH3L	CTL2	HAL	45.0	107.50	35.00	2.0836	ethanolamine
CTL1	CTL1	CTL1	53.35	111.00	8.00	2.561	alkanes (protein)
CTL1	CTL1	CTL2	58.35	113.50	11.16	2.561	glycerol
CTL1	CTL1	CTL3	53.35	108.50	8.00	2.561	alkanes (protein)
CTL1	CTL2	CTL1	58.35	113.50	11.16	2.561	glycerol
CTL1	CTL2	CTL2	58.35	113.50	11.16	2.561	glycerol
CTL1	CTL2	CTL3	58.35	113.50	11.16	2.561	glycerol
CTL2	CTL1	CTL2	58.35	113.50	11.16	2.561	glycerol
CTL2	CTL1	CTL3	58.35	113.50	11.16	2.561	glycerol
CTL2	CTL2	CTL2	58.35	113.60	11.16	2.561	alkanes (protein)
CTL2	CTL2	CTL3	58.0	115.00	8.00	2.561	alkanes (protein)
HOL	OHL	CTL1	57.5	106.00			glycerol
HOL	OHL	CTL2	57.5	106.00			glycerol
HOL	OHL	CTL3	57.5	106.00			glycerol
OHL	CTL1	CTL2	75.7	110.10			glycerol
OHL	CTL2	CTL1	75.7	110.10			glycerol
OHL	CTL2	CTL2	75.7	110.10			glycerol
OHL	CTL2	CTL3	75.7	110.10			glycerol
OHL	CTL1	HAL	45.9	108.89			glycerol
OHL	CTL2	HAL	45.9	108.89			glycerol
OHL	CTL3	HAL	45.9	108.89			glycerol
O2L	SL	O2L	130.0	109.47	35.0	2.45	methylsulfate
O2L	SL	OSL	85.0	98.0			methylsulfate
CTL2	OSL	SL	15.0	109.0	27.00	1.90	methylsulfate
CTL3	OSL	SL	15.0	109.0	27.00	1.90	methylsulfate

[a] K_θ in kcal/mole/rad^2, θ_0 in degrees, K_{UB} in kcal/mole/Å^2 and S_0 in Å.

Table III. Dihedral Angle Parameters[a]

Atom types				K_χ	n	δ	Source
X	CTL1	OHL	X	0.14	3	0.00	glycerol
X	CTL2	OHL	X	0.14	3	0.00	glycerol
X	CTL3	OHL	X	0.14	3	0.00	glycerol
OBL	CL	CTL2	HAL	0.00	6	180.00	acetic acid
OBL	CL	CTL3	HAL	0.00	6	180.00	acetic acid
OSL	CL	CTL2	HAL	0.00	6	180.00	acetic acid
OSL	CL	CTL3	HAL	0.00	6	180.00	acetic acid
OBL	CL	OSL	CTL1	3.85	2	180.00	methyl acetate
OBL	CL	OSL	CTL1	0.965	1	180.00	methyl acetate
OBL	CL	OSL	CTL2	3.85	2	180.00	methyl acetate
OBL	CL	OSL	CTL2	0.965	1	180.00	methyl acetate
OBL	CL	OSL	CTL3	3.85	2	180.00	methyl acetate
OBL	CL	OSL	CTL3	0.965	1	180.00	methyl acetate
X	CL	OSL	X	2.05	2	180.00	methyl acetate
X	CTL2	CL	X	0.05	6	180.00	methyl acetate
X	CTL3	CL	X	0.05	6	180.00	methyl acetate
X	CL	OHL	X	2.05	2	180.00	acetic acid
HAL	CTL2	CL	OHL	0.00	6	180.00	acetic acid
HAL	CTL2	CL	OHL	0.00	6	180.00	acetic acid
HAL	CTL3	CL	OHL	0.00	6	180.00	acetic acid
OSL	PL	OSL	CTL2	0.95	2	0.00	phosphate (nucleic acids)
OSL	PL	OSL	CTL2	0.50	3	0.00	phosphate (nucleic acids)
O2L	PL	OSL	CTL2	0.10	3	0.00	phosphate (nucleic acids)
OSL	PL	OSL	CTL3	0.95	2	0.00	phosphate (nucleic acids)
OSL	PL	OSL	CTL3	0.50	3	0.00	phosphate (nucleic acids)
O2L	PL	OSL	CTL3	0.10	3	0.00	phosphate (nucleic acids)
OHL	PL	OSL	CTL2	0.95	2	0.00	phosphate (nucleic acids)
OHL	PL	OSL	CTL2	0.50	3	0.00	phosphate (nucleic acids)
OHL	PL	OSL	CTL3	0.95	2	0.00	phosphate (nucleic acids)
OHL	PL	OSL	CTL3	0.50	3	0.00	phosphate (nucleic acids)
X	OHL	PL	X	0.30	3	0.00	phosphate (nucleic acids)
X	CTL1	OSL	X	−0.10	3	0.00	phosphate (nucleic acids)
X	CTL2	OSL	X	−0.10	3	0.00	phosphate (nucleic acids)
X	CTL3	OSL	X	−0.10	3	0.00	phosphate (nucleic acids)
CTL3	CTL2	OSL	CL	0.7	1	180.00	ethyl acetate
CTL2	CTL2	OSL	CL	0.7	1	180.00	ethyl acetate
CTL3	CTL1	OSL	CL	0.7	1	180.00	ethyl acetate
CTL2	CTL1	OSL	CL	0.7	1	180.00	ethyl acetate
CTL2	CTL2	CL	OSL	−0.15	1	180.00	methyl propionate
CTL2	CTL2	CL	OSL	0.53	2	180.00	methyl propionate2
CTL3	CTL2	CL	OSL	−0.15	1	180.00	methyl propionate
CTL3	CTL2	CL	OSL	0.53	2	180.00	methyl propionate
X	CTL2	NTL	X	0.26	3	0.00	tetramethylammonium
X	CTL3	NTL	X	0.23	3	0.00	tetramethylammonium
X	CTL2	NH3L	X	0.10	3	0.00	ethanolamine
NH3L	CTL2	CTL2	OHL	0.7	1	180.00	ethanolamine
NH3L	CTL2	CTL2	OSL	0.7	1	180.00	ethanolamine

(continued)

Table III. (Continued)

Atom types				K_χ	n	δ	Source
NTL	CTL2	CTL2	OHL	4.3	1	180.00	choline
NTL	CTL2	CTL2	OSL	3.3	1	180.00	choline
NTL	CTL2	CTL2	OHL	−0.4	3	180.00	choline
NTL	CTL2	CTL2	OSL	−0.4	3	180.00	choline
CTL3	CTL2	CTL2	CTL3	0.15	1	0.00	alkanes (protein)
CTL2	CTL2	CTL2	CTL2	0.15	1	0.00	alkanes (protein)
CTL2	CTL2	CTL2	CTL3	0.15	1	0.00	alkanes (protein)
X	CTL1	CTL1	X	0.200	3	0.00	alkanes (protein)
X	CTL1	CTL2	X	0.200	3	0.00	alkanes (protein)
X	CTL1	CTL3	X	0.200	3	0.00	alkanes (protein)
X	CTL2	CTL2	X	0.195	3	0.00	alkanes (protein)
X	CTL2	CTL3	X	0.160	3	0.00	alkanes (protein)
X	CTL3	CTL3	X	0.155	3	0.00	alkanes (protein)
HAL	CTL3	OSL	SL	0.00	3	0.00	methylsulfate
CTL2	OSL	SL	O2L	0.00	3	0.00	methylsulfate
CTL3	OSL	SL	O2L	0.00	3	0.00	methylsulfate

[a] K_χ in kcal/mole, δ in degrees. X indicates a wild card.

Table IV. Improper Dihedral Angle Parameters[a]

Atom types				K_φ	φ_0	Source
OBL	X	X	CL	100.00	0.00	acetic acid

[a] K_φ in kcal/mole/rad^2, φ_0 in degrees. X indicates a wildcard.

Table V. Lennard-Jones Parameters[a]

Atom type	ε	$R_{min}/2$	$\varepsilon_{1,4}$	$R_{min,1,4}/2$	Source
HOL	−0.046	0.2245			hydroxyl (protein)
HAL	−0.022	1.3200			alkane (protein)
HCL	−0.046	0.2245			ethanolamine
HL	−0.046	0.70			tetramethylammonium
CL	−0.070	2.00			methyl acetate
CTL1	−0.0200	2.275	−0.01	1.9	alkane (protein)
CTL2	−0.0550	2.175	−0.01	1.9	alkane (protein)
CTL3	−0.0800	2.06	−0.01	1.9	alkane (protein)
OBL	−0.12	1.70	−0.12	1.4	carbonyl (protein)
O2L	−0.12	1.70			anionic (nucleic acid)
OHL	−0.1521	1.77			hydroxyl (protein)
OSL	−0.1521	1.77			ester (nucleic acid)
NH3L	−0.20	1.85			ethanolamine
NTL	−0.20	1.85			tetramethylammonium
PL	−0.585	2.15			phosphate (nucleic acid)
SL	−0.47	2.10			methylsulfate

[a] ε in kcal/mole, $R_{min}/2$ in Å, $\varepsilon_{1,4}$ in kcal/mole and $R_{min,1,4}/2$ in Å.

Appendix Figure 1. Partial atomic charges and atom types for the model compounds A) tetramethylammonium, B) ethyltrimethylammonium, C) choline, D) ethanolamine, E) methylacetate, F) methylproprionate, and G) ethylacetate.

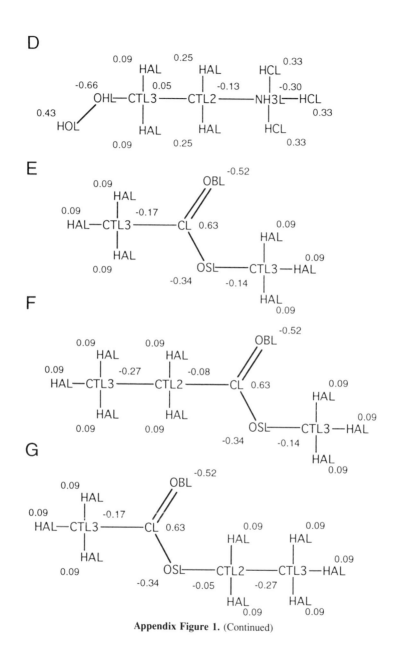

Appendix Figure 1. (Continued)

3

Statistical Mechanics and Monte Carlo Studies of Lipid Membranes

H. LARRY SCOTT

Introduction

Over the past 25 years or so experimental studies of lipid bilayer membranes have progressed to the point at which a wealth of data are available from a large variety of experimental measurements. In particular, NMR experiments (Bloom et al, 1991; Brown et al, 1983; Seelig, 1977), X-ray experiments (McIntosh 1990; Tristram-Nagle et al, 1993), and infrared spectroscopic experiments (Mendelsohn and Senak, 1993) yield data that are related directly to structures and interactions at the molecular level. To aid in the interpretation of this data, and to gain a more complete understanding of lipid bilayers at the molecular level, the next step is to construct theoretical models. To be of use, a theoretical model must be consistent with the data and must contain all of the important atomic level properties as determined from experiment. Ideally, the model will have predictive capabilities. That is, the input to the model will be entirely based upon atomic level properties of the constituent molecules which are independently determined. Then, observable properties of the model will be calculated by the methods of statistical mechanics or from direct computer simulation. Lastly, calculated properties (predictions) of the model will be compared with experimental measurements. Unfortunately, this process is not often so simple in practice. Theoretical models with molecular or atomic level detail are generally too complex to be solved analytically. For this reason, computer simulation methods play a large role in the theoretical analysis of models for complex systems such as lipid bilayers in solution.

   The most commonly used direct computer simulation method is molecular dynamics (MD), in which equations of motion for all of the atoms in a system are integrated numerically. Macroscopic observables are calculated from time

Biological Membranes
K. Merz, Jr. and B. Roux, Editors
© Birkhäuser Boston 1996

averages over calculated trajectories. The connection with equilibrium systems is made through ergodic arguments which connect time averages with configuration averages. The problems and issues surrounding the application of MD to lipid membranes are discussed in several chapters in this volume. The present chapter will describe alternative methods, namely Monte Carlo (MC) methods, which are based on statistical mechanics rather than directly integrated dynamical trajectories. In the next sections of this chapter, both equilibrium and kinetic Monte Carlo methods will be described. This will be followed by a historical review of MC modeling of lipid bilayers. The review will discuss applications of the Monte Carlo method to lipid bilayers at two levels:

- Direct MC simulation of a lipid bilayer with atomic resolution for the study of molecular conformations and interactions.
- MC calculation of the properties of a statistical mechanical model for the study of cooperative, large scale behavior.

In its most basic form, the Monte Carlo method consists of calculating equilibrium thermodynamical averages by a biased sampling of the configuration space of a lipid bilayer, using atomic-resolution models for lipids and water. The aim is to determine equilibrium properties of the constituent molecules as well as thermodynamic averages. Unfortunately, as will be described below, it is difficult in practice to efficiently sample this configuration space if the number of degrees of freedom per molecule is large.

By judiciously reducing the number of degrees of freedom, one can devise a simplified model for the lipid bilayer in terms of a Hamiltonian function which is in some sense a lower resolution approximation to the real system. The loss of information due to the reduction in degrees of freedom is, one hopes, offset by the gain in the scope and accuracy of the simulation. As an example, in the following section a model will be described in which fluid phase type degrees of freedom are left out. While all fluid-phase information is lost in the reduced model, it will be seen that the model is capable of describing a cooperative phase transition in which long range intermolecular correlations are involved. The model is applied to the ripple phase-to-gel phase transition in lipid bilayers. This application serves to illustrate the power of the direct use of the tools of Statistical Mechanics to predict cooperative global molecular phase changes over time and length scales which are not accessible to MD simulations.

Finally, there is a description of a new MC algorithm that has the potential to sample the configuration space of a lipid bilayer far more efficiently than conventional direct MC or MD sampling. Throughout the present chapter, the emphasis will be on use of independently calculated atomic or molecular level interactions to predict structural and thermodynamic properties of lipid bilayers in thermal equilibrium. None of the discussions in any of the sections in this chapter will involve models in which ad hoc input parameters are adjusted to obtain good agreement with experiment.

Monte Carlo Methods for Lipid Bilayers

Equilibrium Monte Carlo Methods

The basic Monte Carlo method which is still widely used today, was developed in 1953 (Metropolis et al, 1953). While part of the early motivation may have been to study explosions, it was quickly realized that this technique could be applied to fundamental problems in condensed matter physics. The foundation for the MC method has been reviewed recently by Binder (1986). Since fast desktop computers have become widely available, MC simulations have been applied in one form or another in many fields of science. Because of this rapid growth in use of the MC method, and for completeness, a synopsis of the basic MC methodology and its underlying rationale will be given below.

The basic problem in statistical mechanics is to evaluate sums of the form:

$$Z = \sum_C \exp[-\mathcal{H}(C)/k_B T] \tag{1}$$

or

$$\langle O \rangle = \sum_C O(C) \exp[-\mathcal{H}(C)/k_B T]/\sum_C \exp[-\mathcal{H}(C)/k_B T] \tag{2}$$

where the sum runs over all configurations of the system, $\mathcal{H}$ is the Hamiltonian for the given configuration, k_B is Boltzmann's constant, T is the absolute temperature, and O is any dynamical variable. It is well known that this sum cannot be evaluated analytically for more than a handful of nontrivial models (Baxter, 1982), and these generally involve formidable mathematical analysis. One is therefore motivated to attempt to carry out the summations numerically. Direct evaluation of Z by numerical summation is, unfortunately, also out of the question. A system of N particles with, say, 2 discrete degrees of freedom per particle (as in a simple spin 1/2 magnetic system) has 2^N configurations, where $N \approx 10^{23}$!

In reality, the large majority of the terms in Equations (1) or (2) contribute only a negligible amount to the sum. Therefore, it is possible to numerically estimate such a sum by including only the important terms. For example, if terms in the sum are generated with probability proportional to their equilibrium probability, that is, proportional to $\exp[-\mathcal{H}(C)/k_B T]$, then the partition sum Z simply equals the number of terms generated, and the average $\langle O \rangle$ is given by an arithmatic average:

$$\langle O \rangle = \frac{\sum_{C_t} O(C_t) P_t^{-1} \exp[-\mathcal{H}(C_t)/k_B T]}{\sum_{C_t} P_t^{-1} \exp[-\mathcal{H}(C_t)/k_B T]} = \frac{1}{N} \sum_{C_t} O(C_t) \tag{3}$$

where C_t is a trial configuration picked with probability given by $P_e(C_t) = Z^{-1} \exp[-\mathcal{H}(C)/k_B T]$ (Binder, 1986).

The question now becomes, how can the trial configurations be generated with probability equal to their equilibrium probabilities? To do this sampling directly requires not only knowledge of $\mathcal{H}(C)$ but also the exact partition function Z which is, as mentioned above, not possible. What is needed is a way to sample from the subset of important states even though exact knowledge of the probabilities of these states is not available. The Importance Sampling procedure developed by Metropolis and co-workers (Metropolis et al, 1953) provides such a method. This is a procedure for generating configurations with probabilities that approach the equilibrium probabilities in the limit of a large number of steps.

The procedure consists of constructing configurations for sampling in a successive manner such that state $n + 1$ depends only upon state n and not upon any of the states preceding state n (a Markov walk in configuration space). If $P(C_t)$ is the probability that trial configuration C_t occurs, and if $w(t, t')$ is the transition probability from a state t to a state t' in the Markov walk, then the rate of change of $P(C_t)$ is given by the master equation:

$$\frac{dP(C_t)}{dt} = \sum_{t' \neq t} \left[w(t', t)P(C_{t'}) - w(t, t')P(C_t) \right] \tag{4}$$

For equilibrium the left hand side of the above equation must vanish and $P(C_t)$ must equal the equilibrium value, $P_e(C_t)$. In this limit the master equation becomes:

$$P_e(C_{t'})w(t', t) = P_e(C_t)w(t, t') \tag{5}$$

which is called the detailed balance criterion. If we rearrange the detailed balance equation, and make use of the fact that the equilibrium probabilities are proportional to Boltzmann factors, we get:

$$\frac{w(t, t')}{w(t', t)} = \frac{P_e(C_t')}{P_e(C_t)} = \exp[-(\mathcal{H}(C_t') - \mathcal{H}(C_t))/k_B T] \tag{6}$$

Note that unknown proportionality factors in the equilibrium probabilities are canceled in the ratio. The key result is that the quotient of the transition probabilities depends only on the difference in energy between the two states. It follows that a sufficient condition for the sampling probabilities to approach their equilibrium values is that the sampling procedure must be a Markov walk through phase space with step-to-step transition probabilities which satisfy Equations (4) and (5). Two typical examples of such probabilities are:

$$w(t, t') = \begin{array}{ll} \frac{1}{m} \exp[-(\Delta\mathcal{H})/k_B T] & \Delta\mathcal{H} > 0 \\ \frac{1}{m} & \Delta\mathcal{H} \leq 0 \end{array} \tag{7}$$

or

$$w(t, t') = \frac{1}{2m} \frac{\exp[-(\Delta\mathcal{H})/k_B T]}{[1 + \exp-(\Delta\mathcal{H})/k_B T]} \tag{8}$$

where m is a factor related to the number of ways to pick a succeeding move in the Markov walk, and $\Delta\mathcal{H} = (\mathcal{H}(C_{t'}) - \mathcal{H}(C_t))$.

A formal proof that the above procedure samples configurations with probabilities that approach the canonical equilibrium values can be constructed using the Central Limit Theorem. An heuristic argument has been given by Binder (1986) who shows that, under the conditions of Equations (4) and (5):

$$\Delta N_{t \to t'} \propto \left(\exp[-(\Delta \mathcal{H})/k_B T] - \frac{N_{t'}}{N_t} \right) \tag{9}$$

so that if the population ratio of states t and t' is greater than the equilibrium ratio (the exponential term) then $\Delta N_{t \to t'}$ is negative. If the population ratio is less than the equilibrium ratio then $\Delta N_{t \to t'}$ is positive. Therefore the system should approach a state where $\Delta N_{t \to t'} = 0$, which is the equilibrium state.

The procedure described above describes the Monte Carlo method for generating equilibrium configurations and calculating equilibrium averages for dynamical variables. It is natural to now ask whether the Markov walk can yield **dynamical** information about the system as equilibrium as approached and after it is attained. In the following section this question is addressed.

Kinetic Monte Carlo Methods

In some sense a Markov Monte Carlo walk in configuration space is also a walk in time. For example, consider the diffusion of a single atom on a surface for which the only energetics consist of the adatom hopping over barriers between neighboring potential wells on the surface. If all wells have the same depth, then all hops must traverse the same barrier. If we know the real-time frequency r with which the adatom attempts to hop to a neighboring site, and if z is the number of nearest neighbors, then the time associated with a single jump to a neighboring site is $t = 1/rz$. Similarly, in any system for which all moves have the same difference in energy, the steps in a MC walk can be related in a linear fashion to time. However, in most cases of interest, such as lipid bilayers, there are many different types of transitions between configurations, with many different energy barriers. In fact, even for fairly simple systems, Weinberg and coworkers (Fichtorn and Weinberg, 1991; Kang and Weinberg, 1989) have shown that Markov walks based on the transition probability given by Equation (8) multiplied by the $\exp[-E_b/k_B T]$ where E_b is the barrier height (this is sometimes called Kawasaki dynamics) does not agree with results calculated by simple barrier hopping except at high temperatures.

Binder (1986) has described alternative MC procedures for systems with many different energy barriers. In one such procedure transitions between states occur at random times according to a Poisson distribution. That is, if a given move occurs with rate r, the probability that exactly n moves will occur in time t is given by:

$$p(n) = \frac{(rt)^n}{n!} \exp(-rt) \tag{10}$$

A great advantage of the Poisson distribution is that it is multiplicative for multiple events, so that if we have a set of events with rates r_i, and a set of occurance numbers n_i for each event, then:

$$p(N) = \prod_i p(n_i) = \frac{R^N}{N!} \exp(-Rt) \tag{11}$$

where $N = \sum_i n_i$ and $R = \sum_i r_i$. If individual rates for all events in a simulation are known, the following algorithm can be used to produce a real-time MC simulation (Binder, 1986):

- Pick an event ν with probability r_ν/R
- Execute the event
- Advance the clock by a random time chosen from the Poisson distribution for the total rate R for all moves.

This method will yield real-time trajectories if one knows the values for the individual rates. Unfortunately in most cases of interest the potential surface is highly complex and rates for individual molecular moves are hard to determine. In some cases these may be estimated from experiment or calculated from molecular dynamics trajectory averaging for individual events.

Monte Carlo Studies of Lipid Chain Properties

In this section and the following section two applications of equilibrium MC calculations to lipids will be described. First, the calculation of conformational properties of individual lipid chains under various circumstances will be described. To this end, consider a lipid bilayer made up of one lipid, say dimyristoyl phosphatidylcholine (DMPC), in an aqueous solution. DMPC consists of 46 atoms (not counting hydrogens) on three distinct chains. Figure 1 shows a stick drawing of a DMPC molecule. The molecule is characterized by two distinct regions:

- Two hydrocarbon chains of 16 carbons each (the hydrophobic region). This portion of the molecule is highly insoluble in water.
- A chain containing a phosphate group (PO_4^-), and a choline group ($N^+(CH_3)_3$) (the head group). This portion of the molecule is highly soluble in water.

All three of the chains in DMPC or in any other biomembrane lipid are flexible. Distortions in bond length and bond angle, and, most importantly, rotations about bonds, all are degrees of freedom which control the conformation of each DMPC. In turn, the interactions between DMPC molecules, between DMPC and water, and between DMPC and other membrane molecules such as cholesterol or proteins, depend strongly upon molecular conformation. Unfortunately for theoreticians, the number of degrees of freedom that can contribute to the conformation of a single DMPC molecule is enormous. An estimate of the number of degrees of freedom that are active in fluid lipids can be made from the

Figure 1. Stick model of dimyristoyl phosphatidylcholine molecule (DMPC). The upper chain is the polar head group and the lower two chains are $(CH_2)_{12}CH_3$ hydrophobic chains. Hydrogen atoms are not shown.

enthalpy change at the lipid bilayer melting phase transition (Nagle and Scott, 1978). The result is around 15–20 degrees of freedom per molecule, largely associated with the hydrocarbon chains.

The problem of sampling in a system with a large number of degrees of freedom is illustrated by considering a single hydrocarbon chain of N carbons. Neglecting bond stretching and C–C bond torsion leaves only dihedral rotations about C–C bonds to consider. Each bond i may be characterized by a dihedral angle ϕ_i, $0 \leq \phi_i \leq 2\pi$. If we quantize the dihedral angle, allowing only n discrete values for each ϕ_j (say, $2\pi/n$) then for each state of bond i there are n states of every other bond. The total number of conformations of one chain based on dihedral rotations alone is then on the order of N^n. In practice many conformations in this set are disallowed by steric overlaps, but it is certain that the configuration space for the single chain due only to dihedral degrees of freedom is very large.

When torsional degrees of freedom for the polar groups of the lipid and for the water are added in, along with translational degrees of freedom for all atoms, it becomes immediately clear that the configuration space for a DMPC molecule and its neighboring waters is enormous compared to that of a simple fluid or magnet. A simple atomic fluid has six degrees of freedom per molecule, associated with position and momentum in the x, y, and z directions, and an Ising ferromagnet has two, corresponding to spin up or down.

Because the configuration space of a DMPC molecule plus its associated water molecules is so large, designing a good statistical sampling procedure is a difficult problem. For this reason the more popular simulation approach to date has been MD, in which molecular conformations evolve according to classical equations of motion. The hope of MD researchers is that simulations are sufficiently long that the most important subregions of configuration space are effectively sampled by the the system. That is, the interactions between molecules force the system into thermodynamic equilibrium conformations in a matter of a few hundred picoseconds. While this is likely to be true for certain fast dynamical properties, such as rotational reorientations, there are other dynamical variables that may require orders of magnitude more time to reach thermal equilibrium. Global rearrangements such as phase changes are far out of reach of MD simulation times.

Clearly, then, there remains a need to examine alternative methods for simulation of lipid bilayers. Since the main lipid phase transition is almost entirely a consequence of the activation of a subset of the total number of degrees of freedom, namely those associated with hydrocarbon chain conformations, one approach is to begin with such a simpler subsystem. The application of the MC method to a system of hydrocarbon chains attached to an planar interface has in fact been carried out by several groups, beginning as early as 1966. In this section MC simulations of systems consisting of only hydrocarbon chains, and some other simple models, will be briefly reviewed. The problems that arise when nonlipids are included in even these simple simulations will be described.

Historically, the application of the MC method to lipids probably began with the lattice-based simulations of Whittington and Chapman (1966). Whittington and Chapman used a two-dimensional lattice to generate a line of chains using a self-avoiding walk algorithm. In general lattice-based simulations have the advantage that the configuration space of the system is greatly reduced, making long simulations of larger systems possible. The problem is that the reduction is a very severe limitation on the accessible molecular conformations in the case of large, flexible molecules such as phospholipids,. As computing power has grown, the need for an underlying lattice in MC simulations has diminished and is mainly used for models that aim to describe long range cooperative phenomena.

Early Monte Carlo continuum simulations of hydrocarbon chains in a lipid bilayer type of environment were carried out by Scott (1978). The simulations utilized rotation matrix chain generation methods developed for polymer chains by Curro (1974). The first such simulations consisted of ten or fewer chains of hard spheres with the topmost sphere attached to an interface. Since that time, models studied by the MC method have progressed to bilayers of 200 chains interacting via 6–12 potentials between atoms, and containing cholesterol (Scott, 1991) or gramicidin (Xing and Scott, 1992). Because the method is limited to equilibrium properties, early MC simulations were utilized mainly to calculate segmental order parameters in the different models. Segmental order parameters are defined as:

$$\langle S_n \rangle = \frac{1}{2} \langle 3 \cos^2 \theta_n - 1 \rangle \tag{12}$$

where θ_n is the angular deviation of bond n from its orientation in the all-trans chain conformation.

The above definition of the angle θ_n was originally used because it is consistent with the classical definition of order parameter, that is, a quantity which is unity in a perfectly ordered phase and which is zero in a perfectly disordered phase. Currently, it is more useful to calculate CD order parameters, as these are directly measurable in Deuterium NMR experiments using selectively or fully deuterated lipids. CD order parameters differ from segmental order parameters in that the angle θ_n represents the deviation of the average carbon-to-deuterium bond vector at carbon position n from its orientation for ordered, all-trans, chains. The connection between segmental and CD order parameters is complex, but may in a first approximation be considered to be a linear scaling (Peterson and Chan, 1977). Since either type of order parameters are primarily influenced by chain packing, MC simulations provided insights into chain packing as a function of area per chain (Scott, 1990). Even in the simplest hard sphere chain simulations, the qualitative profile of S_n vs n plots followed the standard shape of a nearly flat plateau region for small n, followed by a drop in S_n as n approached the ends of the chains. One useful result of the early simulations was that the shape of the order parameter profiles depended critically upon the density of the chains. For low density (high area per molecule) the plateau on the order parameter profiles disappeared.

In order to carry out simulations in reasonable time in the early 1980s, it was necessary to restrict the dihedral angles allowed for $C–C$ bonds to three states per bond, 0, and ± 120 degrees. This restriction is consistent with the locations of three distinct minima in the dihedral energy function (Ryckaert and Bellemanns, 1975), and is commonly called the rotational isomeric approximation. Even within the rotational isomeric approximation, it was not possible to include, for example, lipid head groups or water in MC simulations at that time. A consequence of the restricted dihedral angles was that order parameter plots for the chains in the simulations were not as smooth as experimental plots.

Computing power has advanced by many orders of magnitude since the studies described above were done. A question now is: Is it currently possible to use the standard Metropolis MC method to simulate the equilibrium behavior of a hydrated lipid bilayer, explicitly including water and lipid head groups? This was attempted recently by Taga and Masuda (1995). These authors used the standard MC sampling procedure to calculate equilibrium properties of a lipid bilayer consisting of 36 molecules of dipalmitoyl phosphatidylcholine (DPPC), 18 in each monolayer, with no water of hydration. DPPC molecules were represented in full atomic detail, using 1–6–12 (that is, electrostatic and van der Waals) interactions between atoms, with separate torsional and dihedral potentials for chain atoms. Code was written to take advantage of vector processing. Still, simulations consisted of only 1.2×10^5 steps, which amounts to only about 60 moves for each coordinate in the system (three rotational and three translational coordinates per atom). Taga and Masuda report order parameter profiles and electron density profiles that qualitatively are similar to experimental data, but the results are compromised by the small number of steps. Ideally an MC simulation should incorporate at least 10^3 moves per degree of freedom. The conclusion drawn by this writer is that thorough MC simulation of a hydrated lipid bilayer is not yet feasible if the standard Metropolis sampling procedure is used. Later in this chapter an alternative sampling procedure which promises to greatly improve sampling efficiency will be described.

A natural application for all simulations of lipid bilayers is to study the affect of biologically important nonphospholipid molecules on the physical state of the hydrocarbon region of a bilayer. Using the simple hydrocarbon chains-only models described above, Scott and coworkers carried out MC calculations of the properties of systems of hydrocarbon chains and cholesterol (Scott, 1991; Scott and Kalaskar, 1989), hydrocarbon chains and a cylindrical model polypeptide (Scott, 1986), and hydrocarbon chains and gramicidin A (Xing and Scott, 1992; Xing and Scott, 1989). Perhaps most definitive were the chain-cholesterol simulations, because the structure of cholesterol is simple, and it interacts nearly entirely within the hydrophobic interior of a bilayer. A synopsis of the results of these simulations will be presented below because the work revealed an important problem common to all simulations involving membrane lipids and nonlipid membrane molecules.

The Metropolis Monte Carlo method was used to study interactions between lipid chains and cholesterol at several cholesterol concentrations. The objective of the simulations was to calculate MC average order parameter profiles for chains that are near neighbors to cholesterol, and to compare these profiles with those for chains that are not close to cholesterol (bulk chains). MC simulations were run for the following systems:

- 3 cholesterol molecules and 97 lipid chains of length 14, 16, and 18 carbons, respectively.
- 7 cholesterol molecules and 93 14-carbon lipid chains.
- 13 cholesterol molecules and 87 14-carbon lipid chains.

Chains were limited conformationally by the rotational isomeric approximation, and C atoms interacted with other C atoms through 6–12 potentials. Explicit hydrogen atoms were not included. Simulations were run at fixed area per chain of 30 Å^2, at $T = 300 K$ (sufficient for chain disordering for all chains at the fixed molecular areas used). Equilibration consisted of 40×10^3 configurations per chain, and averages were calculated over a further $30–40 \times 10^3$ configurations per chain. Errors were estimated from fluctuations in values of the bond order parameters calculated at intermediate steps during each run.

At the lower concentrations no lipid chain is a near neighbor to more than a single cholesterol. For the 13 cholesterol plus 87 chain simulations, there were several chains that were near neighbors to two or three cholesterol molecules. In all cases the cholesterols are not allowed to undergo translational motion although the conformations of the tail chains on each cholesterol were allowed to change.

Figure 2 shows selected order parameter profiles calculated for chains that are near neighbor to more than one cholesterol (from the 87 chain–13 cholesterol simulations), chains which are near neighbor to a single cholesterol, and bulk chains (the latter two profiles from 93 chain–7 cholesterol simulations). The profiles in Figure 2 for chains that are neighbors to more than one cholesterol molecule show that these chains on average exist in configurations that are substantially more ordered than the other chains. The error bars on each of the profiles in Figure 2 measure the extent to which the given bonds are able to change dihedral states during the simulation. Smaller error bars for most bonds between 2 and 8 for the profile for chains adjacent to a single cholesterol are a consequence of the restrictions on the rotational mobility of these bonds. This implies that the cholesterol inhibits the rotameric freedom of its neighbor chains, although on average the neighbors are not in all-trans conformations. The profile for chains which are near neighbor to two or more cholesterols shows a sharp dip at bond 3 which reflects this inhibition. The dip exists because, on several of the chains included in this profile, a kink formed which reoriented bond 3 but left bonds 4–6 in trans orientations. The steric presence of the two cholesterol molecules made this kink a relatively stable structure during the simulation. It proved to be very difficult for a chain to undergo additional gauche rotations

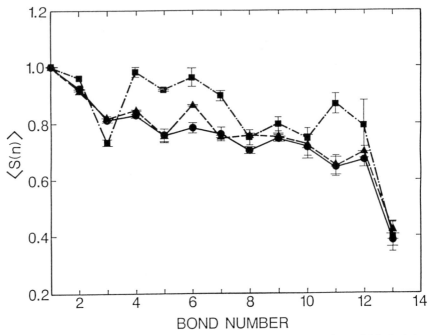

Figure 2. Plot of bond order parameter profiles versus number for bulk chains (solid line), chains neighbor to one cholesterol (dashed line), and chains neighbor to two or more cholesterols (dot-dashed line). Error bars are standard deviations. Reprinted from (Scott, 1991) with permission.

about its upper carbons in the steric presence of two cholesterols as well as other chains after equilibrium was established.

The above conclusions were sufficient to provide a basis for a statistical mechanical model (Scott, 1991). The model was subsequently found to predict a lipid–cholesterol phase diagram which agreed qualitatively with experimental data. The key lesson from these simulations is that the rejection rate for MC moves for a lipid chain near a cholesterol is substantially higher than the rate for a chain whose neighbors are other chains. This implies that the configuration space of a lipid bilayer with 15% or more cholesterol present is very difficult to efficiently sample by standard MC means within the rotational-isomeric model. Similar conclusions were drawn in Monte Carlo simulations of lipid chains adjacent to hard cylinders with hemispherical lumps (Scott, 1986) and of lipid chains adjacent to gramicidin A. In these simulations interpretation of the order parameter profiles was more difficult due to the nonsmooth interface between lipid and nonlipid. In the gramicidin simulations, comparison of data for chains of length 14, 16, and 18 carbons suggested that the hydrophobic length of the gramicidin allowed the shorter chains to pack more efficiently around it than was the case for the long chains (Xing and Scott, 1992).

Relaxing the rotational-isomeric approximation might reduce the steric bottleneck forced on the hydrocarbon chains by nonflexible perturbants in the bilayer, but then one is again faced with the problem of a greatly enlarged configuration space to sample. A novel and more efficient sampling scheme which may alleviate the problem will be discussed later in this chapter. First the use of the MC method in the analysis of a model with a reduced configuration space, with application to long range reordering transitions, will be reviewed in the following subsection.

The Monte Carlo Method in Lattice Statistical Mechanics Modeling

Beginning in 1973 with the pioneering work of Nagle (Nagle, 1973) a number of workers have devised statistical mechanical models for lipid bilayers. This type of modeling differs from direct MD and MC simulations in that one begins with a set of assumptions about lipid bilayers. The assumptions are designed to reduce the number of degrees of freedom of the system to a manageable few. Then, the statistical mechanical partition function is calculated by approximate or exact analysis. Predicted thermodynamic properties of the model, usually related to phase transitions, are compared with available experimental data. A detailed discussion and review of this type of statistical mechanical modeling work was given several years ago by Nagle (1980). For a system as complicated as a lipid bilayer the major obstacle to theoretical modeling lies in the analysis of the model. If the constructed model is realistic in its depiction of intermolecular interactions and degrees of freedom, then it is almost certainly so complex that severe approximations are needed to calculate the predicted thermodynamic behavior. The approximations cloud the picture considerably because the extent to which calculated properties are a result of mathematical artifacts is never clear.

One escape from the above dilemma is to use the Monte Carlo method to calculate these properties. In the field of lipid bilayers this approach has been used extensively by Pink, Mouritsen, and coworkers (Caille et al, 1980; Mouritsen, 1990; Pink, 1990) and by the present author. The work of Pink, Mouritsen, and coworkers on the application of simple lattice models of point molecules to lipid systems is discussed elsewhere in this volume. The application of the MC method to calculate the properties of a statistical mechanical model involving extended and structured molecules will be briefly described here.

At sufficiently low temperatures, below the lipid chain melting phase transition temperature, the hydrocarbon chains of lipid molecules in a bilayer may be considered to be highly inflexible. In this region, then, the number of degrees of freedom for the lipid molecules is dramatically reduced, and it is possible to write a Hamiltonian with atomic detail but which does not include the frozen degrees of freedom. Since lateral diffusive motion is not important in solid-like phases, we can construct an underlying lattice and ignore diffusional degrees of freedom. The total number of degrees of freedom for the system is now reduced from over 20 per molecule to two or three per molecule. Pearce, Scott, and coworkers utilized this simplification to formulate and evaluate a model for the gel-to-ripple phase transition which occurs in some bilayers (McCullough et al, 1990; Scott and Pearce, 1989). After the model was constructed, several powerful analytic techniques were employed to calculate the properties, and, in particular, to determine if a ripple phase was predicted. In the end, it was the Monte Carlo solution of the model that showed the existence of the ripple phase and, later, showed graphically the effect of cholesterol on this phase. To illustrate this application of the MC method, we begin with the definition of the model below.

We start with a two dimensional square lattice, representing a plane of reference for a lipid bilayer. Let two lipid molecules, one in each monolayer of the bilayer, occupy each site of the lattice. The lipid molecules are represented as block L-shaped molecules, with the long segment of the L associated with the hydrocarbon chains and the short segment associated with the head group. Each L-shaped molecule is assigned two degrees of freedom; a height variable, n, and an orientation variable, σ. The height variable is quantized for convenience, $n = 0, \pm 1, \pm 2, \ldots$, and represents the perpendicular displacement of the molecule above or below the plane of reference. The orientation variable $\sigma = \pm 1$ represents the orientation of the lipid head group along one axis in the plane of reference. The basic critical assumptions in this representation of a lipid bilayer **below the main lipid phase transition** are:

- Lipid chain dihedral rotational disordering, while present to a small extent in the ripple phase, is not an important factor the formation of this phase.
- There is a preferred orientation axis for the head groups. This seems to be valid at least in anhydrous lipid lamellar phases (Hauser et al, 1981).
- The two monolayers in the bilayer are, below the main lipid phase transition, very strongly coupled, so that vertical displacements are correlated and no free volume opens up in the bilayer center (it seems obvious that in the ripple phase the two monolayers could not have out-of-phase ripples). This means one need only consider one monolayer.

The Hamiltonian for the model was originally constructed from a numerical calculation of intermolecular interaction energies for a set of close-packed lipid-lipid pair conformations (this is where the atomic details entered into this model). The construction of an analytic form for the Hamiltonian is not required if one is to simply use Monte Carlo simulations. A simple table of energies is all that is needed. But, the mapping is necessary in order to carry out analytical studies of the predicted phases. Since analytical results were a principal goal this work, the calculated energies were mapped onto a Hamiltonian function which not only correctly reproduced the three lowest calculated interaction energies, but also was an extension of a model known to have incommensurate phases with only nearest neighbor interactions. (McCullough et al, 1990; Scott and Pearce, 1989). The Hamiltonian is:

$$\mathcal{H} = -\sum_i^N \sum_j^N \left(J_0 \cos\left\{ \frac{2\pi}{p} \left[n_{i+1,j} - n_{i,j} + \frac{\Delta}{2}(\sigma_{i+1,j} + \sigma_{i,j}) \right] \right\} \right.$$

$$\left. + J_1 \cos\left[\frac{2\pi}{p}(n_{i,j} - n_{i,j+1}) \right] + J_2 \sigma_{i,j}\sigma_{i+1,j} + J_3 \sigma_{i,j}\sigma_{i,j+1} \right) \qquad (13)$$

where p is an integer that was set equal to 5. This is the lowest value for which a limiting case of the model yields a sufficiently broad incommensurate phase region (Scott and Pearce, 1989). The variable $n_{i,j}$ is the height of the molecule at site (i, j), and the variable $\sigma_{i,j} = \pm 1$ denotes the orientation of the head group of the molecule at site (i, j). The sums run over all sites of the lattice. The quantities J_0 and J_2 are the interaction parameters for nearest neighbor pairs in the ripple direction (i), and J_1 and J_3 are the interaction parameters for nearest neighbor pairs in the perpendicular direction. With p set equal to 5, the values for the interaction parameters were chosen so that the lowest energy pair configurations in the Hamiltonian Equation (13) agree precisely with the lowest van der Waals energies for lipid-lipid configurations as determined by independent calculation (Scott and Pearce, 1989).

Statistical mechanical analysis of this model showed that, for the above values of the interaction parameters, the model exhibited a low temperature ordered phase in which chains were tilted by about 27°. The chain tilt was a direct consequence of packing the L-shaped molecules as closely as possible, which meant stacking them so that the headgroup of one molecule was slightly above that of its neighbor if both head groups point in the same direction. The tilt angle followed directly from the molecular dimensions of the L-molecules, which were estimated from CPK models. This angle was remarkably close to the experimental value.

However several approximate methods from statistical mechanics all, when applied to this model, failed to predict any periodic structure or correlations. The best analytical result was an exact solution of the one-dimensional version of the model. The one-dimensional model is incapable of exhibiting long range order, but the pair correlation function had a sinusoidal component which decayed algebraically with interparticle separation (McCullough et al, 1990). This result was sufficiently encouraging that a Monte Carlo simulation of the model was carried out. In the MC simulation, periodic ripple structures were found. When Monte Carlo simulations were done with unwanted periodicities in Equation (13) forbidden, the model exhibited a complex ripple phase with two typical wavelengths of about 160 Å and 220 Å (McCullough et al, 1990). Snapshots of the computer generated ripples, one of which is shown in Figure 3a, are very similar in appearance to scanning tunneling microscopy images (Zasadzinski et al, 1988) and X-ray scattering electron density profiles (Sun et al, 1996) of the ripple phase in lipid bilayers.

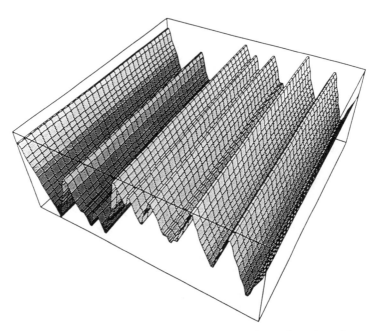

Figure 3. Snapshots of vertical molecular positions in final lattice configurations after annealing and running at the final temperature of $k_B T / J_0 = 0.2$. (a) cholesterol concentration $= 0.00$; (b) cholesterol concentration $= 0.02$; (c) cholesterol concentration $= 0.07$; (d) cholesterol concentration $= 0.10$. Reprinted from Scott and McCullough (1993) with permission.

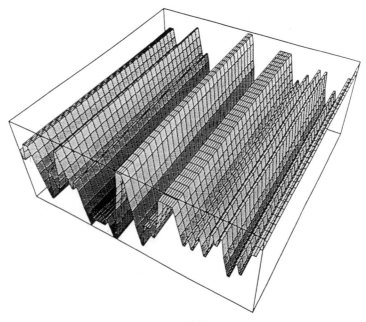

Figure 3(b).

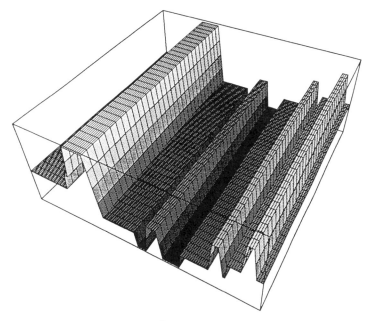

Figure 3(c).

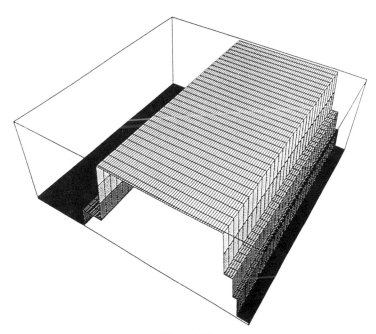

Figure 3(d).

The model was extended to include cholesterol primarily by changing the model from a spin one-half to a spin one model. That is, the σ variable, previously restricted to the values ± 1, now was allowed to have the values $0, \pm 1$. The interpretation for lipid–cholesterol modeling was that a site at which $\sigma_{i,j} = 0$ was occupied by a cholesterol, while a site at which $\sigma = \pm 1$ was occupied by a L-shaped lipid, exactly as in the original model. Both lipid and cholesterol sites still required the state variable $n_{i,j}$ in addition to the spin variable for full specification. An additional term was added to Equation (13), representing the interaction between cholesterol and lipid molecules or between pairs of cholesterol molecules on neighboring sites of the lattice. To gain some guidance for the magnitude of this term, intermolecular interaction energies were numerically calculated for a lipid molecule (DMPC) and an adjacent cholesterol molecule in the same way as was done for pairs of lipid molecules in the original model.

A spin 1 model is considerably more complex analytically than a spin $\frac{1}{2}$ model (this illustrates the effect of adding a single degree of freedom to a simple model)! This fact, plus our experience with the spin $\frac{1}{2}$ model, led us to probe the properties of the extended model by Monte Carlo simulation. In this case simulations were done keeping the cholesterol concentration fixed but allowing for migration of the cholesterol molecules (the spin 0 sites). A simulated annealing procedure was followed, starting at $k_B T / J_0 = 1$ and proceeding to $k_B T / J_0 = 0.2$ in steps of 0.1. At each temperature 50,000 MC steps per site were run for equilibration followed by another 50,000 steps for averaging.

The major new result to emerge from MC simulations of the spin 1 model was that introduction of cholesterol molecules produce linear regions in the system which are flat and which lie at the ripple peaks or troughs. With increasing cholesterol content the flat regions expand, effectively pushing the ripples apart. This is in excellent accord with experimental observations (Copeland and McConnell, 1980; Hicks et al, 1987). Figure 3 shows snapshots of ripple phases for several cholesterol concentrations. The increase in the spacing of the ripples with increasing cholesterol concentration is clearly evident in this figure. The flat areas consist of cholesterol molecules mixed with lipids with antiparallel head group ordering. At higher cholesterol concentrations ($\geq 10\%$) cholesterol molecules tend to aggregate. In all cases nearly all cholesterol molecules lie in the flat areas between the ripples. Detailed comparison of the predicted ripple structures with experimental data show a strong correspondence between experiment and theory (Scott and McCullough, 1993).

The above model, while relatively simple in its formulation, could not have been solved (that is, its predictive properties exactly determined) correctly without use of the Monte Carlo method. This is an example of the interplay between analytical model construction and the use of the computer to carry out the full, exact analysis. For models that attempt to describe large scale molecular reorganization events, with long range correlations and long timescales, direct MC calculation of thermodynamic properties appears to be the best available method for obtaining reliable predictions. Work is now in progress to extend the above model to include lipid chain degrees of freedom, with the aim of describing the chain melting phase transition along with the ripple phase.

Alternative Monte Carlo Sampling for Lipids

If it is true, as discussed earlier in this chapter, that the configuration space of a system of lipids in a bilayer with, or even without, surrounding water is too enormous for straightforward Metropolis MC sampling to be effective, are there alternative sampling algorithms which might work? There are a number of smart Monte Carlo methods that have been applied to problems in liquid physics (Allen and Tildesley, 1987). These methods either sample preferentially (for example, move molecules near a perturbant more often) or bias the MC moves according to the intermolecular force or some other quantity such as the virial. For a system of flexible chain molecules, force-bias MC algorithms would be difficult to apply because the net force will not be constant over the entire chain, and resulting displacements would either violate bonding constraints or would only be optimal for a portion of the chain. While preferential sampling might improve the efficiency of a simulation involving, for example, lipid and cholesterol, the rejection rate for Metropolis MC moves for chains which are boxed in by cholesterol or a membrane protein is so high that little improvement is likely.

A method that holds high promise for efficient Monte Carlo Simulation of flexible chains was recently developed by Siepmann and Frenkel (1992). The algorithm, called the Configuration Bias Monte Carlo Method (CBMC), attempts to choose Monte Carlo steps that have a higher likelihood of producing energetically favorable chain conformations. The procedure has as its basis the early work of Rosenbluth and Rosenbluth (1955) on MC simulation of a long polymer chain on a cubic lattice. Here the method will be described as it is applied to an off-lattice monolayer of lipid hydrocarbon chains. For this system, the CBMC method proceeds as follows.

First, the system is initialized, perhaps in an ordered array of all-trans chains, although this is not necessary. After a chain is picked at random, a set of allowed moves are defined for the each carbon atom in the chain. For the topmost carbon, the allowed moves usually consist of random small translations. For the second carbon the allowed moves should consist of displacement from the top carbon with the bond length fixed. That is, the polar angle may be chosen from $\frac{\pi}{2} \leq \theta \leq \pi$ (to keep the chain from pointing out of the monolayer into the vacuum). The azimuth angle should be picked from $0 \leq \phi \leq 2\pi$. In the CBMC scheme, a set of n trials are made for each carbon atom, so for the second carbon n pairs (θ, ϕ) are picked as possible moves. For the remaining carbons on the chain, the C–C bond orientation is calculated by rotation from the previous C–C bond, and the set of allowed moves consists of n values for the dihedral angle ϕ. These values of the dihedral may be picked from a uniform random distribution or from a distribution biased by the dihedral angle energy function

(Ryckaert and Bellemanns, 1975). Once the set of possible moves is known, the simulation proceeds by the following procedure:

(1) For each carbon j in the chain, calculate the energy before the move, of each possible state i for this atom, $v_{old}(i, j)$. Calculate the old weight $w_{old}(j) = \sum_{i=1}^{n} e^{-v_{old}(ij)/kT}$. Repeat for each j.

(2) Starting at $j = 1$ move carbon j to each of its allowed new states, i. Calculate the new energy $v_{new}(ij)$ and the new weight $w_{new}(j) = \sum_{i=1}^{n} e^{-v_{new}(ij)/kT}$.

(3) Accept the new state for carbon j with probability $e^{-v_{new}(ij)/kT}/w_{new}(j)$.

(4) Repeat the above steps for all j. If it is desired to start the process at a carbon atom below the top of the chain, the new weights for all nonaffected carbons are set equal to their old weights.

(5) Accept the new chain configuration according to a Metropolis sampling procedure with the usual $e^{-\Delta E/kT}$ replaced by

$$\prod_{j=1}^{n} w_{new}(j) / \prod_{j=1}^{n} w_{old}(j).$$

Seipmann and Frenkel (1992) have proven that the above procedure satisfies the detailed balance and step reversibility criteria which are the primary requirements of a Monte Carlo walk.

While each step of the CBMC procedure is lengthy, the end result is (if accepted) a move to an entirely new chain configuration which fits well (in a Monte Carlo sense) with the local environment of that chain. In a traditional MC procedure, in which only one or two C–C bonds are rotated to a new dihedral position, such an optimal **whole-chain** conformation would be generated only very rarely. Smit and Seipmann (1994) have tested the CBMC algorithm against classical Monte Carlo and other techniques for systems of alkane chains of different lengths. In all cases the CBMC method provided much faster convergence and better accuracy, in some cases by many orders of magnitude, than did the other methods.

The author and M. Clark (Scott and Clark, 1995) have applied the CBMC method to a monolayer consisting of 1000 chains, each of which has a spherical head group followed by a 16 carbon chain. Results are very promising. The simulations were started from an all-trans hexagonal array with area 31 Å^2 per chain, similar to lipid fluid phase packing. In only 1000 steps per chain order parameter profiles were obtained which closely match experimental results. The CBMC method offers promise as a way to avoid the sampling problem of chains near a bulky molecule such as cholesterol or a membrane protein. Application of the method to these types of problems is in progress at this writing.

REFERENCES

Allen MP, Tildesley D (1987): *Computer Simulation of Liquids.* New York, NY: Oxford University Press

Baxter RJ (1982): *Exactly Solved Models in Statistical Mechanics.* San Diego, CA: Academic Press

Binder K (1986): *Monte Carlo Methods in Statistical Physics*, 2nd ed. Binder K, ed. Berlin: Springer-Verlag

Bloom M, Evans E, Mouritsen O (1991): Physical properties of the fluid lipid bilayer component of cell membranes: A perspective. *Q Rev Biophys* 24:293–397

Brown MF, Ribeiro AA, Williams GD (1983): New view of lipid bilayer dynamics from ^{2}H and ^{13}C NMR relaxation time measurements. *Proc Nat Acad Sci USA* 80:4325–4329

Caille A, Pink DA, de Verteuil F, Zuckermann MJ (1980): Theoretical models for quasi-two-dimensional mesomorphic monolayers and membrane bilayers. *Can J Phys* 58:581–611

Copeland BR, McConnell HM (1980): The rippled structure in bilayer membranes of phosphatidylcholine and binary mixtures of phosphatidylcholine and cholesterol. *Biochim Biophys Acta* 599:95–109

Curro JG (1974): Computer simulation of multiple chain systems—The effect of density on the average chain dimensions. *J Chem Phys* 61:1203–1207

Fichtorn KA, Weinberg WH (1991): Theoretical foundations of dynamical Monte Carlo simulations. *J Chem Phys* 95:1090–1096

Hauser H, Pascher I, Pearson I, Sundell S (1981): Preferred conformation and molecular packing of phosphatidylethanolamine and phosphatidylcholine. *Biochim Biophys Acta* 650:21–51

Hicks A, Dinda M, Singer MA (1987): The ripple phase of phosphatidylcholines: Effect of chain length and cholesterol. *Biochim Biophys Acta* 903:177–185

Kang HC, Weinberg WH (1989): Dynamic Monte Carlo with a proper energy barrier: Surface diffusion and two-dimensional domain ordering. *J Chem Phys* 90:2824–2830

McIntosh TJ (1990): X-Ray diffraction analysis of membrane lipids. In: *Molecular Description of Biological Membrane Components by Computer Aided Conformational Analysis*, Brasseur R, ed. Boca Raton, FL: CRC Press

McCullough WS, Perk JHH, Scott HL (1990): Analysis of a model for the ripple phase of lipid bilayers. *J Chem Phys* 93:6070–6080

Mendelsohn R, Senak L (1993): Quantitative determination of conformational disorder in biological membranes by FTIR spectroscopy. In: *Biomolecular Spectroscopy*, Clark JR, Heister RE, eds. New York: Wiley

Metropolis N, Rosenbluth N, Rosenbluth A, Teller H, Teller E (1953): Equation of state calculations by fast computing machines. *J Chem Phys* 21:1087–1092

Mouritsen O (1990): Computer simulation of cooperative phenomena in lipid membranes: In: *Molecular Description of Biological Membrane Components by Computer Aided Conformational Analysis*, Brasseur R, ed. Boca Raton, FL: CRC Press

Nagle JF (1980): Theory of the main lipid bilayer phase transition. *Annu Rev Phys Chem* 31:157–192

Nagle JF (1973): Theory of biomembrane phase transitions. *J Chem Phys* 58:252–271

Nagle JF, Scott HL (1978): Biomembrane phase transitions. *Phys Today* 31:38–47

Peterson NO, Chan SI (1977): More on the motional state of lipid bilayer membranes: Interpretation of order parameters obtained from Nuclear Magnetic Resonance experiments. *Biochemistry* 16:2657–2667

Pink D (1990): Computer simulation of biological membranes. In: *Molecular Description of Biological Membrane Components by Computer Aided Conformational Analysis*, Brasseur R, ed. Boca Raton, FL: CRC Press

Rosenbluth MN, Rosenbluth AW (1955): Monte Carlo calculation of the average extension of molecular chains. *J Chem Phys* 23:356–359

Ryckaert JP, Bellemanns A (1975): Molecular dynamics of liquid n-butane near its boiling point. *Chem Phys Lett* 30:123–125

Scott HL (1990): Computer aided methods for the study of lipid chain packing in model membranes and micelles. In: *Molecular Description of Biological Membrane Components by Computer Aided Conformational Analysis*, Brasseur R, ed. Boca Raton, FL: CRC Press

Scott HL (1991): Lipid-cholesterol interactions: Monte Carlo simulations and theory. *Biophys J* 59:445–455

Scott HL (1986): Monte Carlo calculations of order parameters in models for lipid-protein interactions. *Biochemistry* 25:6122–6129

Scott HL (1978): Monte Carlo studies of the hydrocarbon region of lipid bilayers. *Biochim Biophys Acta* 469:264–271

Scott HL, Clark M (1995): Unpublished research

Scott HL, Kalaskar S (1989): Lipid chains and cholesterol in model membranes: A Monte Carlo study. *Biochemistry* 28:3687–3692

Scott HL, McCullough WS (1993): Lipid-cholesterol interactions in the $P_{\beta'}$ phase: Application of a statistical mechanical model. *Biophys J* 64:1398–1404

Scott HL, Pearce PA (1989): Calculation of intermolecular interaction strengths in the $P_{\beta'}$ phase in lipid bilayers. *Biophys J* 55:339–345

Seelig J (1977): Deuterium magnetic resonance: Theory and application to lipid membranes. *Q Rev Biophys* 10:353–418

Siepmann JI, Frenkel D (1992): Configuration bias Monte Carlo: A new sampling scheme for flexible chains. *Mol Phys* 75:59–70

Smit B, Siepmann JI (1994): Simulating the adsorption of alkanes in zeolites. *Science* 264:1118–1120

Sun WJ, Tristam-Nagle S, Suter RM, Nagle JF (1996): Structure of the ripple phase in lecithin bilayers. (preprint)

Taga T, Masuda K (1995): Monte Carlo study of lipid membranes: Simulation of dipalmitoylphosphatadylcholine bilayers in gel and liquid-crystalline phases. *J Comp Chem* 16:235–242

Tristram-Nagle S, Zhang R, Suter RM, Worthington CR, Sun WJ, Nagle JF (1993): Measurement of chain tilt angle in fully hydrated bilayers of gel phase lecithins. *Biophys J* 64:1097–1109

Whittington S, Chapman D (1966): Effect of density on configurational properties of long chain molecules using a Monte Carlo method. *Trans Faraday Soc* 62:62–72

Xing J, Scott HL (1992): Monte Carlo studies of a model for lipid-gramicidin A bilayers. *Biochim Biophys Acta* 1106:227–232

Xing J, Scott HL (1989): Monte Carlo studies of lipid chains and gramicidin A in a model membrane. *Biochem Biophys Res Comm* 165:1–6

Zasadzinski JAN, Schneir J, Gurley J, Elings V, Hansma PK (1988): Scanning tunneling microscopy of freeze-fracture replicas of biomembranes. *Science* 239:1013–1015

4

Strategic Issues in Molecular Dynamics Simulations of Membranes

ERIC JAKOBSSON, SHANKAR SUBRAMANIAM,
AND H. LARRY SCOTT

Introduction

The heterogeneity associated with membrane systems poses a huge challenge for computer simulations of membrane dynamics and structure. Unlike proteins or nucleic acids with well-defined three-dimensional structures, membrane components such as lipid bilayers derive a vast majority of their properties and function from their fluid nature. This introduces the problem of setting up the correct bilayer model system for any realistic computer simulation. The model includes: choice of the system size; interatomic force fields; treatment of short and long-range interactions; and, most important, the macroscopic boundary conditions that best mimic experimental conditions. The simulation is thus an integral part of the model.

In choosing methodologies for the simulation of lipid bilayer membranes, many choices must be made in the area of boundary conditions and in the manner of computing long range forces. No choice is perfect. All involve trade-offs. The trade-offs may be between accuracy and efficiency or sometimes, considering the enormous computer intensiveness of membrane simulations, between accuracy and feasibility of the simulation. The purpose of this paper is to systematically review these choices, considering the positives and negatives of the choices as exemplified in the experiences of both our groups and others, and finally conclude with our best judgment as to optimal methods of membrane simulations.

Size of Simulated System

The lipid membrane is of infinite size by any reasonable standards of molecular simulation. Thus only a partial sample of a membrane can be computed. Any

Biological Membranes
K. Merz, Jr. and B. Roux, Editors
© Birkhäuser Boston 1996

truncation of the size of the system will introduce artefacts. As a smectic liquid crystal (Helfrich and Jakobsson, 1990), the fluid phase membrane will have a continuous spectrum of normal mode frequencies and wavelengths for thermally induced fluctuations in thickness and curvature. When the system is truncated, all wavelengths longer than the size of the truncated system will be eliminated. On the other hand, the normal modes with the same symmetry and periodicity of the truncated system may be artefactually enhanced. Another problem induced by truncation is that the molecules at the system boundary will behave differently (and presumably less realistically) from those away from the boundary, if a nonperiodic system is chosen. In a very large system the boundary molecules are a negligible fraction of the total number of molecules. However, this is not the case in even a large membrane patch (by standards of feasible MD simulations), consisting of 100 lipids in each monolayer. In this case fully 40% of the lipids are at a boundary!

Because of these concerns the best answer to the question, "What size membrane patch should one simulate?", is generally, "As big as you can afford to compute." It should be noted that after the simulated membrane is equilibrated, it is almost as economical to do a sampling run on a large membrane patch as a small one. This is because the large patch gives more data per picosecond than the small one, since one samples the data in a membrane spatially as well as temporally. Thus an equilibrated simulated membrane with 100 phospholipid molecules in each half of the bilayer will provide as much data in a 100 ps simulation as an equilibrated simulated membrane with 50 phospholipid molecules in each half of the membrane will provide in 200 ps of simulation, and the data will be less susceptible to boundary artefacts for the larger system. Thus the big efficiency disadvantage for larger simulated patches is in the equilibration process itself (it will take more computer time to equilibrate) rather than in the production run after equilibration.

In the simulations in our lab that we describe in this paper, we made the judgement that we wished to equilibrate the system and take a reasonable amount of data in a time on the order of months. Based on our initial benchmarks, this led us to a system of 100 phospholipid molecules (DMPC) and 2300 water molecules. This decision was made in the context of a decision to use periodic boundary conditions and electrostatics with a cut-off of 20 Å. The same criteria would lead us now to a somewhat larger system, since availability of computer power has increased significantly in the past year.

Atomic Interaction Force Fields

Choice of interatomic force fields and parameters for macromolecular simulations is an important aspect of the simulation model. There is an increasing convergence on force field parameters representing bonding interactions. However unlike protein simulations, where a torsion term connecting 1–4 interactions suffices to describe the conformational properties of the polypeptide chain, hydrocarbon chains warrant collective torsion interactions. Conformational lability of the hydrocarbon chain has a direct bearing on the membrane fluidity and hence proper treatment of torsion terms is warranted. Ryckaert-Bellemans (1975, 1978) potentials appear to be well-suited for membrane simulations and yield experimentally observed gauche-trans ratio for the hydrocarbon chains in DPPC (Egberts, 1988; Egberts et al, 1994) and DMPC bilayers. In our simulations and those of several other groups, combining the Ryckaert-Bellemans potential with a united atom approach for methylene and methyl groups is found to be adequate to describe the chain conformational and organization properties.

In the computation of nonbonded interactions, a variety of partial charge models, ranging from the semi-empirically derived partial charges to those derived by fitting electrostatic potentials to high-level ab initio calculations, have been employed. In our simulations, we have employed Mulliken charges derived from Hartree-Fock level ab initio calculations using the 6-31G* basis set, and no scaling was found to be necessary owing to appropriate treatment of the solvent force field. A good measure of the validity of the charge distribution is the simulated water dipole orientations and the bilayer order parameters. The choice of the water model also plays an important role in the simulated behavior of the interface. Any model that yields realistic self-diffusion coefficient and dielectric constant for bulk water should be realistic in describing the dynamical and electrostatic behavior of interfacial solvent. In our own simulations, we have employed the SPC/E model developed by Berendsen et al (1987).

Periodic or Nonperiodic Boundary Conditions? Method of Calculating Electrostatics

The issues of using periodic or nonperiodic boundary conditions and of the method of calculating electrostatics need to be considered together, because there are trade-offs between them.

In periodic boundary conditions the system being simulated is considered to be replicated in space at its boundary, so that it is adjacent in each direction to an identical system. For systems that are homogenous in chemical composition and essentially infinite in extent on the molecular size scale, such as solutions and membranes, periodic boundary conditions are the most natural and have been used for many membrane simulations, for example Alper et al, 1993a,b; Berendsen et al, 1992; Chiu et al, 1995; Chiu et al, 1992; Damodaran and Merz, 1994; Egberts et al, 1994; Egberts, 1988; Huang et al, 1994; Marrink and Berendsen, 1994; Marrink et al, 1993; Robinson et al, 1994; Venable et al, 1993. The great advantage of the periodic boundary conditions is that the molecules at the boundaries do not behave differently in any systematic way from the molecules in the interior of the simulated system. Other than this, other disadvantages of truncation of systems pertain to periodic boundary conditions, such as the reinforcement of normal modes of the wavelength of the size of the central simulation cell. In addition the membrane simulated with periodic boundary conditions is precluded from exhibiting bending or splay modes, because such modes would prevent the matching of the membrane with its image at the lateral boundaries of the central simulation cell. Bending and splay modes are almost certainly characteristic of real lipid bilayer membranes, but the timescale for these modes is quite long, probably beyond the reach of current MD simulations.

The most straightforward way of calculating electrostatic interactions is to simply evaluate the Coulomb interactions between each pair of charged atoms in the system. But for a large system such as a membrane the computational cost of including all the atom pair interactions is prohibitive. Also, for periodic boundary conditions it is not even correct. In order to prevent double counting of the interactions between the atoms in the central simulation cell and those in the images, the calculated electrostatic interactions must be cut off at a distance somewhat shorter than half the size of the central simulation cell. For atom-based cutoffs (the electrostatics are computed for each atom pair within a specified distance), substantial distortion of the situation occurs when the cutoff distance from an atom (e.g., A) includes one of a pair of bonded atoms of opposite charge (B or C), but not the other. This will cause an exaggerated net attraction between A and B–C and an exaggerated charge-dipole orientation of B–C relative to A. To eliminate these effects it is common to use a group-based cutoff. In this tactic the atoms are clustered into bonded charge groups. If any atom pair, one in each of the two groups, is within the cutoff distance, then all the atom pairs in the two groups are included in the electrostatic computation. Sometimes each group is a whole molecule, and sometimes it is part of a molecule. If the groups are parts of a molecule and the molecule is neutral, it is common to make each group neutral. If the molecule is carrying a net charge, it is common to make one or more of the groups charged.

One interesting question is: How long must the cutoff be not to introduce serious errors? One systematic study of this question in the context of a membrane simulation is by Alper et al (1993a,b). This study suggested that a group-based cutoff longer than 18 Å would have no significant effect on water orientations at the interface, a factor judged to be a particularly sensitive indicator of the consequences of a cutoff.

If one strongly desires not to use a cutoff, the electric field may be computed efficiently to essentially arbitrary accuracy by a fast multipole expansion method (Board et al, 1992). This has been done on membrane simulations (Heller et al, 1993; Zhou and Schulten, 1995). The problem with applying the fast multipole method to membranes is that this method for calculating the electrostatics is incompatible with the periodic boundary conditions that are otherwise the most natural and correct for membrane simulations, as discussed above. So one is faced with a question: Which leads to more serious artefacts, cutoffs in doing the electrostatics, or the artificial behavior at the boundary molecules in nonperiodic boundary conditions?

In our studies we made a decision, partly based on the results of Alper et al (1993a,b), to combine periodic boundary conditions with a 20 Å cutoff. This has produced generally satisfactory results, but one would like to devise a way to keep the periodic boundary conditions and improve further the accuracy of the electrostatics calculations.

Possible Ensembles for the Central Simulation Cell

NVE (Constant Volume, Constant Energy)

Constant-volume, constant energy seems clearly unsuitable for these simulations since the experimental conditions are characterized by a constant temperature rather than a constant energy. Thus simulations using this ensemble does not correspond appropriately to experiment.

NVT (Constant Volume, Constant Temperature)

A number of membrane simulations have been done with this ensemble (Alper et al, 1993a,b; Venable et al, 1993; Damodaran and Merz, 1994; Robinson et al, 1994). It is better than NVE because it does provide constant temperature, but it does have the problem that one must make assumptions about the total molecular density of the system that are to some degree uncertain and to some degree arbitrary. To produce a fluid phase membrane one must, for example, initially make a good estimate of the dimensions and density of the fluid membrane. If this estimate, which must be built into the starting configuration, is good, then some experimentally observed features of membrane structure will be replicated well. If the estimate is very bad, the results will have features quite unlike experimentally observed membranes. For example, if the density is too low in a fluid phase simulation, a gap can develop between the monolayers or the two monolayers may bend, because of their inherent curvature. At the other extreme, if the packing is too tight in a gel phase simulation, the chains in the monolayers spontaneously tilt in opposite directions, producing a herringbone pattern for the bilayer in a side view. However, one must be concerned about the larger danger using the NVT method that the results will be spuriously good; i.e., the results will look good even if there are underlying flaws in the calculation. For example, if the system density is adjusted to a fairly correct value, there is a concern that the simulation might produce a rather good looking fluid phase membrane even if the force fields are not right, because the constraints of the constant volume would prevent the system from going to an incorrect density.

NPT (Constant Pressure, Constant Temperature)

In constant-pressure simulations, one computes the pressure from the internal virial and the kinetic energy according to the following expression:

$$P = \frac{2\left(E_k + .5\sum_{i<j}(\vec{r}_i - \vec{r}_j) \cdot \vec{F}_{ij}\right)}{3V} \tag{1}$$

Where P is the pressure, E_k is the kinetic energy, $(\vec{r}_i - \vec{r}_j)$ is the vector of the position difference between between atom i and atom j, $\vec{F}_{ij}$ is the force between atom i and atom j, and V is the volume of the system.

Note that the first term within the outer parenthesis of the numerator on the right hand side of Equation (1), the kinetic energy, is the contribution to the pressure according to the perfect gas law. The second term, the virial, comprises the deviations from the perfect gas law. Whereas the first term is always positive, the second term may be either positive or negative. In fact, for biological materials such as macromolecules in electrolyte the second term will always be negative (except possibly deep under water). The negative virial reflects the fact that the net intermolecular forces in electrolyte solution are attractive.

In the simulations the computed pressure is fed back as a control on the size of the system. If the instantaneous computed pressure is higher (lower) than the set point the system is expanded (contracted) slowly as the simulation proceeds (Berendsen et al, 1984). Thus the system can adjust its density to whatever is appropriate for the molecular force fields used and the experimental conditions emulated. NPT simulations of lipid bilayer membranes with one atmosphere set point pressure have been done in the lab of Berendsen (Egberts, 1988; Egberts et al, 1994). The results are problematic. These simulations for a PC membrane produced a gel phase even at temperatures where the membrane should be fluid. In order to produce an appropriate fluid phase the partial charges on the phospholipid molecules were reduced by a factor of two.

The results of the NVT and the NPT computations pose a conundrum. The NVT method produces better results for phospholipid molecules that have reasonably correct force fields, yet the NPT method seems to have a better theoretical basis. The resolution of this conundrum may lie in considering a characteristic of interfaces that is not taken into account in the standard NPT method. This is the surface tension, considered below.

$N\gamma T$ (Constant Surface Tension, Constant Temperature)

This variant of the constant pressure ensemble takes account of the fact that at the interface of a fluid there is a negative tangential component to the pressure that gives rise to the macroscopic measurable phenomenon known as surface tension. This negative tangential component will cause the pressure component in the membrane plane to be different from the pressure component imposed by the environment normal to the membrane plane. To derive the pressure components we start with a definition of the surface tension in terms of the normal and tangential pressure components. Because surface tension is a macroscopic phenomenon its operational definition is in terms of an integral expression including the interface (Bakker, 1911; White, 1980):

$$\gamma = \int_{Z_1}^{Z_2} [P_N(Z) - P_T(Z)]dZ \tag{2}$$

where γ is the surface tension, the Z axis is normal to the plane of the membrane, P_N is the pressure normal to the membrane surface, P_T is the pressure tangent to the membrane surface and Z_2 and Z_1 are any positions along the Z-axis that bracket the interface. In the experimental conditions normally simulated, the mean value of P_N is about one atmosphere.

Rearranging Equation (2), and assuming the pressure normal to the membrane is one atmosphere, gives an expression for the set point for the lateral pressure:

$$\frac{\int_{Z_1}^{Z_2} P_T(Z)dZ}{(Z_2 - Z_1)} = 1atm = \frac{\gamma}{(Z_2 - Z_1)} \tag{3}$$

The left-hand side of Equation (3) provides the mean tangential pressure set point. Thus by using this expression for the set point, the $N\gamma T$ method can be implemented in a fashion analogous to the NPT method. In this method the normal and tangential spatial components of the virial (the vector dot product on the right hand side of Equation (1)) are calculated separately. The molecular dynamics program adjusts the dimensions of the simulation cell independently in the normal and tangential directions. The method is shown schematically in Figure 1.

In order to implement this method, one must know a numerical value for the surface tension to insert into Equation (3). One method is to assume that a bilayer is almost exactly equivalent to two monolayers and infer the bilayer surface tension from Langmuir trough data on monolayers. The logic of the inference is that the surface tension in the monolayer data that produces a phase change at the same temperature as it does in the bilayer should be half the surface tension of the bilayer. From the monolayer data of Albrecht et al (1978) on DPPC monolayers, the bilayer surface tension would be estimated to be 56 dynes/cm. An independent method, reported by MacDonald and Simon (1987) is to determine the surface tension of a monolayer in equilibrium with a suspension of vesicular bilayers. For DMPC this method results in an estimate of 46 dynes/cm for the bilayer. The reasonably close agreement between these two independent determinations suggests that we would not be far wrong in using them to estimate the bilayer surface tension in implementing the $N\gamma T$ boundary conditions for membrane simulations.

The $N\gamma T$ ensemble with periodic boundary conditions has been implemented in a simulation of a DMPC membrane (Chiu et al, 1995). The implementation was essentially completely successful, producing a realistic fluid phase membrane (a bit under 60 Å^2 surface area per phospholipid molecule) from a starting X-ray crystal structure (about 38 Å^2 surface area per phospholipid molecule). It is especially noteworthy that the fluid phase was produced from a starting configuration that was ordered in a crystalline arrangement. This has previously been thought not to be feasible because of long equilibration times. The key to the rapid fluidization and approach to equibilibrium appears

1 atm.

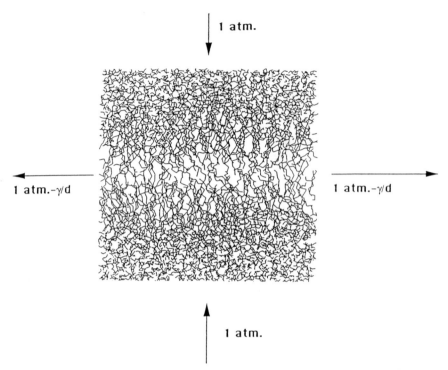

1 atm.$-\gamma/d$ 1 atm.$-\gamma/d$

1 atm.

Figure 1. Boundary conditions for computing the NγT ensemble. Functionally the boundary conditions are implemented like the NPT ensemble, but the lateral pressure is a function of the surface tension as shown, where d is the thickness of the central computation cell in the direction normal to the membrane plane.

to be the magnitude of the lateral pressure set point as computed from Equation (3). For the system simulated in our laboratory the amplitude of that lateral pressure was about -100 atmospheres, a rather large force pushing the system towards the correct equilibrated state.

In summary the advantages of the NγT method are: (1) it moves towards equilibration relatively rapidly; (2) it shows what state is predicted by the force fields used; and (3) when used with reasonable force fields it produces an accurate simulation of fluid phase membranes. It seems to be a good method for simulating a membrane in excess water, where surface tension is known fairly well and can thus can be used to determine boundary conditions.

Using Simulations to See Details Not Experimentally Accessible: Electrical Potentials in the Interfacial Region of Membranes

A disadvantage of simulations compared to experiments is that simulations are not as real. Therefore simulations can not be justified as a simple emulation of nature. One must ask: "What can be learned from simulations that cannot be learned without them?" The answer lies in the fact that in simulations one can observe the results at a level of detail far beyond that achievable experimentally. Thus, if the simulations are close to reality, it may be possible to use simulations to learn about some aspects of mechanism and structure in a more detailed way than is possible without the simulations.

One phenomenon whose basis our labs have explored with simulations is the dipole potential. This potential is clearly seen across phospholipid monolayers at an air-water interface (for original results and a review see Hladky and Haydon, 1973; McLaughlin, 1977). It is of somewhat different magnitude for different types of phospholipids but is always hundreds of millivolts, with the air positive relative to the electrolyte. If a bilayer is similar to two monolayers, the existence of the dipole potential implies that the hydrocarbon interior of the membrane should be substantially positive relative to the surrounding electrolyte. This cannot be measured directly, but it has been inferred from the experimental fact that negatively charged organic ions have a higher permeation rate through lipid bilayer membranes than do positively charged organic ions of comparable size (Flewelling and Hubbell, 1986).

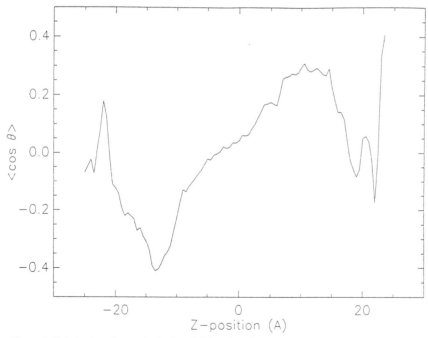

Figure 2. Polarization of water in the interfacial region of electrolyte-lipid bilayer system. Vertical axis is mean cosine of the angle between the water dipole moment vector and the axis normal to the membrane plane. Horizontal axis is position normal to the membrane plane. Origin is between two opposing membrane faces, so positive z-values are in one interfacial region, negative values in the opposing face. Values are time-averaged for the 50 ps of simulated time. (Reprinted with permission, from Chiu et al, 1995).

It is notable that the magnitude of the dipole potential for uncharged membranes that have been explored is completely independent of ionic strength (Hladky and Haydon, 1973). This is remarkable because in general the intensity of electrostatic interactions among macromolecules is a strong function of ionic strength, so the dipole potential stands as an intriguing exception. The absence of ionic influence is serendipitous for molecular dynamics studies of the basis for the dipole potential because this makes the relevant molecular dynamics simulations much more tractable, since we need include in the simulation neither explicit ions nor any provision for screening effects of ions. The question of the basis of the dipole potential comes down to the question of how the polar entities at the membrane electrolyte interface (the water molecules and the polar regions of the phospholipids) interact to produce the potential. We have explored this issue by analysis of NγT molecular dynamics simulations of DMPC membranes (Chiu et al, 1995), as shown in Figures 2 through 6 (from Chiu et al, 1995). Figure 2 shows the 50 ps time-averaged orientation of water dipole moments in the interfacial region as a function of the depth of penetration of the water into the interface. It is seen that there is a clear tendency for the water molecules to orient their dipole moments with the positive charges (hydrogens) towards the interior

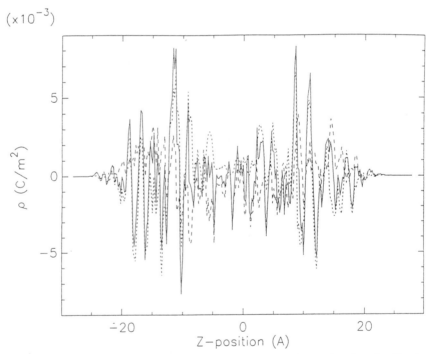

Figure 3. From the same data set as Figure 2, shows the charge density as a function of position in the dimension normal to the membrane plane for the water (short dashes), the lipid (long dashes) and the total (water plus lipid). (Reprinted with permission, from Chiu et al, 1995).

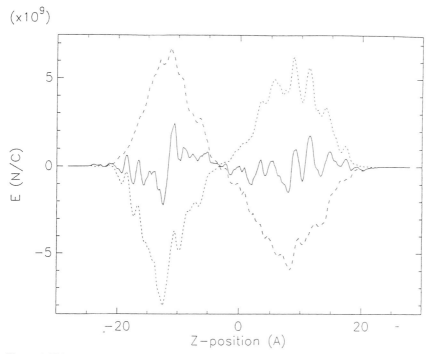

Figure 4. This curve shows the negative of the first integral of the curves in Figure 3, which gives the contribution to the electric field normal to the membrane contributed by the water (short dashes), the lipid (long dashes), and the total field (solid). (Reprinted with permission, from Chiu et al, 1995).

hydrocarbon region of the membrane, in the direction appropriate to account for the dipole potential. To assess the contribution of the water molecules and the phospholipid in producing the dipole potential, the time-averaged charge was computed as a function of the depth in the membrane (Figure 3), integrated once to produce the mean electric field (Figure 4), and integrated a second time to produce the potential (Figure 5), according to Poisson's equation, $\partial^2 V / \partial x^2 = -\rho / \varepsilon_0$.

From Figures 2 through 5 the pattern of the dipole potential production emerges. The water in the interfacial region is selectively oriented to make the interior of the membrane postive relative to the electrolyte. The phospholipid in the interfacial region is selectively oriented in the opposite direction to the water. Each of these orientations is extreme enough to produce a potential of several volts across the interface (see Figure 5). Thus the simulations show us that the dipole potential, although it is large compared to ionic diffusion potentials such as the trans-membrane potential and the Gouy-Chapman double layer at the membrane surface (McLaughlin, 1989), is in fact due to a relatively small difference between two much larger and oppositely directed potentials, one due to the water orientations and the other to the phospholipid polar regions. The

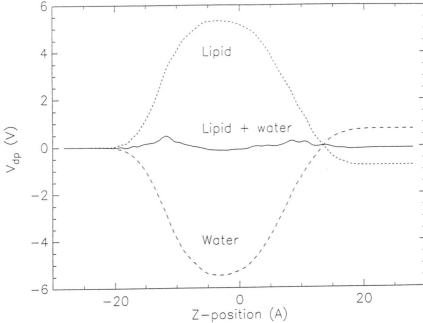

Figure 5. These curves are the integrals of the curves in Figure 4, which give the contribution to the potential profile for the water, the lipid, and the total potential profile. (Reprinted with permission, from Chiu et al, 1995).

net potential is positive because the net water polarization is larger in magnitude than the net phospholipid polarization.

Figure 6 shows the total potential curve of Figure 5 enlarged to a larger scale. From this curve one sees an additional feature of the potential profile. In addition to a net dipole potential, in which the hydrocarbon region is positive relative to the electrolyte, there is a positive potential peak in the interfacial region. This feature emerged from the simulations, and was not anticipated prior to the analysis of the simulated trajectories. It seems possible that the peak may contribute to the mechanism for the positive-inside rule of membrane protein folding (von Heijne, 1994). This rule refers to the fact that positively charged amino acids are disproportionately found in intracellular loops of membrane proteins, suggesting that it is particularly difficult for positive residues to cross the interface between the cytoplasm and the membrane interior.

In summary the results of Figures 2 through 6 show the general basis for a well-known membrane electrical phenomenon, the dipole potential, and reveal the existence of a previously unobserved electrical phenomenon, a positive potential barrier in the interfacial region. By being able to show such a relatively subtle phenomenon as the balance between water and phospholipid polarization, these results that the force fields and methodologies employed are adequate to be useful in exploring the details of the organization

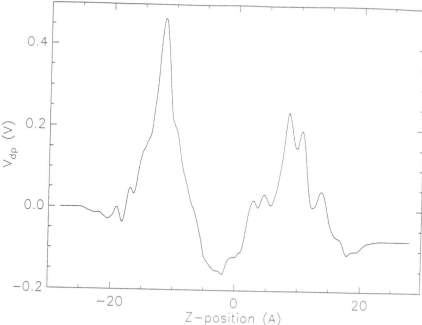

Figure 6. Total potential profile from Figure 5 with an expanded scale, so that the detail can be seen. Note that there is a positive potential positive to the membrane, in agreement with experiment, and that there is a potential barrier in the interfacial region. (Reprinted with permission, from Chiu et al, 1995).

of the membrane-electrolyte interface. (However it should be noted that the net dipole potential shown in Figure 6 is somewhat smaller than that seen experimentally, so there is still some refinement that should be considered for the methodology.)

More details still need to be explored. Experimentally the dipole potential always has the same sign, whatever the species of the phospholipid. This suggests that the water potential always overbalances phospholipid potential. Why should this be so? Perhaps a partial explanation lies in results from simulation of water at an idealized hydrophobic surface (Lee et al, 1984). In that simulation it is seen that water at the surface orients in such a way as to leave hydrogens dangling towards the surface, so that the potential would be substantially positive at the surface compared to the bulk water. This surface polarization arises from the asymmetric charge distribution in water (two positive charge locii but only one negative charge locus) plus the energetic imperative to maximize the water-water hydrogen bonding. It may be that this mechanism for water polarization is the essential underlying cause of the dipole potential, with the magnitude of the induced potential being modulated by the phospholipid. More extensive analysis of the predominant hydrogen-bonding patterns in the interfacial region should illuminate this issue.

Possible New Directions in Methodology

In this section we consider some possible new directions for improving and extending simulation methodology for membrane dynamics.

More Efficient Methods for Equilibrating Membrane Simulations

As stated in the section on the size of the simulated system, a key limitation in the size of the simulated system is the time to equilibrate the system. A possible way to speed equilibration for a large system is to first equilibrate a smaller system that is a subunit of the large system. Then after the small system is equilibrated, a larger system can be generated by copying the subunits next to each other. Then a final equilibration of the larger system can be carried out. For example, one can equilibrate a membrane model with 25 lipid molecules in each half membrane with different initial conditions, then put four of these together to make a membrane with 100 lipid molecules in each monolayer. A variant of this method would be to first equilibrate several smaller systems that are subunits of the large system concurrently on a parallel computer with different initial conditions. This case would give greater randomization than assembling the large system out of microscopically identical subunits. Putting the system together out of nonidentical subunits would cause the boundaries of the subunits not to match with each other, necessitating another energy minimization before starting the equilibration of the larger system. Then the larger system is equilibrated again, but the second equilibration should take much less computer time, since the configuration is made up from an ensemble of equilibrated configurations. The logic is that most of the equilibration is done on the smaller subunit. Such a strategy is ideal for large periodic systems with high degree of fluidity. The authors do not know of any laboratories using this strategy for equilibration of simulated membranes, but it seems as though it should be a computationally elegant approach.

Improved Electrostatics, Combined with Periodic Boundary Conditions

A possible way to optimize the trade-off between periodic boundary conditions and fast, accurate electrostatics is to consider an alternative transformation method to the fast multipole method for computing the electrostatics accurately. The alternative method would be Ewald sums. Ewald sums are consistent with the use of periodic boundary conditions. Recently improvements have been made in the efficiency with which Ewald sums can be calculated (Darden et al, 1993). A drawback to Ewald sums compared to direct computation of the Coulomb forces is that the direct computations automatically produce the terms needed for computing the virial contribution to the internal system pressure, while Ewald sums do not. Therefore some thought and computer experimentation will be necessary to decide how to do the virial computation efficiently when they must be done independently of the electrostatics.

Hybrid Simulations

In some cases of interest, the timescale for molecular relaxation may be in the nanoseconds domain. In this case molecular dynamics can only be done with enormous computing resources or by using membrane patches which are too small to provide much useful information. An example is a lipid bilayer containing cholesterol. In the case of cholesterol the lipids that are nearest neighbor to a cholesterol are blocked by the rigid fused rings of the cholesterol from many otherwise accessible molecular conformations.

One alternative way to gain information about such a system is to run a hybrid simulation which mixes molecular dynamics and Monte Carlo methods. The Monte Carlo methods have the potential, if carefully designed, to hop over barriers in the potential surface of a lipid next to a nonlipid molecule. This can potentially allow for better sampling of configuration space for lipids which are near neighbors to cholesterol or to a membrane protein. A reasonable scenario for this procedure is as follows:

(1) Initialize a system and run molecular dynamics using the $N\gamma T$ procedure described above to equilibrate all lipids which are not adjacent to the nonlipid molecules (but, of course, run MD on these lipids too).

(2) Run a Monte Carlo program, using a smart sampling algorithm (see the chapter by Scott in this volume) to move the lipids adjacent to nonlipids, and, perhaps, their neighboring lipids as well, for sufficient steps until these lipids equilibrate as measured, say, by convergence of order parameters.

(3) Run more MD on the entire system.

(4) Repeat the Monte Carlo sampling for the lipids adjacent to nonlipids.

(5) Continue until entire system is equilibrated and all desired averages are calculated.

The disadvantage of the above procedure is that a specific timescale is lost. What is gained is information about interactions between lipids and nonlipid molecules in a model membrane environment under equilibrium conditions.

Choosing the Optimal Computer Architecture for Membrane Simulations

Because of the size of the system that must be simulated, simulating membranes is of necessity done on supercomputers or powerful dedicated workstations. One suitable large machine is the vector supercomputer, such as the Cray C90. It would be desirable to have even more power via parallel machines. It has proven to be difficult to port molecular dynamics code to run efficiently on massively parallel distributed memory machines such as the Thinking Machines CM-5. At the present writing there has been much better performance from moderately parallel (8-way) simulations on machines based on clusters of powerful microprocessors, such as the Silicon Graphics Power Challenge and the Convex Exemplar. The individual processors on these machines are not quite as powerful as those on the Cray. However the relatively efficient parallelization of molecular dynamics on these machines compared to that for distributed memory machines, and the large memory available in the shared memory paradigm, makes these machines suitable for molecular dynamics on large systems.

More generally, it appears that there will be a trend in the next few years for data-intensive and high-performance biological computing to move towards distributed computing within a World-Wide-Web environment. This has been demonstrated to be feasible by recent developments in interactive computing in the Web environment, such as Java. At the National Center for Supercomputing Applications, the Computational Biology Group has developed the Biology Workbench. To the user of the Biology Workbench, the interface is a Web page exhibited by a browser such as Netscape or Mosaic. The machine on the scientist's desk can be anything that will support such a networked browser, and may in the future be a stripped-down special purpose Web terminal that will cost significantly less than today's workstations. Instructions for computing are done by clicking on hypertext or pushbuttons, or typing in specified blanks. Computations are shifted to appropriate remote machines by links to URL's that underly the interface. The software underlying the interface contains translators that account for machine-specific or application-specific aspects of the analysis or simulation programs, or databases, to which the links are being made. In this way, all the important protein and gene sequence and protein structure databases are at the biologist's fingertips, as well as powerful analysis programs that permit drawing inferences about relationships, and validating and making predictions from the raw data. The scientist need not learn any of the relevant specific interfaces or operating systems, nor even have any idea of the geographic or physical locations of the data bases or the applications programs.

Although the Biology Workbench is specific to molecular biology, it provides a paradigm for the future of high performance computing in any area of biology, for example neuroscience, or in any information-rich field in which interconnectivity between databases and high-performance computers is critical. Indeed, this may be the future paradigm for high-performance computing in general. There is no fundamental reason why "high-performance computing" should be synonymous with "technically difficult computing." The Biology Workbench may presage a day when the working scientist who needs high performance computing may not need any specific knowledge about computer operating systems or specifications. Rather the scientist will make known his or her computing needs to a Web-browser interface backed by an intelligent system that will use the information provided by the scientist to run the appropriate application on the appropriate computing platform, using the appropriate archival data from databases that may be anywhere in the country or perhaps even the world. This is just what the Biology Workbench is doing today for a wide range of problems in molecular biology. Because these developments are in progress in a rapidly changing technical environment, it is more useful to look on the Web than in archival literature to learn about this type of computing and its application to molecular simulations as well as other aspects of molecular biology. For a good starting point on biological applications refer to the Web pages for this book at http://www.birkhauser.com/books/. On these pages you will find an up-to-date list of computational biology resources on the Internet.

Summary and Conclusions

Membrane simulations have been done that are sufficiently accurate to discover new details for the mechanisms of lipid bilayer membrane organization and dynamics, such as the development of electrical potentials at the membrane-electrolyte interface. The available computer hardware will become more powerful and the methodology will continue to improve in the foreseeable future. These improvements will lead to more accurate simulations of larger and more heterogenous membrane systems. Such simulations, playing a complementary role to experimental studies, will contribute significantly to elucidating mechanisms for the interactions between lipid bilayer membranes and organic molecules, peptides, and proteins.

ACKNOWLEDGMENTS

Our work was supported by a grant from the National Science Foundation. Computations were done on machines at the Pittsburgh Supercomputing Center and the National Center for Supercomputing Applications. We presented the basic idea of the $N\gamma T$ method, plus early simulations and analysis, at the 1994 Biophysical Society meetings, the 1994 Jerusalem Conference on Biochemistry, and the 1994 meeting on numerical algorithms in biomolecular simulations at the University of Kansas. By discussions at meetings we have become aware that a similar algorithm is being developed by R. Pastor and B. Brooks. We also had useful discussions with many other participants at those meetings and individually with S. Simon. S. White pointed out to us the definition of the surface tension in terms of the integrated lateral pressure in Equation 2.

REFERENCES

Albrecht O, Gruler H, Sackman E (1978): Polymorphism of phospholipid bilayers. *J Physique* 39:301–313

Alper HE, Bassolino D, Stouch TR (1993a): Computer simulation of a phospholipid monolayer-water system. The influence of long range forces on water structure and dynamics. *J Chem Phys* 98:9798–9807

Alper HE, Bassolino-Klimas D, Stouch TR (1993b): The limiting behavior of water hydrating a phospholipid monolayer: A computer simulation study. *J Chem Phys* 99:5547–5559

Bakker G (1911): *Theorie de la Couche Capillaire Plane dans les Corps Purs.* Paris: Gauthier-Villars

Berendsen HJC, Egberts B, Marrink S-J, Ahlstrom P (1992): Molecular dynamics simulations of phospholipid membranes and their interaction with phospholipase A_2. In: *Membrane Proteins: Structures, Interactions and Models*, Pullman A, Jortner J, Pullman B, eds. Dordrecht, The Netherlands: Kluwer Academic Publishers

Berendsen HJC, Grigera JR, Straatsma TP (1987): The missing term in effective pair potentials. *J Chem Phys* 91:6289–6291

Berendsen HJC, Postma JPM, van Gunsteren WF, DiNola A, Haak JR (1984): Molecular dynamics with coupling to an external bath. *J Chem Phys* 81:3684–3689

Board JA Jr, Causey JW, Leathrum JR Jr, Windemuth A, Schulten K (1992): Accelerated molecular dynamics simulation with the parallel fast multipole algorithm. *Chem Phys Lett* 198:89

Chiu SW, Gulukota K, Jakobsson E (1992): Computational approaches to understanding the ion channel-lipid system. In: *Membrane Proteins: Structures, Interactions, and Models*, Pullman A, Jortner J, Pullman B, eds. Dordrecht, The Netherlands: Kluwer Academic Publishers

Chiu SW, Clark M, Balajiv V, Subramaniam S, Scott HL, Jakobsson E (1995): Incorporation of surface tension into molecular dynamics simulation of an interface: A fluid phase lipid bilayer membrane. *Biophys J* 69:1230–1245

Damodaran KV, Merz KM (1994): A comparison of DMPC- and DLPE-based lipid bilayers. *Biophys J* 66:1076–1087

Darden T, York D, Pedersen L (1993): Particle mesh Ewald: An $N \cdot \cdot \log(N)$ method for Ewald sums in large systems. *J Chem Phys* 98:10089–10092

Egberts E (1988): Molecular dynamics simulations of multibilayer membranes (dissertation). University of Groningen, Groningen, The Netherlands

Egberts E, Marrink SJ, Berendsen HJC (1994): Molecular dynamics simulation of a phospholipid membrane. *Eur Biophys J* 22:423–436

Flewelling RF, Hubbell WL (1986): The membrane dipole potential in a total membrane potential model. Applications to hydrophobic ion interactions with membranes. *Biophys J* 49:541–552

Helfrich P, Jakobsson E (1990): Calculation of deformation energies and conformations in lipid membranes containing gramicidin channels. *Biophys J* 57:1075–1084

Heller H, Schaefer M, Schulten K (1993): Molecular dynamics simulation of a bilayer of 200 lipids in the gel and in the liquid-crystal phases. *J Phys Chem* 97:8343–8360

Hladky SB, Haydon DA (1973): Membrane conductance and surface potential. *Biochim Biophys Acta* 318:464–468

Huang P, Perez JJ, Loew GH (1994): Molecular dynamics simulations of phospholipid bilayers. *J Biomol Struct Dyn* 11:927–956

Lee CY, McCammon JA, Rossky PJ (1984): The structure of liquid water at an extended hydrophobic surface. *J Chem Phys* 80:4448

MacDonald RC, Simon SA (1987): Lipid monolayer states and their relation to bilayers. *Proc Natl Acad Sci USA* 84:4089–4093

Marrink SJ, Berendsen HJC (1994): Simulation of water transport through a lipid membrane. *J Phys Chem* 98:4155–4168

Marrink SJ, Berkowitz M, Berendsen HJC (1993): Molecular dynamics simulation of a membrane water interface—the ordering of water and its relation to the hydration force. *Langmuir* 9:3122–3131

McLaughlin S (1989): The electrostatic properties of membranes. *Ann Rev Biophys Biophys Chem* 18:113–136

McLaughlin S (1977): Electrostatic potentials at membrane-solution interfaces. *Curr Top Membr Transp* 9:71–144

Robinson AJ, Richards WG, Thomas PJ, Hann MM (1994): Head group and chain behavior in biological membranes: A molecular dynamics computer simulation. *Biophys J* 67:2345–2354

Ryckaert JP, Bellemans A (1978): Molecular dynamics of liquid alkanes. *Far Disc Chem Soc* 66:95–106

Ryckaert JP, Bellemans A (1975): Molecular dynamics of liquid n-butane near its boiling point. *Chem Phys Lett* 30:123–125

Venable RM, Zhang Y, Hardy BJ, Pastor RW (1993): Molecular dynamics simulations of a lipid bilayer and of hexadecane: An investigation of membrane fluidity. *Science* 262:223–226

von Heijne G (1994): Membrane proteins: From sequence to structure. *Annu Rev Biophys Biomol Struct* 23:167–192

White SH (1980): Small phospholipid vesicles: Internal pressure, surface tension, and surface free energy. *Proc Natl Acad Sci USA* 77:4048–4050

Zhou F, Schulten K (1995): Molecular dynamics study of a membrane-water interface. *J Phys Chem* 99:2194–2208

Part II

EXPERIMENTAL PROBES OF BIOMEMBRANE STRUCTURE AND DYNAMICS

This section outlines basic experimental techniques that are used to obtain insights into biomembrane structure and dynamics.These methods, while providing important insights into biomembranes, also provide information by which the quality of a molecular simulation can be assessed. The first chapter in Section II (Chapter 5) emphasizes the use of diffraction based techniques to establish the structure of fluid lipid bilayers. Stephen White and Michael Weiner have carried out several seminal X-ray and neutron diffraction studies over the last five years that have established a low resolution picture of how peptides interact with lipid bilayers and that have established as well the relative positioning of the various regions of a bilayer (i.e., headgroup, alkyl chain, and glycerol backbone). Rich Mendelsohn and Robert Snyder describe, in Chapter 6, how Infra Red (IR) spectroscopy techniques can be used to determine the populations of the various conformers of the alkyl chains in lipid bilayers. The final chapter in this section by Mike Brown (Chapter 7) gives a very detailed account of how NMR techniques can be applied to the study of the structure of biomembranes and assesses the role dynamics plays in determining biomembrane properties.

5

The Liquid-Crystallographic Structure of Fluid Lipid Bilayer Membranes

STEPHEN H. WHITE AND MICHAEL C. WIENER

Introduction

Knowledge of the structure of lipid bilayers in the fluid liquid-crystalline state is important for understanding the permeability and stability of membranes and the insertion and folding of membrane proteins. Quantitative structural models are especially important at the present time for validation of Monte Carlo and molecular dynamics simulations of lipid bilayers (Pastor, 1994). Diffraction studies of phospholipid crystals at low hydrations can provide atomic-resolution images of the phospholipid molecules of membranes (Pascher et al, 1992) but such images are of marginal value for understanding membrane bilayers for the obvious reason that the phospholipids are in a noncrystalline state. Fluid bilayers present special problems to the structural biologist because their inherent thermal motion and disorder exclude entirely the possibility of obtaining three-dimensional structural information. The only kind of structural image that can be obtained by diffraction methods is a one-dimensional one consisting of the time-averaged transbilayer distributions of the multiatom submolecular groups comprising the lipids such as the phosphate, carbonyl groups, double-bonds, etc. (Figure 1). Such projections have become a standard method for describing the results of bilayer simulations (Damodaran and Merz, 1994; Egberts et al, 1994; Fattal and Ben-Shaul, 1994; Heller et al, 1993). The changes in mean intergroup distances obtainable from the images (called bilayer profiles) are invaluable for understanding how lipid composition and proteins affect membrane organization (White and Wimley, 1994). However, because the profiles represent long-time averages projected onto a line normal to the membrane plane, important information such as instantaneous three-dimensional atomic positions and trajectories are lost from view.

Biological Membranes
K. Merz, Jr. and B. Roux, Editors
© Birkhäuser Boston 1996

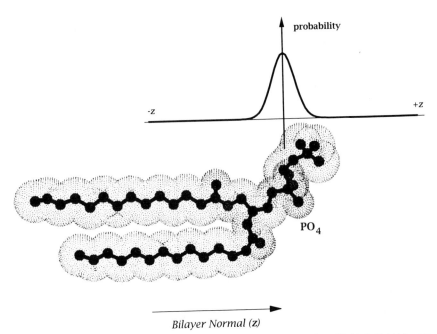

Bilayer Normal (z)

Figure 1. Concept of quasimolecular structure. The inherent thermal motion of lipids in fluid bilayers precludes the possibility of three-dimensional atomic-level crystallographic structures like that of dimyristoylphosphatidylcholine (DMPC) (Pearson and Pascher, 1979) illustrated in the molecular graphics image. However, one can obtain by combined X-ray and neutron diffraction measurements useful structures that consist of time-averaged transbilayer probability distributions for the principal structural groups such as the phosphate group shown here (Wiener and White, 1991b,c).

Monte Carlo and molecular dynamics simulations of membranes can, in principle, provide this information and are thus important endeavors. The evolution of computer hardware and software technologies in the coming years will make simulations increasingly useful and practical tools. Direct bilayer structural information obtained by diffraction methods combined with simulations thus have the potential for providing comprehensive descriptions of complex membrane systems.

We consider in this chapter the structure of a fluid bilayer membrane obtained by what we call liquid-crystallography in which X-ray and neutron diffraction measurements are combined through joint-refinement procedures (Wiener and White, 1992a,b; 1991a,b,c). We have recently summarized the liquid crystallographic method in a comprehensive review (White and Wiener, 1995). Here we provide an overview of the nature of the fluid bilayer structure determination problem, our composition-space structural refinement method for combining the X-ray and neutron diffraction data, and the fully resolved structural image of fluid bilayers formed from dioleoylphosphatidylcholine (DOPC).

Nature of the Fluid Bilayer Structure Problem

Diffraction studies of fluid bilayers are generally accomplished using multilamellar bilayer arrays (multilayers) formed from phospholipids by either dispersal in water or deposition on glass substrates. The resulting one-dimensional lattice of thermally disordered bilayer unit cells typically yields five to ten orders of lamellar diffracted intensities from which the bilayer profile can be constructed (Blaurock, 1982; Franks and Levine, 1981; Levine and Wilkins, 1971). The profiles can be expressed as transbilayer electron density, scattering-length density, or probability density depending upon the scale factors used in the Fourier transformation of the phased structure factors obtained from the diffracted intensities. Examples of X-ray and neutron scattering-length density profiles for bilayers formed from dioleoylphosphocholine (DOPC) bilayers are shown in Figure 2. These profiles leave the impression that they contain only modest amounts of information. In fact, they contain a great deal of information because all of the atoms in the unit cell contribute to the profile. The object of liquid-crystallography is to decompose these rather smooth profiles into subprofiles that represent the transbilayer distributions of submolecular groupings such as the phosphates, cholines, and carbonyl groups.

To be useful, distributions of the submolecular groups must represent fully resolved images. These can be obtained only when: (1) the one-dimensional lattice is perfect; and (2) all h_{max} of the observable diffracted intensities are recorded. The canonical resolution of the experiment is d/h_{max} where d is the one-dimensional Bragg spacing. The Bragg spacing for bilayer systems is typically 50 Å with $h_{max} = 5$ to 10 so that the canonical resolution is 5–10 Å. This resolution is frequently assumed to be the limit on the accuracy with which the separation of structural features can be determined. That assumption is incorrect; the positions of resolvable features can be determined with a precision that greatly exceeds the canonical resolution (Wiener and White, 1991b). For example, the high electron density peaks in the bilayer profiles obtained from X-ray diffraction (Figure 2B) are assigned to the phosphate moieties, and the distance between them (d_{p-p}) is frequently cited to a precision of 1 Å or better (Inoko and Mitsui, 1978; Janiak et al, 1979; McIntosh and Simon, 1986; Ranck et al, 1977). We call this aspect of resolution, resolution precision. The apparent conflict between the canonical resolution and the resolution precision can be resolved by a careful consideration of the nature of the disorder found in multilamellar fluid bilayer systems and its effects on the images obtained by Fourier transformation of the phased structure factors.

Resolution and resolution precision are controlled by three types of disorder (Blaurock, 1982; Hosemann and Bagchi, 1962; Schwartz et al, 1975) in diffraction experiments. Disorder of the first kind is thermal disorder in which the atoms or molecular fragments oscillate about well-defined positions within the unit cell. A sample with only this type of disorder will have a unit cell of well-defined composition and a lattice with a high degree of long-range order. All of the diffraction peaks will be perfect images of the incident beam so that

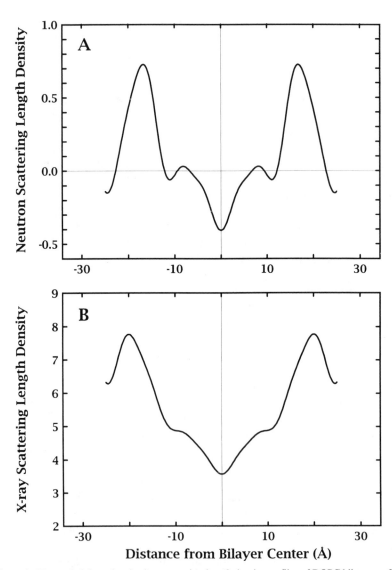

Figure 2. Observed eight-order absolute scattering-length density profiles of DOPC bilayers at 23°C and 66% RH (Wiener and White, 1992a). A. Profile determined by neutron diffraction. Hydrogen atoms have negative values of scattering length so that the average scattering length density is close to zero. B. Profile determined by X-ray diffraction. Notice the difference in the locations of the dominant headgroup peaks in two profiles. This is because the dominant structural feature for neutron scattering is the carbonyl group whereas for X-ray scatttering the phosphate group dominates.

the widths of diffraction peaks will be independent of diffraction order h. The intensities of the peaks, however, will decrease more rapidly with increasing values of h compared to unit cells with less thermal disorder. Disorder of the second kind, lattice disorder, results when long-range order and/or uniform unit cell composition are lacking. This is typically the case for compacted stacks of biological membranes. The membranes have constant composition but the water spacing between membranes varies and thus causes a loss of spatial coherence. When lattice disorder is present, the decreased long-range order causes the diffracted peaks to increase in width as h increases. A third type of disorder is orientational disorder which is related solely to the macroscopic features of a particular sample. For example, a single crystal of salt will produce discrete diffraction spots at well-defined angular positions relative to the X-ray beam. If the crystal is broken up to form a powder, the numerous small crystallites will be oriented at various angles with respect to one another and the X-ray beam so that ring-like diffraction spots are produced. The diffracting lattices can be nearly perfect in both cases.

The nature and quality of images obtained in diffraction experiments is governed by: (1) the thermal disorder of the unit cell, which determines the maximum number of diffraction orders h_{max} that can be obtained under ideal conditions; (2) the number h_{obs} of diffraction orders which are observable as result of the disorder of the lattice or other experimental conditions; and (3) the number of diffraction orders h_{for} actually used in the Fourier reconstruction of the image (Wiener and White, 1991b). Thermal disorder sets the ultimate upper limit on the image obtainable. If thermal disorder is very low, atomic-level structures, referred to as high-resolution structures, can be obtained. If thermal disorder is high, then only polyatomic-level structures can be obtained. These are called low-resolution structures. In either case, with an excellent lattice and good experimental technique, $h_{obs} = h_{max}$, and Fourier transformation with $h_{for} = h_{max}$ yields a fully resolved image of the unit cell. If $h_{for} < h_{max}$, the resulting image will be a partially resolved image of the high- or low-resolution structure. A fluid bilayer structure is inherently a low-resolution structure because thermal motion causes the atoms of the molecules to be broadly distributed over distances of 5–10 Å so that h_{max} is limited to 5 to 10 diffraction orders regardless of the care with which samples are prepared, the sensitivity of the detector, or the intensity of the source. In such a case, Fourier transformation using $h_{for} = h_{max}$ yields a fully resolved image of the low-resolution structure.

The number of diffraction orders observed from crystalline and liquid-crystalline phases is a direct consequence of the spatial distribution of matter resolvable over the time course of a diffraction experiment. Individual atoms or small groups of atoms are discernible in the high-resolution structure of a crystal while the thermal disorder of the liquid crystal causes these distributions to overlap, producing a low-resolution structure. As discussed below, the physically appropriate structural subunits of the liquid crystal are these overlapping multiatomic quasimolecular pieces which have Gaussian shapes (King and White, 1986; Wiener and White, 1991b,c; 1992a). For both crystalline and liquid-crystalline materials, the intensities of the diffracted X-rays can be accurately measured, and in both cases models of appropriate structural resolution can be constructed that allow one to refine the structural image with a resolution precision which is considerably better than the canonical resolution. Thus, it is not correct to assume that the low canonical resolution of the bilayer diffraction experiment makes it impossible to determine distances and distributions to better than d/h_{max}. On the contrary, if thermal motion is the only cause of disorder, very accurate fully resolved images of the low-resolution structure can be obtained. This frequently is the case for oriented multilamellar bilayer arrays (Franks and Lieb, 1979; Smith et al, 1987; Wiener and White, 1991b).

Determination of Fully-Resolved Images of Fluid Bilayers

Joint-Refinement of X-ray and Neutron Data

The structural resolution of bilayer diffraction experiments can be increased using neutron diffraction and specific labeling with deuterium at various positions within a lipid molecule (Blasie et al, 1975; Schoenborn, 1975; Worcester, 1975). The transbilayer position and distribution of labels can be determined with a precision of better than 1 Å (Büldt et al, 1979; 1978; Worcester and Franks, 1976; Zaccai et al, 1979). The general difficulty with such experiments is the heroic amount of chemical and diffraction work that must be done (Büldt et al, 1979; 1978; Zaccai et al, 1979). To avoid heroism, we developed the so called composition-space joint refinement method that combines X-ray and neutron data in a way that minimizes the amount of neutron data required to obtain a detailed image of a fluid bilayer.

The refinement method is possible because of the significant differences in the neutron and X-ray scattering density profiles (Figure 2) observed for phospholipid bilayers (Franks and Lieb, 1979). The neutron scattering-length density profile is different from the X-ray scattering-length density profile because neutrons interact with nuclei whereas X-rays interact with electrons. Because there is no specific relation between X-ray scattering length (determined by atomic number) and neutron scattering length (determined by nuclear scattering), X-ray and neutron diffraction data sets are independent of each other. Thus, the use of both kinds of diffraction doubles the amount of data available for structure refinement. Each experimental method sees a different representation of the molecule in its own scattering space, and each method has different sensitivities to various regions of the molecule. Neutrons scatter most strongly from the carbonyl groups of phospholipids because this part of the molecule lacks hydrogens whereas X-rays scatter most strongly from the electron-dense phosphate moiety.

Quasimolecular Models and the Multi-Gaussian
Representation of Bilayers

The principal objective of molecular modeling in bilayer diffraction studies is to construct a real-space model for the distribution of matter across the bilayer. The one-dimensional projection of a perfect crystalline lipid structure along the bilayer normal will result in a series of sharp (approximately δ-function) peaks representing the individual atoms. Thermal disorder, described by the Debye-Waller factors of the atoms in crystallographic refinements, broadens these peaks to produce a disordered crystalline model. The Gaussian quasi-molecular model is a logical extension of the disordered crystal model in that Debye-Waller factors for small crystals are rigorously derived by considering the Gaussian-distributed deviations of atoms from their equilibrium positions (Warren, 1969). If the thermal disorder is very high, the broadened adjacent atomic peaks will overlap and thus make it impossible to resolve them individually. These overlapping atomic distributions merge into a single Gaussian function representing a multiatomic grouping. The quasimolecular model thus consists appropriately of a family of Gaussians that accounts for all of the atomic mass of the unit cell. It should be emphasized that the use of Gaussians is more than a mathematical convenience because direct structural determinations demonstrate that the transbilayer profiles of specific multiatomic groupings are in fact Gaussian (Wiener and White, 1991a; Wiener et al, 1991). The positions of the Gaussians represent the time-averaged positions of the submolecular pieces while their widths describe the range of thermal motion of the pieces (Willis and Pryor, 1975). Because the quasimolecular model accounts for thermal motion from the start, Debye-Waller terms are not included in the transform. The use of Gaussian distributions implies that the motions of these multiatomic distributions are primarily harmonic. In crystal structures, some atoms probably undergo anharmonic motion, but molecular dynamics calculations suggest that these regions are best described by a series of Gaussians rather than a single non-Gaussian distribution (Kuriyan et al, 1986).

The number of Gaussians necessary for modeling the bilayer is related to the number of observable diffraction orders (Wiener and White, 1991b). The canonical resolution, d/h_{max}, is the most appropriate length scale for describing the bilayer because it represents the characteristic sizes of molecular subunits that are discernible in the diffraction experiment. If ten diffraction orders are observable from a bilayer with a d-spacing of 50 Å, the principal scattering centers will be about 5 Å wide. Therefore, appropriate quasimolecular models will have Gaussian distributions with $1/e$-halfwidths of about 2.5 Å. Because of the importance of the canonical resolution in the determination of the appropriate length scale, all of the observable diffraction orders must be recorded, as noted earlier. Furthermore, the experimental errors of the structure factors must be carefully estimated so that the limits on spatial resolution can be determined (Wiener and White, 1991b). A model based upon an imperfect data set, i.e., one that excludes significant higher order structure factors, will result in an incorrect model of the bilayer.

Composition Space

Our joint refinement procedure is based upon the obvious fact that, for thermally disordered liquid-crystalline bilayers, there is a single time- and space-averaged bilayer structure that is invariant with respect to the type of beam used in the diffraction experiment. We therefore used a composition-space representation in which the quasimolecular Gaussian distributions describe the number or probability of occupancy per unit length across the width of the bilayer of each component (Wiener and White, 1991c). This representation permits the joint refinement of neutron and X-ray lamellar diffraction data by means of a single quasimolecular structure that is fit simultaneously to both diffraction data sets (Figure 3). Scaling of each component by the appropriate neutron or X-ray scattering length maps the composition space profile to the appropriate scattering-length space for comparison to experimental data. Other extensive properties, such as mass, can also be obtained by an appropriate scaling of the refined composition-space structure. Based upon simple bilayer models involving crystal and liquid crystal structural information (Wiener and White, 1991c), we estimate that a fluid bilayer with h_{max} observed diffraction orders will be accurately represented by a structure with approximately h_{max} quasimolecular components.

The time-averaged Gaussian probability distribution of each quasimolecular piece projected onto the bilayer normal can be described by:

$$n_i(z) = \left(\frac{N_i}{A_i \sqrt{\pi}}\right) \exp\left[-\left(\frac{z - Z_i}{A_i}\right)^2\right] \tag{1}$$

where $n_i(z)$ is the fraction of the piece located at position Z_i with $1/e$-halfwidth A_i (Figure 3). The distribution can be viewed as the convolution of the hard-sphere or steric distribution of the quasimolecular fragment with an envelope of thermal motion (Wiener and White, 1991a; Wiener et al, 1991). In general, each piece i consists of $N_i \geq 1$ identical subpieces. The $n_i(z)$ include the water molecules associated with the lipid and any other molecules contained within the unit cell. The distribution of matter across the bilayer can be represented in terms of neutron scattering length or X-ray scattering length by multiplying Equation (1) by, respectively, the neutron scattering length b_{ni} or X-ray scattering length b_{xi} of piece i so that the scattering length per unit length is:

$$\rho_{ji}^*(z) = b_{ji} \cdot n_i(z) \tag{2}$$

where $j = n$ or x. Thus, the neutron or X-ray scattering length per unit length at any point in the bilayer is given by:

$$\rho_j^*(z) = \sum_{i=1}^{p} \rho_{ji}^*(z) \tag{3}$$

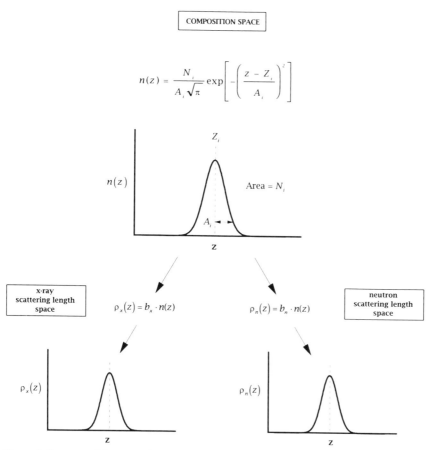

Figure 3. Summary of the composition-space refinement method. The basic strategy is to determine probability or occupancy functions $n_i(z)$ which describe the time-averaged transbilayer distribution of various parts of the hydrated lipid molecule. Scaling the functions by their neutron and X-ray scattering-lengths yield scattering length profiles consistent with diffraction measurements. This approach recognizes the simple and obvious fact there is a bilayer structure that is independent of the diffraction method used to determine it. Because X-ray and neutron scattering lengths are not related, the use of both diffraction methods effectively doubles the amount of data available for the construction of quasimolecular models. Redrawn and modified from Wiener and White (1991c). Used with permission.

where p is the number of quasimolecular pieces per lipid. The neutron and X-ray structure factors $F_j(h)$ of the model consisting of the set of p quasimolecular pieces are then given analytically by the Fourier transform of Equation (2) summed over all of the pieces:

$$F_j(h) = 2\sum_{i=1}^{p} b_{ji} N_i \cdot \exp\left[-\left(\frac{\pi A_i h}{d}\right)^2\right]\cos\left(\frac{2\pi Z_i h}{d}\right) \qquad (4)$$

The rule-of-thumb for the number of quasimolecular components required is that the number of Gaussians p is approximately equal to the number h_{max} of observed lamellar diffraction orders (Wiener and White, 1991c). There are many ways to divide a lipid molecule into p fragments, but two important guidelines simplify the process. The first is to parse the hydrated molecule into $p \approx h_{max}$ pieces that have widths $2A_i \approx d/h_{max}$. The second guideline is inherent in the composition-space refinement method. Namely, the positions Z_{ji} of the pieces must be the same in both X-ray and neutron scattering-length spaces. This entails parsing the atoms among the pieces so that the weighting by the scattering lengths, Equation (2), leads to model scattering-length profiles consistent with the observed ones. The appropriate parsing is ultimately determined by experimental sensitivity and the relative widths and scattering lengths of the distributions (Wiener and White, 1991b). The parsing must be done largely by trial and error in specific cases.

Model Refinement

The structural model is obtained by finding the set of composition-space models that yield the best agreement to both the neutron and X-ray data (Wiener and White, 1991c; 1992a). Nonlinear minimization with the standard Levenberg-Marquardt algorithm (Bevington, 1969; Press et al, 1989) is carried out to determine the parameters Z_i and A_i of Equation (4) which minimizes the joint crystallographic R-factor defined here as:

$$R = \sum_{j=n,x} R_j \tag{5}$$

where:

$$R_j = \frac{\sum_h \left| |F_j(h)| - |F_j^*(h)| \right|}{\sum_h |F_j^*(h)|} \tag{6}$$

$F_j^*(h)$ are the experimental structure factors scaled to the appropriate relative absolute scale (Jacobs and White, 1989; Wiener and White, 1991c; Wiener et al, 1991). A composition-space structure is judged to be satisfactory if it provides fits to both the neutron and X-ray data sets that were below the experimental noise or "self-R" (Wiener and White, 1991b). The robustness of the structure determination and the uncertainties in the parameters are examined by introducing Gaussian-distributed noise into the data sets. Each of the absolute neutron and X-ray structure factors has an associated uncertainty which is used to define the width of a normal distribution centered at the best value of the structure factor. Monte Carlo methods (Press et al, 1989) are used to select data sets from these distributions which are in turn used as the input for the structural calculations (Wiener and White, 1991a; Wiener et al, 1991). Generally, about fifty different data sets are selected in this way, which result in an equal number of suitable models. The mean values of the positions and widths of the Gaussians provide the best estimates of the actual positions and widths. The standard deviations give the estimated experimental uncertainties of the final model.

The Structure of a Fluid Bilayer

Quasimolecular Model

We obtained the complete structure of 1,2-dioleoyl-sn-glycero-3-phosphocholine (DOPC) in the liquid-crystalline phase (66% RH, 23°C) by the joint refinement of neutron and X-ray lamellar diffraction data (Wiener and White, 1992a). The requirement that a successful quasimolecular model fit two independent sets of data strongly constrains the ways in which the molecule can be divided. Figure 4 depicts the quasimolecular model of DOPC and its associated water molecules that was used in the structural determination. This one was chosen initially because it logically identified the obvious molecular fragments. Subsequent examination of more than 30 other parsing schemes did not lead to successful refinements, suggesting that the chosen model may be unique. The methylene region (part 2) in Figure 4 is represented by three Gaussians so that ten quasimolecular fragments are required to obtain the complete structure of the DOPC bilayer. Each piece requires three parameters: position Z_i, $1/e$-halfwidth A_i, and area N_i. The water and double-bond distributions were determined independently from neutron diffraction experiments (Wiener et al, 1991) which reduced the number of parameters from thirty to twenty-four. The terminal methyl distribution was determined from a direct combination of neutron and X-ray data prior to the full joint refinement so the parameter set was further reduced to twenty-one (Wiener and White, 1992b). The contents of each of the remaining pieces of the model, except for the methylene envelope, were fixed by the parsing so that only the positions and $1/e$-halfwidths were determined during the nonlinear minimization. Specifically, the contents of the carbonyl, glycerol, phosphate, and choline fragments were fixed so that the number of parameters was reduced to seventeen. We estimated previously that a fluid bilayer yielding h_{max} diffraction orders would require $p \approx h_{max}$ quasimolecular Gaussian distributions to describe it adequately (Wiener and White, 1991c). We had $h_{max} = 8$ orders for both X-rays and neutrons and $p = 10$ Gaussian fragments consistent with the approximation.

The Complete Structure of DOPC Bilayers at 66% RH

The complete and fully-resolved image of a DOPC bilayer is shown in Figure 5. The positions and widths of the Gaussians and their experimental uncertainties may be found in the original publication (Wiener and White, 1992a). The positions were determined with precisions ranging from 0.02 Å for the carbonyl group (the most strongly scattering feature in neutron experiments) to 0.77 Å

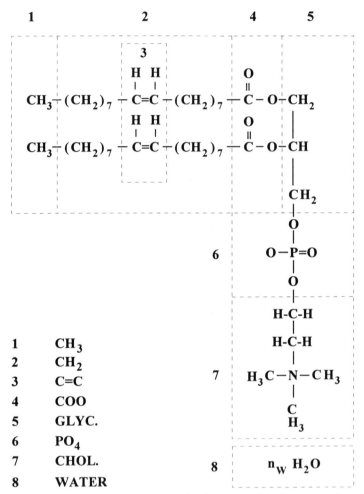

Figure 4. The parsing of DOPC into the quasimolecular parts used in the structure determination by the joint refinement of X-ray and neutron data. Redrawn from Wiener and White (1992a). Used with permission.

for the water. Because this real-space image is physically meaningful, there is useful and interesting information in the widths of the Gaussian distributions that characterize each quasimolecular fragment. The positions of the distributions denote the most likely place to locate the center of scattering of each fragment whereas the widths describe the range of thermal motions projected onto the bilayer normal assuming undulatory motions are insignificant in our system. The $1/e$-halfwidth A_i of a quasimolecular Gaussian fragment can be viewed as the convolution of a hard-sphere of van der Waals radius D_H located at Z_i with a Gaussian envelope of thermal motion describing the range over which that piece moves within the bilayer. The observed $1/e$-halfwidth is

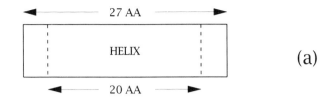

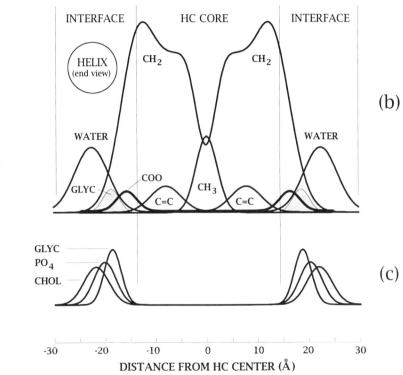

Figure 5. The structure of a fluid dioleoylphosphocholine (DOPC) bilayer determined by the joint refinement of X-ray and neutron diffraction data (Wiener and White, 1992a). The structure consists of the time-averaged distributions of the principal structural groups of the lipid projected onto an axis normal to the bilayer plane. The distributions are Gaussians whose areas equal the number of structural groups represented by them; the distributions therefore represent the probability of finding a structural group at a particular location. A Gaussian distribution is appropriate because that distribution is invariably observed in direct structural determinations of the individual fragments (Wiener and White, 1991a; Wiener et al, 1991). The positions and widths of the distributions can generally be determined with a precision of better than 0.5 Å. (a) A representation of the length of a 27 amino acid transbilayer helix. (b) The distributions of the methyl (CH_3), methylene (CH_2), double-bonds (C=C), carbonyls (COO), glycerol (GLYC), and water. The interfaces of the bilayer are defined as the regions occupied by water. Notice that an α-helix that is parallel to the bilayer can be comfortably accommodated in the interfaces. (c) The distributions of the glycerol, choline (CHOL), and phosphate (PO_4) groups. Note that the interfaces of the bilayer, each about 15 Å thick, account for 50% of the bilayer thickness. These regions are highly heterogeneous chemically and can therefore host a wide variety of non-covalent interactions with peptides and proteins. The figure is slightly modified from White (1994) and used with permission.

given approximately by $\sqrt{D_H^2 + D_T^2}$ where D_T describes the envelope of thermal motion (Wiener and White, 1991a). Because of the approximate nature of this crude expression and the ambiguity in estimating hard-sphere widths of each of the quasimolecular fragments, we did not explicitly repeat the calculation for all of the fragments. The narrowest thermal distribution is that of the glycerol region ($A_{GLYC} = 2.46 \pm 0.38$Å). The $1/e$-halfwidths of the quasimolecular pieces on either side of the glycerol backbone increase as shown graphically in Figure 5c. The general image is a gradient of thermal motion within the interface zone in which the regions bounding the relatively rigid glycerol backbone undergo increasing ranges of motion that are roughly proportional to the distance from the glycerol fulcrum. This is consistent with NMR results (Braach-Maksvytis and Cornell, 1988; Strenk et al, 1985) and crystallographic measurements (Elder et al, 1977; Hitchcock et al, 1975) which indicate that the glycerol backbone is the most rigid portion of the liquid-crystalline phospholipid bilayer on DMPC.

The glycerol region is also interesting because it is at the extreme boundaries of both the methylene and water distributions and thus marks the water-methylene interface (shaded Gaussian in Figure 5b). The net thermal motions within the hydrocarbon region, compared to the interface zone, are qualitatively different in that the $1/e$-halfwidth of the terminal methyl groups (2.95 Å) is about the same as the carbonyl (2.77 Å) or phosphate groups (3.09 Å) while the width of the double-bond distribution is significantly larger (4.29 Å). This situation makes it appear as though the flexible acyl chain is tethered at one end to the interface by the carbonyls and at other end to the bilayer center by the terminal methyls. Because the half-thickness of the hydrocarbon is considerably shorter than the length of the fully extended chain, this apparent tethering allows the double bonds to diffuse over a relatively large volume of space relative to the carbonyls and terminal methyls. Tethering also permits some of the methylenes to venture beyond the C(2) carbons into the interfacial zone (Figure 5b). The concept of tethering, used here strictly in a spatial sense, seems to violate the NMR observation that there is a smooth increase in motion along the chain such that the terminal methyl is the most disordered part of the chain (Seelig and Seelig, 1977; 1974). However, deuterium NMR orientational order-parameter measurements are made in the time domain whereas the diffraction measurements are made in the spatial domain. The diffraction results simply indicate that the motion of the terminal methyl groups reported by the NMR experiment occurs in a limited region of the bilayer thickness.

An approximate average tilt-angle of the phosphocholine dipole with respect to the bilayer surface can be estimated from the distance between the centers of the phosphate and choline pieces along the bilayer normal. Assuming that the phosphorus and nitrogen atoms are the centers-of-scattering of each of these roughly spherical fragments and a phosphate-nitrogen distance of 4.5 Å obtained from the crystal structure of DMPC (Hauser et al, 1981), the dipole is calculated to be canted with an angle of $22 \pm 4°$ with respect to the bilayer surface. This compares favorably with the values obtained from crystal structures (Hauser et al, 1981) and neutron diffraction of oriented multilayers (Büldt et al, 1979) and is in reasonable agreement with the recent value of 18° obtained from ^{2}H-NMR and Raman spectroscopic studies of DPPC (Akutsu and Nagamori, 1991). Similar orientations are observed in several molecular dynamics simulations (Egberts et al, 1994; Heller et al, 1993; Zhou and Schulten, 1995).

Structural Implications for Peptide-Bilayer Interactions

The structure shown in Figure 5 reveals the complexity of the bilayer as the nonpolar phase for the partitioning of peptides and proteins. The structure has been divided into interface and hydrocarbon (HC) core regions based upon the water distribution. Note that the total interfacial thickness is about the same as the HC core thickness. This means that it is not correct to think of the bilayer as a thin slab of hydrocarbon separating two aqueous phases. Furthermore, the interfacial region is a complex mixture of phosphocholine, glycerol, carbonyl, and methyl groups. The interactions of peptides with these region are equally complex. Finally, as shown in Figure 5b, the thermal thickness of the interface is sufficient to accommodate comfortably an α-helix parallel to the bilayer surface. This emphasizes the likely importance of the interface in protein folding and insertion (Jacobs and White, 1989; White and Wimley, 1994).

ACKNOWLEDGMENTS

We acknowledge the support of grants from the National Institute of General Medical Sciences (GM-37291 and GM-46823) and the National Science Foundation (DMB-887043) which made the work described in this review possible.

REFERENCES

Akutsu H, Nagamori T (1991): Conformational analysis of the polar head group in phosphatidylcholine bilayers—A structural change induced by cations. *Biochemistry* 30:4510–4516

Bevington PR (1969): *Data Reduction and Error Analysis for the Physical Sciences.* New York: McGraw-Hill Book Company

Blasie JK, Schoenborn BP, Zaccai G (1975): Direct methods for the analysis of lamellar neutron diffraction from oriented multilayers: A difference Patterson deconvolution approach. *Brookhaven Symp Biol* 27:III58–III67

Blaurock AE (1982): Evidence of bilayer structure and of membrane interactions from X-ray diffraction analysis. *Biochim Biophys Acta* 650:167–207

Braach-Maksvytis VLB, Cornell BA (1988): Chemical shift anisotropies obtained from aligned egg yolk phosphatidylcholine by solid state 13C nuclear magnetic resonance. *Biophys J* 53:839–843

Büldt G, Gally HU, Seelig J, Zaccai G (1979): Neutron diffraction studies on phosphatidylcholine model membranes. I. Head group conformation. *J Mol Biol* 134:673–691

Büldt G, Gally HU, Seelig A, Seelig J, Zaccai G (1978): Neutron diffraction studies on selectively deuterated phospholipid bilayers. *Nature* 271:182–184

Damodaran KV, Merz KM (1994): A comparison of DMPC- and DLPE-based lipid bilayers. *Biophys J* 66:1076–1087

Egberts E, Marrink SJ, Berendsen HJC (1994): Molecular dynamics simulation of a phospholipid membrane. *Eur Biophys J* 22:423–436

Elder M, Hitchcock PB, Mason R, Shipley GG (1977): A refinement analysis of the crystallography of the phospholipid 1,2-dilauroyl-DL-phosphatidylethanolamine, and some remarks on lipid-lipid and lipid-protein interactions. *Proc R Soc Lond* 354:157–170

Fattal DR, Ben-Shaul A (1994): Mean-field calculations of chain packing and conformational statistics in lipid bilayers: Comparison with experiments and molecular dynamics studies. *Biophys J* 67:983–995

Franks NP, Levine YK (1981): Low-angle X-ray diffraction. In: *Membrane Spectroscopy,* Grell E, ed. Berlin: Springer-Verlag

Franks NP, Lieb WR (1979): The structure of lipid bilayers and the effects of general anesthetics: An X-ray and neutron diffraction study. *J Mol Biol* 133:469–500

Hauser H, Pascher I, Pearson RH, Sundell S (1981): Preferred conformation and molecular packing of phosphatidylethanolamine and phosphatidylcholine. *Biochim Biophys Acta* 650:21–51

Heller H, Schaefer M, Schulten K (1993): Molecular dynamics simulation of a bilayer of 200 lipids in the gel and in the liquid-crystal phases. *J Phys Chem* 97:8343–8360

Hitchcock PB, Mason R, Shipley GG (1975): Phospholipid arrangements in multilayers and artificial membranes: Quantitative analysis of the X-ray data from a multilayer of 1,2-dimyristoyl-DL-phosphatiylethanolamine. *J Mol Biol* 94:297–299

Hosemann R, Bagchi SN (1962): *Direct Analysis of Diffraction by Matter.* Amsterdam: North-Holland

Inoko Y, Mitsui T (1978): Structural parameters of dipalmitoylphosphatidylcholine lamellar phases and bilayer phase transitions. *J Phys Soc Japan* 44:1918–1924

Jacobs RE, White SH (1989): The nature of the hydrophobic binding of small peptides at the bilayer interface: Implications for the insertion of transbilayer helices. *Biochemistry* 28:3421–3437

Janiak MJ, Small DM, Shipley GG (1979): Temperature and compositional dependence of the structure of hydrated dymyristoyl lecithin. *J Biol Chem* 254:6068–6078

King GI, White SH (1986): Determining bilayer hydrocarbon thickness from neutron diffraction measurements using strip-function models. *Biophys J* 49:1047–1054

Kuriyan J, Petsko GA, Levy RM, Karplus M (1986): Effect of anisotropy and anharmonicity on protein crystallographic refinement. An evaluation by molecular dynamics. *J Mol Biol* 190:227–254

Levine YK, Wilkins MHF (1971): Structure of oriented lipid bilayers. *Nature New Biol* 230:69–72

McIntosh TJ, Simon SA (1986): Hydration force and bilayer deformation: A reevaluation. *Biochemistry* 25:4058–4066

Pascher I, Lundmark M, Nyholm P-G, Sundell S (1992): Crystal structures of membrane lipids. *Biochim Biophys Acta* 1113:339–373

Pastor RW (1994): Molecular dynamics and Monte Carlo simulations of lipid bilayers. *Curr Opin Struct Biol* 4:486–492

Pearson RH, Pascher I (1979): The molecular structure of lecithin dihydrate. *Nature* 281:499–501

Press WH, Flannery BP, Teukolsky SA (1989): *Numerical Recipes. The Art of Scienctific Computing.* Cambridge, UK: Cambridge University Press

Ranck JL, Keira T, Luzzati V (1977): A novel packing of the hydrocarbon chains in lipids: The low temperature phases of dipalmitoylphosphatidylglycerol. *Biochim Biophys Acta* 488:432–441

Schoenborn BP (1975): Advantages of neutron scattering for biological structure analysis. *Brookhaven Symp Biol* 27:110–117

Schwartz S, Cain JE, Dratz EA, Blasie JK (1975): An analysis of lamellar X-ray diffraction from disordered membrane multilayers with application to data from retinal rod outer segments. *Biophys J* 15:1201–1233

Seelig A, Seelig J (1977): Effect of a single cis double bond on the structure of a phospholipid bilayer. *Biochemistry* 16:45–50

Seelig A, Seelig J (1974): The dynamic structure of fatty acyl chains in a phospholipid bilayer measured by deuterium magnetic resonance. *Biochemistry* 13:4839–4845

Smith GS, Safinya CR, Roux D, Clark NA (1987): X-ray studies of freely suspended films of a multilamellar lipid system. *Mol Cryst Liq Cryst* 144:235–255

Strenk LM, Westerman PW, Doane JW (1985): A model of orientational ordering in phosphatidylcholine bilayers based on conformational analysis of the glycerol backbone region. *Biophys J* 48:765–773

Warren BE (1969): *X-ray Diffraction.* Reading, MA: Addison-Wesley

White SH (1994): Hydropathy plots and the prediction of membrane protein topology. In: *Membrane Protein Structure: Experimental Approaches,* White SH, ed. New York: Oxford University Press

White SH, Wiener MC (1995): Determination of the structure of fluid lipid bilayer membranes. In: *Permeability and Stability of Lipid Bilayers,* Disalvo EA, Simon SA, eds. Boca Raton: CRC Press

White SH, Wimley WC (1994): Peptides in lipid bilayers: Structural and thermodynamic basis for partitioning and folding. *Cur Opinion Struc Biol* 4:79–86

Wiener MC, King GI, White SH (1991): Structure of a fluid dioleoylphosphatidylcholine bilayer determined by joint refinement of x-ray and neutron diffraction data. I. Scaling of neutron data and the distribution of double-bonds and water. *Biophys J* 60:568–576

Wiener MC, White SH (1991a): Transbilayer distribution of bromine in fluid bilayers containing a specifically brominated analog of dioleoylphosphatidylcholine. *Biochemistry* 30:6997–7008

Wiener MC, White SH (1991b): Fluid bilayer structure determination by the combined use of X-ray and neutron diffraction. I. Fluid bilayer models and the limits of resolution. *Biophys J* 59:162–173

Wiener MC, White SH (1991c): Fluid bilayer structure determination by the combined use of X-ray and neutron diffraction II. "Compostion-Space" refinement method. *Biophys J* 59:174–185

Wiener MC, White SH (1992a): Structure of a fluid dioleoylphosphatidylcholine bilayer determined by joint refinement of x-ray and neutron diffraction data. III. Complete structure. *Biophys J* 61:434–447

Wiener MC, White SH (1992b): Structure of a fluid dioleoylphosphatidylcholine bilayer determined by joint refinement of x-ray and neutron diffraction data. II. Distribution and packing of terminal methyl groups. *Biophys J* 61:428–433

Willis BTM, Pryor AW (1975): *Thermal Vibrations in Crystallography.* Cambridge, UK: Cambridge University Press

Worcester DL (1975): Structural analysis of hydrated egg lecithin and cholesterol bilayers. *Brookhaven Symp Biol* 27:III37–III57

Worcester DL, Franks NP (1976): Structural analysis of hydrated egg lecithin and cholesterol bilayers. II. Neutron diffraction. *J Mol Biol* 100:359–378

Zaccai G, Büldt G, Seelig A, Seelig J (1979): Neutron diffraction studies on phosphatidylcholine model membranes. II. Chain conformation and segmental disorder. *J Mol Biol* 134:693–706

Zhou F, Schulten K (1995): Molecular dynamics study of a membrane-water interface. *J Phys Chem* 99:2194–2207

6

Infrared Spectroscopic Determination of Conformational Disorder and Microphase Separation in Phospholipid Acyl Chains

RICHARD MENDELSOHN AND ROBERT G. SNYDER

Introduction

Infrared (IR) spectroscopy offers unique advantages for the study of phospholipid acyl chain structures and interactions. As is well known, the technique monitors molecular vibrations that produce dipole moment oscillations at infrared frequencies. The observed frequencies, intensities, and band shapes are dependent on molecular conformation, configuration, and chain packing. The method does not depend on probe molecules, is effective in sampling all relevant phases adopted by phospholipids (crystal, gel, liquid crystal, inverted hexagonal, cubic, etc.), and provides structural information not directly accessible to many other methods, e.g., the membranes of intact cells in normal and pathological states (Mantsch et al, 1988; Moore and Mendelsohn, 1994; Moore et al, 1995; Moore et al, 1993), and monolayer films (Dluhy, 1986; Flach et al, 1994; Gericke and Hühnerfuss, 1993; Mitchell and Dluhy, 1988). A major advantage for the study of phospholipids lies in the extensive IR spectra-structure correlations derived from alkanes. Assignments of the observed frequencies to specific normal modes, and the sensitivities of these modes to conformational change are in large part understood.

Prior to the widespread availability of Fourier transform IR (FT-IR) technology in the mid 1970s, studies of biological molecules in aqueous phases were severely limited because the high absorbance of water obscures many vibration bands of interest. The experimental sensitivity gained from the multiplex and throughput advantages of FT-IR (Griffiths and de Haseth, 1986) permits accurate spectral subtraction of water bands and has resulted in extensive applications to the study of biological membrane models. This topic has been reviewed (Mantsch and McElhaney, 1991; Mendelsohn and Mantsch, 1986). In

Biological Membranes
K. Merz, Jr. and B. Roux, Editors
© Birkhäuser Boston 1996

the current article, we first provide a brief summary of the main conformation-sensitive regions of phospholipid IR spectra, then detail three recent applications of new FT-IR methods that provide quantitative information about phospholipid microphase separation and acyl chain conformation.

Conformation-Sensitive Regions of the IR Spectrum

Several regions of the IR spectrum have proven useful for studies of phospholipid structures and phase transitions. The overall sensitivity of the spectrum to phase and conformational change is illustrated in Figure 1 for the disaturated phospholipid, 1,2 distearoylphosphatidylcholine (diC$_{18}$PC) in its gel ($-19°C$) and liquid-crystalline ($58°C$) phases. The approximate frequencies, assignments, and structural association for some conformation marker bands arising from the acyl chains are listed in Table 1.

A typical application of FT-IR to study phospholipid phase transitions involves monitoring the thermotropic responses of various spectral parameters. As shown in Figure 1, the strongest bands arise from the symmetric and antisymmetric CH$_2$ stretching vibrations near 2850 cm^{-1} and 2920 cm^{-1} respectively. The exact wavenumber positions change in response to the introduction of con-

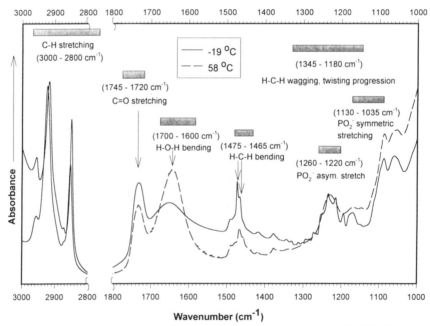

Figure 1. FT-IR spectra of fully hydrated 1,2 distearoylphosphatidylcholine (1,2 diC$_{18}$PC) in the gel ($-19°C$, solid line) and liquid crystal ($58°C$, dashed line). The spectral regions where particular chain and head group modes appear are marked.

Table 1. IR Modes used for Analysis of Hydrocarbon Chain Conformation

Mode	Frequency range (cm^{-1})	Comment
CH$_2$ sym. stretch	2849–2854	The frequencies of these bands are qualitative
CH$_2$ antisym. stretch	2916–2924	measures of conformational disorder
CD$_2$ sym. stretch	2090–2100	
CD$_2$ antisym. stretch	2195–2200	
CH$_2$ scissoring	1462,1473	orthorhombic phase doublet
	1468	hexagonal or triclinic phase
	1473	triclinic phase
CD$_2$ scissoring	1086,1094	orthorhombic phase doublet
	~1089	hexagonal or triclinic phase
CH$_2$ wagging (ordered phases)	1180–1350	For the all-trans chain, these modes couple and split to produce a series of bands characteristic of the number of CH$_2$ groups
CH$_2$ wagging in (conformationally disordered phases)		For disordered phases, particular 2- or 3-bond conformational sequences have their wagging modes as noted:
	1368	gtg sequence
	1353	double gauche sequence
	1341	end gauche sequence
Isolated CD$_2$ rocking modes of an isolated CD$_2$ group	622	tt sequence
	645	$g\underline{t}gt$ or $g'\underline{t}gt$ (see text)
	≈650	tg conformation in isolated g states
	660–680	gg conformation

formational disorder (trans-gauche isomerization) into the chains. The symmetric stretching frequency increases from ≈2850 cm^{-1} for an all-trans chain near room temperature to ≈2854 cm^{-1} for the liquid-crystalline state. Similarly, the antisymmetric stretching frequency increases from 2918 cm^{-1} to 2924 cm^{-1}. Since modern FT-IR equipment can readily detect of frequency shifts of 0.1 cm^{-1}, the observed increases provide a sensitive probe of conformational disordering.

Other features in the spectrum (Figure 1) respond to phospholipid structure changes. When the chains are in the all-trans conformation, the wagging motions of the methylenes couple to form a progression of bands between 1180 cm^{-1} and 1330 cm^{-1} (Cameron et al, 1980; Chia and Mendelsohn, 1992; Schachtschneider, 1963; Senak et al, 1992; Snyder, 1960). The existence of the all-trans conformation in the gel phase of diC$_{18}$PC is therefore immediately evident. In addition, the CH$_2$ scissoring band (≈1460 cm^{-1}) splits into a doublet through interchain coupling when ordered acyl chains pack in a perpendicular orthorhombic subcell (Snyder, 1961). Thus from Figure 1, it may be immediately deduced that the unit subcell of crystalline diC$_{18}$PC at low temperatures is orthorhombic.

While the spectral parameters enumerated in Table 1 have proven useful in studies of phospholipid assemblies, lipid-protein interaction, and living cells, the structural information forthcoming has for the most part been qualitative. For example, increases observed in the CH_2 stretching frequencies have not been quantitatively correlated with either the number or location of gauche bonds in the acyl chains. Similarly, although the intensity of wagging mode progression decreases as the chain packing loosens and disorder is introduced, few quantitative correlations between the intensity and lipid structural parameters have been determined.

Three FT-IR experiments to be described below, have recently been developed to yield quantitative information about phospholipid domains and conformational order. First, we describe how the splitting or broadening of the CH_2 scissoring band is used to detect and measure microdomains in the size range of 1–100 molecules. Next, the CD_2 rocking bands of specifically deuterated alkanes and phospholipids are used to quantitatively determine the extent of conformational disorder at particular positions (depths) in the bilayer. Finally, the bands associated with localized CH_2 wagging modes are used to monitor specific 2- and 3-bond conformational sequences in disordered chains. This approach is applied to disordered bilayer-hexagonal ($L_\alpha \rightarrow H_{II}$) interconversion in 1-palmitoyl, 2-oleoylphosphatidylethanolamine (POPE).

Microaggregation and Domain Demixing in Alkanes and Phospholipid Bilayers

In this section we discuss a sensitive, quantitative, and structure-oriented infrared method useful for measuring microaggregation in gel-state phospholipids. Aggregates ranging in size from 1 to 100 acyl chains can be detected and their size estimated. We will describe the method, discuss its application to binary n-alkane mixtures, and end by summarizing recent applications to mixed phospholipid bilayers.

Method

The method employs the interchain splitting or broadening of the methylene scissors infrared bands to detect aggregation (Snyder et al, 1993; 1992). It is applicable to systems whose spectra exhibit such splitting, namely to crystalline assemblies in which the polymethylene chains are packed in an orthorhombic perpendicular unit subcell. This requirement is not unduly selective since orthorhombic packing is generally preferred by trans polymethylene chains. Such is the case for the n-alkanes and phospholipids at low temperature. A second requirement, discussed below, is that the chains of one of the components of the mixture must be proteated, while those of the other component(s) must be deuterated, or vice versa.

The method is based on the fact that the magnitude of the band splitting is related to the average size of the aggregates. This relation can derived from a simple model, since interchain interaction between the scissors vibrations is lateral (perpendicular to the chain direction) and short range. Because of this interaction, the scissors vibrations of the chains in the crystal are coherently coupled. Because there are two chains in the unit subcell of the orthorhombic crystal, the intense infrared methylene scissors band near 1465 cm^{-1} is split into two components of nearly equal intensity (Snyder, 1961). As noted, the magnitude of the splitting depends on the lateral dimensions of the crystal. This dependence becomes significant only when the lateral size of the crystal is very small, e.g., less than 100 chains. The effect of crystal size on the magnitude of the splitting is not normally observed because the lateral dimensions of the crystals of even a finely powdered sample are much greater than 100 molecules. In this case, the scissors band is maximally split, component separation being about 10 cm^{-1}.

A phase separated mixture may consist of very small domains, with lateral dimensions of perhaps only a few chains. If the chains of one component are of an isotope different from those of the other components, the domains act as isolated single crystals, in so far as their scissors vibration are concerned. This happens because the frequencies of the scissors vibrations of proteated and deuterated n-alkyl chains are so disparate (1465 cm^{-1} versus 1090 cm^{-1}) that vibrational coupling between them is negligible. As a result, the scissors vibrations associated with a given domain do not interact with those of the chains surrounding the domain. The magnitude of the scissors band splitting is then dependent on the average size of the domains and therefore can be determined from the spectrum. A binary mixture is a special case in that, with one component proteated and the other deuterated, the average domain sizes of both components can be monitored simultaneously.

The scissors band for the minority component of a partially demixed crystal often has a characteristic triplet shape, as shown in Figure 2, for particular example is the CH$_2$ scissors band of the C$_{30}^{H}$ chains in a C$_{30}^{H}$/C$_{36}^{D}$ (n-C$_{30}$H$_{62}$/n-C$_{36}$D$_{74}$) mixture at a molar concentration ratio of 1:4. The two outside bands constitute the scissors band doublet associated with domains of the minor component. Their separation is a measure of the average size of C$_{30}^{H}$ domains. Between these bands is a third band representing C$_{30}^{H}$ chains that are isolated in the matrix of the majority component, C$_{36}^{D}$.

To determine domain size in a quantitative way, we need the relation between domain size and band splitting. This has been carried out for orthorhombic n-alkane crystals on the basis of a simple model (Snyder et al, 1992). The scissors mode interaction between pairs of neighboring chains is described by a single force constant. (Which can be evaluated from the splitting observed for a pure n-alkane.) To a good approximation, we can assume that interchain interaction affects neither the form of the normal coordinate nor the intrinsic infrared intensities. Since the domains may assume a great variety of shapes, the value of the splitting associated with a domain of a given number of chains is taken as the average of the splittings calculated for an ensemble of random shapes. This average is then weighted somewhat towards values associated with domains having compact shapes.

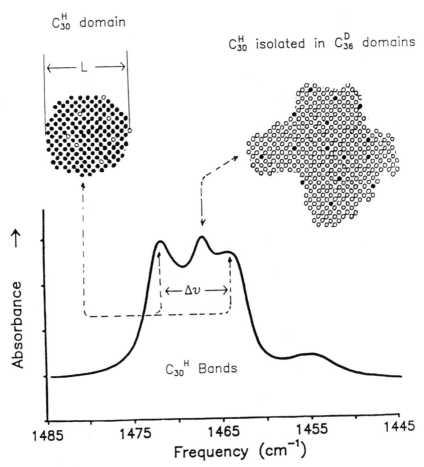

Figure 2. Assignment of the components of the triplet scissors band of C_{30}^H observed in the infrared spectrum of partially demixed 1:4 C_{30}^H/C_{36}^D. Reprinted with permission from Snyder et al (1992).

The sensitivity of the method increases as the domains become smaller, a fact that makes the method of special interest. As the domains become larger, the splitting asymptotically approaches its maximum value, normally in the range 9–13 cm^{-1}. For example, when the lateral domain size reaches 50 chains, the splitting is about 85% of its maximum value. Estimates of domain size therefore becomes increasingly uncertain for domains larger than about 100 chains.

Additional information is provided from band intensities. The intensity of the doublet band relative to that of the central band represents the ratio of the number of chains involved in aggregates to the number isolated or nearly isolated chains within the domains of the other component(s).

n-Alkane Mixtures

The infrared isotope method has been used to study the spontaneous demixing that occurs for some binary n-alkane mixtures after being melt-quenched to room temperature (Snyder et al, 1992). Figure 3 shows the time evolution of the scissors band of the minority component in a C_{30}^H/C_{36}^D mixture at a molar concentration ratio of 1:4. Just after the quench, the central band has shoulders that are just discernable. Without its shoulders, it closely resembles the C_{36}^H band observed for randomly mixed 1:4 C_{36}^H/C_{36}^D. It is evident that the C_{30}^H/C_{36}^D mixture, immediately after the quench, is highly mixed. With time, the demixing accelerates. The shoulders grow, and the band assumes the pronounced triplet shape that characterizes aggregation.

Domain growth is informatively and conveniently displayed in (kinetic) curves consisting of log/log plots in which the average lateral dimension of the domains is plotted against the time elapsed after the quench. Figure 4 shows kinetic curves for the minority component in the 4:1 mixtures C_{30}^H/C_{36}^D, C_{29}^H/C_{36}^D, and C_{28}^H/C_{36}^D. The rate of demixing increases dramatically with the chain-length difference between the components. An increase of one carbon increases the demixing rate by a factor of about 30. Clearly, the driving force for demixing is predominately the chain-length difference.

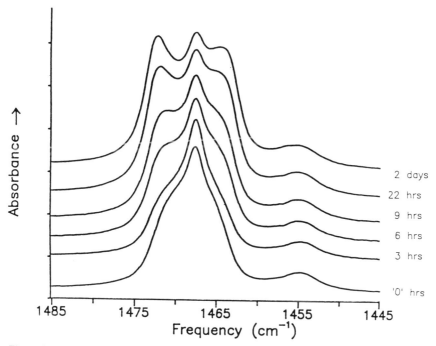

Figure 3. Time evolution of the infrared scissors band of C_{30}^H during the spontaneous microphase segregation of a 1:4 C_{30}^H/C_{36}^D mixture at 22°C. Reprinted with permission from Snyder et al (1992).

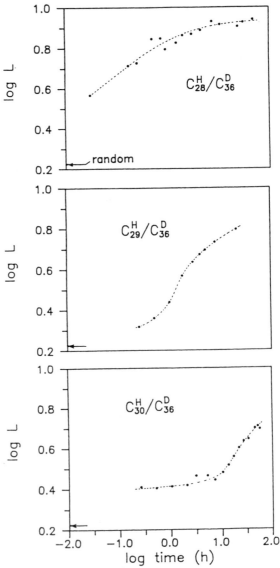

Figure 4. Kinetic curves (log L versus log t) for the minority components in the 4:1 mixtures C_{28}^H/C_{36}^D, C_{29}^H/C_{36}^D, and C_{30}^H/C_{36}^D at room temperature. The estimated value of log L for random mixing is indicated. L is the average lateral dimension of the minority component domains in terms of number of alkyl chains. Reprinted with permission from Snyder et al (1992).

The shapes of the kinetic curves in Figure 4 change systematically in going from one mixture to the next. These curves correspond to different time regions of a common kinetic curve that has an S-shape, the origin of which is discussed by Snyder et al (1992).

The isotopic (protonated vs deuterated) composition of a mixture can also affect the degree and rate of microphase separation (Snyder et al, 1994). This effect results from the small molar-volume difference between CH_2 and CD_2 groups, and can be summarized as follows. For the mixture $C_{n^i}/C_{n'^{i'}}$ (with $n < n'$ and with i and i' being either H or D), there are four possible isotopic combinations. These differ more or less in their tendency to demix. Listed in order of increasing tendency to demix, they are: $C_n^H/C_{n'}^D$, $C_n^H/C_{n'}^H \cong C_n^D/C_{n'}^D$, and $C_n^D/C_{n'}^H$. For the n-alkane mixture C_{30}/C_{36} switching from H/D to D/H is roughly equivalent to an increase in the chain-length difference of about one-half of a carbon atom.

Mixed Diacylphosphatidylcholine Bilayers

This method has been extended to diacylphosphatidylcholine bilayers in the gel state. It has also been applied to binary mixtures of alkyl methyl esters (Snyder et al, 1995), whose complexity is intermediate between the n-alkanes and the phospholipids. The demixing in ester mixtures was found to resemble that for n-alkanes.

Comparing the phospholipids and n-alkanes, we note that the acyl chains in the gel-state phospholipids at low temperatures also pack in an orthorhombic subcell, so the infrared CH_2 scissors band is normally split. Splitting comparable to that for the n-alkanes has been observed for aqueous dispersions of diacylphosphatidylcholine, diacylphosphatidylethanolamine, and diacylphosphatidic acid (Cameron et al, 1981). However, the splitting for a phospholipid dispersion is highly temperature dependent. It diminishes rapidly with increasing temperature and vanishes at a temperature some $30°$ to $40°$ below the gel-to-liquid crystal transition (T_m). In contrast, the splitting for orthorhombic n-alkanes is fairly temperature independent.

For the longer pure diacylphosphatidylcholines, as well as for their mixtures, there are two types of orthorhombic chain-packing in the gel state whose scissors band splittings are significantly different (Mendelsohn et al, 1995). The two types of packing are the normal orthorhombic (G_d) phase and a new highly ordered orthorhombic (G_o) phase (their scissors-band splitting being about 11.6 cm^{-1} and 9.2 cm^{-1}, respectively). The G_d phase is ordinarily associated with even-numbered acyl chains with lengths in the range $12-16$ carbons. The G_o form, has only recently been identified, occurs for homologues with even-numbered acyl chains of lengths in the range of $18-24$ carbons (Mendelsohn et al, 1995b). The G_d and G_o phases can be distinguished from the frequency and shape of their scissors bands.

The relation between domain size and splitting derived earlier for n-alkanes mixtures (Snyder et al, 1992) is also appropriate for the phospholipid mixtures.

Very small domain sizes represents an exception because there are two chains per molecule for the lipid. In that case a small correction may be required (Snyder et al, 1995).

Aggregation in the series of mixtures $diC_{18}^D PC/diC_{20}^H PC$, $diC_{18}^D PC/diC_{22}^H PC$, and $diC_{18}^D PC/diC_{24}^H PC$ has been measured using the infrared method (Snyder et al, 1995; Mendelsohn et al, 1995b). These mixtures have in common the majority component ($diC_{18}^D PC$) whose acyl chains are deuterated. The minor components form a homologous series having chain-length differences of 2, 4, and 6 carbons. This ensures that a wide range of miscibility is sampled. Spontaneous demixing was not observed for these mixtures.

AGGREGATION AT $-19°C$

As the concentration of one component increases, the average size of its domains also increases. This happens whether or not there is aggregation, since the size of randomly formed domains necessarily increases with concentration. Real aggregation can be detected by comparing the average size of the domains of the a component of a given mixture with the average size observed for the corresponding component in a random mixture at the same concentration ratio. Aggregation of $diC_{22}^H PC$ has been demonstrated in this way in $diC_{18}^D PC/diC_{22}^H PC$ at $-19°C$ and at a number of different concentration ratios (Snyder et al, 1995). In Figure 5, the separation and the relative intensities of the two outer bands of the scissoring band triplet for $diC_{22}^H PC$ are shown to increase as the concentration of this component is increased. Figure 6 compares the size of the $diC_{22}^H PC$ domains, measured for the mixture at each concentration, with those associated with a random mixture. The size is expressed in terms log L, with L being the average lateral dimension of the $diC_{22}^H PC$ domains. The average sizes of the $diC_{22}^H PC$ domains in the $diC_{18}^D PC/diC_{22}^H PC$ mixture are larger than for those for $diC_{18}^D PC$ in the random mixture, $diC_{18}^D PC/diC_{18}^H PC$. The disparity increases markedly with increasing $diC_{22}^H PC$ concentration, indicating increasing aggregation.

The chain-length difference between the components in binary mixtures is the main factor in determining miscibility, as it was for the n-alkanes. The relation between component miscibility and chain-length difference for diacylphosphatidylcholine mixtures is well known, having been previously mapped out, mainly from calorimetric studies (Cevc and Marsh, 1987). (We note that the calorimetry was done primarily on phospholipids acyl chains in the range 12–18 carbons, shorter than those reported here.) Calorimetry indicates that a chain-length difference of two carbons leads to near ideal mixing and that there is little if any aggregation, while a difference of six carbons leads to nearly complete phase separation. The case of a four carbon difference, being intermediate, is less well-defined. These generalities seem to be fairly independent of average chain length. Our infrared measurements of miscibility in the 4:1 $diC_{18}^D PC/diC_{n'}^H PC$ ($n' = 20, 22, 24$) mixtures are qualitatively in line with those indicated by calorimetry.

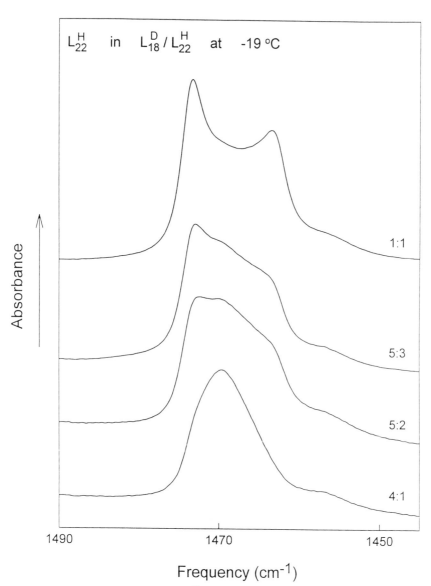

Figure 5. The CH$_2$ scissors band of diC$_{22}^{H}$PC for the diC$_{18}^{D}$PC/diC$_{22}^{H}$PC mixture at $-19°$C at concentration ratios of 4:1, 5:2, 5:3, and 1:1. (L$_n$ = diC$_n$PC). Reprinted with permission from Snyder et al (1992).

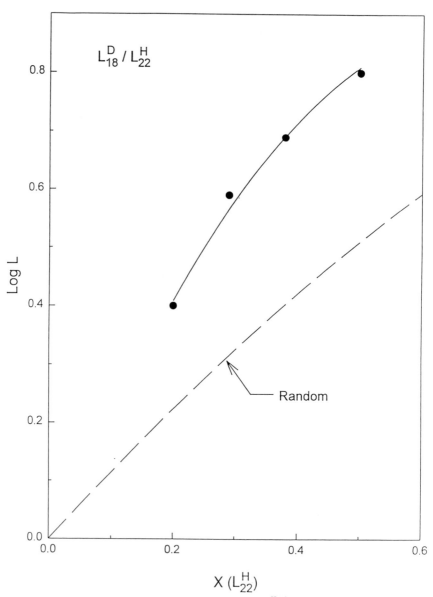

Figure 6. Log of L (the average lateral dimension of the $diC_{22}^H PC$ domains in terms of number of acyl chains) plotted against $X(diC_{22}^H PC)$ (the molar concentration of $diC_{22}^H PC$) for a series of $diC_{18}^D PC/diC_{22}^H PC$ mixtures ($-19°C$) at various concentration ratios. ($L_n = diC_n PC$)

The high sensitivity of the infrared method to low-level aggregation is well displayed in application to the 4:1 $diC_{18}^D PC/diC_{20}^H PC$ mixture at $-19°C$. While this mixture, with its chain-length difference of two carbons, would be classified as an essentially ideal solid solution by calorimetry, infrared measurements clearly indicate aggregation. This is revealed in the observation that the width of

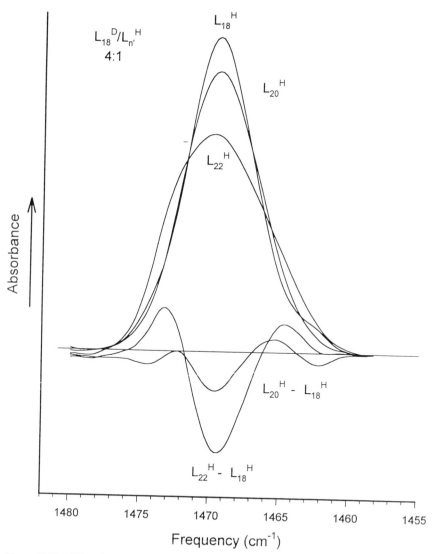

Figure 7. The CH_2 scissors band for the minority components in the mixtures $diC_{18}^D PC/diC_{18}^H PC$, $diC_{18}^D PC/diC_{20}^H PC$, and $diC_{18}^D PC/diC_{22}^H PC$, all at $-19°C$ and at a concentration ratio 4:1. The increase in widths indicates increasing aggregation of the minority phospholipids. In this figure, the differences in the band shapes are expressed as difference spectra. $(L_n = diC_n PC)$

the scissors band of the minority component (diC$_{20}^H$PC) is significantly greater than that observed for the corresponding component (diC$_{18}^H$PC) in randomly mixed 4:1 diC$_{18}^D$PC/diC$_{18}^H$PC.

The difference in widths, apparent in Figure 7, can be utilized in a quantitative way to estimate the fraction of minority-component chains involved in aggregation. This was done by subtracting the scissors band of the minority component (diC$_{18}^H$PC) of randomly mixed 4:1 diC$_{18}^D$PC/diC$_{18}^H$PC from the corresponding band for the minority component of diC$_{18}^D$PC/diC$_{20}^H$PC and, in addition, for diC$_{18}^D$PC/diC$_{22}^H$PC.

This analysis shows that about 4% of the diC$_{20}^H$PC molecules in 4:1 diC$_{18}^D$PC/ diC$_{20}^H$PC at $-19°C$ are involved in aggregates. For 4:1 diC$_{18}^D$PC/diC$_{22}^H$PC at $-19°C$, a mixture with a four carbon chain-length difference, the width for the minority component (diC$_{22}^H$PC) is significantly greater than that for diC$_{20}^H$PC in

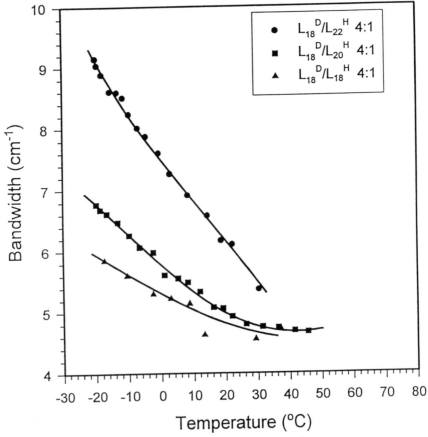

Figure 8. Temperature dependence of aggregation as indicated by the width (full width at half maximum) of the CH$_2$ scissors band for the minority component in 4:1 binary mixtures of diC$_{18}^D$PC paired with diC$_{18}^H$PC(▲), diC$_{20}^H$PC(■), or diC$_{22}^H$PC(●). (L$_n$ = diC$_n$PC)

4:1 $diC_{18}^{D}PC/diC_{20}^{H}PC$. Difference spectra indicate about 11% of the $diC_{22}^{H}PC$ molecules in $diC_{18}^{D}PC/diC_{22}^{H}PC$ at $-19°C$ are involved in aggregates.

The 4:1 $diC_{18}^{D}PC/diC_{24}^{H}PC$ mixture at $-19°C$ is drastically different from the other mixtures considered here in that its components are nearly immiscible.

The aggregates of the minority component undergo dissolution as the temperature of the mixture is increased. This process can be monitored via the temperature dependence of the scissors-band width. Figure 8 shows width plotted against temperature for the 4:1 mixtures, $diC_{18}^{D}PC/diC_{20}^{H}PC$ and $diC_{18}^{D}PC/diC_{22}^{H}PC$, and also for the random mixture $diC_{18}^{D}PC/diC_{18}^{H}PC$, which is used as a reference. The dissolution of the aggregates is essentially complete at the temperature at which the width equals that for the random mixture. Minority-aggregate dissolution temperatures for 4:1 $diC_{18}^{D}PC/diC_{20}^{H}PC$, $diC_{18}^{D}PC/diC_{22}^{H}PC$, and $diC_{18}^{D}PC/diC_{24}^{H}PC$ are estimated to be $15 \pm 5°$, $40 \pm 3°$, and $>47°C$, respectively. For the $diC_{18}^{D}PC/diC_{24}^{H}PC$ mixture, the **onset** temperature for dissolution can be measured. It is estimated to be $24 \pm 3°C$.

CD$_2$ Rocking Modes as Quantitative Probes of Chain Conformational Disorder

To test the predictions of statistical models of lipid assemblies as well as to provide quantitative details for the fluid mosaic model of membrane structure, experimental techniques that quantitatively determine acyl chain conformational order must be developed. Several physical methods have been applied to this end. Fluorescence and electron paramagnetic resonance spectroscopies provide highly sensitive probes of lipid conformational changes and dynamics. However, these techniques require incorporation of probe molecules (fluorophores or spin labels) into the bilayer, that introduce uncertainties concerning the location of the probe and the degree of perturbation of membrane order and dynamics by the probe. In addition, transferring the measured spectral properties of the probe into structural and dynamic information requires formulating some sort of model.

In contrast, IR spectroscopy, with its characteristic time scale of $<10^{-12}$ s, presents a snapshot of the acyl chains in specific conformations. The spectra are not sensitive to slower motions. It is thus possible to associate bands with particular kinds of conformational states, providing that band separation is greater than the bandwidth. The IR experiments described in this and the following sections satisfy this condition and are applied to quantitatively address the following questions:

(1) What are the concentrations of trans and gauche bonds as a function of depth in the bilayer?

(2) What is the population of particular two- or three-bond conformational sequences such as *gg* (double gauche) or *gtg* states in these chains?

(3) What effects do membrane components such as cholesterol have on the extent and distribution of conformational disorder?

Snyder and Poore (1973) described a method for measuring gauche bond concentrations at specific sites in hydrocarbon chains. Their approach is based upon the sensitivity of the rocking mode frequency of an isotopically isolated CD_2 group to the conformation of the two C–C bonds flanking the CD_2 group. The CD_2 rocking frequencies associated with trans-trans (tt), trans-gauche (tg), and gauche-gauche (gg) conformations of this three-carbon fragment appear at $\approx$622 cm^{-1}, $\approx$650 cm^{-1}, and $\approx$680 cm^{-1}, respectively.

The CD_2 rocking mode, which is largely localized, is conformationally sensitive because its normal coordinate involves some rocking motion of the adjoining CH_2 groups (Snyder and Poore, 1973; Maroncelli et al, 1985a,b). The participation of these groups makes the CD_2-mode frequency dependent on the conformation of the C–C bonds that connect the CD_2 group to its CH_2 neighbors.

We illustrate the utility of this spectral region for the small molecule n-heptane. CD_2 rocking mode data for liquid n-heptane-4-d_2 over the temperature range $-90°$ to $-20°$C are shown in Figure 9. The intense band at 622 cm^{-1} (Figure 9A) arises from tt sequences (the tt designation in the current case represents the 3-4,4-5 C–C bond pair). Conformational disorder is manifest by the appearance of a pair of tg bands (near 649 cm^{-1} and 644 cm^{-1}), instead of the single band that would be expected on the basis of the above discussion. There are two bands because the next nearest CH_2 neighbors of the CD_2 group are involved to a sufficient extent to affect the CD_2 rocking frequency and thus provide conformational information farther away from the tagged methylene. As a result, we can identify the sequences $g\underline{t}\underline{g}t$ or $g'\underline{t}\underline{g}t$ (where the CD_2 group lies between the underlined bond pair) which appears at 644 cm^{-1}, and the $t\underline{t}\underline{g}t$ sequence, which appears at 649 cm^{-1}. When the temperature is decreased, the relative intensity of the tt marker band increases compared to the tg band, as anticipated. As shown in Figure 9B, the 644 cm^{-1} intensity increases more rapidly with temperature than the 649 cm^{-1} band. Spectra-structure correlations for these modes are included in Table 1. The 680 cm^{-1} gg marker band is not easily detected.

As noted above, the strongest bands in IR spectra of phospholipids arise from the CH_2 stretching vibrations. However, it is difficult a priori to correlate the frequencies of these bands with chain order in a quantitative way. IR data for n-heptane-4-d_2 enable us to relate the conformational information from the CD_2 rocking modes to the temperature-induced variation in the methylene stretching frequencies. In Figure 10 the CD_2 antisymmetric stretching frequencies near 2170 cm^{-1} are plotted against $\{I(644) + I(649)\}/\{I(622)\}$, e.g., the gauche/trans population ratio. A linear least-squares line is drawn through the data. The observed correlation persists over a wide temperature range ($-90°$ to $+50°$C).

The main problem in using the CD_2 rocking modes to study acyl chain conformations in phospholipids is the experimental one of overcoming spectral interference from the intense H_2O (D_2O) libration band near 800 (600) cm^{-1} in hydrated lipid dispersions which overlaps the CD_2 bands. The intensity of this water band is about 50 times stronger than the tt marker band, which

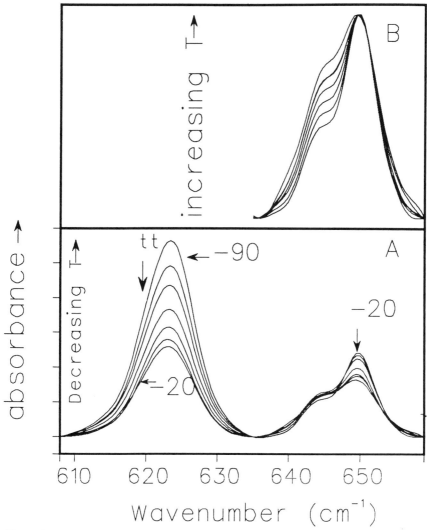

Figure 9. (A) Spectra of the CD_2 rocking region of n-heptane-4d$_2$ at $\approx$10 degree intervals from $-90°$ to $-20°C$. The *tt* (see text) marker band appears at 622 cm^{-1} and increases in intensity with decreasing temperature compared to the *tg* marker bands at 644 cm^{-1} and 649 cm^{-1}. (B) The *tg* marker bands are isolated as a function of temperature over the same range as in the bottom figure. Band assignments are given in the text. With increasing temperature, the 644 cm^{-1} band increases in intensity relative to the 649 cm^{-1} band.

in turn (for phospholipids in the L_α phase) is 5–10 times stronger than the *tg* marker bands. Increasing IR cell path lengths simply increases the solvent absorption to the point where insufficient radiation is transmitted at the desired frequencies. Path lengths for transmission experiments are thus limited to 6 μM. The best spectral data are acquired with D_2O suspensions. Spectra of fully

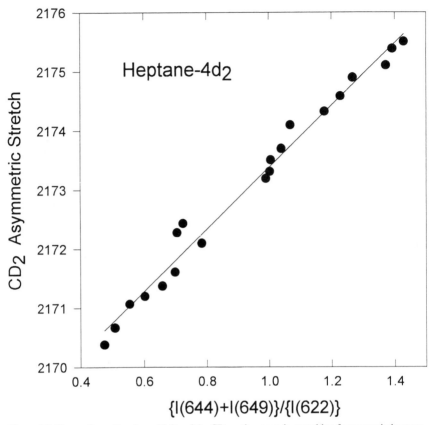

Figure 10. The conformational sensitivity of the CD_2 antisymmetric stretching frequency in heptane-4-d_2 is shown. This parameter (ordinate scale) is plotted against I{(644) + I(649)}/I(622) where the numerator is the sum of the integrated intensities of the tg marker bands and I(622) is the integrated intensity of the tt marker band. The linear regression line is shown. The data plotted encompass the temperature interval $-90°$ to $+50°$C.

proteated phospholipids at the same path length, temperature, and D_2O content are used as references for spectral subtraction (Mendelsohn et al, 1989).

It is necessary to convert the measured integrated intensities of the CD_2 rocking bands to relative populations of the trans and gauche bonds. To do this, the relative extinction coefficients must be established. This has been done three different ways. All give essentially the same result, namely that extinction coefficients for the bands associated with tt, tg, and gg sequences are nearly equal. (1) One method involves computing the relative intensities on the basis of a group moment model (Snyder, 1965). This model assumes that the transition moment is the sum of localized dipole derivatives from each group, methylene groups in our case, and that the direction of the local moment is determined by local symmetry. Normal coordinates are needed to indicate how much each group contributes to the sum. (2) Another method, somewhat more experimentally oriented, is to use the rotational isomeric state model (Flory, 1960) to estimate the relative concentrations of the conformational pairs. From the experimentally observed intensity ratios and a reasonable value for the gauche-trans energy difference, the intensity ratios for the pairs can be estimated. (3) The final method is the most direct. Cyclohexadecane in the crystalline state has a square conformation that may be described as $(ggttg'g'tt)_2$. In this ring the tt, tg, and gg pairs occur in the ratio 1:2:1. The infrared spectrum of a crystalline sample of cyclohexadecane-d_2 (each ring contains one CD_2 group) shows CD_2 rocking bands whose intensities closely follow the 1:2:1 ratios obtained by counting (Shannon et al, 1989).

Applications of the CD_2 rocking mode method to determine conformational disorder in 1,2 dipalmitoylphosphatidylcholine (diC$_{16}$PC) and in diC$_{16}$PC/cholesterol mixtures were reported by Mendelsohn and coworkers (Davies et al, 1990a; Mendelsohn et al, 1991; 1989). A series of specifically deuterated derivatives of diC$_{16}$PC was synthesized. Six derivatives, with CD_2 groups at the 3(3'), 4(4'), 6(6'), 10(10'), 12(12'), and 13(13') positions, respectively, were examined. IR spectra, as typified by the data for 6,6,6,6-d_4 diC$_{16}$PC in the gel (20°) and liquid crystal (50°) states were acquired (Figure 11). The latter spectra are remarkably similar to that of liquid n-heptane-4-d_2 at low temperatures (compare with Figure 9). Conversion of the measured intensities to conformational states is straightforward since the extinction coefficients calculated for alkanes apply to phospholipids. The data are analyzed according to Equation 1 below,

$$f_g(m-1, m) = \left[\frac{1}{r_g(m-1, m)} + 1 \right]^{-1} \qquad (1)$$

where r_g is the average gauche ratio for bonds m, $m-1$ so that $r_g(m-1, m) = [r_g(m-1) + r_g(m)]/2$ with $r_g(m-1, m) = n_{gt}/n_{tt} = I(650)/2(622)$.

The position dependence of the conformational disorder for 1,2 diC$_{16}$PC in its gel (33°C) and liquid-crystalline (50°C) phases is given in Figure 12, along with data for DPPC in a 1:1 binary mixture with cholesterol at 50°C. The results for phospholipids are essentially halved from earlier reports (Davies et al, 1990b; Mendelsohn et al, 1991; 1989), in which the factor of 2 required in the definition of $r_g(m-1, m)$ in equation (1) was overlooked.

The gel phase of DPPC is characterized by high conformational order ($\approx$2% gauche bonds) at the 6 and 10 positions, with a little more disorder ($\approx$5% gauche bonds) at position 4. Conformational disorder in the liquid-crystalline phase is $\approx$9%–11% at positions 4, 10, and 12, and rises to about 16% disorder at position 6. The disorder increases to about 40% at the penultimate C–C bond near the bilayer center (measured as described for end-gauche conformational states earlier). Included in Figure 12 is the position-independent value of $\approx$36% calculated from the RIS model for an isotropic alkane at 50°C. This value corresponds to $\approx$5 gauche bonds for liquid hexadecane (C$_{16}$ chain, the same length as diC$_{16}$PC) at 50°C.

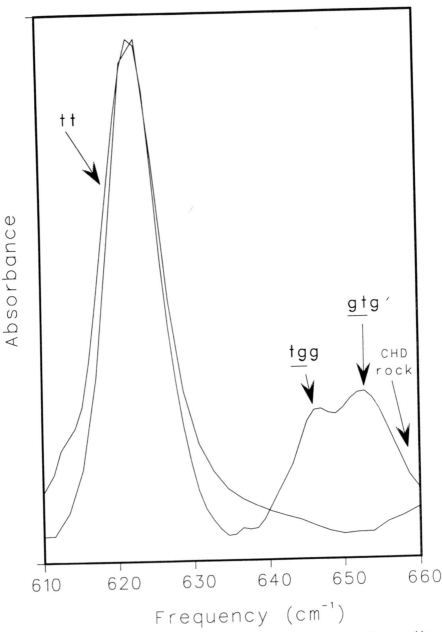

Figure 11. Spectra of the CD_2 rocking region for $6,6,6',6'$-d_4 DPPC below the phase transition (33°C) and above the phase transition (50°C). Note the emergence of the disorder bands (646 cm^{-1}, 652 cm^{-1}) above T_m.

Gauche Concentration in DPPC and 1/1 DPPC/Cholesterol Mixtures

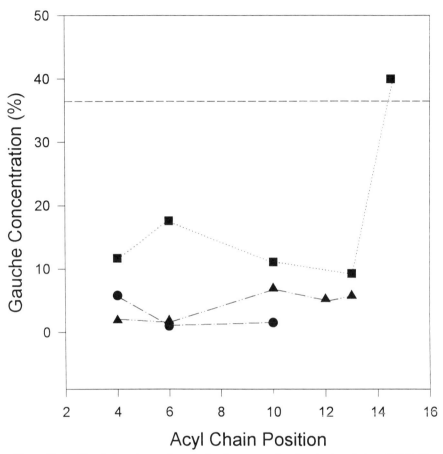

Figure 12. Positional dependence of gauche bond concentration disorder in gel phase DPPC (●, _._.), liquid crystalline diC$_{16}$PC at 50°C (■, ...), and 1:1 mol ratio diC$_{16}$PC/cholesterol mixture at 50°C (▲, _._.._.). The horizontal dashed line represents the disorder (36%) for liquid n-alkanes at 41°C calculated from the RIS model.

The CD$_2$ rocking data thus reveal substantial constraints to conformational disordering imposed by the bilayer structure. About 25%–30% (4, 10, and 12 positions) to 45% (6 position) of the disorder expected for an isotropic liquid is observed. Incorporation of cholesterol into the bilayer further reduces the extent and depth distribution of gauche bonds (Figure 12). At 50°C, conformational disorder at the 4 and 6 positions is reduced from 10% and 16% to about 2%–3% respectively, a fivefold diminution. The reduction is not as significant at positions

10 and 12. Cholesterol reduces conformational disorder at these locations from 10% to about 6%, a factor of ≈ 1.7.

Current models for cholesterol insertion into bilayers provide a rationale for these results. The location of the rigid sterol nucleus in the bilayer extends from acyl chain position 2 in the interfacial region to about position 8 or 9, while the aliphatic side chain of cholesterol extends from positions 9 or 10 towards the bilayer center. The IR data lend quantitative support to the idea that the rigid sterol nucleus is more effective in suppressing acyl chain conformational disorder by providing a more rigid environment than the aliphatic side chain of the sterol.

The above results demonstrate the advantages of the CD_2 rocking mode experiment to monitor conformational order. The method provides a direct, depth-dependent probe of trans-gauche isomerization in the acyl chains and is currently the only direct means available for extracting this information.

General use of the CD_2 rocking bands for the determination of conformational disorder is constrained by the extensive and tedious syntheses required to prepare specifically deuterated phospholipid derivatives and the relatively low intensity of the CD_2 rocking bands. These problems preclude extension of this experiment to systems containing a large number of membrane components. The CH_2 wagging modes, as discussed below, provide a complementary albeit less specific approach to measuring conformational disorder.

CH_2 Wagging Modes as Probes of Two- and Three-Bond Conformational States

The localized CH_2 wagging vibrations provide a convenient means to determine the contribution of particular two- and three-bond conformational states to acyl chain disorder. Snyder (1967) assigned wagging vibrations in the 1300–1370 cm^{-1} region of liquid n-alkanes spectra to particular nonplanar conformers. A wagging band at 1306 is associated with gtg' and possibly other sequences. The wagging bands observed at 1341 cm^{-1} and 1353 cm^{-1}, arise from the CH_2 wagging modes of end-gauche (eg) and double gauche (gg) conformers, respectively. The eg conformation refers to the penultimate C–C bond (i.e. the 14–15 bond in a C_{16} chain).

A note on the assignment of the wagging band at 1368 cm^{-1} is appropriate since this band appears to be associated with the gtg sequence rather than with the kink sequence, gtg', as has been generally assumed by most authors. The frequencies of the wagging band associated with gtg and gtg' are the same within a few wavenumbers (Snyder, 1967) so they are difficult to distinguish on that basis alone. Originally, it was argued from local symmetry considerations that only the gtg band would be infrared active, but with the caveat that the gtg' band might appear if the dihedral angle of the central trans bond underwent displacement from its equilibrium value. Recent calculations indicate that the intensity generated in that way would be negligible (Cates et al, 1994).

Spectra of the CH_2 wagging region (1330–1390 cm^{-1}) for the L_α (disordered bilayer) state of diC$_{16}$PC and for diC$_{16}$PC in CHCl$_3$ solution (monomeric state) are shown in Figure 13. The assignment of bands to particular conformers are noted. The most intense band in this spectral region arises from the symmetric methyl deformation (methyl umbrella) mode at 1378 cm^{-1}. This normal mode is highly localized so its associated band is suitable as an internal intensity reference standard.

One qualitative conclusion may be deduced directly from Figure 13. Since the relative contributions of the 1353 cm^{-1} band to the intensity of the contour is less for the bilayer than for the monomeric solution, the number of gg sequences is severely restricted by the bilayer structure, presumably a consequence of the severe disruption to acyl chain packing, that these sequences would entail.

Quantitative analysis of the two- and three-bond disorder (Casal and McElhaney, 1990; Senak et al, 1991) rests on comparison of the measured band intensities in phospholipids to those in liquid alkanes, whose conformational states can be estimated from the RIS model. The various integrated wagging band intensities are determined by curve-fitting the measured contour. The resultant bands are normalized to the intensity of the methyl umbrella mode. The concentration of specific sequences in a phospholipid acyl chain is then calculated from the ratio of the normalized intensity of the marker band for the phospholipids to that for alkanes.

The RIS model predicts that the number of gg and gtg conformers increases with temperature. When the theoretical predictions are compared with the experimentally observed intensity ratios, the alkane experimental data for the temperature dependence of the gg and eg modes for hexadecane are consistent with the predictions of the RIS model. However, the measured relative integrated intensities for the 1368 cm^{-1} band decrease with temperature, in disagreement with the theoretical predictions. This behavior was observed for all alkanes studied over the range C_{12}–C_{28} (Senak et al, 1991). Thus, the estimated number of gtg sequences in phospholipids, must be considered semiquantitative.

Conformational disorder for the L_α and H_{II} states of some phospholipids is summarized in Table 2. The restrictions (compared with the values calculated for alkanes) on the number of gg forms are substantial; 1.1 gg states/chain are calculated from the RIS model for a C_{16} chain at 25°C while 0.4 gg states/chain are observed for diC$_{16}$PC at 45°C. In contrast, the number of gtg states is only slightly reduced (from 1.2 to 1.0) in going from a liquid alkane to a bilayer.

Significant differences in two- and three-bond conformational states between lipid classes are also revealed in Table 2. Phosphatidylethanolamines (PEs) in bilayers are expected to have a more ordered liquid crystalline state than PCs, a consequence of their smaller headgroup. The magnitude of the effect is significant, as revealed in Table 2. The number of gg states is reduced to 0.2/chain in diC$_{16}$PE from 0.4/chain for diC$_{16}$PC, and the number of eg states is reduced to 0.1 for diC$_{16}$PE from 0.45 for diC$_{16}$PC. In contrast, the number of gtg states is virtually unaltered.

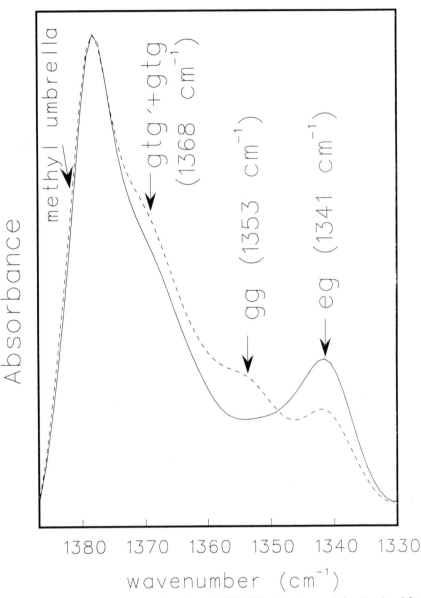

Figure 13. Spectra of the CH$_2$ wagging region for diC$_{16}$PC in chloroform solution (- - - -) and for liquid-crystalline diC$_{16}$PC (——) at 50°C. The various conformation marker bands at 1341 cm^{-1} (end-gauche), 1353 cm^{-1} (*gg*) and 1368 cm^{-1} (*gtg* states) are indicated. The diminution in intensity of the *gg* marker band in going from the chloroform solution (monomers) to the phospholipid vesicles (bilayers) is evident. Reprinted with permission from Senak et al (1991).

Table 2. Two- and Three-Bond Conformational Order in Some Phospholipids

Molecule	State	T(°C)	# gg	# eg	# gtg
$C_{16}H_{34}$	liquid	25	1.1^a	0.6^a	1.2^a
$diC_{16}PC$	L_α	45	0.4	0.4	1.0
$diC_{16}PC$	L_α	80	0.6	0.6	1.1
$diC_{16}PE$	L_α	65	0.2	0.1	1.0
$POPE^b$	L_α	50	0.2	0.05	0.9
POPE	H_{II}	75	0.4	0.1	1.0
DOPE	H_{II}	50	0.4	0.05	1.0
DOPC	L_α	50	0.9	0.6	1.1
$diC_{20:1}PC^c$	L_α	50	1.1	0.8	1.2

^{a} As calculated from the rotational isomeric state model
^{b} The numbers for all molecules containing C=C bonds are rendered uncertain by the lack of information about the conformations adopted by the C–C bonds adjacent to the C=C bond.
^{c} The C=C bond is in the *cis* configuration

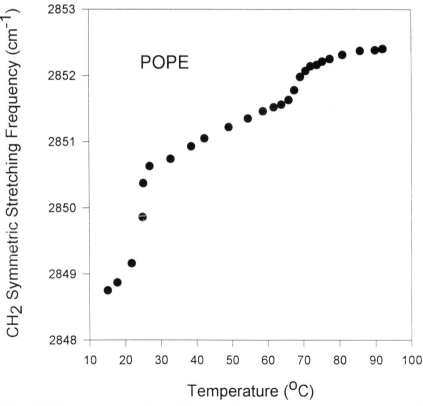

Figure 14. The temperature dependence of the symmetric CH_2 stretching frequency near 2850 cm^{-1} for 1-palmitoyl, 2-oleoyl PE (POPE) is shown. The gel-L_α phase transition for this molecule is evident near 25°C, while the L_α-H_{II} transition is evident near 70°C. Increasing disorder is indicated by increasing frequency.

Finally, the wagging modes also provide a convenient means to probe the conformation of phospholipids containing unsaturated acyl chains. Little structural information exists for such systems although they are the dominant species in the lipids of most membranes. Some preliminary results are included in Table 2. The primary trends are that the populations of *gg* and *eg* sequences are higher than in saturated acyl chains of the corresponding length.

A useful application of the CH_2 wagging experiment is to the study of the bilayer-inverted hexagonal (H_{II} phase) transition that occurs in phospho-

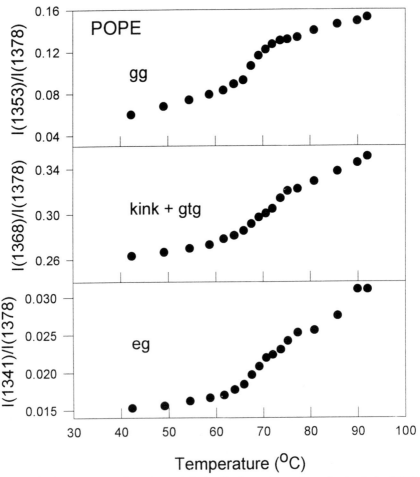

Figure 15. The temperature dependence of the localized wagging mode marker bands for POPE shown in the interval surrounding the L_α-H_{II} transition at $\approx 70°C$, as follows: (top), the *gg* conformation, 1353 cm^{-1}; (middle), *gtg* conformations, 1368 cm^{-1}; (bottom) the *eg* conformation, 1341 cm^{-1}. The sensitivity of IR to structural change is manifest for the *gg* mode, where the intensity increase during the transition corresponds to an additional 0.05–0.1 *gg* states.

lipids with small head groups and unsaturated acyl chains (Cevc and Marsh, 1985). Inverted hexagonal (H_{II}) phases are under current intense investigation because of their putative transient role in membrane fusion. The effects of the gel $\rightarrow$ L_α $\rightarrow$ H_{II} sequence of structure changes in 1-palmitoyl, 2-oleoyl PE (POPE) on the symmetric CH_2 stretching frequency is shown in Figure 14. The effects of the L_α-H_{II} interconversion on the three wagging spectral parameters are shown in Figure 15. This molecule has its gel-liquid crystal transition at 25°C and its bilayer-hexagonal transformation at 70°–75°C. Increases in the symmetric CH_2 at each transition (Figure 14) are indicative of increased acyl chain conformational disorder. The methylene wagging modes (Figure 15) permit a more quantitative evaluation of changes in the conformer distribution during the second transition. The gg conformation marker shows a well defined sigmoidal behavior at the L_α-H_{II} transition. The increased 1353 cm^{-1} band intensity corresponds to an additional 0.05–0.1 gg states during the transition. The energy involved for this increase is about 100–150 calories, depending on the value assumed for the energy of trans-gauche interconversion.

Structurally, the formation of gg states in the H_{II} phase indicates some relaxation of the packing constraints that inhibit the formation of such forms in bilayers. The intensity of the eg marker bands and the gtg conformational marker bands show little if any significant cooperative increase during the transition.

The CH_2 wagging modes offer several advantages for studies of acyl chain conformation. Unlike the CD_2 rocking mode approach, no synthesis is required. In addition, the wagging modes are relatively intense and appear in a spectral region free of interference from water absorption. Two limitations are: The position dependence of the conformational disorder cannot be deduced; other membrane components cannot be incorporated, since these may have absorption bands that interfere with the wagging bands.

Trends in Membrane IR Spectroscopy

The application of the three experiments described in the current chapter have been extended in two cases to encompass protein-containing systems. A study of Gramicidin/DPPC interaction (Davies et al, 1990a) used the CD_2 rocking modes to demonstrate that this peptide disordered the lipid gel phase while ordering the L_α phase. Holloway and coworkers (Doebler et al, 1995) used the CH_2 wagging modes to study lipid conformational changes upon the binding of native and mutant cytochrome b5 to PC vesicles. Future progress along these lines will depend upon improved methods for accurately removing spectral backgrounds.

The versatility of IR sampling ensures that future applications will require the acquisition of spectra from more challenging samples. Technology has been developed to obtain spectra in the external reflectance-absorbance (IR-RAS) mode from monolayers of lipids and proteins in situ at the A/W interface (Mendelsohn et al, 1995a,b). Spectra of the membranes from intact human erythrocytes have been observed (Moore et al, 1995) in spite of the fact that the vast majority of cell weight (excluding water) is derived from hemoglobin. Thus the possibility arises of examining changes to cell membranes during pathological states or upon treatment with therapeutic agents. The availability of IR microscopes permits the acquisition of structural information with a spatial resolution of ≈ 10 microns. Evaluation of the membrane structure in single cells (or at least in small groups of cells) is feasible.

ACKNOWLEDGMENTS

This work was supported in part by the Public Health Service through grants GM 29864 to Rutgers University at Newark and GM 27690 to the University of California at Berkeley.

REFERENCES

Cameron DG, Gudgin EF, Mantsch HH (1981): Dependence of acyl chain packing of phospholipids on the head group and acyl chain length. *Biochemistry* 20:4496–4500

Cameron DG, Casal HL, Mantsch HH (1980): Characterization of the pretransition in 1,2-dipalmitoyl-sn-glycero-3-phosphocholine by Fourier transform infrared spectroscopy. *Biochemistry* 19:3665–3672

Casal HL, McElhaney RN (1990): Quantitative determination of hydrocarbon chain conformational order in bilayers of saturated phosphatidylcholines of various chain lengths by Fourier transform infrared spectroscopy. *Biochemistry* 29:5423–5427

Cates DA, Strauss HL, Snyder RG (1994): Vibrational modes of liquid n-alkanes. *J Phys Chem* 98:4482–4488

Cevc G, Marsh D (1987): *Phospholipid Bilayers: Physical Principles and Models* New York: Wiley Interscience

Chia N-C, Mendelsohn R (1992): CH_2 wagging modes of unsaturated acyl chains as IR probes of conformational order in methyl alkenoates and phospholipid bilayers. *J Phys Chem* 96:10543–10547

Davies MA, Brauner, JW, Schuster HF, Mendelsohn R (1990a): A quantitative infrared determination of acyl chain conformation in gramicidin/dipalmitoylphosphatidylcholine mixtures. *Biochem Biophys Res Comm* 168:85–90

Davies MA, Schuster HF, Brauner JW, Mendelsohn R (1990b): Effects of cholesterol on conformational disorder in dipalmitoylphosphatidylcholine bilayers. A quantitative IR study of the depth dependence. *Biochemistry* 29:4368–4373

Dluhy RA (1986): Quantitative external reflection infrared spectroscopic analysis of insoluble monolayers spread at the air-water interface. *J Phys Chem* 90:1373–1379

Doebler RW, Steggles AW, Holloway, PW (1995): FTIR analysis of acyl chain conformation and its role in the free energy of binding of native and mutant cytochrome B5 to lipid vesicles. *Biophys J* 68:A429

Flach CR, Brauner JW, Taylor JW, Baldwin RC, Mendelsohn R (1994): External reflection FTIR of peptide monolayer films in situ at the air/water interface: Experimental design, spectra-structure correlations, and effects of hydrogen-deuterium exchange. *Biophys J* 67:402–41

Flory PJ (1989): Statistical Mechanics of Chain Molecules. Hanser, Munich

Gericke A, Hühnerfuss H (1993): In situ investigation of saturated long-chain fatty acids at the air/water interface by external infrared reflection-absorption spectroscopy. *J Phys Chem* 97:12899–12908

Griffiths PR, de Haseth JA (1986): Fourier transform infrared spectrometry. In: *Chemical Analysis: A Series of Monographs on Analytical Chemistry and Its Applications*, Vol 83, Elving PJ, Winefordner JD, eds.New York: Wiley Interscience

Mantsch HH, McElhaney RN (1991): Phospholipid phase transitions in model and biological membranes as studied by infrared spectroscopy. *Chem Phys Lipids* 57:213–226

Mantsch HH, Yang PW, Martin A, Cameron DG (1988): Infrared spectroscopic studies of Acholeplasma laidlawii B membranes. Comparison of the gel to liquid-crystal phase transition in intact cells and isolated membranes. *Eur J Biochem* 178:335–341

Maroncelli M, Strauss HL, Snyder RG (1985a): On the CD_2 probe infrared method for determining polymethylene chain conformation. *J Phys Chem* 89:4390–4395

Maroncelli M, Strauss HL, Snyder RG (1985b): The distribution of conformational disorder in the high-temperature phases of the crystalline n-alkanes. *J Chem Phys* 82:2811–2824

Mendelsohn R, Mantsch HH (1986): Fourier transform infrared studies of lipid-protein interaction. In: *Progress in Protein-Lipid Interactions 2* Watts A, De Pont JJHHM, eds. Amsterdam: Elsevier-North Holland

Mendelsohn R, Brauner JW, Gericke A (1995a): External infrared reflection absorption spectrometry of monolayer films at the air/water interface. *Ann Rev Phys Chem* 46:305–334

Mendelsohn R, Davies MA, Schuster HF, Xu Z, Bittman R (1991): CD_2 rocking modes as quantitative infrared probes of one-, two-, and three-bond conformational disorder in dipalmitoylphosphatidylcholine and dipalmitoylphosphatidylcholine/cholesterol mixtures. *Biochemistry* 30:8558–8563

Mendelsohn R, Davies MA, Brauner JW, Schuster HF, Dluhy RA (1989): Quantitative determination of conformational disorder in the acyl chains of phospholipid bilayers by infrared spectroscopy. *Biochemistry* 28:8934–8939

Mendelsohn R, Liang GL, Strauss HL, Snyder RG (1995b): IR spectroscopic determination of gel state miscibility in long-chain phosphatidylcholines mixtures. *Biophys J*: submitted

Mitchell ML, Dluhy RA (1988): In situ FT-IR investigation of phospholipid monolayer phase transitions at the air-water interface. *J Am Chem Soc* 110:712–718

Moore DJ, Mendelsohn R (1994): Adaptation to altered growth temperatures in Acholeplasma laidlawii B: Fourier transform infrared studies of acyl chain conformational order in live cells. *Biochemistry* 33:4080–4085

Moore DJ, Sills RH, Mendelsohn R (1995) Peroxidation of erythrocytes: FTIR spectroscopy studies of extracted lipids, isolated membranes, and intact cells. *Biospectroscopy*: in press

Moore DJ, Wyrwa M, Reboulleau CP, Mendelsohn R (1993): Quantitative IR studies of acyl chain conformational order in fatty acid homogeneous membranes of live cells of Acholeplasma laidlawii B. *Biochemistry* 32:6281–6287

Schachtschneider JM, Snyder RG (1963): Vibrational analysis of the n-paraffins-II. Normal co-ordinate calculations. *Spectrochim Acta* 19:117–168

Senak L, Davies MA, Mendelsohn R (1991): A quantitative IR study of hydrocarbon chain conformation in alkanes and phospholipids. CH_2 wagging modes in disordered bilayer and H_{II} phases. *J Phys Chem* 95:2565–2571

Senak L, Moore D, Mendelsohn R (1992): CH_2 wagging progressions as IR probes of slightly disordered phospholipid acyl chain states. *J Phys Chem* 96:2749–2754

Shannon VL, Strauss HL, Snyder RG, Elliger CA, Mattice WL (1989): Conformation of the cycloalkanes $C_{14}H_{28}$, $C_{16}H_{32}$, and $C_{22}H_{44}$ in the liquid and high-temperature crystalline phases by vibrational spectroscopy. *J Am Chem Soc* 111:1947–1958

Snyder RG (1967): Vibrational study of the chain conformation of the liquid n-paraffins and molten polyethylene. *J Chem Phys* 47:1316–1360

Snyder RG (1965): Group moment interpretation of the infrared intensities of crystalline n-paraffins. *J Chem Phys* 42:1744–1763

Snyder RG (1961): Vibrational spectra of crystalline n-paraffins. Part II. Intermolecular effects. *J Mol Spec* 7:116–144

Snyder RG (1960): Vibrational spectra of crystalline n-paraffins Part I. methylene rocking and wagging modes. *J Mol Spec* 4:411–434

Snyder RG, Poore MW (1973): Conformational structure of polyethylene chains from the infrared spectrum of the partially deuterated polymer. *Macromolecules* 6:708–712

Snyder RG, Conti G, Strauss HL, Dorset DL (1993): Thermally-induced mixing in partially microphase segregated binary n-alkane crystals. *J Phys Chem* 97:7342–7350

Snyder RG, Goh MC, Srivatsavoy VJP, Strauss HL, White JW, Dorset DL (1992): Measurement of the growth kinetics of microdomains in binary n-alkane solid solutions by infrared spectroscopy. *J Phys Chem* 96:10008–10019

Snyder RG, Srivatsavoy VJP, Cates DA, Strauss HL, Dorset DL (1994): Hydrogen/deuterium isotope effects on microphase separation in unstable crystalline mixtures of binary n-alkanes. *J Phys Chem* 98:674–684

Snyder RG, Strauss HL, Cates DA (1995): Detection and measurement of microaggregation in binary mixtures of esters and of phospholipid dispersions. *J Phys Chem*: in press

7

Membrane Structure and Dynamics Studied with NMR Spectroscopy

Michael F. Brown

Motivation and Problem Addressed

Biomembranes mediate the diverse functions of life and comprise mainly lipids and proteins, together with carbohydrates associated with the cellular and organelle surfaces. Present knowledge indicates that the lipids typically form a bilayer containing proteins that span the membrane or are attached to its surface. The lipid and protein moieties are amphiphilic, i.e., part of the molecule is polar and preferentially associated with water, whereas part is nonpolar and only sparingly soluble in aqueous media. The hydrophobic effect (Tanford, 1980) is thus an important determinant of the self-assembly of the lipids and proteins into biological membranes. In the liquid-crystalline state, as found in native membranes, the polar head groups of the lipids are on the exterior of the bilayer; whereas the nonpolar hydrocarbon chains are sequestered away from water, within the membrane interior. The lipid bilayer represents the fundamental permeability barrier to the passage of ions and polar molecules into or out of a cell or organelle. In addition the bilayer lipids play a role in the vectorial organization of membrane components. On the other hand, the distinctive functions of biological membranes are largely due to proteins, which may be influenced by lipid-protein interactions.

Functions of biomembranes that may be sensitive to physicochemical properties of the bilayer include the triggering of visual excitation (Gibson and Brown, 1993; Wiedmann et al, 1988), active transport of ions (Michelangeli et al, 1991; Navarro et al, 1984), nerve impulse generation (Fong and McNamee, 1987), and prokaryotic and eukaryotic cellular regulation and growth (Epand, 1990; Jensen and Schutzbach, 1984; Kinnunen et al, 1994; Lindblom and Rilfors, 1989; Newton, 1993). In addition, the activities of membrane bound

Biological Membranes
K. Merz, Jr. and B. Roux, Editors
© Birkhäuser Boston 1996

peptides such as gramicidin A (Lundbæk and Andersen, 1994) and alamecithin (Hall et al, 1984) depend on the lipid environment in planar bilayers; other biologically active peptides such as the magainins are membrane bound. There are also membrane proteins that have a predominantly structural capacity, such as the viral coat proteins. One can summarize existing knowledge by stating that in some cases there is evidence for an influence of the membrane lipid bilayer on the biological activity of membrane proteins, as exemplified by rhodopsin (Brown, 1994). In other instances it is plausible that the bilayer physicochemical properties may be similarly implicated in membrane function.

Since a lipid bilayer is typically a liquid-crystalline material, being intermediate between the liquid and solid states of matter, it is the properties of the liquid-crystalline state that may be functionally relevant. Knowledge of both structure and dynamics is needed. Nematic and smectic liquid crystals are characterized by the presence of orientational order accompanied by long range positional order of the constituents; yet the molecules undergo rapid rotational and translational diffusion as in the case of a simple fluid. The presence of motion is associated with characteristic properties of liquid-crystalline materials, which in turn may be implicated in the distinctive functions of biomembranes. In the past, considerable emphasis has been devoted to the influences of electrostatics on membrane functions, including the surface potential and transmembrane potential (McLaughlin, 1989; Seelig et al, 1987). However, it is becoming increasingly apparent that nonelectrostatic properties associated with the presence of neutral membrane lipids also may be quite important (Brown, 1994; Jensen and Schutzbach, 1984; Lindblom and Rilfors, 1989; Navarro et al, 1984; Wiedmann et al, 1988). Examples of equilibrium properties that may be biologically relevant include the thickness of the bilayer and the associated interfacial area per molecule, together with lateral and curvature deformation of the membrane as formulated using a flexible surface model (Gibson and Brown, 1993). In this regard, the presence of lipids close to a L_α-H_{II} phase boundary in biomembranes (Deese et al, 1981a) may yield a characteristic spontaneous curvature and/or bending rigidity associated with the individual apposed monolayers of the membrane bilayer. The resulting curvature free energy of the two monolayers (frustration) can provide a driving force, i.e., a source of work, for conformational changes of membrane proteins linked to key biological functions (Brown,

---

Abbreviations: COSY, correlation spectroscopy; DLPC–d_{46}, 1,2-diperdeuteriolauroyl-*sn*-glycero-3-phosphocholine; DMPC–d_{54}, 1,2-diperdeuteriomyristoyl-*sn*-glycero-3-phosphocholine; DMPC, 1,2-dimyristoyl-*sn*-glycero-3-phosphocholine; DPH, 1,6-diphenyl-1,3,5-hexatriene; DPPC, 1,2-dipalmitoyl-*sn*-glycero-3-phosphocholine; EPR, electron paramagnetic resonance; EFG, electric field gradient; FT IR, Fourier transform infrared; MAS, magic angle sample spinning; NMR, nuclear magnetic resonance; NOE, nuclear Overhauser effect; NOESY, nuclear Overhauser effect spectroscopy; PAS, principal axis system; PDPC–d_{31}, 1-perdeuteriopalmitoyl-2-docosahexaenoyl-*sn*-glycero-3-phosphocholine; PMMA–d_8, poly(methyl methacrylate)-d_8; REDOR, rotational-echo double-resonance; RR, rotational resonance; SANS, small angle neutron scattering; SAXS, small angle x-ray scattering; TOCSY, total correlation spectroscopy.

1994). Clearly the above properties can be modulated by additional components such as cholesterol, alkanes, and general anesthetics.

Besides such equilibrium aspects of the membrane bilayer, dynamical properties associated with motions of the lipid molecules may also be important (McConnell, 1976). In the liquid-crystalline state, these may include local motions of the flexible molecular segments having correlation times τ_c of $\approx 5-20$ picoseconds; whole molecule motions with correlation times ranging from ≈ 100 picoseconds to several nanoseconds for rotational diffusion and ≈ 50 nanoseconds for lateral diffusion; and finally collective excitations of the bilayer spanning a broad range of time scales, from ≈ 100 picoseconds to milliseconds or longer depending on the sample dimensions. Properties of the lipid/water interface are associated with the polar head groups, which together with the acyl chains may govern the membrane permeability to water and nonpolar solutes. Molecular motions include axial rotation and translational diffusion of the molecular center of mass, and may influence formation of transient complexes of membrane enzymes or the activities of transport proteins. Interestingly, collective undulatory excitations involving the membrane surface and the bilayer interior plus vibrational out-of-plane motions of the individual molecules can give rise to repulsive forces between the lamellae (Israelachvili and Wennerström, 1992). These repulsive forces are of inherent physical interest, and may be connected to biological functions such as adhesion and membrane fusion associated with exo- or endocytosis.

How can one acquire further information regarding the equilibrium and dynamical properties of membrane lipid bilayers and membrane proteins? Clearly both theoretical and experimental studies are needed as a means of integrating existing knowledge with new concepts. Experimental avenues at present include small angle neutron scattering (SANS) (Zaccai et al, 1979), small angle X-ray scattering (SAXS) (McIntosh and Simon, 1986; Wiener and White, 1992), infrared spectroscopy (Davies et al, 1990; Siminovitch et al, 1988b), and nuclear magnetic resonance (NMR) spectroscopy (Brown and Chan, 1995; Davis, 1983; Seelig, 1977). With the exception of NMR spectroscopy, however, the above methodologies mainly provide knowledge pertinent to the structures of membrane constituents averaged over an appropriate time scale. Deuterium (^{2}H) NMR spectroscopy has the considerable advantage in that detailed information is obtainable regarding the fluctuations that give rise to the average structural properties. With respect to dynamics, the most promising experimental methods at present comprise NMR relaxation (Bloom et al, 1992; Brown and Chan, 1995) and dynamic neutron scattering (König et al, 1992). Theoretical approaches involving Monte Carlo methods, as well as molecular and Langevin (Brownian) dynamics simulations of bilayers, have also yielded significant insights (Pastor et al, 1991; Peters et al, 1995; Scott, 1986; Scott and Kalaskar, 1989; van der Ploeg and Berendsen, 1982; Venable et al, 1993). Such computer simulations represent an important means of integrating conceptually the various experimental findings with regard to the underlying force fields associated with the equilibrium and dynamical bilayer properties.

Our goal in this chapter is thus to briefly outline the conceptual formalism for application of NMR lineshape and relaxation methods to membrane constituents in general. Simple approaches in analytical closed form are described, that may help to shed light on the types of more detailed models that can be developed in future work. Finally, the results of experimental NMR studies of lipid bilayers and membrane proteins are described selectively, which provide a basis for comparison to future theoretical molecular and Brownian dynamics studies of membranes.

NMR Spectroscopy in Relation to Membrane Biophysics

For illustrative purposes, and as a means of focussing attention of the problem at hand, a representative biomembrane comprising both integral and peripheral membrane proteins is depicted schematically in Figure 1. The results of various biophysical studies, including NMR lineshape and relaxation studies (Brown and Chan, 1995), small angle X-ray and neutron scattering, and computer simulations (Pastor et al, 1991) can be described briefly as follows. In the biologically relevant liquid-crystalline phase, the lipid constituents of membranes are highly disordered and a wide range of motions occurs. The Zwitterionic and anionic lipid polar head groups are hydrated and in contact with water on either side of the bilayer, which has a molecularly rough lipid/water interface undergoing rapid fluctuations. The acyl chains are closely packed, usually perpendicularly to the aqueous interface, yet they have substantial disorder with considerable entanglement. The acyl segments are broadly distributed along the perpendicular to the membrane surface, as is evident with regard to the terminal methyl groups. NMR relaxation studies suggest that various collective bilayer excitations exist in which the microviscosity of the bilayer interior is comparable to that of a liquid hydrocarbon such as n-hexadecane (Brown et al, 1983). Only small effects of integral membrane proteins on the bilayer properties are evident in the fluid phase (Brown et al, 1982, 1977; Ellena et al, 1986; Scott, 1986; Seelig and Seelig, 1980). Because the membrane properties involve a balance of long range attractive van der Waals interactions, together with short range repulsive forces (Israelachvili, 1992), one can consider the force balance using a flexible surface model (Gibson and Brown, 1993). We have proposed elsewhere that this balance of forces involving the lipid polar head groups and hydrocarbon region is a key determinant of membrane protein function in the case of rhodopsin (Wiedmann et al, 1988). Such a formulation in terms of curvature free energy involving fluid membranes represents a very useful concept in biophysics (Brown, 1996, 1994).

Yet in spite of the above knowledge, some of the most intriguing questions remain currently unanswered and await future research. Such problems are particularly amenable to study at this time, due to development of the appropriate biophysical and biochemical tools together with conceptual advances. For example, what are the physicochemical properties that characterize the liquid-

a) biomembrane

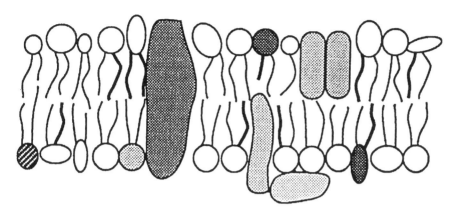

b) lipid bilayer

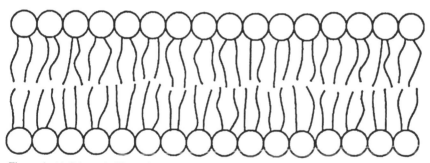

Figure 1. (a) Schematic illustration of a representative biomembrane, depicting integral proteins that span the lipid bilayer, together with peripheral proteins that are associated with the surface. (b) Illustration of a model bilayer as employed for investigations of membrane lipid properties. Studies of model lipid systems are of fundamental importance since they deal with self-assembling liquid-crystalline materials. In addition, they comprise a necessary foundation for biological applications which involve membrane lipid-protein interactions.

crystalline state of the membrane lipids, and how do these depend on the polar head group and acyl chain compositions? How do cholesterol and integral membrane proteins influence the properties of lipid bilayers? What are the effects of bilayer lipids on membrane proteins, and is there a single key property that governs or influences protein-mediated functions of biological membranes? One hypothesis is that equilibrium properties of the membrane lipid bilayer are related to the energetics of the conformational states of membrane proteins, whereas dynamical properties associated with the amplitudes and rates of the lipid motions influence stochastic events involving membrane proteins. Inasmuch as NMR spectroscopy provides knowledge at the microscopic level regarding the average and dynamic properties of bilayer membranes, it plays a unique and vital role in their investigation. A corollary is that NMR relaxation techniques afford a powerful means of experimentally obtaining information about the correlation functions and associated motion spectral densities of the membrane lipids. Here the decay of the orientational correlations involving the various molecular segments provides knowledge of time-dependent processes; whereas the equilibrium or long time values yield information regarding the properties averaged over the relevant time scale. By contrast, with reference to disordered systems, including liquid-crystalline materials and biological membranes, X-ray and neutron scattering techniques do not yield such atomic level information.

Equilibrium Properties of Membrane Constituents Obtained from NMR Lineshapes

How can one investigate the equilibrium and dynamical properties of membrane constituents in the liquid-crystalline phase? In NMR spectroscopy, the various observable quantities correspond to magnetic or electrical interactions involving the nuclei which depend on their orientation and/or relative spatial position. The relevant interactions are described as a rule by second-rank Cartesian tensors, which are most usefully formulated in a spherical basis, rather than in the more familiar rectangular Cartesian coordinates (see Appendix). In what follows, we shall mainly emphasize deuterium (^{2}H) NMR spectroscopy, which constitutes a general paradigm for the application of NMR methods to liquid-crystalline systems and molecular solids. Because a single electric quadrupolar interaction is dominant, the influences of magnetic dipolar interactions and chemical shielding are minimal. Moreover, the quadrupolar interaction is *intra*molecular, so that the effects of *inter*molecular interactions can be largely neglected.

Coupling Hamiltonians in NMR Spectroscopy

In what follows the case of spin $I = 1$ systems is considered; examples include the quadrupolar interaction of the ^{2}H nucleus having $I = 1$ with the electric field gradient of its chemical bond, and the direct ^{1}H-^{1}H and ^{13}C-^{1}H magnetic dipolar interactions involving spin $I = 1/2$ nuclei. The coupling of a spin $I = 1$ system is described by the contraction or scalar product of irreducible tensors of second-rank, the first corresponding to the nuclear spin angular momentum operators, and the second representing a specific coupling mechanism as discussed in the classic book of Häberlen (1976). Thus the Hamiltonian operator is given by:

$$\hat{H}_\lambda = C_\lambda \hat{T} \cdot V = C_\lambda \sum_{m=-2}^{2} (-1)^m \hat{T}_{-m}^{(2)\text{lab}} V_m^{(2)\text{lab}}, \tag{1}$$

where the hat (^) denotes an operator, C_λ is a constant, and λ refers to a specific interaction, e.g., the quadrupolar (Q) or dipolar (D) coupling. In the case of second-rank tensors the summation over the projection index m runs from $-2, \ldots, 2$ in integral steps. All symbols have their familiar meanings, where $\hat{T}_m^{(2)}$ and $V_m^{(2)}$ are second-rank irreducible spherical tensors corresponding to the nuclear spin angular momentum operators and the coupling tensor V, respectively.

Clearly the spin angular momentum is quantized in the main external magnetic field (denoted lab, with coordinate axes X, Y, Z). The corresponding irreducible tensor operators are given by Häberlen (1976):

$$\hat{T}_0^{(2)\text{lab}} = \frac{1}{\sqrt{6}}(3\hat{I}_Z\hat{S}_Z - \hat{\boldsymbol{I}} \cdot \hat{\boldsymbol{S}}), \tag{2a}$$

$$\hat{T}_{\pm 1}^{(2)\text{lab}} = \mp \frac{1}{2}(\hat{I}_Z\hat{S}_\pm + \hat{I}_\pm\hat{S}_Z), \tag{2b}$$

$$\hat{T}_{\pm 2}^{(2)\text{lab}} = \frac{1}{2}\hat{I}_\pm\hat{S}_\pm, \tag{2c}$$

in which $\hat{I}_\pm \equiv \hat{I}_X \pm i\hat{I}_Y$ and likewise for $\hat{S}_\pm$. Here $\hat{\boldsymbol{I}}$ denotes the angular momentum operators for the nuclear spin in units of $\hbar$, i.e., $\hat{\boldsymbol{I}} = \hat{\boldsymbol{J}}/\hbar$ where $\hat{\boldsymbol{J}}$ is the angular momentum. (Note that for the electric quadrupolar coupling or the homonuclear dipolar coupling $\hat{\boldsymbol{I}} = \hat{\boldsymbol{S}}$; whereas for the heteronuclear magnetic dipolar coupling $\hat{\boldsymbol{I}} \neq \hat{\boldsymbol{S}}$.) By contrast, the irreducible components of the coupling tensor $V_m^{(2)}$ are expressed in a molecule-fixed principal axis system (PAS, with coordinate axes x, y, z) as described by Häberlen (1976):

$$V_0^{(2)\text{PAS}} = \sqrt{\frac{3}{2}}\delta_\lambda, \tag{3a}$$

$$V_{\pm 1}^{(2)\text{PAS}} = 0, \tag{3b}$$

$$V_{\pm 2}^{(2)\text{PAS}} = -\frac{1}{2}\delta_\lambda\eta_\lambda. \tag{3c}$$

In the above formulas $\delta_\lambda \equiv V_{zz}$ indicates the largest Cartesian principal component, and the asymmetry parameter $\eta_\lambda \equiv (V_{yy} - V_{xx})/V_{zz}$ characterizes the deviation of the Cartesian principal values from axial symmetry, where $\eta_\lambda \in [0, 1]$. The coupling parameters for the electric quadrupolar interaction and the magnetic dipolar interaction are the following: quadrupolar coupling ($I = 1$), $C_Q = eQ/2\hbar$, $\delta_Q = eq$, and η_Q; dipolar coupling, $C_D = -2\gamma_I\gamma_S\hbar$, $\delta_D = \langle r_{IS}^{-3}\rangle$, and $\eta_D = 0$. For the dipolar coupling the internuclear distance cubed is averaged over faster vibrational degrees of freedom, such that $\langle r_{IS}^{-3}\rangle^{-1/3}$ can be greater than the equilibrium internuclear distance (Brown, 1984b; Söderman, 1986); analogous considerations apply to the quadrupolar coupling (Vold and Vold, 1991).

Deuterium NMR: Calculation of Expectation Values
and Quadrupolar Frequencies

Recall that the above irreducible tensors are expressed in different reference frames; the spin operators $\hat{I}$ and $\hat{S}$ correspond to the laboratory coordinate system defined by the main magnetic field, whereas the quadrupolar and dipolar couplings involve the molecule-fixed frame. To express the Hamiltonian in a single frame, e.g., that of the laboratory, the transformation properties of irreducible tensors under rotations are utilized (Brink and Satchler, 1968). The irreducible components of the coupling tensor in the laboratory frame and the principal axis system are related by (see Appendix):

$$V_m^{(2)\text{lab}} = \sum_{s=-2}^{2} V_s^{(2)\text{PAS}} D_{sm}^{(2)}(\Omega_{PL}).\tag{4}$$

Here $\Omega_{PL} = (\alpha_{PL}, \beta_{PL}, \gamma_{PL})$ are the Euler angles which describe active transformation (Brink and Satchler, 1968) of the irreducible components from the principal axis system (P) to the laboratory frame (L). The convention of Brink and Satchler (1968) is used for the Wigner rotation matrices, $D_{m'm}^{(j)}(\Omega) = e^{-im'\alpha} d_{m'm}^{(j)}(\beta) e^{-im\gamma}$; where (m', m) are generic projection indices, j is the angular momentum, $\Omega \equiv (\alpha, \beta, \gamma)$ are the Euler angles, and the reduced rotation matrix elements are indicated by $d_{m'm}^{(j)}(\beta)$. [One should also recognize that $D_{00}^{(j)}(\Omega) = d_{00}^{(j)}(\beta) = P_j(\cos\beta)$, where P_j is a Legendre polynomial of order j.]

In discussing NMR lineshapes, let us restrict ourselves to deuterium (^{2}H) NMR spectroscopy ($I = 1$ nucleus), which is illustrative of the principles in general. The coupling due to the direct magnetic dipolar interaction, as well as the anisotropic chemical shift (see above), can be similarly considered. Expressed within the laboratory frame, the quadrupolar Hamiltonian takes the form:

$$\hat{H}_Q = \delta_Q C_Q \sum_m (-1)^m \hat{T}_{-m}^{(2)\text{lab}} \left\{ D_{0m}^{(2)}(\Omega_{PL}) - \frac{\eta_Q}{\sqrt{6}} [D_{-2m}^{(2)}(\Omega_{PL}) + D_{2m}^{(2)}(\Omega_{PL})] \right\},\tag{5}$$

which is in dimensions of angular frequency. The expectation values are calculated by making the conventional high-field approximation, in which the shifting of the energy levels due to the coupling interaction $\hat{H}_Q$ is small in comparison to the much larger Zeeman interaction, given by $\hbar\hat{H}_Z = -\gamma\hbar B_0\hat{I}_Z$. We thus restrict ourselves to the secular ($m = 0$) term in Equation (5), also referred to as truncating the Hamiltonian. This is equivalent to the conventional first order perturbation theory employed in quantum mechanics. The truncated internal Hamiltonian in dimensions of energy then becomes:

$$\hbar\hat{H}_Q = \frac{e^2 q Q}{4}(3\hat{I}_Z^2 - \hat{I}^2)\left\{ D_{00}^{(2)}(\Omega_{PL}) - \frac{\eta_Q}{\sqrt{6}}[D_{-20}^{(2)}(\Omega_{PL}) + D_{20}^{(2)}(\Omega_{PL})] \right\}.\tag{6}$$

The quadrupolar frequencies are thus related to the expectation values of the secular part of the quadrupolar Hamiltonian, which commutes with $\hat{H}_Z$ and causes first order energy level shifts involving the unperturbed wavefunctions.

Let us label the spin wavefunctions ψ by the total spin angular momentum quantum number I and its projection m onto the magnetic field axis, in units of $\hbar$, which in the Dirac bra-ket notation is denoted by $|I, m\rangle$. The coupling energies due to the Zeeman and quadrupolar interactions in the $|I, m\rangle$ basis are given to first order by the Schrödinger equation (Atkins, 1990):

$$\hbar\hat{H}|I, m\rangle = E_m|I, m\rangle, \tag{7}$$

where $\hbar\hat{H} = \hbar\hat{H}_Z + \hbar\hat{H}_Q$ and $\hbar\hat{H}_Z = -\gamma\hbar B_0\hat{I}_Z$. The eigenvalues of the angular momentum operators correspond to the following equations (Atkins, 1990):

$$\hat{I}^2|I, m\rangle = I(I + 1)|I, m\rangle, \tag{8a}$$

$$\hat{I}_Z|I, m\rangle = m|I, m\rangle. \tag{8b}$$

From these relations, the following results for the three energy levels of an $I = 1$ system are then obtained, which is left as an exercise for the reader:

$$E_{+1} = -\gamma\hbar B_0 + \left(\frac{e^2qQ}{4}\right)\left\{D_{00}^{(2)}(\Omega_{PL}) - \frac{\eta_Q}{\sqrt{6}}[D_{-20}^{(2)}(\Omega_{PL}) + D_{20}^{(2)}(\Omega_{PL})]\right\}, \tag{9a}$$

$$E_0 = \left(\frac{-e^2qQ}{2}\right)\left\{D_{00}^{(2)}(\Omega_{PL}) - \frac{\eta_Q}{\sqrt{6}}[D_{-20}^{(2)}(\Omega_{PL}) + D_{20}^{(2)}(\Omega_{PL})]\right\}, \tag{9b}$$

$$E_{-1} = \gamma\hbar B_0 + \left(\frac{e^2qQ}{4}\right)\left\{D_{00}^{(2)}(\Omega_{PL}) - \frac{\eta_Q}{\sqrt{6}}[D_{-20}^{(2)}(\Omega_{PL}) + D_{20}^{(2)}(\Omega_{PL})]\right\}. \tag{9c}$$

Here the first term is the Zeeman energy, $E_m = -\gamma\hbar B_0 m$, whereas the second is the shift of the energy levels to first order due to the electric quadrupolar interaction of the ^{2}H nucleus with the electric field gradient of its chemical bond.

Finally, the frequencies of the two allowed single-quantum ($\Delta m = \pm 1$) transitions of the ^{2}H nucleus are given in terms of the Bohr frequency condition by $E_{-1} - E_0 = h\nu_+$ and $E_0 - E_{+1} = h\nu_-$. Here $\nu_{\pm} = \nu_0 + \nu_Q^{\pm}$, where $\nu_0 = \gamma\hbar B_0$ is the Larmor frequency, and $\nu_Q^{\pm}$ are the quadrupolar frequencies. This leads to the following results for the first-order quadrupolar frequencies:

$$\nu_Q^{\pm} = \pm\frac{3}{4}\chi_Q\left\{D_{00}^{(2)}(\Omega_{PL}) - \frac{\eta_Q}{\sqrt{6}}[D_{-20}^{(2)}(\Omega_{PL}) + D_{20}^{(2)}(\Omega_{PL})]\right\}, \tag{10}$$

where $\chi_Q \equiv e^2qQ/h$ is the *static* quadrupolar coupling constant. Now substitution of the Wigner rotation matrix elements (see Appendix) yields $[D_{-20}^{(2)}(\Omega) +$

$D_{20}^{(2)}(\Omega)] = \sqrt{\frac{3}{2}} \sin^2 \beta \cos(2\alpha)$, where Ω are generalized Euler angles. We then obtain for the quadrupolar splitting $\Delta \nu_Q$:

$$\Delta \nu_Q \equiv \nu_Q^+ - \nu_Q^-, \tag{11a}$$

$$= \frac{3}{2} \chi_Q \left[\frac{1}{2}(3\cos^2 \beta_{PL} - 1) - \frac{\eta_Q}{2} \sin^2 \beta_{PL} \cos(2\alpha_{PL}) \right]. \tag{11b}$$

Furthermore, it turns out that the static EFG tensor of the C–^{2}H bond is nearly axially symmetric ($\eta_Q \approx 0$), which leads to the simpler result:

$$\nu_Q^{\pm} = \pm \frac{3}{4} \chi_Q D_{00}^{(2)}(\Omega_{PL}). \tag{12}$$

The quadrupolar splitting is thus:

$$\Delta \nu_Q = \frac{3}{2} \chi_Q D_{00}^{(2)}(\Omega_{PL}), \tag{13a}$$

$$= \frac{3}{2} \chi_Q \left(\frac{3\cos^2 \beta_{PL} - 1}{2} \right). \tag{13b}$$

The above expression describes the dependence of the quadrupolar splitting on the (Euler) angles that rotate the coupling tensor from its principal axes system to the laboratory frame defined by the main magnetic field.

To gain a simple physical feel for these results, let us consider as a simple illustrative example the case that $\eta_Q \approx 0$. Take an imaginary single crystal and start with an alignment of the magnetic field along one of the principal axes of the coupling tensor, e.g., the z-axis, which corresponds to the largest principal value of the electric field gradient. (Here the reader may wish to refer to Figure A1 of the Appendix.) According to Equations (13) the quadrupolar splitting then takes its maximum value of $\Delta \nu_Q = \frac{3}{2}\chi_Q$, where $\chi_Q = 170$ kHz for an aliphatic C–^{2}H bond. Rotating the single crystal about the z-axis of the electric field gradient (associated with the orientation of the C–^{2}H bonds within the crystal axes system) by the Euler angle α_{PL} ($\equiv \phi$ in spherical polar coordinates) does nothing, since the tensor is axially symmetric ($\eta_Q \approx 0$). However, for a nonaxially symmetric tensor the behavior is more complicated as described by Equations (11). Next the single crystal is rotated about the new y'-axis by the Euler angle β_{PL} ($\equiv \theta$ in spherical polar coordinates). The quadrupolar splitting now becomes progressively smaller and reaches zero at the so-called magic-angle [where $\beta_{PL} = \cos^{-1}(1/\sqrt{3}) = 54.7°$], beyond which the sign of the splitting changes. When $\beta_{PL} = 90°$, the magnetic field is now aligned within the x-y plane of the axially symmetric coupling tensor, such that $\Delta \nu_Q = -\frac{3}{4}\chi_Q$. Rotation about the third Euler angle (γ_{PL}), i.e., about the z''-axis, is again immaterial on account of the cylindrical symmetry around the main magnetic field direction. The effect of these rotations is described by Equations (13) for an axially symmetric tensor ($\eta_Q = 0$), and more generally by Equations (11) for the case of a nonaxially symmetric tensor ($\eta_Q \neq 0$).

The reader should note that when motion is present the various rotation angles become time-dependent as discussed below.

The above discussion indicates how the orientation of a single crystal within the magnetic field can be altered to generate a rotation pattern for the quadrupolar splittings. It turns out that randomly oriented samples are often investigated, giving rise to what are called powder or powder-type spectra. A random dispersion of membranes in water involves a spherical distribution, analogous to grinding a single crystal to form a powder with a random distribution of crystallite axes. In such powder-type spectra the three principal axes of the coupling tensor, i.e., static or residual, have an arbitrary orientation relative to the external magnetic field. It follows that a distribution of resonance intensity $S(\nu)$ is present on account of the orientation dependence of the quadrupolar frequencies $\nu_Q^\pm$, due to the single quantum $|1, 0\rangle \rightarrow |1, -1\rangle$ and $|1, +1\rangle \rightarrow |1, 0\rangle$ transitions. The axially symmetric powder-type spectrum for a spin $I = 1$ system is known as a Pake doublet, and the process of deconvoluting such a powder pattern to obtain the subspectrum due to a particular orientation is called de-Pakeing (Davis, 1983).

Generalized Transformation of Coordinates

A major advantage of NMR spectroscopy in membrane structural biology is that information is obtained pertinent to both average structural properties and molecular motions. Let us now consider the effects of fluctuations of the coupling tensor on the ^{2}H NMR lineshapes. In general, the presence of motion leads to averaging of the *static* coupling tensor to yield a *residual* coupling tensor, whose principal values depend on both the static coupling parameters as well as the type of motion (Brown and Söderman, 1990). When motion occurs on a time scale comparable to or less than the inverse static splitting, Equation (13) for the case of an axially symmetric *static* EFG tensor ($\eta \approx 0$) is modified, leading to:

$$\Delta\nu_Q = \frac{3}{2}\chi_Q \langle D_{00}^{(2)}(\Omega_{PL})\rangle. \tag{14}$$

Here the brackets indicate an average over the possible tensor orientations with respect to the laboratory frame sampled on the NMR time scale (ergodic hypothesis of statistical mechanics).

Now in fluid bilayer membranes the motions of the constituent molecules are usually cylindrically symmetric about the bilayer normal, an axis known as the director. The transformation of the various coordinate frames under rotations is most easily handled using irreducible tensor methods (Brink and Satchler, 1968; Rose, 1957). Treatment of the effects of rotations is facilitated by application of a simple principle from group theory known as closure (see Appendix). This in effect states that any overall rotation can be considered as the result of a sequence of rotations involving various intermediate coordinate frames. For the case of a lipid bilayer, the overall rotation from the principal axis system of the coupling tensor to the laboratory, described by the Euler angles Ω_{PL}, can be represented by the effect of two consecutive rotations. The first, with Euler angles Ω_{PD}, represents rotation from the principal axis system to the director frame (whose z-axis is the bilayer normal); whereas the second, with Euler angles Ω_{DL}, represents the rotation from the director to the laboratory frame:

$$D_{00}^{(2)}(\Omega_{PL}) = \sum_n D_{0n}^{(2)}(\Omega_{PD}) D_{n0}^{(2)}(\Omega_{DL}). \tag{15}$$

Because of the cylindrical symmetry of the motions about the director, the Wigner rotation matrix elements containing the index n are averaged to zero. (Recall that Euler's formula states that $e^{\pm i\phi} = \cos\phi \pm i\sin\phi$, where in the present case $\phi \equiv n\alpha_{DL}$ and all angles α_{DL} are equally probable.) This gives the result:

$$\langle D_{00}^{(2)}(\Omega_{PL})\rangle = \langle D_{00}^{(2)}(\Omega_{PD})\rangle D_{00}^{(2)}(\Omega_{DL}). \tag{16}$$

Finally, substituting Equation (16) into Equation (14) yields the final expression for the quadrupolar splitting:

$$\Delta\nu_Q = \frac{3}{2}\chi_Q S_{CD}\left(\frac{3\cos^2\beta_{DL} - 1}{2}\right). \tag{17}$$

Here S_{CD} denotes the C–^{2}H segmental order parameter, which is defined as (Seelig, 1977):

$$S_{CD} = \langle D_{00}^{(2)}(\Omega_{PD})\rangle = \frac{1}{2}\langle 3\cos^2\beta_{PD} - 1\rangle. \tag{18}$$

The case of a lipid hexagonal phase (H_I or H_{II}) can also be considered through a second application of closure (Thurmond et al, 1993). There is now an additional rotation of coordinates, which can be imagined to involve curling of a hypothetical lipid monolayer to form a cylindrical aggregate. Thus applying closure twice gives:

$$D_{00}^{(2)}(\Omega_{PL}) = \sum_n \sum_{n'} D_{0n}^{(2)}(\Omega_{PD}) D_{nn'}^{(2)}(\Omega_{DC}) D_{n'0}^{(2)}(\Omega_{CL}). \tag{19}$$

Assuming there is axial symmetry about *both* the normal to the surface of the cylinder, corresponding to the director axis of a planar bilayer, as well as the long axis of the cylinder leads to:

$$\langle D_{00}^{(2)}(\Omega_{PL})\rangle = \langle D_{00}^{(2)}(\Omega_{PD})\rangle D_{00}^{(2)}(\Omega_{DC}) D_{00}^{(2)}(\Omega_{CL}). \tag{20}$$

In general for a lipid hexagonal phase (H_I or H_{II}) one can assume that $\beta_{DC} = 90°$ in which case $D_{00}^{(2)}(\Omega_{DC}) = -1/2$. Substitution into Equation (14) for the quadrupolar splitting then yields the result:

$$\Delta\nu_Q = -\frac{3}{4}\chi_Q S_{CD}\left(\frac{3\cos^2\beta_{CL} - 1}{2}\right). \tag{21}$$

Here S_{CD} is the order parameter relative to the normal to the cylinder surface, and is analogous to that in the lamellar phase (Thurmond et al, 1993). Assuming that S_{CD} is the same, Equation (21) shows that the quadrupolar splitting $\Delta \nu_Q$ is reversed in sign and reduced in absolute magnitude by a factor of one-half versus the lamellar phase, Equation (17). In the case of ^{2}H NMR spectroscopy, only the absolute value of the splitting $|\Delta \nu_Q|$ is measured. However, for $I = 1/2$ nuclei such as ^{31}P this reversal in sign and reduction by one-half is seen in the chemical shift anisotropy $\Delta \sigma$ (Thurmond et al, 1993). A similar formalism can be applied to the homonuclear dipolar interaction in the case of ^{1}H NMR, and the heteronuclear ^{13}C–^{1}H dipolar interaction in the case of ^{13}C NMR spectroscopy.

Let us next consider a more general formulation that is capable of dealing with the various possible types of motions of the molecules within the bilayer in a unified fashion. In the liquid-crystalline state, a hierarchy of motions exists as indicated by the frequency dependence of the ^{1}H, ^{2}H, and ^{13}C nuclear relaxation rates (Brown et al, 1990, 1983; Kroon et al, 1976; Rommel et al, 1988; Sefcik et al, 1983). It follows that one can expect the presence of contributions from segmental motions, as well as molecular motions and possibly collective fluctuations of the bilayer itself. The general types of motion models that can be considered are indicated schematically in Figure 2. In the case of segmental motions, the C–^{2}H bond vector reorientation is described by the angles $\Omega_{PD}(t)$ between the frame of the coupling tensor and the director; the remaining transformations are collapsed. Likewise molecular motions are considered in terms of the angles $\Omega_{MD}(t)$ between the molecular frame and the director. Finally, collective motions are formulated in terms of fluctuations of an instantaneous director relative to the average director, and are represented by the angles $\Omega_{ND}(t)$.

As discussed in the Appendix, one can use the closure property of the rotation group (Brink and Satchler, 1968) to express the Wigner rotation matrix elements for the overall transformation, $D_{sm}^{(2)}(\Omega_{PL})$, in terms of the various intermediate frames (Figure 2). Given such a general formulation one can write (Trouard et al, 1994):

$$D_{sm}^{(2)}(\Omega_{PL}; t) = \sum_r \sum_q \sum_p \sum_n D_{sr}^{(2)}(\Omega_{PI}; t) D_{rq}^{(2)}(\Omega_{IM}; t) D_{qp}^{(2)}(\Omega_{MN}; t)$$
$$\times D_{pn}^{(2)}(\Omega_{ND}; t) D_{nm}^{(2)}(\Omega_{DL}), \tag{22}$$

where all the summations run from -2 to 2 in integral steps. The Euler angles Ω_{PI} specify transformation of the irreducible components from the principal axis system (P) of the static tensor to the internal (I) motional frame; the angles Ω_{IM} correspond to rotation from the internal frame to the molecular (M) axis system; Ω_{MN} to transformation from the molecular frame to the *instantaneous* director $n(t)$, an axis of local cylindrical symmetry; Ω_{ND} to rotation from the instantaneous director to the *average* director n_0, that is the macroscopic bilayer normal; and finally the angles Ω_{DL} describe the fixed transformation from the average director (D) to the laboratory frame (L) corresponding to the static external magnetic field. In the latter transformation, the spherical polar angles characterizing the orientation of the magnetic field within the frame of the bilayer are indicated by $(\theta, \phi) \equiv (\beta_{DL}, \alpha_{DL})$. The Euler angles describing the

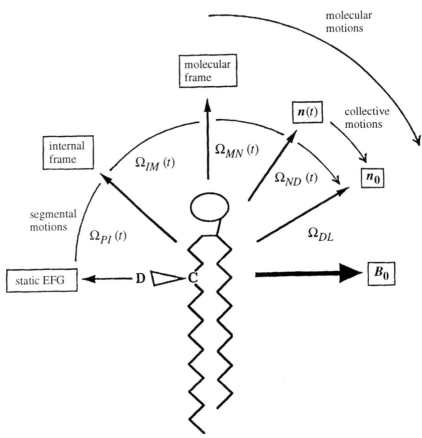

Figure 2. Schematic depiction of transformations employed in generalized dynamical models for membrane constituents. The Euler angles describing the various rotations are explained in further detail in the text.

various rotations in Figure 2 may be either time-dependent or fixed; by contrast the Ω_{DL} transformation is fixed. The expansion, Equation (22), can be further generalized to include an arbitrary number of coordinate transformations; alternatively closure enables those transformations not included in a specific motion model to be collapsed (Trouard et al, 1994). Thus, for the hexagonal phases (H_I or H_{II}), closure can be used to further expand the transformation involving the Euler angles Ω_{DL} into an additional summation $\sum_{n'}$ involving rotation from the normal to the cylinder surface to the cylinder long axis, with Euler angles Ω_{DC}, followed by rotation from the cylinder long axis to the magnetic field, having Euler angles Ω_{CL}.

An additional important aspect is the following: as noted by Brown and Söderman (1990), the *static* coupling tensor can be preaveraged by rapid fluctuations to yield a *residual* coupling tensor. Furthermore, even if the static tensor is axially symmetric ($\eta_Q = 0$), the residual or effective tensor can be nonaxially symmetric ($\eta_Q^{\text{eff}} \neq 0$). The possibility of such preaveraging accounts for the introduction of an internal coordinate frame in Figure 2. Within this frame the *residual* coupling tensor is diagonal in terms of a Cartesian basis, and the Euler angles that relate the PAS of the *static* tensor to the residual tensor are designated Ω_{PI}. The residual coupling parameters are then given by (Brown and Söderman, 1990; Trouard et al, 1994):

$$\chi_\lambda^{\text{eff}} \equiv \langle \chi_\lambda \rangle, \tag{23a}$$

$$\equiv \chi_\lambda \langle D_{00}^{(2)}(\Omega_{PI}) \rangle, \tag{23b}$$

and

$$\eta_\lambda^{\text{eff}} = -\frac{\sqrt{6} \langle D_{0\pm2}^{(2)}(\Omega_{PI}) \rangle}{\langle D_{00}^{(2)}(\Omega_{PI}) \rangle}, \tag{24a}$$

$$= -\frac{3}{2} \frac{\langle \sin^2 \beta_{PI} \cos(2\gamma_{PI}) \rangle}{\langle D_{00}^{(2)}(\Omega_{PI}) \rangle}, \tag{24b}$$

where $\lambda = Q$ in the case of the quadrupolar interaction. Thus one can consider in a unified fashion various motion models in which a static tensor is averaged by local motions, or alternatively a residual tensor (preaveraged by faster motions) is further averaged by slower molecular motions or collective fluctuations of the aggregate. The usefulness of this approach will become further apparent with regard to the treatment of motion models for NMR relaxation.

Results for Quadrupolar Frequencies of Oriented Samples

We shall now return to the case of ^{2}H NMR spectroscopy involving $I = 1$ nuclei. Using closure, Equation (22), some general expressions can be derived for various limiting motion cases which are pertinent to organized assemblies of amphiphiles in aqueous dispersions, such as membrane bilayers (Seelig, 1977). As a means of generalizing the results to either a static or residual coupling tensor, we shall modify our notation somewhat at this point and introduce the following definitions: $\chi \equiv \chi_\lambda$ or $\chi_\lambda^{\text{eff}}$; $\eta \equiv \eta_\lambda$ or $\eta_\lambda^{\text{eff}}$; and $\Omega_{XD} \equiv \Omega_{PD}$ or Ω_{ID}. A summary of expressions for the angular-dependent quadrupolar frequencies $\nu_Q^{\pm}$, derived analytically in closed form, are presented in Table 1 for a number of simplified motion cases. (Results are also included for the corresponding powder-type ^{2}H NMR spectra, as discussed below.) In Table 1 the following cases are considered: (*i*) The principal axis system of the coupling tensor is immobile, representing cases where motion is largely absent. (*ii*) A cylindrical distribution with an immobile principal axis system of the coupling tensor, corresponding to membrane proteins with no axial rotation about the director axis. (*iii*) Axial rotation of the coupling tensor about the director axis, representing membrane bilayers in the fluid, lamellar phase. (*iv*) Axial rotation of the

Table 1. Angular-Dependent Quadrupolar Frequencies and Discontinuities in Powder-Type ^{2}H NMR Spectra for Various Motion Cases[a,b]

(*i*) *Immobile Principal Axis System*

$$v_Q^{\pm} = \pm \frac{3}{4} \chi \left\{ D_{00}^{(2)}(\Omega_{XL}) - \frac{\eta}{\sqrt{6}} [D_{-20}^{(2)}(\Omega_{XL}) + D_{20}^{(2)}(\Omega_{XL})] \right\}$$

$$(v_Q^{\pm})_x = \mp \frac{3}{8} \chi (1 + \eta)$$

$$(v_Q^{\pm})_y = \mp \frac{3}{8} \chi (1 - \eta)$$

$$(v_Q^{\pm})_z = \pm \frac{3}{4} \chi$$

(*ii*) *Immobile Principal Axis System having Cylindrical Distribution*

$$v_Q^{\pm} = \pm \frac{3}{4} \chi \sum_n \left\{ D_{0n}^{(2)}(\Omega_{XD}) - \frac{\eta}{\sqrt{6}} [D_{-2n}^{(2)}(\Omega_{XD}) + D_{2n}^{(2)}(\Omega_{XD})] \right\} D_{n0}^{(2)}(\Omega_{DL})$$

limit of $\eta = 0$:

$$(v_Q^{\pm})_{\perp} = \mp \frac{3}{8} \chi$$

$$(v_Q^{\pm})_{\theta_{\min}} = \pm \frac{3}{8} \chi [3 \cos^2(\beta_{DL} - \beta_{XD}) - 1]$$

$$(v_Q^{\pm})_{\theta_{\max}} = \pm \frac{3}{8} \chi [3 \cos^2(\beta_{DL} + \beta_{XD}) - 1]$$

(*iii*) *Axial Rotation about Director*

$$\langle v_Q^{\pm} \rangle = \pm \frac{3}{4} \chi \left\{ \langle D_{00}^{(2)}(\Omega_{XD}) \rangle - \frac{\eta}{\sqrt{6}} [\langle D_{-20}^{(2)}(\Omega_{XD}) \rangle + \langle D_{20}^{(2)}(\Omega_{XD}) \rangle] \right\} D_{00}^{(2)}(\Omega_{DL})$$

$$\langle v_Q^{\pm} \rangle_{\perp} = \mp \frac{3}{8} \chi \left\{ \langle D_{00}^{(2)}(\Omega_{XD}) \rangle - \frac{\eta}{\sqrt{6}} [\langle D_{-20}^{(2)}(\Omega_{XD}) \rangle + \langle D_{20}^{(2)}(\Omega_{XD}) \rangle] \right\}$$

$$\langle v_Q^{\pm} \rangle_{\parallel} = \pm \frac{3}{4} \chi \left\{ \langle D_{00}^{(2)}(\Omega_{XD}) \rangle - \frac{\eta}{\sqrt{6}} [\langle D_{-20}^{(2)}(\Omega_{XD}) \rangle + \langle D_{20}^{(2)}(\Omega_{XD}) \rangle] \right\}$$

(*iv*) *Axial Rotation about Director and Cylinder Axis*

$$\langle v_Q^{\pm} \rangle = \pm \frac{3}{4} \chi \left\{ \langle D_{00}^{(2)}(\Omega_{XD}) \rangle - \frac{\eta}{\sqrt{6}} [\langle D_{-20}^{(2)}(\Omega_{XD}) \rangle + \langle D_{20}^{(2)}(\Omega_{XD}) \rangle] \right\} \langle D_{00}^{(2)}(\Omega_{DC}) \rangle D_{00}^{(2)}(\Omega_{CL})$$

$$\langle v_Q^{\pm} \rangle_{\perp} = \mp \frac{3}{8} \chi \left\{ \langle D_{00}^{(2)}(\Omega_{XD}) \rangle - \frac{\eta}{\sqrt{6}} [\langle D_{-20}^{(2)}(\Omega_{XD}) \rangle + \langle D_{20}^{(2)}(\Omega_{XD}) \rangle] \right\} \langle D_{00}^{(2)}(\Omega_{DC}) \rangle$$

$$\langle v_Q^{\pm} \rangle_{\parallel} = \pm \frac{3}{4} \chi \left\{ \langle D_{00}^{(2)}(\Omega_{XD}) \rangle - \frac{\eta}{\sqrt{6}} [\langle D_{-20}^{(2)}(\Omega_{XD}) \rangle + \langle D_{20}^{(2)}(\Omega_{XD}) \rangle] \right\} \langle D_{00}^{(2)}(\Omega_{DC}) \rangle$$

---

[a] See text for definition of the various Euler angles.

[b] The above expressions correspond to a generalized electric field gradient (EFG) tensor. Either a *static* EFG tensor or a *residual* EFG tensor preaveraged by faster motions is considered. The coupling parameters are $\chi \equiv \chi_Q$ or χ_Q^{eff} and $\eta \equiv \eta_Q$ or η_Q^{eff}. Generalized Euler angles are denoted by $\Omega_{XD} \equiv \Omega_{PD}$ or Ω_{ID}. The correspondence between the averaged Wigner rotation matrix elements and the order parameters S_{ii} is: $\langle D_{00}^{(2)}(\Omega_{XD}) \rangle = S_{33}$; and $\langle D_{-20}^{(2)}(\Omega_{XD}) \rangle + \langle D_{20}^{(2)}(\Omega_{XD}) \rangle = \sqrt{\frac{3}{2}} \langle \sin^2 \beta_{XD} \cos(2\alpha_{XD}) \rangle = \sqrt{\frac{2}{3}} (S_{11} - S_{22})$. Note that the definition of Häberlen (1976) is used for the asymmetry parameter η, rather than the definition of Abragam (1961).

coupling tensor about the director axis plus additional rotation about a cylinder axis, as found in hexagonal phases (H_I or H_{II}) of lipids or surfactants in water.

Note that in Table 1 the third (iii) and fourth (iv) of the above motion cases involve axial rotation about the director on the ^{2}H NMR time scale. The averaged values of the Wigner rotation matrix elements are thus referred to a single coordinate system, i.e., that of the director frame. In such cases, the averaged Wigner rotation matrix elements are related to the order matrix S introduced by Saupe (1964) and applied extensively by Seelig (1977) to lipid systems, having the diagonal elements $S_{ii} = \frac{1}{2}\langle 3\cos^2\beta_i - 1\rangle$, where $i = x, y,$ or z. Here the angles β_i represent the direction cosines between the Cartesian axes of the coupling tensor and the director. On account of the orthonormality of the direction cosines, we have that trace $S \equiv \sum_i S_{ii} = 0$; hence only two order parameters are independent. The correspondence between the averaged Wigner rotation matrix elements in Table 1 and the order parameters S_{ii} is summarized below:

$$\langle D_{00}^{(2)}(\Omega_{XD})\rangle = S_{33}, \tag{25a}$$

$$\langle D_{-20}^{(2)}(\Omega_{XD})\rangle + \langle D_{20}^{(2)}(\Omega_{XD})\rangle = \sqrt{\frac{3}{2}}\langle \sin^2\beta_{XD}\cos(2\alpha_{XD})\rangle, \tag{25b}$$

$$= \sqrt{\frac{2}{3}}(S_{11} - S_{22}). \tag{25c}$$

We use the definition of Häberlen (1976) for the asymmetry parameter η rather than the convention of Abragam (1961).

Perhaps at this juncture, some brief mention should also be made of the rates of the fluctuations which produce the motion averaging, i.e., the ^{2}H NMR time scale. As noted above, one of the significant advantages of NMR spectroscopy is that information regarding both average (or equilibrium) and dynamic properties of the membrane constituents are obtained. For motion averaging to occur in NMR spectroscopy, the rate of the motion must be greater than the difference in the resonance frequencies, here corresponding to the angular anisotropy of the quadrupolar interaction. A somewhat general constraint in the case of ^{2}H NMR spectroscopy is that the motion rate must be greater than $9\chi_Q/8 = 191$ kHz, corresponding to a time-scale less than about 5×10^{-6} s, as in cases (iii) and (iv) of Table 1. However, if smaller anisotropies are involved then significantly slower motions can also yield averaging of the nonhomogeneous NMR lineshapes.

Spectral Lineshapes and Quadrupolar Frequencies
for Powder-Type Samples

In the preceding sections, we have outlined how the quadrupolar frequencies $\nu_Q^{\pm}$ are obtained as a function of orientation for some limiting motion cases. One can also consider various orientational distributions, in which the motions are not sufficiently rapid to yield averaging of the frequencies due to the principal values of the coupling tensor, i.e., essentially static distributions are considered. In such powder-type spectra, the principal values of the coupling tensor (static or residual) correspond to characteristic features or discontinuities (singularities or edges). For a spherical distribution, these represent the frequencies of the NMR transitions for the main magnetic field aligned along one of the three orthogonal principal axes of the coupling tensor. Thus the principal values can be read off the NMR spectra directly (Häberlen, 1976). Information regarding the orientations of the principal axes relative to the frame of the aggregate (analogous to the crystal frame) is clearly lost in such powder-type samples. Here we shall discuss both spherical and cylindrical distributions of the coupling tensor as depicted in Figure 3. A spherical distribution is applicable to a random dispersion of bilayer membranes, or randomly dispersed lipid hexagonal phase aggregates in water. Likewise a cylindrical distribution represents immobilized membrane proteins having uniaxial symmetry about the bilayer normal, where the latter is in turn randomly oriented, or alternatively hexagonal phase (H_I or H_{II}) aggregates with the cylinder axes oriented randomly between glass plates (or DNA fibers or fibrous proteins for that matter).

The results for the quadrupolar frequencies corresponding to the discontinuities (singularities or edges) in the powder-type ^2H NMR spectra are also summarized in Table 1 for the various motion cases. The first two cases, (i) and (ii), correspond to the absence of motion in which the results for the ^2H NMR lineshapes are derived immediately below. When motion occurs, as in cases (iii) and (iv) the quadrupolar frequencies are averaged to $\langle \nu_Q^{\pm} \rangle$, yielding a reduction in the quadrupolar splitting $\Delta \nu_Q$. The presence of axial rotation about a preferred director is assumed, so that the quadrupolar frequencies transform as $\pm D_{00}^{(2)}(\Omega_{DL})$ or $\pm D_{00}^{(2)}(\Omega_{CL})$, respectively. It follows that axially symmetric powder-type ^2H NMR spectra are obtained, in which the spectral discontinuities $\langle \nu_Q^{\pm} \rangle_{\perp}$ and $\langle \nu_Q^{\pm} \rangle_{\parallel}$ are appropriately scaled. The discontinuities in the distribution of the quadrupolar frequencies are thus proportional to those for the spherically symmetric static distribution, case (i) with $\eta = 0$, as summarized in Table 1.

SPHERICAL PROBABILITY DISTRIBUTION—
AXIALLY SYMMETRIC COUPLING TENSOR

Let us now treat the case of a spherical distribution function, also known as a random distribution. This situation corresponds to lipids in the L_α or hexagonal (H_I or H_{II}) phases, or to membrane proteins undergoing axially symmetric rotation about the bilayer normal, such as rhodopsin. Let $(\theta, \phi) \equiv (\beta_{XL}, \alpha_{XL})$ which correspond to the spherical polar coordinates of the laboratory frame, defined by the main magnetic field, relative to the principal axis system of the coupling tensor (part a of Figure 3). We shall first consider the situation in which there is axial symmetry about a preferred direction. For example, this may correspond to an axially symmetric static or residual coupling tensor ($\eta = 0$); or alternatively to an asymmetric tensor ($\eta \neq 0$), which in turn is modulated by axially symmetric motion about the director or cylinder axis. One should note that in

a) spherical distribution

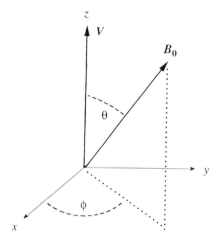

b) cylindrical distribution

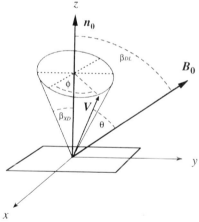

Figure 3. Illustration of spherical and cylindrical distributions employed in calculating NMR line-shapes of membrane constituents. (a) Spherical distribution in which the orientation of the main magnetic field B_0 relative to the frame of the coupling tensor V (either static or residual; see text) is described by the spherical polar angles (θ, ϕ); all angles are possible. Such a distribution is applicable to aqueous dispersions of membrane lipids and biomembranes having the bilayer normal (director) randomly oriented with respect to the main magnetic field. (b) Cylindrical distribution in which the z-axes of the individual coupling tensors fall on the rim of a cone making a semi-angle β_{XD} with respect to the axis of cylindrical symmetry n_0. The cylinder axis n_0 in turn makes an angle β_{DL} with respect to the main magnetic field B_0 and is randomly oriented. Due to the uniaxial immobilized distribution not all angles are possible. Such a cylindrical distribution is applicable to membrane proteins and other membrane constituents having a lack of axial motion about the bilayer normal.

the latter case the axial symmetry about the director is imposed on the final averaged tensor. The quadrupolar frequencies are then given by:

$$v_Q^{\pm} \propto \pm D_{00}^{(2)}(\theta), \tag{26}$$

where the characteristic scaling of Equation (26) accounts for the axially symmetric powder-pattern lineshape.

Consider a sphere of radius r having surface area $A = 4\pi r^2$ and area element $dA = dV/dr = r^2 \sin\theta \, d\theta \, d\phi$; all symbols have their customary meanings. Then the probability $p(\theta) \, d\theta$ for the number fraction of the N nuclei between θ and $\theta + d\theta$ is given by the area of a zone of the sphere divided by the total area:

$$\frac{dN(\theta)}{N} = \frac{2\pi r^2 \sin\theta \, d\theta}{4\pi r^2} = \frac{1}{2} \sin\theta \, d\theta. \tag{27}$$

It follows that the probability density in terms of the angle θ is $p(\theta) = \frac{1}{2} \sin\theta$. The next step is to introduce a dimensionless reduced resonance frequency $\xi_{\pm}$ for each of the two allowed transitions, given by:

$$\xi_{\pm} = \pm D_{00}^{(2)}(\theta). \tag{28}$$

For example, with an axially symmetric static or residual coupling tensor ($\eta = 0$), the quadrupolar frequencies $v_Q^{\pm}$ are given by Equation (12), and thus:

$$\xi_{\pm} = \frac{v_Q^{\pm}}{(3/4)\chi}. \tag{29}$$

For the other motion cases involving axial rotation about a director or cylinder axis, see cases (iii) and (iv) of Table 1, the denominator of Equation (29) is multiplied accordingly, thus yielding the $\pm D_{00}^{(2)}(\theta)$ scaling implied by Equation (28).

Our task is now to go from the probability density for the fraction of spins as a function of angle, $p(\theta)$, to the probability density for the nuclei as a function of the reduced frequency $p(\xi_{\pm})$. Applying the chain rule gives:

$$\frac{dN(\xi_{\pm})}{N \, d\xi_{\pm}} = \frac{dN(\theta)}{N \, d\theta} \frac{d\theta}{d\xi_{\pm}}, \tag{30}$$

which leads to:

$$p(\xi_{\pm}) = p(\theta) \frac{d\theta}{d\xi_{\pm}} = \frac{1}{2} \sin\theta \frac{d\theta}{d\xi_{\pm}}. \tag{31}$$

Finally, taking the derivative in Equation (31) yields the probability density for the case of a spherical distribution of the coupling tensor:

$$|p(\xi_{\pm})| \propto (1 \pm 2\xi_{\pm})^{-1/2}. \tag{32}$$

Equation (32) reveals that weak singularities (having infinite but integrable intensities) exist in the distribution function $p(\xi_{\pm})$ where the reduced frequency

$\xi_\pm = \mp\frac{1}{2}$, corresponding to $\theta = \beta_{XL} = 90°$. This orientation corresponds to the equator of the spherical probability distribution, where the solid angle is largest and the intensity is clearly maximal. Moreover, there is a cutoff for the spectral intensity at the limiting values of $\xi_\pm = \pm 1$, yielding the edges in the powder-type spectrum, where $\theta = \beta_{XL} = 0°$. This orientation represents the poles of the spherical probability distribution where the solid angle is smallest and the intensity minimal.

Now, as discussed above, in ^{2}H NMR spectroscopy there are two resonances corresponding to the reduced frequencies $\xi_\pm$, so that the total probability density is $p(\xi) = p(\xi_-) + p(\xi_+)$, where ξ is a generic reduced frequency variable. The absorption intensity $S(\xi)$ plotted on a reduced frequency scale ξ is proportional to $|p(\xi)|$, where the individual probability densities are convoluted with a line broadening function, e.g., corresponding to a Lorentzian or Gaussian distribution. Given an axially symmetric coupling tensor, the discontinuities (singularities or edges) in the powder-type spectra correspond to the following quadrupolar frequencies:

$$(\nu_Q^\pm)_\perp = \mp\frac{3}{8}\chi, \tag{33a}$$

$$(\nu_Q^\pm)_\| = \pm\frac{3}{4}\chi, \tag{33b}$$

where $\chi \equiv \chi_Q$ or χ_Q^{eff} is the static or residual quadrupolar coupling constant. The discontinuities in the powder-type spectra are summarized in Table 1. In such powder-type ^{2}H NMR spectra, the largest peaks correspond to weak singularities (infinite but integrable) in the lineshape distribution function, where $\beta_{XL} = 90°$, yielding $\nu_Q^\pm = \mp\frac{3}{8}\chi$. The $\pm$ limits of the lineshape distribution function correspond to orientations where $\beta_{XL} = 0°$, such that $\nu_Q^\pm = \pm\frac{3}{4}\chi$ for the spectral edges. Finally, at the so-called magic angle $D_{00}^{(2)}(\Omega_{XL}) = \frac{1}{2}(3\cos^2\beta_{XL} - 1) = 0$, yielding that $\beta_{XL} = 54.7°$, which corresponds to the spectral center. An example of an experimental spectrum of a solid polymer, PMMA-d_8 (plexiglass) is shown in Figure 4, together with a theoretical powder pattern lineshape calculated using Equation (32).

SPHERICAL PROBABILITY DISTRIBUTION—
NONAXIALLY SYMMETRIC COUPLING TENSOR

Let us next consider the case of a nonaxially symmetric coupling tensor. It turns out that for an asymmetric coupling tensor (static or residual with $\eta \neq 0$) the situation is somewhat more complicated (Häberlen, 1976; Slichter, 1990). Here the angular dependence of the resonance frequencies $\nu_Q^\pm$ is given by Equation (10); see also Table 1. As in the symmetric case, we again introduce a reduced resonance frequency scale, defined as:

$$\xi_\pm = \frac{\nu_Q^\pm}{(3/4)\chi} = \pm D_{00}^{(2)}(\theta) \mp \frac{\eta}{2}\sin^2\theta\cos(2\phi). \tag{34}$$

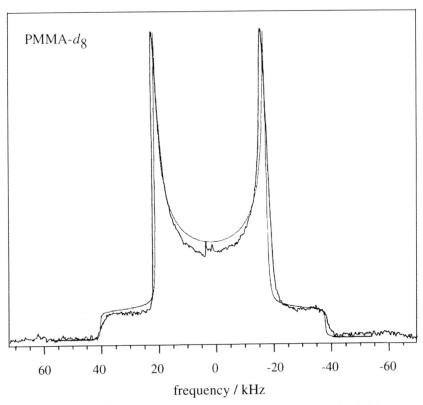

PMMA-d_8

frequency / kHz

Figure 4. Experimental ^{2}H NMR spectrum of solid poly(methyl methacrylate)-d_8 (plexiglass) ($\nu_0 = 46.1$ MHz; T = 20°C). A theoretical lineshape simulation based on Equation (32) of the text is indicated by the smooth line (Moltke S, Shumway JA, Brown MF, unpublished). The observed powder pattern arises from the C^{2}H$_3$ groups which undergo rapid 3-fold rotation in the solid state, such that $\chi_Q^{\mathrm{eff}} = -\frac{1}{3}\chi_Q$ and $\chi = 170$ kHz. The singularities at $(\nu_Q^{\pm})_\perp = \mp\frac{3}{8}\chi_Q^{\mathrm{eff}} = \pm21$ kHz originate from those regions of the sample where the main magnetic field is aligned perpendicular to the symmetry axis of the coupling tensor (the methyl rotor axis). The weaker edges or shoulders at $(\nu_Q^{\pm})_\parallel = \pm\frac{3}{4}\chi_Q^{\mathrm{eff}} = \mp42$ kHz arise from regions where the magnetic field is parallel to the tensor symmetry axis (see text).

The general result for the probability density in terms of the reduced frequency $p(\xi_\pm)$ involves three characteristic frequencies, given by $\xi_i^{\pm} = \pm(3\chi/4)\nu_i^{\pm}$ where $i = x, y, z$, corresponding to the principal axes of the coupling tensor in a Cartesian basis:

(i) for $\xi_x^{\pm} \leq \xi_\pm \leq \xi_y^{\pm}$,

$$|p(\xi_\pm)| \propto [(\xi_z^{\pm} - \xi_\pm)(\xi_y^{\pm} - \xi_x^{\pm})]^{-1/2} F(\phi, k), \tag{35a}$$

$$k = \left[\frac{(\xi_z^{\pm} - \xi_y^{\pm})(\xi_\pm - \xi_x^{\pm})}{(\xi_z^{\pm} - \xi_\pm)(\xi_y^{\pm} - \xi_x^{\pm})} \right]^{1/2} ; \tag{35b}$$

(*ii*) for $\xi_y^\pm \leq \xi_\pm \leq \xi_z^\pm$,

$$|p(\xi_\pm)| \propto [(\xi_z^\pm - \xi_y^\pm)(\xi_\pm - \xi_x^\pm)]^{-1/2} F(\phi, k), \tag{35c}$$

$$k = \left[\frac{(\xi_z^\pm - \xi_\pm)(\xi_y^\pm - \xi_x^\pm)}{(\xi_z^\pm - \xi_y^\pm)(\xi_\pm - \xi_x^\pm)} \right]^{1/2}. \tag{35d}$$

Here $F(\phi, k)$ is a complete elliptic integral of the first kind:

$$F(\phi, k) = \int_0^{\pi/2} \frac{da}{\sqrt{1 - k^2 \sin^2 a}}, \tag{36}$$

in which $\phi = \pi/2$. Now in the case of a nonaxially symmetric tensor ($\eta \neq 0$), each of the two allowed transitions yields three discontinuities in the powder-type ^{2}H NMR spectra; as a result there are six characteristic peaks. For such a spherical powder-type distribution with $\eta \neq 0$, the three discontinuities (singularities or edges) correspond to reduced frequencies of $\xi_x^\pm = -\frac{1}{2}(1 + \eta)$, $\xi_y^\pm = -\frac{1}{2}(1 - \eta)$, and $\xi_z^\pm = 1$. As noted above, these correspond to the magnetic field aligned along the three principal axes of the coupling tensor (x, y, z). However, knowledge of the orientation of the principal axes system within the frame of the aggregate is lost in such powder-type samples, as mentioned above. The discontinuities in the powder-type ^{2}H NMR spectra for the general case that $\eta \neq 0$ are summarized in Table 1, case (*i*).

CYLINDRICAL PROBABILITY DISTRIBUTION— AXIALLY SYMMETRIC COUPLING TENSOR

In addition, other distributions also can be readily considered in terms of the above formalism, e.g., a cylindrical distribution corresponding to a uniaxial immobilized sample, as in case (*ii*) of Table 1. As noted above, this may be applicable to immobilized membrane proteins having a cylindrically symmetric probability distribution about the membrane normal, such as bacteriorhodopsin (part *b* of Figure 3). To keep things simple, we shall only consider the case of an axially symmetric coupling tensor (static or residual with $\eta = 0$), with quadrupolar frequencies $v_Q^\pm$ given in Table 1. The reduced frequency scale is again defined by Equations (28) and (29). As in the spherical case, $\theta \equiv \beta_{XL}$ is the polar angle (colatitude) of the magnetic field direction relative to the frame of the coupling tensor. However, the static average is now about the orientation axis involving the azimuthal angle $\phi \equiv \gamma_{XD}$. The problem can be approached using closure as described above, where the details are left as an exercise for the reader. The probability density in terms of reduced frequency $p(\xi_\pm)$ is related to the probability density in terms of the angle ϕ by:

$$p(\xi_\pm) = p(\phi) \frac{d\phi}{d\xi_\pm}, \tag{37}$$

where $p(\phi) = 1/2\pi$. Finally, one needs to take the derivative $d\phi/d\xi_\pm$, which after some algebra and trigonometry yields the following result:

$$|p(\xi_\pm)| \propto \left\{ \left(\frac{1 \pm 2\xi_\pm}{3} \right) \left[-\cos(\beta_{DL} - \beta_{XD}) + \left(\frac{1 \pm 2\xi_\pm}{3} \right)^{1/2} \right] \right.$$
$$\left. \times \left[\cos(\beta_{DL} + \beta_{XD}) - \left(\frac{1 \pm 2\xi_\pm}{3} \right)^{1/2} \right] \right\}^{-1/2}. \qquad (38)$$

Equation (38) predicts three weak singularities (infinite intensity, but integrable), representing reduced frequencies of $\xi_\pm = \mp\frac{1}{2}$, corresponding to $\theta = \beta_{XL} = \pm 90°$; $\xi_\pm = \pm\frac{1}{2}[3\cos^2(\beta_{DL} - \beta_{XD}) - 1]$, corresponding to $\theta = \theta_{\min} = \beta_{DL} - \beta_{XD}$; and lastly $\xi_\pm = \pm\frac{1}{2}[3\cos^2(\beta_{DL} + \beta_{XD}) - 1]$, corresponding to $\theta = \theta_{\max} = \beta_{DL} + \beta_{XD}$. Because the various tensors are distributed with their z-axes on the rim of a cone, the singularities do not correspond to a single orientation of the magnetic field, as in the case of a spherical or random distribution. Inspection of Table 1 indicates that the first discontinuity, $(v_Q^\pm)_\perp$, is independent of orientation and corresponds to the main magnetic field perpendicular to the z-axis of the coupling tensor. Similarly, $(v_Q^\pm)_{\theta_{\min}}$ corresponds to the magnetic field aligned along the minimum value of θ for a given orientation β_{DL}; whereas $(v_Q^\pm)_{\theta_{\max}}$ corresponds to the maximum value (see Figure 3, part b).

In passing, the reader may have noticed that a similar formalism can also be applied to the analysis of discrete jumps or hopping of the coupling tensor due to chemical exchange among the various possible orientations (Wittebort et al, 1987). Typically a Cartesian basis is employed; however, as described here a generalized approach in terms of an irreducible representation is an alternative. As noted above, for a spherical distribution the discontinuities (singularities or edges) in the powder-type NMR spectra occur where the external magnetic field B_0 is aligned along the various principal axes of the coupling tensor, i.e., static or residual. Such jump models have been applied to analysis of the motion averaging of the quadrupolar, dipolar, and chemical shift tensors of bilayer lipids, as well as membranous peptides and proteins (Auger et al, 1990; Huang et al, 1980; Siminovitch et al, 1988a).

Motion Averaging in Liquid-Crystalline Membranes

One can also treat more specifically the various motion models that involve axial rotation about the director in the liquid-crystalline state, as in case (iii) of Table 1. With regard to L_α phase bilayers, three broad classes of motions can give rise to a reduction of the quadrupolar frequencies from their maximal values (Brown, 1982). The averaged quadrupolar frequencies corresponding to the segmental, molecular, and collective models are summarized in Table 2. Several points are noteworthy, among which are the following. First, in the case of *segmental* motions χ and η (≈ 0) refer to the *static* coupling tensor and $\Omega_{XD} = \Omega_{PD}$. The order parameter for the frame undergoing the motion is

Table 2. Summary of Quadrupolar Frequencies for Models Involving Axial Rotation about Director[a]

(i) Segmental Motions

$$\langle \nu_Q^{\pm} \rangle = \pm \frac{3}{4} \chi_Q D_{00}^{(2)}(\Omega_{PI}) \langle D_{00}^{(2)}(\Omega_{ID}) \rangle D_{00}^{(2)}(\Omega_{DL})$$

$$= \pm \frac{3}{4} \chi_Q \langle D_{00}^{(2)}(\Omega_{PD}) \rangle D_{00}^{(2)}(\Omega_{DL})$$

(ii) Molecular Motions[b]

$$\langle \nu_Q^{\pm} \rangle = \pm \frac{3}{4} \chi \left\{ D_{00}^{(2)}(\Omega_{XM}) - \frac{\eta}{\sqrt{6}} [D_{-20}^{(2)}(\Omega_{XM}) + D_{20}^{(2)}(\Omega_{XM})] \right\} \langle D_{00}^{(2)}(\Omega_{MD}) \rangle D_{00}^{(2)}(\Omega_{DL})$$

(iii) Collective Motions

$$\langle \nu_Q^{\pm} \rangle = \pm \frac{3}{4} \chi_Q^{\text{eff}} D_{00}^{(2)}(\Omega_{IN}) \langle D_{00}^{(2)}(\Omega_{ND}) \rangle D_{00}^{(2)}(\Omega_{DL})$$

$$= \pm \frac{3}{4} \chi_Q^{\text{eff}} \langle D_{00}^{(2)}(\Omega_{ND}) \rangle D_{00}^{(2)}(\Omega_{DL})$$

---

[a] See text for definition of the various Euler angles.
[b] The expression corresponds to a generalized EFG tensor. Either a *static* EFG tensor or a *residual* EFG tensor preaveraged by faster motions is considered. The coupling parameters are $\chi \equiv \chi_Q$ or χ_Q^{eff} and $\eta \equiv \eta_Q$ or η_Q^{eff}. Generalized Euler angles are denoted by $\Omega_{XM} \equiv \Omega_{PM}$ or Ω_{IM}.

given by $\langle D_{00}^{(2)}(\Omega_{PD}) \rangle$ in the case of segmental motions. Secondly, in the case of *molecular* motions there are two possibilities: for rigid molecules such as cholesterol the *static* coupling tensor is modulated as described above; whereas for flexible phospholipids χ_{eff} and η_{eff} correspond to the *residual* coupling tensor and $\Omega_{XD} = \Omega_{ID} \rightarrow \Omega_{MD}$. The order parameters are given by $\langle D_{00}^{(2)}(\Omega_{PD}) \rangle$ for rigid molecules and by $\langle D_{00}^{(2)}(\Omega_{MD}) \rangle$ for flexible molecules. Finally, for *collective* motions χ_{eff} represents the *residual* coupling tensor, where $\eta_{\text{eff}} = 0$ and $\Omega_{XD} = \Omega_{ID} \rightarrow \Omega_{ND}$; i.e., the internal frame (I) corresponds to an instantaneous director axis (N). The order parameter for the motion frame is given by $\langle D_{00}^{(2)}(\Omega_{ND}) \rangle$ for collective motions.

In addition, closure can be used to include the orientation of the internal frame relative to the frame undergoing the motion, e.g., the segmental (P) or molecular (M) frame or the instantaneous director (N) in the case of collective motions. This is useful, e.g., for the analysis of relaxation due to rotational diffusion, and enables the isomorphism (equivalence) of the results for the segmental and molecular models to be easily seen. Alternatively, the orientation of the internal frame relative to the diffusion axis system can be bypassed (Trouard et al, 1994), as in case (iii) of Table 1. According to the results in Table 2 the averaged quadrupolar frequencies $\langle \nu_Q^{\pm} \rangle$ for the segmental, molecular, and collective motion models all transform as $D_{00}^{(2)}(\Omega_{DL})$, thus yielding axially symmetric ²H NMR spectra. Clearly lineshape studies cannot be used to distinguish among such models involving rotational symmetry about a preferred director. Rather it is desirable to carry out relaxation studies (see below). In principle, these offer a powerful means of dissecting the types of motions that lead to averaging of the coupling tensors in NMR spectroscopy, including the amplitudes and time-scales of the characteristic fluctuations.

Moments of Deuterium NMR Lineshapes

Another approach in the case of powder-type spectra involves calculation of the moments of the ^{2}H NMR lineshapes, e.g., for lipids having perdeuterated acyl chains (Davis, 1983). The moments represent a general means of characterizing a distribution function, in the present case corresponding to the various quadrupolar splittings or the segmental order parameters. It follows that the theory of distributions can be used to further interpret the NMR lineshapes. The kth spectral moment is defined as:

$$M_k = \frac{\int_0^\infty v^k S(v)dv}{\int_0^\infty S(v)dv}, \tag{39}$$

where only one-half of the symmetric spectrum $S(v)$ is considered, thereby enabling consideration of both the even and odd moments. This leads to the following expressions for the first and second moments (Davis, 1983; Jansson et al, 1992):

$$M_1 = \frac{\pi}{\sqrt{3}}\chi_Q\langle|S_{CD}|\rangle, \tag{40a}$$

$$M_2 = \frac{9\pi^2}{20}\chi_Q^2\langle|S_{CD}|^2\rangle. \tag{40b}$$

Also, the width of the distribution of order parameters about the mean can be included in terms of the parameter Δ_2, defined as:

$$\Delta_2 = \frac{\langle|S_{CD}|^2\rangle - \langle|S_{CD}|\rangle^2}{\langle|S_{CD}|\rangle^2} = \frac{20}{27}\frac{M_2}{M_1^2} - 1. \tag{41}$$

Here Δ_2 is the fractional mean-squared width, which is related to the variance of the distribution of order parameters. The above expressions indicate how the moments of the ^{2}H NMR lineshapes are related to the distribution of quadrupolar splittings, i.e., order parameters. It is also possible to calculate the moments corresponding to the various motion models considered in Tables 1 and 2 (not shown). The distribution of quadrupolar splittings as described by the ^{2}H NMR spectral moments is related to equilibrium or average properties of the membrane lipids (see below).

Average Structures of Membrane Constituents

In NMR spectroscopy the second-rank tensors associated with the quadrupolar, dipolar, or chemical shift interactions are related to equilibrium or averaged properties of the system. (Dynamical properties are accessible from relaxation studies and are discussed below.) The correspondence of the motion-averaged tensors to equilibrium properties thus yields a conceptual link to the results of other biophysical methods, including SAXS and SANS (Wiener and White, 1992; Zaccai et al, 1979), as well as FT IR studies (Davies et al, 1990; Siminovitch et al, 1988b). Let us confine our attention at present mainly to the lipid constituents of the membrane bilayer. With regard to the acyl chain region of the bilayer, such equilibrium properties include the thickness, mean cross-sectional or interfacial area per molecule, isobaric coefficient of thermal expansion, and the isothermal compressibility. Additional properties are related to the curvature free energy of a monolayer film, including the spontaneous curvature and binding rigidity (Olsson and Wennerström, 1994). The latter account for the existence of nonlamellar phases, e.g., hexagonal (H_I and H_{II}) and cubic phases (Gruner, 1989; Lindblom and Rilfors, 1989; Seddon, 1990). We have described elsewhere how the properties of nonlamellar forming lipids may be important with regard to the functions carried out by membrane proteins such as rhodopsin (Brown, 1994).

In discussing the relationship of NMR observables to equilibrium and dynamical properties of lipid bilayers, ^{2}H NMR spectroscopy provides a general paradigm (Brown and Chan, 1995; Davis, 1983; Seelig, 1977). Here the electric quadrupolar interaction is by far the largest coupling, and dominates over the chemical shift or homo- or heteronuclear dipolar interactions. The value of ^{2}H NMR is that one can study the properties of an isolated spin $I = 1$ system without the complications of additional interactions of comparable magnitude. Moreover, as noted above the quadrupolar interaction is isomorphous to the dipolar coupling, so that the formalism can be readily extended to other nuclei such as ^{1}H and ^{13}C. A representative ^{2}H NMR spectrum of a lipid bilayer having perdeuterated acyl chains in the fluid, liquid-crystalline state is shown in Figure 5 (Salmon et al, 1987). For significant motion averaging to occur, the rate of the orientational fluctuations of the acyl segments must be greater than about 10^5 s^{-1}. If all the segments were equivalent, then a single quadrupolar splitting would be observed. However, this is clearly not the case; rather a family of splittings is found which indicate variations in the degree of motion averaging along the chain (Figure 5). Although the static quadrupolar coupling is everywhere the same, the residual quadrupolar couplings of the various segments are clearly inequivalent. It is remarkable that well-defined residual quadrupolar interactions are observed, corresponding to the order parameters, $S_{CD}^{(i)}$, of the various acyl chain segments (index i) (Seelig, 1977). Although the system is quite fluid, the lipid molecules still give rise to distinct quadrupolar splittings in the ^{2}H NMR spectra. Thus one obtains information at the atomic level in the disordered, liquid-crystalline state.

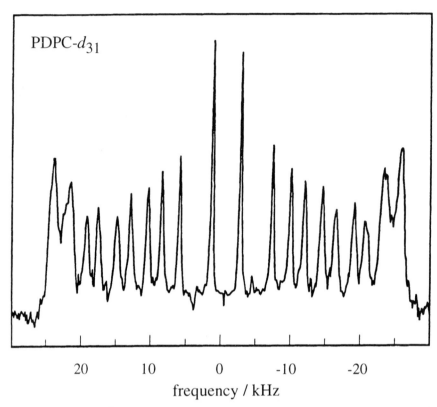

Figure 5. Illustrative ^{2}H NMR spectrum (de-Paked) of a random multilamellar dispersion of a phospholipid bilayer having perdeuterated acyl chains, comprising 1-perdeuteriopalmitoyl-2-docosahexaenoyl-*sn*-glycero-3-phosphocholine (PDPC-d_{31}), in the liquid-crystalline (L_α) phase ($\nu_0 = 55.4$ MHz; T = 30°C). Although the *static* electric field gradient tensor is the same for all the deuterated positions, the *residual* tensor is different, leading to a profile of the orientational order parameters $S_{CD}^{(i)}$ along the chain. (Salmon A, Brown MF, unpublished.)

PROFILE OF SEGMENTAL ORDER PARAMETERS VERSUS CHAIN POSITION

With regard to ^{2}H NMR spectroscopy of fluid bilayers, profiles of the second-rank order parameters $S_{CD}^{(i)} \equiv S^{(2,i)}$ as a function of acyl chain position (index i) represent one of the primary experimental observables for modeling of the equilibrium bilayer properties (Pastor et al, 1991). The latter include the average projected acyl length and mean cross-sectional area per molecule, as well as properties pertinent to the force balances in lamellar and nonlamellar phases of membrane lipids. Such order profiles were first obtained from studies of bilayer lipids with specifically deuterated chains by Seelig and coworkers (Seelig, 1977), and have since been extended to include lipids having perdeuterated acyl chains (Davis, 1983; Jansson et al, 1992; Salmon et al, 1987; Thurmond et al, 1991). A representative plot of the order parameters versus the acyl chain position is

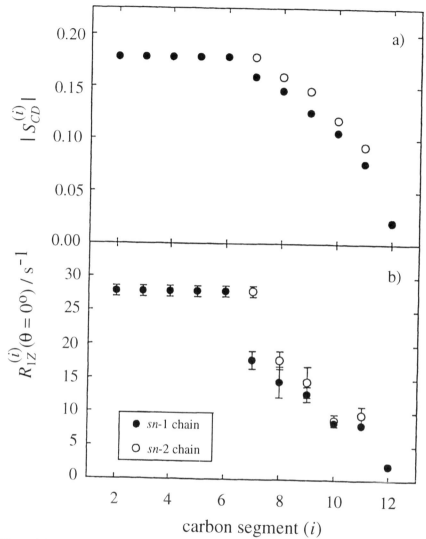

Figure 6. Profiles of (a) segmental order parameters, $S_{CD}^{(i)}$, and (b) spin-lattice relaxation rates, $R_{1Z}^{(i)}$, versus acyl segment position (i) for DLPC-d_{46} oriented on planar glass substrates in the liquid-crystalline (L_α) phase ($\nu_0 = 46.1$ MHz; T = 40°C). In both cases a plateau is observed for the initial part of the chains, followed by a progressive decrease towards the hydrocarbon interior of the bilayer. (Trouard et al, 1994. Used by permission.)

shown for 1,2-diperdeuteriolauroyl-sn-glycero-3-phosphocholine (DLPC-d_{46}) in part a of Figure 6. (The profile of the NMR relaxation rate R_{1Z} is also shown in part b as discussed subsequently.) The order profile comprises a plateau in the $S_{CD}^{(i)}$ values over the first part of the chains, followed by a decrease as the chain terminus is approached. Analogous ^{2}H NMR studies have been carried out for lipids in other physical states, including the normal hexagonal (H_I) and reverse hexagonal (H_{II}) phases (Lafleur et al, 1990; Thurmond et al, 1994, 1990). In the case of ^{13}C NMR spectroscopy, order parameters, S_{CH}, for the phospholipid acyl chains and polar head groups are obtained by comparing the residual dipolar or chemical shift tensors to their static values, and yield good agreement with previous ^2H NMR results (Brown and Chan, 1995).

A physical feeling for the order profiles of bilayer lipids (part a of Figure 6) can be gained from considering several limiting cases that are derivable in closed form. First, one can consider an all-trans rotating chain, for which $\beta_{PD}^{(i)} = 90°$ and $S_{CD}^{(i)} = -\frac{1}{2}$. One can thus deduce immediately that the chains are substantially disordered, e.g. due to rotational isomerism. On the other hand, the classical oil drop model of Overton implies that $S_{CD}^{(i)} \approx 0$. The presence of a plateau in the order profile might suggest a crankshaft model involving coupled isomerizations, such as kink and jog configurations. For a polymethylene chain saturated with kinks take the repeating unit as trans-gauche$^\pm$-trans-gauche$^\mp$, where the pentane rule of Flory is assumed. Applying a three-site exchange model together with axial rotation yields $S_{CD}^{(i)} = -\frac{1}{6}$ for the gauche sites and $\langle S_{CD} \rangle = -\frac{1}{3}$ for the entire chain. The acyl chain ordering with values of $|S_{CD}^{(i)}| \lesssim 0.2$ thus falls in between the limiting crankshaft and oil drop models. Consideration of additional chain configurations is necessary, which requires detailed enumeration using a computer. The plateau in the order profile is associated with tethering of the acyl chains to the aqueous interface, whereas beyond the plateau region, the segments are substantially more disordered to fill in the free volume that would otherwise be present due to the chain terminations. In other words the chains are entangled to maintain the density constant at essentially that of liquid hydrocarbon.

More generally, one can expect that the order parameters from ^2H NMR spectroscopy may include contributions from local segmental motions, e.g., dihedral angle isomerizations, as well as slower molecular motions and possibly collective excitations of the bilayer. For the present purposes, we can roughly separate these putative contributions into those having a segmental order parameter $S_f^{(2,i)}$ which depends on the segment position (index i) in the chain, and those due to additional slower motions, which are roughly independent of chain position, having an order parameter $S_s^{(2)}$. The former originates largely from trans-gauche isomerizations of the acyl chain segments together with librational motions within the various conformational states; whereas the latter may represent molecular motions and/or collective disturbances of the bilayer. As a first approximation, let us assume that the above motions comprise statistically independent Markov processes, e.g., associated with their time-scale separation. This enables a simple product approximation to be made, namely that:

$$S^{(2,i)} = S_f^{(2,i)} S_s^{(2)}, \tag{42}$$

where the fast and slow motions are decoupled. One possibility is that the equilibrium bilayer thickness and mean interfacial area per molecule arise mainly from local isomerizations; the influences of the slower motions as described as by the order parameter $S_s^{(2)}$ are small. In this case the observed order parameter $S^{(2,i)}$ mainly reflects the local order parameter $S_f^{(2,i)}$ due to trans-gauche rotational isomerizations of the chains.

MEAN PROJECTED LENGTH AND CROSS-SECTIONAL AREA PER LIPID

From the above it follows that simple statistical mechanical models for the local segmental fluctuations of the lipids can be employed (Jansson et al, 1992; Salmon et al, 1987; Schindler and Seelig, 1975). One approach involves distribution of the C–C and C–^{2}H bonds of the methylene segments on a diamond lattice, together with rotational disorder of the chains about their reference long axis (the bilayer normal or director). Now in the fluid, liquid-crystalline phase, the presence of rotational isomerization of the chains yields a reduction of the mean bilayer thickness, accompanied by an increase in the mean area per molecule relative to the low temperature (gel) state. Given such a diamond-lattice model, the mean cross-sectional or interfacial area per molecule $\langle A \rangle$ can then be estimated in two different ways. These treatments differ depending on whether the entire chain is considered and the length-area fluctuations are assumed to be uncorrelated, i.e., the covariance of the length and area distributions is zero, or whether one considers only the initial part of the chain (giving rise to the order plateau) and assumes the length-area fluctuations to be inversely correlated (Jansson et al, 1992). Knowledge of the mean area per lipid is important with regard to both Monte Carlo and molecular dynamics simulations of lipid bilayers.

In the former instance the mean acyl length $\langle L \rangle$ projected along the bilayer normal is given by the sum of the average projections of the individual segments. It can be shown that the average distance traveled along the bilayer normal for an individual methylene segment is:

$$\langle l_i \rangle = l_0 \left[\frac{1}{2} - S^{(2,i)} \right]. \tag{43}$$

Here $l_0 = 1.25$ Å is the length of a C–C bond projected onto the all-trans reference axis, and the variance of the length distribution is given by $\sigma^2 = \langle l_i^2 \rangle - \langle l_i \rangle^2 = -l_0^2[\frac{1}{2}S^{(2,i)} + S^{(2,i)2}]$. As noted above it is assumed that $S^{(2,i)} \equiv S_{CD}^{(i)} \approx S_f^{(2,i)}$, where the calculation of the order parameters assumes that $\chi_Q = 170$ kHz. For an all-trans rotating chain $S^{(2,i)} = -\frac{1}{2}$ and thus $\langle l_i \rangle = l_0$ as expected. One can then calculate the mean acyl length $\langle L \rangle$ projected along the bilayer normal from the profile of the order parameters $S^{(2,i)}$ as a function of chain position (Salmon et al, 1987; Schindler and Seelig, 1975). Alternatively the first moment M_1 of the ^{2}H NMR spectra for lipids having perdeuterated acyl chains can be used (Jansson et al, 1992). Besides the rotational isomeric approximation, the above model assumes that folding-back of the chains towards the interface is negligible, so that only three of the six possible diamond lattice orientations are considered, i.e., $\beta_{ID} = 0°, 60°, 90°$; the same result is obtained if only $\beta_{ID} = 0°$ and $60°$ are considered.

In addition, statistical correlations in the distributions of the projected length and cross-sectional areas of the chains in the bilayer are neglected, and the possible influences of larger amplitude slow motions (see above and below) of the lipids are assumed to be relatively minor. One simply parcels up the available volume among the lipids into cells that represent the average molecular shape for a given aggregate geometry. Because of the softness of the effective potential, significant entanglement of the chains can occur, so that the above approach is clearly an approximation. With these caveats in mind, the mean projected segmental lengths $\langle l_i \rangle$ can be summed to yield the mean acyl length projected along the bilayer normal, given by $\langle L \rangle = \sum_i \langle l_i \rangle$. In the case of a bilayer the formula $\langle A \rangle = V_{\text{chain}}/\langle L \rangle$ is used, where the segmental volumes $V_{\text{CH}_2} = 27.6$ Å^3 and $V_{\text{CH}_3} = 2V_{\text{CH}_2}$ are obtained from dilatometry studies. This approach gives a value of $\langle A \rangle = 71.8$ Å^2 for DPPC-d_{62} at 50°C and $\langle A \rangle = 67.0$ Å^2 for DPPE-d_{62} at 60°C in the liquid-crystalline state (Thurmond et al, 1991); whereas the corresponding values from SAXS studies are 57.6 Å^2 to 70.9 Å^2 for the case of DPPC (Nagle and Wiener, 1988). Finally, the temperature dependence of $\langle L \rangle$ and $\langle A \rangle$ enables one to estimate the thermal expansion coefficients perpendicular and parallel to the bilayer normal, yielding values of about $\alpha_\parallel = -2 \times 10^{-3}$ K^{-1} and $\alpha_\perp = 1 \times 10^{-3}$ K^{-1} for DPPC (Jansson et al, 1992).

The values of $\langle A \rangle$ estimated using the above cell model for the DPPC bilayer are in agreement with the high end of the range of SAXS values; whereas the lower values are significantly overestimated, by as much as 25% (Nagle and Wiener, 1988). However, the trend towards a smaller interfacial or cross-sectional chain area for PE versus PC at similar absolute temperatures is reproduced. The above approximations have been criticized by Nagle (1993) who has argued that such a formalism is incorrect, and rather should only be applied to the segments giving rise to the order plateau (Thurmond et al, 1991). Such an approach gives a value of $\langle A \rangle = 63.6$ Å^2 for DPPC-d_{62} in the fluid phase at 50°C, which is closer to the median of the SAXS values. However, Euclidean geometry leads one to expect that for a bilayer the projected lengths and cross-sectional areas of the cells are inversely correlated for the initial chain segments, a point apparently overlooked by Nagle (1993). It follows that the mean cross-sectional area for the plateau segments is given by $\langle A \rangle = V_{\text{plat}}\langle 1/l \rangle_{\text{plat}}$; whereas the erroneous use of $\langle l \rangle_{\text{plat}}^{-1}$ in place of $\langle 1/l \rangle_{\text{plat}}$ yields a lower bound to the area obtained with the diamond-lattice model (Jansson et al, 1992). One can then calculate the value of $\langle 1/l_i \rangle$ for a methylene group, using the above diamond-lattice model with rotational disorder, by assuming that $\beta_{ID} = 0°$ and $60°$ only, i.e., only two of the six possible diamond-lattice orientations are significantly populated. The following result is obtained:

$$\left\langle \frac{1}{l_i} \right\rangle = \frac{2}{l_0}[1 + S^{(2,i)}]. \tag{44}$$

Note that in the limiting case of an all-trans chain $S^{(2,i)} = -\frac{1}{2}$ leading to $\langle 1/l_i \rangle = 1/l_0$; thus Equation (44) has the correct behavior.

Moreover, one can consider the product of the quantities in Equations (43) and (44):

$$p = \langle 1/l_i \rangle \langle l_i \rangle, \tag{45a}$$

$$= 1 + \frac{2\sigma^2}{l_0^2}. \tag{45b}$$

Assuming that only the $\beta_{ID} = 0°$ and $60°$ orientations are populated significantly, the interfacial or cross-sectional area occupied per molecule is:

$$\langle A \rangle = V_{CH_2} \langle 1/l_i \rangle_{plat}, \tag{46a}$$

$$= \frac{p V_{CH_2}}{\langle l_i \rangle_{plat}}, \tag{46b}$$

where:

$$p = 1 \quad \text{(uncorrelated)}, \tag{47a}$$

$$= 1 - S^{(2,i)} - 2S^{(2,i)2} \quad \text{(inversely correlated)}. \tag{47b}$$

For DPPC in the fluid L_α phase at 50°C, using Equation (44) a value of $\langle A \rangle = 71.7$ Å^2 is obtained, which is again at the high end of the range of SAXS results (Nagle, 1993). It is also possible to consider the order profiles with regard to more detailed mean-field (Pastor et al, 1988) and lattice models for bilayers (Dill et al, 1988), as well as Monte Carlo (Scott and Kalaskar, 1989) and Brownian dynamics simulations (Pastor et al, 1991).

At this juncture, the reader should note that NMR relaxation studies (see below) suggest the presence of significant contributions from relatively slow motions, e.g., due to molecular fluctuations or collective bilayer disturbances. These modulate the residual coupling tensor remaining from faster motions such as trans-gauche rotational isomerization of the chains. The neglect of such slow motions must be regarded as an inherent deficiency of the above treatment of average bilayer properties using a rotationally disordered diamond-lattice model. Such calculations of the equilibrium projected acyl length and cross-sectional area should be regarded as approximations, which may overestimate the mean interfacial area on account of the neglect of slow motions. However, as discussed by Jansson et al (1992), to some extent there may be a compensation of opposing effects. On the one hand, the inclusion of slow motions increases the calculated value of $\langle L \rangle$ and decreases $\langle A \rangle$ due to the larger values of the local order parameters. On the other hand, the projection of $\langle L \rangle$ onto the average bilayer normal is reduced due to the chain tilt, and accordingly the calculated value of $\langle A \rangle$ is increased.

Comparative Studies of Phospholipids and Influences of Ions and Cholesterol

^{2}H NMR spectroscopy also has been used to investigate various phospholipids having perdeuterated acyl chains, including a homologous series of disaturated phosphatidylcholines, as well as mixed-chain saturated-polyunsaturated phosphatidylcholines (Barry et al, 1991; Rajamoorthi and Brown, 1991). Investigations of the effects of the acyl chain length for a homologous series of 1,2-diacyl-sn-glycero-3-phosphocholines having perdeuterated acyl chains indicate that at the same absolute temperature the length of the plateau in the order profile increases with the number of acyl carbons, whereas the quadrupolar splittings due to the end segments of the acyl chains remain almost constant (Dodd and Brown, 1989). In regard to the above formalism, the projected length of the acyl chains increases with the number of acyl carbons, accompanied by an increase in the average bilayer thickness. It follows that the mean cross-sectional areas are rather similar for the homologous disaturated phosphatidylcholines is the fluid, L_α phase at the same absolute temperature. With regard to the influences of polyunsaturation, for the mixed chain phosphatidylcholines in the fluid L_α phase, the saturated acyl chain at the sn-1 position of the glycerol backbone next to the sn-2 polyunsaturated acyl chain has an increase in configurational freedom versus the corresponding disaturated phosphatidylcholines, as evinced by the quadrupolar splittings at the same absolute temperature. However, the differences in the calculated values of the mean projected length and cross-sectional area are relatively minor compared to the corresponding disaturated phosphatidylcholines (Salmon et al, 1987). As noted above, the effects of the phosphoethanolamine and phosphocholine head groups have been investigated in the case of diacyl phospholipids (Thurmond et al, 1991).

Cholesterol is a ubiquitous component of eukaryotic cell membranes, comprising up to 50% of the weight of the erythrocyte membrane; it is also of considerable medical interest due to its implication in atherosclerosis (Altbach et al, 1991; Pearlman et al, 1988) Rather extensive studies of the effects of cholesterol on the properties of phospholipids have been previously carried out using various biophysical techniques, including differential scanning calorimetry, Fourier transform infrared spectroscopy, spin-label EPR spectroscopy, and ^{13}C NMR and ^{2}H NMR spectroscopy. The influences of cholesterol can be briefly summarized as follows (Ipsen et al, 1987; Recktenwald and McConnell, 1981; Vist and Davis, 1990). ^{2}H NMR studies have revealed that the well-known condensing effect of cholesterol, which yields a reduction of the mean cross-sectional area per lipid molecule, is due to an increase in the acyl chain ordering on account of the greater population of trans acyl chain configurations (Gally et al, 1976). The above findings are corroborated by quantitative FT IR studies indicating that the population of trans acyl configurations is substantially increased in DPPC/cholesterol mixtures (Davies et al, 1990). Analogous ^{2}H NMR studies of lysophosphatidylcholine/cholesterol mixtures have been conducted, yielding similar results (Jansson et al, 1992). Moreover, an extensive study of the phase diagram of DPPC-d_{62}/cholesterol mixtures has been carried out, involving ^{2}H NMR spectroscopy in conjunction with differential scanning calorimetry (Vist and Davis, 1990). In the absence of cholesterol, a lipid bilayer undergoes a thermotropic transition between a low temperature gel phase with conformationally ordered acyl chains to a conformationally disordered liquid-crystalline state. Cholesterol is analogous to a crystal breaker because it interferes with the interactions between the phospholipid molecules in both the solid and liquid states. Consequently an intermediate liquid-ordered phase is formed at high cholesterol mol fractions, in which the phospholipid chains are nearly all-trans, but having rapid axial rotation and lateral diffusion as in the L_α phase of pure phospholipid bilayers.

Besides the above investigations, comprehensive ^{2}H and ^{31}P NMR studies of the polar head groups of the major phospholipid classes have been carried out (Seelig and Seelig, 1980) as well as glycolipids (Auger et al, 1990). An interesting discovery is that the ^{2}H NMR quadrupolar splittings of the various phospholipid polar head groups are quite sensitive to small changes in the head group torsion angles due to the binding of ions (Brown and Seelig, 1977). As a result, ^{2}H NMR is an ideal method to monitor the binding of ions to phospholipid membrane surfaces. By contrast, ^{2}H NMR studies reveal that cholesterol has only minor influences on the average conformations of the phosphocholine and phosphoethanolamine head groups; it acts mainly as a spacer moiety decreasing interactions between the lipids (Brown and Seelig, 1978). Subsequent ^{2}H and ^{31}P NMR studies of the binding of ions and local anesthetics to membrane bilayers (Akutsu and Seelig, 1981; Kelusky and Smith, 1983) have contributed to the development of ^{2}H NMR spectroscopy as a molecular potentiometer for assessing the membrane surface charge density (Seelig et al, 1987).

Dynamical Properties of Membrane Constituents Studied with NMR Relaxation

At this point the reader should recall that NMR spectroscopy allows information to be obtained regarding both the average structural properties and dynamics of membrane constituents, including both lipids and proteins. What are the types, rates, and amplitudes of the motions that lead to averaging of the NMR lineshapes? Can one distinguish among different reorientational mechanisms? For bilayers in the liquid-crystalline state, the motion-averaged ^{2}H NMR lineshapes, having a well-defined profile of quadrupolar splittings (part *a* of Figure 5), enable one to conclude that the bulk of the quadrupolar interaction is fluctuating at frequencies in excess of about 10^5 s^{-1}. However, from such fast limit NMR spectra it is not possible to obtain detailed information regarding the time-scales and amplitudes of the characteristic motions. For example, the reorientational motions of the C–^{2}H vectors can represent largely the internal chain dynamics, or alternatively significant influences due to molecular motions or collective dynamics of the bilayer can be important. Interpretation of the relaxation properties enables one to begin to disentangle the various degrees of freedom that lead to the averaging of the quadrupolar, dipolar, or chemical shift tensors. It is the sensitivity of NMR to motions that confers unique advantages in membrane biophysics, e.g., compared to conventional SAXS and SANS studies (Nagle and Wiener, 1988; Wiener and White, 1992).

As originally shown by Brown, Seelig, and Häberlen (1979) for bilayers in the fluid L_α phase, a profile is seen in the ^{2}H NMR relaxation rates as a function of acyl chain position, which resembles the profile seen for the ^{2}H NMR order parameters. A typical ^{2}H NMR relaxation profile for DLPC-d_{46} is shown in part b of Figure 6. It can be seen that the R_{1Z} profile corresponds to the order profile in that a plateau is found over the initial part of the chain, followed by a decrease in the R_{1Z} relaxation rates towards the acyl terminus. Analogous relaxation profiles are obtained for the ^{13}C R_{1Z} values of small vesicles (Brown et al, 1983), as well as the R_{1Z} and $R_{1\rho}$ values of unsonicated membrane bilayers (Sefcik et al, 1983). In general, only small differences exist between the R_{1Z} rates of unsonicated bilayers and small vesicles (Brown et al, 1979). Because the relaxation rates in NMR spectroscopy depend on both the rates and amplitudes of the motions, the relaxation profile can be due to variation of the motion amplitudes along the chain; alternatively it can be due to an increase in the rate of motion as a function of depth within the bilayer (Brown, 1979). Which of these interpretations is correct? Moreover, another inherent feature of the nuclear spin relaxation is that there is a characteristic dependence on the resonance frequency (magnetic field strength) (Brown and Davis, 1981; Brown et al, 1990, 1983; Kroon et al, 1976). The frequency dependence of the relaxation points to the additional influences of relatively slow motions of the lipids within the bilayer, which are not found in simple fluids.

Density Matrix Formalism for NMR Relaxation

Let us again consider the different coupling mechanisms in NMR spectroscopy, such as the quadrupolar and dipolar interactions ($\lambda = Q, D$). One can apply the density matrix formalism of Bloch-Wangsness-Redfield (Slichter, 1990) to the relaxation of lipid bilayers (Brown, 1979), in which the fluctuating nuclear coupling with its environment yields a time-dependent perturbation. The time-averaged or residual Hamiltonian $\langle \hat{H}_\lambda \rangle$ is subtracted from the time-dependent Hamiltonian to yield the fluctuating part $\hat{H}'_\lambda(t)$, given by (Brown et al, 1979):

$$\hat{H}'_\lambda(t) \equiv \hat{H}_\lambda(t) - \langle \hat{H}_\lambda \rangle. \tag{48}$$

One can then write for the correlation functions of the coupling Hamiltonian:

$$G_{\alpha\beta\alpha'\beta'}(t) = \langle\langle\alpha|\hat{H}'_\lambda(0)|\beta\rangle\langle\beta'|\hat{H}_\lambda(t)|\alpha'\rangle\rangle_{\text{avg}}. \tag{49}$$

In the above expression the Dirac bra-ket notation is used, where the symbols α, α', β, and β' correspond to the eigenstates of the spin system. (The bra-kets should not be confused with the brackets $\langle \ \rangle_{\text{avg}} \equiv \langle \ \rangle$ used to indicate a time or ensemble average.) Substituting for the Hamiltonian $\hat{H}'_\lambda(t)$ gives the correlation functions which describe the fluctuating part of the coupling interaction:

$$G_{\alpha\beta\alpha'\beta'}(t) = \frac{3}{2}\delta_\lambda^2 C_\lambda^2 \sum_m \langle\alpha|\hat{T}_{-m}^{(2)\text{lab}}|\beta\rangle^* \langle\alpha'|\hat{T}_{-m}^{(2)\text{lab}}|\beta'\rangle G_m(t). \tag{50}$$

Equation (50) contains products of the matrix elements of the irreducible tensor operators corresponding to the angular momentum, together with the irreducible correlation functions $G_m(t)$ which characterize the fluctuations of the coupling tensor. Here the matrix elements encompass the selection rules and characterize the allowed transitions in accord with the Wigner-Eckart theorem, whereas the correlation functions $G_m(t)$ describe the time dependence of the fluctuating irreducible components about their mean values within the laboratory frame.

The irreducible correlation functions and corresponding spectral densities of motion characterize the dependence on the motion amplitudes and rates. In the laboratory frame, the correlation functions $G_m(t)$ can be written as:

$$G_m(t) = \langle [V_m^{(2)\text{lab}}(0) - \langle V_m^{(2)\text{lab}} \rangle]^*[V_m^{(2)\text{lab}}(t) - \langle V_m^{(2)\text{lab}} \rangle] \rangle / [V_0^{(2)\text{PAS}}]^2. \quad (51)$$

For the limit that $\eta_\lambda = 0$ only the $V_0^{(2)\text{PAS}}$ irreducible component is nonzero, so that one obtains (Brown, 1984, 1982):

$$G_m(t) = \langle D_{0m}^{(2)*}(\Omega_{PL}; 0) D_{0m}^{(2)}(\Omega_{PL}; t) \rangle - |\langle D_{0m}^{(2)}(\Omega_{PL}) \rangle|^2. \quad (52)$$

However, in general cross-terms involving the symmetric (δ_λ) and asymmetric (η_λ) irreducible components of the coupling tensor may need to be considered (Brown and Söderman, 1990; Trouard et al, 1992). The Fourier transform partners of the irreducible correlation functions, $G_m(t)$, are called the spectral densities, and are given by:

$$J_m(\omega) = \text{Re} \int_{-\infty}^{\infty} G_m(t) e^{-i\omega t} \, dt. \quad (53)$$

The irreducible spectral densities $J_m(\omega)$ describe the amplitudes of the fluctuations of the irreducible components of the coupling tensor at the frequency ω, which produce the observed nuclear spin relaxation. Here the following coupling constants are introduced: quadrupolar coupling ($I = 1$), $\chi_Q \equiv e^2 q Q / h$; and dipolar coupling, $\chi_D \equiv (\gamma_I \gamma_S h / 2\pi^2) \langle r_{IS}^{-3} \rangle$.

Expressions for Nuclear Spin Relaxation Rates

Using the density matrix formalism, equations of motion for the nuclear magnetization can be written in terms of a basis operator representation, Equation (2). First we shall consider the case that $\hat{I} = \hat{S}$, e.g., due to the pairwise magnetic dipolar interaction between ^{1}H nuclei or the electric quadrupolar interaction of the ^{2}H nucleus. The following results are obtained for the various relaxation rates (Brown, 1979):

$$R_{1Z} \equiv \frac{1}{T_{1Z}} = \frac{3}{4}\pi^2 \chi_\lambda^2 [J_1(\omega_I) + 4J_2(2\omega_I)], \quad (54a)$$

$$R_{1\lambda} \equiv \frac{1}{T_{1\lambda}} = \frac{9}{4}\pi^2 \chi_\lambda^2 J_1(\omega_I), \quad (54b)$$

$$R_2 \equiv \frac{1}{T_2} = \frac{3}{8}\pi^2 \chi_\lambda^2 [3J_0(0) + 3J_1(\omega_I) + 2J_2(2\omega_I)], \quad (54c)$$

where $\lambda \equiv D, Q$ so that χ_λ is the coupling constant for the quadrupolar or dipolar interaction, i.e., $\chi_\lambda \equiv \chi_Q$ or χ_D; ω_I is the nuclear Larmor frequency; and $J_m(\omega_I)$ are the spectral densities of motion at frequency ω_I. The relaxation rate for the Zeeman or longitudinal magnetization (i.e., the spin-lattice relaxation rate) is indicated by R_{1Z}, where T_{1Z} is the relaxation time. The corresponding rates for the decay of quadrupolar order and the decay of dipolar order within the two spin $I = 1/2$ approximation are designated by R_{1Q} and R_{1D}, respectively, with relaxation times T_{1Q} and T_{1D}. Finally, the relaxation rate for the decay of transverse single-quantum coherence (i.e., the spin-spin relaxation rate) is given by R_2, where the relaxation time is T_2. In systems undergoing isotropic averaging, e.g., phospholipid vesicles and random multilamellar dispersions (Brown and Davis, 1981), the dependence on the m index vanishes on account of the spherical symmetry, leading to $J_m(\omega_I) \rightarrow \langle J_m(\omega_I) \rangle \equiv J(\omega_I)$. The transverse cross-relaxation rate must then be included, yielding:

$$R_2 \equiv \frac{1}{T_2} = \frac{3}{8}\pi^2 \chi_\lambda^2 [3J(0) + 5J(\omega_I) + 2J(2\omega_I)]. \tag{55}$$

Moreover, the case that $\hat{I} \neq \hat{S}$ can be considered, e.g., corresponding to the ^{13}C–^{1}H direct magnetic dipolar interaction. The various observable relaxation quantities are then given by (Brown, 1984):

$$R_{1Z} \equiv \frac{1}{NT_{1Z}} = \frac{3}{2}\pi^2 \chi_D^2 \left[\frac{1}{6} J_0(\omega_S - \omega_I) + \frac{1}{2} J_1(\omega_I) + J_2(\omega_S + \omega_I) \right], \tag{56a}$$

$$\eta_{IS} = \left(\frac{\gamma_S}{\gamma_I} \right) \frac{-\frac{1}{6} J_0(\omega_S - \omega_I) + J_2(\omega_S + \omega_I)}{\frac{1}{6} J_0(\omega_S - \omega_I) + \frac{1}{2} J_1(\omega_I) + J_2(\omega_S + \omega_I)}, \tag{56b}$$

$$R_2 \equiv \frac{1}{NT_2} = \frac{3}{4}\pi^2 \chi_D^2 \left[\frac{1}{6} J_0(\omega_S - \omega_I) + \frac{1}{2} J_1(\omega_I) + J_2(\omega_S + \omega_I) \right.$$

$$\left. + \frac{2}{3} J_0(0) + J_1(\omega_S) \right]. \tag{56c}$$

Here R_{1Z} is the Zeeman (spin-lattice) relaxation rate with relaxation time T_{1Z}; η_{IS} is the heteronuclear Overhauser effect arising from the fluctuating dipolar interactions; R_2 is the transverse (spin-spin) relaxation rate with relaxation time T_2; and N is the number of directly bonded ^1H nuclei. If orientational averaging occurs (see above), then the dependence on the m index vanishes on account of the spherical symmetry, yielding $J_m(\omega_I) \rightarrow \langle J_m(\omega_I) \rangle \equiv J(\omega_I)$.

Correlation Functions and Spectral Densities

As a rule, one can apply a restricted rotational diffusion model to describe either internal segmental motions or molecular motions; alternatively jump models for the tensor reorientation can be considered. For the case of a rectangular potential with infinite reflecting barriers, the restricted diffusion model corresponds to wobbling in a cone. However, it is preferable to use a more realistic approach, in which the potential of mean torque is expanded in Legendre polynomials (Brown, 1982). Generalized Euler angles are defined as $\Omega \equiv (\alpha, \beta, \gamma)$; that is to say $\Omega = \Omega_{ID}$ for segmental diffusion or $\Omega = \Omega_{MD}$ for molecular diffusion, having projection indices $(m', m) \equiv (r, n)$ or (q, n). The corresponding spectral densities for rotational diffusion $F_{m'm}^{(2)}(\Omega; \omega)$ are Fourier transform partners of the correlation functions $G_{m'm}^{(2)}(\Omega; t)$, as given below:

$$F_{m'm}^{(2)}(\Omega; \omega) = \mathrm{Re} \int_{-\infty}^{\infty} G_{m'm}^{(2)}(\Omega; t) e^{-i\omega t}\, dt, \tag{57a}$$

$$= [\langle |D_{m'm}^{(2)}(\Omega)|^2 \rangle - |\langle D_{m'm}^{(2)}(\Omega) \rangle|^2 \delta_{m'0}\delta_{m0}] j_{m'm}^{(2)}(\Omega; \omega), \tag{57b}$$

where $G_{m'm}^{(2)}(\Omega; t) = G_{m'm}^{(2)}(\Omega; 0) g_{m'm}^{(2)}(\Omega; t)$. Now $F_{rn}^{(2)}(\Omega_{ID}; \omega)$ for segmental diffusion and $F_{qn}^{(2)}(\Omega_{MD}; \omega)$ in the case of molecular diffusion are functions of both even and odd rank order parameters (Halle, 1991; Trouard et al, 1994, 1992). Given a value of the second-rank order parameter $\langle D_{00}^{(2)}(\Omega_{ID}) \rangle$ describing the segmental motions or $\langle D_{00}^{(2)}(\Omega_{MD}) \rangle$ for molecular motions, the parameters characterizing the potential of mean torque can be determined.

The next step is to solve the rotational diffusion equation (Nordio and Segre, 1979) in the presence of a uniaxial potential of mean torque, yielding for the reduced correlation functions that:

$$g_{m'm}^{(2)}(\Omega; t) = e^{-(D_\| - D_\perp)m'^2 t} \sum_{k=1}^{\infty} c_{m'm}^{(2,k)} e^{-\alpha_{m'm}^{(2,k)} D_\perp t}. \tag{58}$$

Here $\alpha_{m'm}^{(2,k)}$ denotes the kth eigenvalue of the diffusion operator, and $c_{m'm}^{(2,k)}$ and $\tau_{m'm}^{(2,k)}$ are the relative normalized weight and correlation time for the kth relaxation component. Relatively simple results in closed form can be obtained by making a single exponential approximation in terms of a short time expansion (Szabo, 1984), giving for the mean of the eigenvalues:

$$\alpha_{m'm}^{(2)} = \sum_{k=1}^{\infty} c_{m'm}^{(2,k)} \alpha_{m'm}^{(2,k)} = \frac{\mu_{m'm}}{G_{m'm}^{(2)}(0)}. \tag{59}$$

In the above formula, the moments $\mu_{m'm}$ are given by their expansions (Trouard et al, 1994) in terms of the order parameters $\langle D_{00}^{(j)}(\Omega) \rangle$. Within the single exponential approximation, the correlation times are denoted by:

$$\frac{1}{\tau_{m'm}^{(2)}} = [\alpha_{m'm}^{(2)} + (\eta_{\mathrm{diff}} - 1)m'^2] D_\perp, \tag{60}$$

where $\eta_{\mathrm{diff}} \equiv D_\|/D_\perp$ and $\mathbf{D}$ is the rotational diffusion tensor. The reduced correlation functions are then:

$$g_{m'm}^{(2)}(\Omega; t) = e^{-t/\tau_{m'm}^{(2)}}, \tag{61}$$

and the corresponding reduced spectral densities are given by:

$$j_{m'm}^{(2)}(\Omega, \omega) = \frac{2\tau_{m'm}^{(2)}}{1 + [\omega \tau_{m'm}^{(2)}]^2}. \tag{62}$$

In the liquid-crystalline state the segmental fluctuations can be rather complicated, such that the dynamics involve dihedral angle isomerizations together with any relatively low frequency vibrational or torsional motions, e.g., coupled to rattling of the molecules in the cage of nearest neighbors. As noted above, one can adopt a restricted rotational diffusion model (Brown, 1982) involving the angles $\Omega_{PD}(t)$ between the frame of the coupling tensor and the director; alternatively jump models can be considered (Auger et al, 1990; Siminovitch et al, 1988a; Torchia and Szabo, 1982). The irreducible spectral densities $J_m(\omega)$ in the laboratory frame calculated for the restricted segmental diffusion model (Brown, 1982; Trouard et al, 1992) are summarized in Table 3. The products of the various Wigner rotation matrix elements $|D_{m'm}^{(2)}(\Omega)|^2$ are given by their Clebsch-Gordan series expansions (Brink and Satchler, 1968) (see Equation A11 of Appendix). Note that the dependence of the relaxation on the bilayer orientation is described by the factor $|D_{nm}^{(2)}(\Omega_{DL})|^2$.

Table 3. Summary of Spectral Densities for Models Involving Axial Rotation about Director[a]

(i) *Segmental Motions*

$$J_m(\omega) = \sum_r \sum_n \left| D_{0r}^{(2)}(\Omega_{PI}) - \frac{\eta_\lambda}{\sqrt{6}} [D_{-2r}^{(2)}(\Omega_{PI}) + D_{2r}^{(2)}(\Omega_{PI})] \right|^2$$

$$\times F_{rn}^{(2)}(\Omega_{ID}; \omega) |D_{nm}^{(2)}(\Omega_{DL})|^2$$

(ii) *Molecular Motions*[b]

$$J_m(\omega) = |\langle D_{00}^{(2)}(\Omega_{PX})\rangle|^2 \sum_q \sum_n \left| D_{0q}^{(2)}(\Omega_{XM}) - \frac{\eta}{\sqrt{6}} [D_{-2q}^{(2)}(\Omega_{XM}) + D_{2q}^{(2)}(\Omega_{XM})] \right|^2$$

$$\times F_{qn}^{(2)}(\Omega_{MD}; \omega) |D_{nm}^{(2)}(\Omega_{DL})|^2$$

(iii) *Collective Motions*

three-dimensional:

$$J_m(\omega) = C |\langle D_{00}^{(2)}(\Omega_{PI})\rangle|^2 |\omega|^{-1/2} |D_{\pm 1m}^{(2)}(\Omega_{DL})|^2$$

two-dimensional:

$$J_m(\omega) = C' |\langle D_{00}^{(2)}(\Omega_{PI})\rangle|^2 |\omega|^{-1} |D_{\pm 1m}^{(2)}(\Omega_{DL})|^2$$

[a] See text for definition of the various Euler angles.
[b] Either a *static* EFG tensor or a *residual* EFG tensor preaveraged by faster motions is considered. The coupling parameters are $\chi \equiv \chi_\lambda$ or $\chi_\lambda^{\text{eff}}$ and $\eta \equiv \eta_\lambda$ or $\eta_\lambda^{\text{eff}}$. Generalized Euler angles are denoted by $\Omega_{XM} \equiv \Omega_{PM}$ or Ω_{IM}. The correspondence of the residual coupling parameters to the static values is given by $\chi_\lambda^{\text{eff}} \equiv \langle \chi_\lambda \rangle = \chi_\lambda \langle D_{00}^{(2)}(\Omega_{PI})\rangle$ and $\eta_\lambda^{\text{eff}} = -\sqrt{6}\langle D_{0\pm2}^{(2)}(\Omega_{PI})\rangle/\langle D_{00}^{(2)}(\Omega_{PI})\rangle = -3\langle \sin^2 \beta_{PI} \cos(2\gamma_{PI})\rangle/2\langle D_{00}^{(2)}(\Omega_{PI})\rangle$. Note that $\lambda = Q, D$ for the quadrupolar and dipolar interactions, respectively.

Similarly, the rotational diffusion model can be applied to restricted molecular rotations of the flexible phospholipids within the bilayer. In this approach the motions involving the molecular long axis are formulated in terms of reorientation of a single molecule in the mean field due to its various neighbors. The slow motions may be noncollective, i.e., molecular in nature, due to wobbling or rattling of the lipids within the potential of mean torque, as described by the angles $\Omega_{MD}(t)$ between the molecular frame and the director. One can show (Brown, 1982) that the spectral densities of motion are formally equivalent or isomorphous to those for segmental diffusion (Brown, 1979), with the substitutions $\chi_\lambda \to \chi_\lambda^{\text{eff}}$ and $\eta_\lambda \to \eta_\lambda^{\text{eff}}$. It follows that the irreducible spectral densities for restricted molecular diffusion are scaled by the local order parameter squared, as summarized in Table 3. The two coupling parameters are $\chi_\lambda^{\text{eff}}$ and $\eta_\lambda^{\text{eff}}$ as given by Equations (23) and (24). Because the spectral densities are multiplied by $|D_{00}^{(2)}(\Omega_{PI})|^2$, the local order parameter squared, the static quadrupolar coupling constant $\chi_\lambda^2 \to (\chi_\lambda^{\text{eff}})^2$ in the relaxation rate expressions, Equations (54)–(56). Note that for rigid molecules such as cholesterol the segmental and molecular models are equivalent as indicated in Table 3.

Another possibility is to adopt a three-dimensional picture for the bilayer fluctuations in terms of a continuum of twist, splay, and bend deformations (Brown, 1982). Such a model describes locally correlated motions of segments of the flexible molecules packed within the bilayer. Here it is assumed that reorientation of the various acyl segments is intrinsically collective; a key distinction versus the above molecular picture is that a broad distribution of elastic modes is considered (Brown, 1982). The shorter wavelengths approach the molecular or chain diameter, where, of course, the continuum description for the local correlations breaks down. On the other hand, for longer wavelengths the fluctuations approach the dimensions of the bilayer (taken as ∞), and fall outside the range considered. In this picture the slow motions are described by the angles $\Omega_{ND}(t)$ between the instantaneous (or local) director and the average director (the bilayer normal); a small angle approximation is made. The irreducible spectral densities for such a three-dimensional collective model are summarized in Table 3, in which C is a collection of various constants (a single elastic constant approximation is made). The angular anisotropy is described by the factor $|D_{\pm 1m}^{(2)}(\Omega_{DL})|^2$ and thus reflects the assumption of small-angle director fluctuations. It should be noted that if the amplitude of the slow motions is nearly uniform along the chain, then a square-law dependence of the spectral densities on the observed order parameter results as for the molecular model (see above). An alternative is to adopt a formulation in which the collective fluctuations are described in terms of a flexible surface model involving the lipid/water interface; here the finite thickness of the bilayer is essentially disregarded (Marqusee et al, 1984; Stohrer et al, 1991). Such a picture of the bilayer deformations is analogous to a smectic undulation model, involving a large increase in the elastic constants for the twist and bend modes (Blinc et al, 1975). As a result only splay is considered in terms of a single elastic constant K_c. The irreducible spectral densities for such a two-dimensional collective model are also included in Table 3, where C' is a collection of elastic and other constants.

Limiting Results for Observed Spectral Densities

Further simplification of the results in Table 3 is possible if one assumes that *orientational averaging* of the relaxation takes place, such as occurs for vesicles and also multilamellar dispersions (Brown and Davis, 1981; Brown et al, 1979). In this case the factor $|D_{nm}^{(2)}(\Omega_{DL})|^2 \rightarrow \langle|D_{nm}^{(2)}(\Omega_{DL})|^2\rangle = 1/5$ and the dependence on the projection index m vanishes for each of the models in Table 3. Let us modify our notation and denote the observed order parameter by $S^{(2)} \equiv S_{CD}$, where the segment index (i) is absorbed, such that $S^{(2)} = S_f^{(2)} S_s^{(2)}$. Here $S_f^{(2)} \equiv \langle D_{00}^{(2)}(\Omega_{PI})\rangle$ is the order parameter for the relatively fast motions, and $S_s^{(2)} \equiv \langle D_{00}^{(2)}(\Omega_{ID})\rangle$ is the order parameter for the relatively slow motions. In terms of the above formulations, the internal fast motions of the chains correspond to the segmental model, whereas the slow motions correspond to either the molecular or collective model. One can then write for the observed spectral densities:

$$J_m(\omega) \rightarrow J(\omega) = J^f(\omega) + J^s(\omega). \tag{63}$$

The assumption of statistically independent fluctuations means that cross-correlations are neglected, that is to say the motions are decoupled. In principle both the fast and slow contributions depend on the degree of ordering, the measuring frequency, and the bilayer orientation.

SEGMENTAL MOTIONS

From the above development it follows that the spectral density due to internal fast motions of restricted orientational amplitude is given by (Brown, 1982):

$$J^f(\omega) = \frac{1}{5}[1 - S_f^{(2)2}]j_f^{(2)}(\omega), \tag{64}$$

where:

$$j_f^{(2)}(\omega) = \frac{2\tau_f^{(2)}}{1 + [\omega\tau_f^{(2)}]^2}. \tag{65}$$

In the above expression $\tau_f^{(2)}$ is the correlation time for isomerizations and other local motions with respect to the local director axis (internal frame). It is noteworthy that the order parameter $S_f^{(2)}$ represents a relatively small correction to the spectral density obtained for isotropic rotational diffusion in the absence of ordering.

MOLECULAR ROTATIONAL DIFFUSION

In the case of relatively slow fluctuations due to molecular motions of restricted amplitude, the result, assuming a single correlation time for the orientationally averaged spectral density, is (Brown, 1982):

$$J^s(\omega) = \frac{1}{5} S_f^{(2)2} [1 - S_s^{(2)2}] j_s^{(2)}(\omega),\qquad(66)$$

where:

$$j_s^{(2)}(\omega) = \frac{2\tau_s^{(2)}}{1 + [\omega\tau_s^{(2)}]^2}.\qquad(67)$$

Notice that there is now a strong dependence (squared) of the spectral densities on the fast order parameter $S_f^{(2)}$. The above general formulation (Brown, 1982, 1979) corresponds to the results for various limiting cases that can be found in the literature. For example, imagine that the slow motion is unrestricted as occurs in globular proteins (Lipari and Szabo, 1982) or detergent micelles (Wennerström et al, 1979). Then there is no preferred symmetry axis (corresponding to the average director), so that the slow order parameter $S_s^{(2)} \rightarrow 0$ giving:

$$J^s(\omega) \rightarrow \frac{1}{5} S_f^{(2)2} j_s^{(2)}(\omega).\qquad(68)$$

The above limiting case is formally equivalent to the results obtained by Wennerström and coworkers (1979) for surfactant micelles as the two-step model and independently by Lipari and Szabo (1982) for globular proteins as the model-free approach. However, because the treatment described here (Brown, 1984a,b, 1982, 1979) is more general it is referred to as the *generalized approach*.

COLLECTIVE BILAYER EXCITATIONS

In a similar fashion, one can consider the case of slow motions due to a continuum of bilayer disturbances arising from correlated segmental or molecular motions (Brown, 1982). For such a collective model the result for the orientationally averaged spectral density due to slow motions is:

$$J^s(\omega) = \frac{1}{5} S_f^{(2)2} j_s^{(2)}(\omega),\qquad(69)$$

where:

$$j_s^{(2)}(\omega) = C\omega^{-1/2}.\qquad(70)$$

Here C includes a combination of elastic and other constants, where a single elastic constant approximation is assumed for the twist, splay, and bend modes. As mentioned above a smectic-like formulation can also be considered in terms of splay only (Blinc et al, 1975; Marqusee et al, 1984), which gives:

$$j_s^{(2)}(\omega) = C'\omega^{-1},\qquad(71)$$

where C' is similarly a collection of elastic and other constants. As in the case of the molecular model; the spectral densities due to slow motions are proportional to the fast order parameter squared, i.e., $S_f^{(2)2}$, as discussed above.

Deuterium and Carbon-13 NMR: Comparison of Theory and Experiment

The above conceptual development indicates that the spectral densities $J_m(\omega)$ and associated relaxation rates depend on the segmental ordering, frequency (magnetic field strength), temperature, and bilayer orientation (Brown, 1982). Here we shall mainly consider the interior of the bilayer comprising the nonpolar fatty acyl chains. Let us return to the ^2H NMR order and relaxation profiles depicted in Figure 6 (Trouard et al, 1994). As noted above, one interpretation is that the static interaction constant is modulated by the motions, and gives rise to the relaxation. Because the residual or effective quadrupolar coupling (≈ 30–40 kHz) is smaller than the static quadrupolar coupling (170 kHz), it is possible that the mean-squared amplitudes provide a relatively minor influence. Hence there is only a minor influence of the order parameter, as in Equation (64), and it follows that the relaxation profile is governed by the local motion rates along the chain. Yet another possibility is that the strength of the interaction varies along the acyl chain, corresponding to an *effective or residual* coupling tensor due to preaveraging by faster local motions such as dihedral angle isomerizations (Brown and Söderman, 1990). Instead of the static coupling being constant and the dynamical parameters varying along the chain, the opposite holds. The residual coupling parameters as reflected in the mean-squared amplitudes then vary with the acyl position, and govern the shape and magnitude of the relaxation profile; whereas the influence of the dynamical parameters is correspondingly smaller.

How are the relaxation rates and derived spectral densities relevant to the equilibrium and dynamical properties of bilayers in the fluid phase? Clearly it is necessary to introduce various motion models that encompass the key physics of interest as a means of further interpreting such relaxation profiles. The angular anisotropy of the relaxation for $I = 1$ nuclei, i.e., the dependence on the angle between the bilayer normal and the main magnetic field, is related to the orientation and symmetry of the coupling tensor within the frame undergoing the fluctuations, corresponding to the segment, molecule, or bilayer (Brown, 1982). By comparing the orientation of the coupling tensor that undergoes the fluctuations to the structural features of the molecule or bilayer, one can gain an idea of the types of motions that yield the observed relaxation in fluid membranes. The experimentally observed angular dependencies of the ^2H R_{1Z} rates of bilayers of disaturated phosphatidylcholines in the fluid (L_α) phase can be explained by each of the segmental, molecular, and collective models to a greater or lesser degree of success, depending on the number of adjustable parameters (Trouard et al, 1994). How can one further distinguish among the above conceptual frameworks? One possibility is to combine the results of ^2H NMR spectroscopy with ^{13}C NMR spectroscopy of lipid bilayers, so that one can obtain more information than with either method alone. The magnitude of the ^2H and ^{13}C relaxation times together with their dependence on the resonance frequency and temperature make it improbable that a single type of local motion can explain the results. Such a formulation has been suggested as an initial approximation for describing the dynamics in fluid L_α phase bilayers (Brown et al, 1979), but appears to be an oversimplification. Subsequent studies have indicated that a fundamentally different physical picture may be involved (Brown, 1982; Brown et al, 1990, 1983), in which the influences of slow motions, which modulate the residual coupling left over from faster local motions, are considered more explicitly.

In the case of segmental motions, the mean-squared amplitudes are formulated in terms of the fast order parameter $S_f^{(2)}$ and represent a minor correction to the overall spectral density, Equation (64). On the other hand, in the case of slow motions the spectral density, Equation (66), is strongly influenced by the fast order parameter $S_f^{(2)}$. For such order fluctuations, the dependence of the relaxation rate on the acyl chain position is largely governed by the two quantities $\chi_\lambda^{\text{eff}}$ and $\eta_\lambda^{\text{eff}}$, assuming that the residual tensor principal axes, together with the mean-squared amplitudes and reduced spectral densities of the slow motions, are nearly constant along the chain. If it is also assumed that the effective asymmetry parameter $\eta_\lambda^{\text{eff}}$ is nearly zero or invariant, then a characteristic square-law functional dependence of the observed spectral densities on the segmental order parameters, $S^{(2,i)} = S_{CD}^{(i)}$, results (Brown, 1982; Brown et al, 1983; Williams et al, 1985). Figure 7 shows that for disaturated lipid bilayers, such as 1,2-diperdeuteriolauroyl-sn-glycero-3-phosphocholine (DLPC-d_{46}), if the ^{2}H $R_{1Z}^{(i)}$ rates are plotted versus the corresponding order parameters, $S_{CD}^{(i)}$, squared, then a nearly straight line results with a positive slope. In fact, the presence of such a square-law dependence (Brown, 1982) appears to be a hallmark of the fluid, liquid-crystalline state of disaturated lipid bilayers. It is consistent with a picture in which the observed order parameters $S^{(2,i)} = S_{CD}^{(i)}$ are given by Equation (42) and the R_{1Z} relaxation rates are given by Equations (54a) and (63) together with a dominant contribution from Equations (66) or (69), where the slow order parameter does not depend strongly on chain position. By contrast, if the data are plotted as $R_{1Z}^{(i)}$ versus $[1 - S_{CD}^{(i)2}]$ according to Equation (64) then the slope is negative, which is physically implausible because the correlation times are positive! This experimentally observed *square-law functional dependence* of $R_{1Z}^{(i)}$ and $S_{CD}^{(i)}$ does not depend on any detailed motion model. Yet it can be simply explained by Fermi's Golden Rule, in which the strength of the interaction depends on the matrix elements squared, and thus the relaxation rates vary with the chain position, whereas the dynamical parameters are relatively constant. Hence although the *slow* motions provide the dominant contribution to the relaxation, the internal *fast* motions govern the profile along the chains.

Further knowledge is obtained from the frequency dependence of the observed relaxation rates. Interpretation of the R_{1Z} relaxation dispersion of phospholipids in the liquid-crystalline state has been somewhat controversial, however. At present the most extensive frequency dependent R_{1Z} relaxation data comprise ^{2}H NMR studies of vesicles of specifically deuterated 1,2-dimyristoyl-sn-glycero-3-phosphocholine (DMPC). The results encompass nine different magnetic field strengths, spanning 2.5 MHz (0.38 T) to 62.4 MHz (9.55 T),

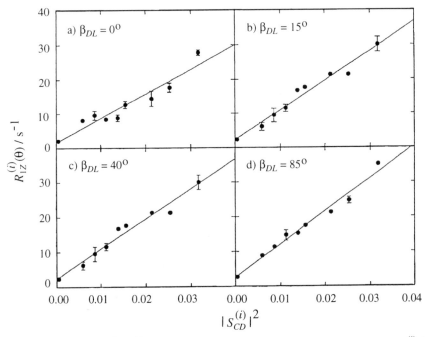

Figure 7. Experimental ^{2}H $R_{1Z}^{(i)}$ relaxation rates plotted versus the corresponding values of $|S_{CD}^{(i)}|^2$ for acyl segments of DLPC-d_{46} in the liquid-crystalline (L_α) phase ($\nu_0 = 46.1$ MHz; T = 40°C). Panels (a)–(d) show data for bilayers on planar supports with the director aligned at different angles ($\theta \equiv \beta_{DL}$) relative to the laboratory frame defined by the main magnetic field. Theory predicts a square-law functional dependence for the case of relaxation governed by relatively slow order fluctuations (see text). (Trouard et al, 1994. Used by permission.)

in which both electromagnets and superconducting NMR solenoids have been utilized (Brown et al, 1990). Figure 8 summarizes the results of fitting the dispersion data for DMPC deuterated at the C-3 position to (part a) the three-dimensional collective model having twist, splay, and bend deformations, leading to an $\omega_I^{-1/2}$ dependence; (part b) the interfacial collective model with splay only, yielding an ω_I^{-1} dependence; and finally (part c) the noncollective molecular model having a single Lorentzian plus a constant to account for internal motions. We have proposed that the isotropic vesicle tumbling is too slow to influence substantially the relaxation dispersion in the MHz regime (Brown et al, 1990). As can be seen, the $\omega_I^{-1/2}$ dependence (Brown, 1982; Brown et al, 1986, 1983) yields an acceptable fit to the data; whereas the fits to an ω_I^{-1} dependence in terms of splay only as proposed by Marqusee et al (1984) and Rommel et al (1988) or to a Lorentzian dispersion (Petersen and Chan, 1977) are less satisfactory. A somewhat different interpretation of these R_{1Z} relaxation rate experiments has also been proposed by Halle (1991).

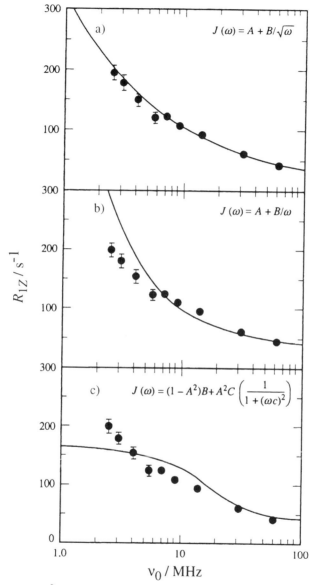

Figure 8. Experimental ^{2}H R_{1Z} relaxation rates plotted as a function of frequency for vesicles of DMPC deuterated at the C-3 acyl segment in the liquid crystalline state (T = 30°C). Data acquired at nine different frequencies (magnetic field strengths) are shown, together with theoretical fits to various motion models. (a) Three-dimensional collective model; (b) interfacial collective model; and (c) noncollective molecular model. The results enable distinction among the various limiting relaxation laws. (Brown et al, 1990. Used by permission.)

Structure and Dynamics of Membrane Bilayers

How can bilayers of membrane lipids in the liquid-crystalline state be studied at the molecular level? With regard to biomembranes and liquid-crystalline materials, the knowledge provided by NMR spectroscopy encompasses the motion-averaged NMR line shapes due to the dipolar, quadrupolar, or chemical shift tensors, as well as the relaxation rates which characterize fluctuations of the coupling tensors about their equilibrium values. The information from NMR spectroscopy is in principle contained in the correlation functions $G_m(t)$ corresponding to the various second-rank quantities, i.e., the dipolar, quadrupolar, or chemical shift tensors, or equivalently their Fourier transform partners, the spectral densities $J_m(\omega)$. The spectral densities contain the complete information that can be obtained from NMR spectroscopy about the equilibrium and dynamical properties of membrane bilayers. Recall that the relaxation rates include both the mean-squared amplitudes of the orientational fluctuations, as well as the reduced spectral densities, whose correlation times describe the motion rates. The mean-squared amplitudes of the orientational fluctuations are related to the bilayer properties averaged over the appropriate time scale. These are given by $\langle |D_{m'm}^{(2)}(\Omega)|^2 \rangle - |\langle D_{m'm}^{(2)}(\Omega)\rangle|^2$, and are related to the area under the appropriate region of the spectral density curve as a function of the frequency ω (see Equation 57b); the total area is constant due to conservation of energy. By contrast, the reduced spectral densities characterize the frequency range of the fluctuations, whose reciprocal correlation times correspond to the cutoff frequency for the power spectrum. Through a combination of line shape and relaxation studies, one can thus obtain valuable insight regarding both the equilibrium and dynamical properties of bilayers.

Summary of Deuterium and Carbon-13 Relaxation Measurements

How can we begin to extract the knowledge of membrane equilibrium and dynamical properties from experimental NMR measurements? Clearly this is a challenging task involving development of an appropriate conceptual framework in membrane biophysics. A summary of currently available NMR results for the fatty acyl chains of multilamellar dispersions and small vesicles of DPPC is given in Figure 9 (Brown, 1982; Brown et al, 1983). The square-law functional dependence of the ^2H R_{1Z} relaxation rates and ^2H NMR order parameters along the acyl chains is depicted in the case of DPPC multilamellar dispersions in part a. The $\omega_C^{-1/2}$ frequency dispersion found in ^{13}C R_{1Z} relaxation studies of small DPPC vesicles is shown in part b. Both dependencies are indicative of the fluid L_α phase of membrane lipid bilayers (Brown, 1982). As noted above, ^2H and ^{13}C R_{1Z} studies of both small vesicles of phosphatidylcholines as well as the corresponding multilamellar dispersions evince little difference in their relaxation rates (Brown et al, 1983; Sefcik et al, 1983). By contrast, part b of Figure 9 illustrates that the ^{13}C R_{1Z} rates of liquid paraf-

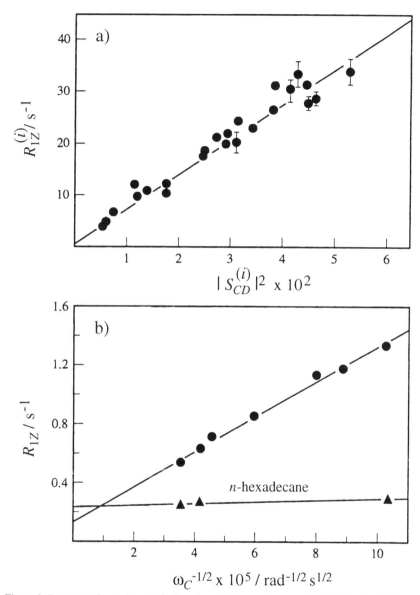

Figure 9. Summary of experimental findings from NMR spectroscopy for DPPC in the liquid crystalline state at 50°C. (a) Plot of ^{2}H spin-lattice relaxation rates $R_{1Z}^{(i)}$ as a function of $|S_{CD}^{(i)}|^2$ for acyl segments of multilamellar dispersions; (b) plot of ^{13}C spin-lattice relaxation rates, R_{1Z}, for $(CH_2)_n$ resonance against $\omega_C^{-1/2}$ for small vesicles. Theory indicates that both dependencies are hallmarks of a contribution to the relaxation arising from relatively slow order fluctuations due to collective bilayer excitations. (Brown et al, 1983. Used by permission.)

fins such as *n*-hexadecane are approximately independent of frequency, i.e., the magnetic field strength, over the entire range studied.

Lipid Bilayers versus Hydrocarbon Fluids

Now in the case of simple liquids and hydrocarbon fluids, there is no residual or time-averaged interaction since the reorientation occurs over all space; local fast motions of the molecules are dominant so that the spectral densities extend to potentially high frequencies. For lipid bilayers, however, the existence of orientational ordering as manifested in the NMR spectral lineshapes represents a fundamental distinction versus liquid hydrocarbons. This orientational ordering is accompanied by long range positional ordering of the lipids, as is readily evident from SAXS or SANS studies (McIntosh and Simon, 1986; Wiener and White, 1992) or electron microscopy (Siegel et al, 1994). Hence there is the possibility of slower motions not found in simple hydrocarbon fluids, associated with collective properties of the assembly, which are manifested in the NMR relaxation rates (Brown et al, 1983). The local orientational ordering of the lipids, e.g., due to segmental motions, is modulated by order fluctuations of larger orientational amplitude. Assuming that significant contributions from both the fast and slow motions are present in lipid bilayers, the spectral density due to internal rapid chain fluctuations can extend to relatively high frequencies, as in the case of liquid hydrocarbons ($\tau_c \approx 5$–20 ps). By contrast the spectral density due to slow motions is concentrated at lower frequencies in lipid bilayers, and can provide a dominant contribution near the resonance frequency. Relative to simple hydrocarbon fluids (see part *b* of Figure 9), there is an enhancement of the relaxation in lipid bilayers that depends on both the frequency, i.e., magnetic field strength, as well as the ordering or mean-squared amplitudes of the fluctuations. The presence of orientational and positional ordering of the lipid chains (Seelig, 1977) yields order fluctuations as found in liquid crystals, thus distinguishing membrane lipid bilayers from simple hydrocarbon fluids (Brown, 1982).

Modeling of Relatively Slow Bilayer Motions

Clearly the constituents of the bilayer have both internal degrees of freedom as well as those characteristic of the entire collection of molecules. One avenue is to approximate the complicated bilayer assembly in a manner sufficiently realistic for insight into the physicochemical behavior to be acquired, yet which can be solved in closed form yielding a relatively small number of free parameters. By analogy to simple fluids, a noncollective molecular picture involving distinct rotational modes can be considered, as discussed above (Brown, 1982; Halle, 1991; Rommel et al, 1988). An alternative viewpoint emphasizes the collective properties of the assembly of bilayer lipids. By analogy to nematic liquid crystals, the bilayer order fluctuations are modeled in terms of a continuum of twist, splay, and bend deformations in the high frequency (or free membrane) limit, or alternatively by smectic undulation waves involving splay only in the low frequency (or strongly coupled) limit (Brown, 1982; Brown et al, 1983). Extension of such a continuum approach to the microscopic level, comprising distances greater than the acyl segmental diameter but less than the bilayer thickness, constitutes an effort to render transparent the complex many body problem that exists in actuality. Within this framework, our goal is thus to capture the most important features of the dynamics leaving the details for subsequent refinement.

Figure 10 shows a schematic illustration of the types of collective excitations that may explain the frequency and ordering dependent relaxation enhancement of membrane bilayers; for clarity only very long wavelength modes are depicted (Brown et al, 1983). In terms of such a continuum approximation, the higher frequency modes have smaller amplitudes, whereas the lower frequency

a)

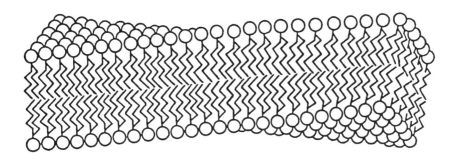

b)

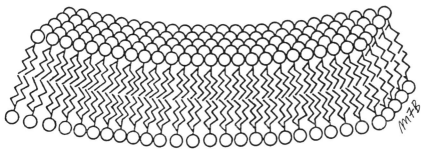

Figure 10. Examples of the types of collective bilayer excitations leading to an $\omega^{-1/2}$ frequency dependence; long wavelength modes are depicted for heuristic reasons only. (Brown et al, 1983. Used by permission.)

modes have larger amplitudes; the cutoffs at high and low frequencies are disregarded as a simplifying approximation. As noted above, there are two limits for a smectic A mesophase such as a membrane lipid bilayer, depending on the repeat distance d between the layers relative to the fluctuation wavelength λ. First, there is the high frequency regime where $\lambda \ll d$, representing the *free membrane limit,* where coupling between the bilayers is negligible; in this case the relaxation is analogous to that of a nematic mesophase. Secondly, there is the low frequency regime where $\lambda \gg d$, representing the *strongly coupled limit,* involving interactions between the various lamellae. Here there is the possibility of smectic undulation waves having a constant interlamellar distance (Blinc et al, 1975; Marqusee et al, 1984), in which the restriction to splay only yields an ω^{-1} frequency dependence over the entire range; other fluctuation modes may also be present (Pfeiffer et al, 1993).

On the one hand, at the higher frequencies in the MHz regime characteristic of the free membrane limit, a nematic-like continuum picture involving relatively short wavelength twist, splay, and bend modes describes the experimentally observed $\omega^{-1/2}$ dependence (Blinc et al, 1975; Brown, 1982; Jeffrey et al, 1979). Over such a distance scale, no simple correspondence exists between the microscopic elastic constants which describe the NMR relaxation and the macroscopic elastic constants characteristic of the bulk material. Because the R_{1Z} rates for vesicles and multilamellar dispersions are nearly the same (Brown et al, 1979), the dispersion in the MHz range most likely does not arise from two-dimensional surface fluctuations (Marqusee et al, 1984). On the other hand, at lower kHz frequencies, the characteristics of the membrane lipid/water interface become increasingly important; so that a transition occurs to the strongly coupled limit involving interactions among the various layers. Here a correspondence with the bulk elastic constants can exist; additional research is needed. The presence of splay only in the case of the lower frequency modes yields an ω^{-1} frequency dependence in the kHz range (Blinc et al, 1975).

Local Motions and the Bilayer Microviscosity

With regard to the internal or local motions, the contribution for a fluid lipid bilayer can be found by extrapolating the ordering and frequency dependent relaxation enhancement to zero ordering or infinite frequency (Figure 9). The contribution from local fast motions, obtained in such a model-dependent fashion, closely matches the relaxation rates of a hydrocarbon liquid having the same chain length, e.g., *n*-hexadecane, which do not depend on frequency. Consequently there may be some physical validity to the analysis. It follows that the *microviscosity* of the bilayer hydrocarbon interior, where a bulk viscosity cannot be measured, corresponds to that of a liquid *n*-alkane having a macroscopic viscosity on the order of a few centipoise (cP). By contrast, earlier fluorescence depolarization studies employing probes such as 1,6-diphenyl-1,3,5-hexatriene (DPH) (Shinitzky and Barenholz, 1978) indicate substantially higher values for the bilayer microviscosity (about 50 to 100-fold greater). An alternative analysis of the ^{2}H NMR relaxation dispersion of membrane lipid vesicles by Halle (1991) also leads to a substantially higher microviscosity (about 10-fold greater), and is probably incorrect. These earlier conclusions (Brown, 1982) now appear substantiated by detailed molecular dynamics computer simulations of lipid bilayers in the fluid phase (Venable et al, 1993). The currently available ^{2}H and ^{13}C NMR results for lipid bilayers thus provide strong supporting evidence for the fluid-like interior of the membrane. Moreover, on account of the liquid-like nature of the bilayer, one is led to consider the importance of the balance of forces involving both the polar head groups and the hydrocarbon chains as an organizing principle, which by analogy to microemulsions can be formulated in terms of a flexible surface model (Brown, 1994).

Influences of Cholesterol and Proteins

Besides the above studies of phospholipids in the liquid-crystalline (L_α) phase, membrane bilayers containing cholesterol and integral proteins such as rhodopsin have also been investigated using NMR relaxation methods. At present, the most extensive knowledge is from ^{2}H NMR studies of bilayers containing cholesterol aligned on planar glass supports (Bonmatin et al, 1990; Morrison and Bloom, 1994; Trouard et al, 1992). The results are indicative of the type of microscopic dynamical information that can be obtained from NMR relaxation studies of membrane constituents in the fluid state. Both the phospholipids, with the acyl chains deuterated, as well as cholesterol deuterated at the various ring positions, have been studied in bilayer membranes. Information is thus obtained regarding the influences of both the rigid cholesterol molecule and the flexible phospholipids on their mutual dynamics in the fluid state.

Figure 11 summarizes the results of combined R_{1Z} and R_{1Q} relaxation measurements of bilayers of 1,2-diperdeuteriomyristoyl-sn-glycero-3-phosphocholine (DMPC-d_{54}) containing cholesterol (1:1 mol ratio) at 40°C (Trouard et al, 1992). The irreducible spectral densities of motion, $J_m(\omega)$, obtained using Equations (54a)–(54b), are plotted as a function of the angle between the bilayer normal (director) and the main magnetic field. Several points are worth noting which include the following. Clearly there is a substantial angular anisotropy of the relaxation rates and derived spectral densities of motion when cholesterol is present in the DMPC-d_{54} bilayer. The angularly anisotropic behavior is more pronounced when cholesterol is present than for phospholipid bilayers in the absence of cholesterol, on account of the increased orientational ordering of the chains. It follows that cholesterol containing bilayers represent a crucial test case for the success of relaxation theories in accounting for experimental NMR data (Bonmatin et al, 1990). As can be seen, the angular dependencies of the individual spectral densities $J_1(\omega)$ and $J_2(\omega)$ are opposite in direction, thereby indicating a rich behavior. By contrast, if only the bulk relaxation rate R_{1Z} is considered, then both spectral densities govern the orientational anisotropy of the relaxation, as described by Equation (54a). In Figure 11 both the direction and magnitude of the anisotropies of the indi-

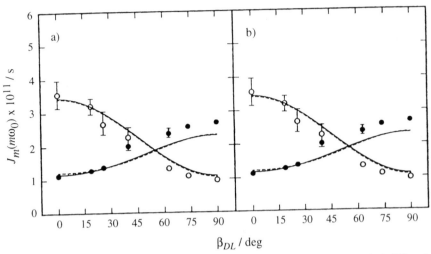

Figure 11. Experimentally determined ^{2}H NMR spectral densities of motion, $J_1(\omega_0)$ (●) and $J_2(2\omega_0)$ (○), for DMPC-d_{54}/cholesterol bilayers (1:1 mol ratio) as a function of angle between the lamellar normal and the main magnetic field ($\nu_0 = 76.8$ MHz; T = 40°C). Note that the experimentally measured spectral densities have opposite angular dependencies. Theoretical fits to the segmental (—) and molecular (----) models using the expressions in Table 3 are indicated by the smooth curves: (a) Maier-Saupe potential; and (b) symmetric top approximation. Both the magnitude and direction of the angular anisotropies of the individual spectral densities are described by a segmental or molecular diffusion model. (Trouard et al, 1992. Used by permission.)

vidual spectral densities $J_1(\omega)$ and $J_2(\omega)$ are reproduced satisfactorily using a restricted rotational diffusion model (Table 3). Due to the rather large order parameters, which approach the maximum value of $-1/2$ for all-trans rotating chains, the segmental and molecular models are nearly equivalent in this case. From the fits to the experimental data in Figure 11, values for the diffusion constants of about $D_\perp = 0.1$–1×10^9 s^{-1} and $D_\parallel = 4$–5×10^9 s^{-1} are obtained (Trouard et al, 1992).

In general, one can expect that many of the internal degrees of freedom of the flexible phospholipids (see above) are significantly reduced by the presence of cholesterol, thereby accounting for the increase in orientational ordering of the acyl chains (Gally et al, 1976; Vist and Davis, 1990). The rather different influences of cholesterol on the conformations of the polar head groups (Brown and Seelig, 1978) and the acyl chains alluded to above can be explained by a spacer effect of the rigid sterol moiety. Interactions among the lipids are more decoupled in the presence of cholesterol, leading to a suppression or damping out of the motion modes, in which case a molecular diffusion model is perhaps appropriate. For heuristic purposes only, one can apply a similar formalism to bilayer lipids in the absence of cholesterol as in its presence (but see above). Regarding DLPC-d_{46} at 40°C, in the liquid-crystalline phase, the angularly anisotropic R_{1Z} relaxation rates for the top part of the chain can be fit to a molecular model, yielding rotational diffusion constants of about $D_\perp \approx D_\parallel = 4\text{--}5 \times 10^7 \text{ s}^{-1}$ (Trouard et al, 1994). Comparison to the corresponding values for DMPC-d_{54} in the presence of cholesterol (1:1 mol ratio) (Trouard et al, 1992) then implies that the rigid sterol moiety may increase somewhat the motion rates of the phospholipid acyl groups, corresponding to a two-fold or greater increase in $D_\perp$ and about a 80-fold increase in $D_\parallel$. A simple physical interpretation is possible, in which the above diffusion constants indicate less entanglement of the chains in the presence of cholesterol, which although having less internal flexibility are nonetheless able to rotate more freely as a unit (Trouard et al, 1992). The above interpretation also appears consistent with NMR pulsed field gradient measurements of the rates of phospholipid lateral diffusion in the absence and presence of cholesterol (Lindblom and Orädd, 1994). Hence the rotational and lateral diffusion are indicative of a liquid-like state, whereas the large increase in the segmental ordering is perhaps more indicative of a molecular solid. The different influences of cholesterol on the ordering and dynamics of the acyl chains characterize in microscopic terms the liquid ordered state present in the phase diagram of phosphatidylcholine/cholesterol mixtures (Ipsen et al, 1987).

In addition, ^{2}H NMR relaxation studies of the cholesterol moiety deuterated at various positions of the sterol ring in DPPC bilayers (1:1 mol ratio) have been carried out (Bonmatin et al, 1990). The various C–^{2}H bonds of cholesterol have different orientations relative to the molecular frame, which leads to a rich anisotropic behavior of the relaxation rates. Because of the fewer internal degrees of freedom of the rigid cholesterol molecule, i.e., relative to the flexible phospholipids, a molecular model involving restricted diffusion within the anisotropic potential of the membrane bilayer is applicable. Figure 12 shows R_{1Z} relaxation rate data for cholesterol deuterated at the various ring positions as a function of the angle $\theta \equiv \beta_{DL}$ between the bilayer normal and the main magnetic field (Bonmatin et al, 1990). As can be seen in parts a–d, the relaxation anisotropy depends on the position of the C–^{2}H label, and thus the orientation β_{PM} of the bond relative to the molecular frame. It is noteworthy that the data for the various C–^{2}H bonds have all been fit simultaneously using the molecular diffusion model (Brown, 1990). Both the magnitudes of the relaxation rates as well as the angular anisotropy are reproduced, yielding diffusion constants of about $D_\perp = 8(0.1) \times 10^6 \text{ s}^{-1}$ and $D_\parallel = 4(0.2) \times 10^8 \text{ s}^{-1}$, depending on whether the short (long) correlation side of the $T_{1Z} \equiv 1/R_{1Z}$ minimum is considered. The diffusion constants obtained for the rigid cholesterol molecule can then be compared to the diffusion constants obtained for the phospholipid acyl chains, adopting a molecular diffusion model. The values imply that the rotational motions of the acyl chains in the mixed bilayers (1:1 mol ratio) are considerably faster than those of the cholesterol molecule, perhaps 10-fold or more. It follows that there may be considerable internal mobility of the acyl groups of the phospholipids in the presence of cholesterol (Trouard et al, 1992).

Finally, the influences of integral membrane proteins such as rhodopsin have been previously studied in native membrane vesicles using ^{1}H and ^{13}C NMR spectroscopy (Brown et al, 1982, 1977). Analogous ^{2}H NMR investigations

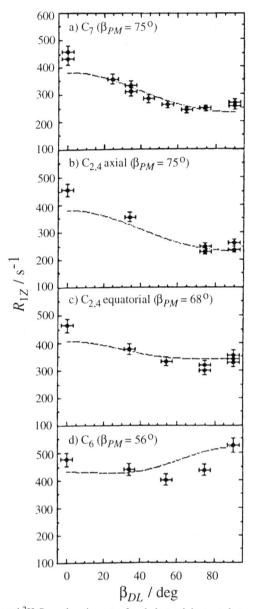

Figure 12. Experimental ^{2}H R_{1Z} relaxation rates for cholesterol deuterated at various ring positions in DPPC bilayers (1:1 mol ratio) as a function of angle between the lamellar normal and the main magnetic field ($\nu_0 = 30.7$ MHz; T $= 30°$C) (Bonmatin et al, 1990). The angle β_{PM} between the C–^{2}H bond vector and the molecular frame defined by the cholesterol long axis is indicated in parts (a)–(b). Theoretical fits of the entire set of experimental R_{1Z} values to the molecular model using Equation (54a) and Table 3 are indicated by the smooth curves. The magnitudes and directions of the angular anisotropies for the various ring positions are described to within experimental error by a rotational diffusion model. (Brown et al, 1990. Used by permission.)

of rhodopsin in recombinant membrane systems have been carried out (Deese et al, 1981b). One should note that rhodopsin (348 amino acids; $M_r = 40,000$) and other integral membrane proteins are substantially larger than cholesterol ($M_r = 387$), and may have greater internal flexibility. It appears that the presence of rhodopsin yields only small or negligible influences on the ^{1}H, ^{2}H, and ^{13}C NMR lineshapes of the lipid components, as also found in ^{31}P NMR studies (Ellena et al, 1986). However, a reduction of the T_{1Z} times is found due to exchange of the annular or boundary lipids on and off the hydrophobic surface on the protein; a single average liquid-crystalline environment is found. An alternative viewpoint in terms of short-range interactions yielding immobilized boundary lipids has been proposed (Yeagle and Kelsey, 1989), but the results cannot be repeated independently (Ellena et al, 1986). Thus in analyzing the influences of the lipid bilayer environment on the function of integral membrane proteins, a flexible surface model involving a balance of forces may be appropriate (Brown, 1994).

Membrane Proteins

It is well known that membrane proteins are difficult to crystallize (Deisenhofer and Michel, 1989; Roth et al, 1989) leading to a search for alternative structural methods. At present, NMR spectroscopy is the only physicochemical method which is capable of yielding structural information regarding proteins comparable to that obtained in X-ray crystallography. Below we summarize the various approaches involving NMR spectroscopy which are applicable to the study of membrane proteins and hydrophobic peptides, either solubilized in detergent micelles, or in membrane lipid bilayers. More extensive discussions of these methods can be found in the literature (Bax, 1989; Clore and Gronenborn, 1991; Smith and Griffin, 1988; Wüthrich, 1989).

High-Resolution Multidimensional NMR Spectroscopy

For proteins in solution, high-resolution multidimensional NMR spectroscopy is in principle capable of allowing the determination of an average structure for molar masses in the 15–30 kD range. Such three- and even four-dimensional NMR methods are widely acknowledged as one of the primary methods in structural biology for the investigation of proteins in solution (Wüthrich, 1989). In this case the dipolar and chemical shift tensors due to the various functional groups are averaged to their traces by rotational tumbling of the molecules. One can then observe the much smaller differences in the isotropic chemical shifts (trace of the chemical shift tensors), as well as the various indirect dipolar couplings. The latter are also known as spin-spin or J-couplings, in which the magnetic dipolar interaction between the nuclei is transmitted indirectly via the electrons of the chemical bond, rather than through space as for the direct dipolar coupling. It is widely appreciated that the interpretation of such high resolution NMR spectra provides a rich source of structural information in organic chemistry and biochemistry.

The application of such high-resolution multidimensional NMR methods to proteins first entails the assignment of the proton (^{1}H) NMR spectra by various through-bond NMR correlation techniques, e.g., total correlation spectroscopy (TOCSY). The next step involves effective measurement of the proton cross-relaxation rates, which characterize the continual exchange of magnetization between dipolar coupled protons, e.g., as manifested in the nuclear Overhauser effect (NOE). The most convenient technique for such measurements is a two-dimensional NMR method called nuclear Overhauser effect spectroscopy (NOESY). Clearly the physical information in any relaxation experiment is contained in the irreducible spectral densities of motion. In the case of the nuclear Overhauser effect, these are due to the dipolar couplings among the various protons, which depend on both the mean-squared amplitudes and reduced spectral densities of the motions, i.e., the correlation times for molecular reorientation. This enables average distances between pairs of dipolar coupled protons to be estimated in NOESY. In practice one only needs to classify the average distances into those giving rise to a strong, medium, or weak NOE. Because the problem is vastly overdetermined, the numerous constraints on the average interproton distances enable an average structure to be determined. A brute force approach is applied involving metric matrix distance geometry, dynamical simulated annealing, and restrained molecular dynamics methods (Bax, 1989; Clore and Gronenborn, 1991). Comparison with the corresponding crystal structures determined via X-ray scattering yields rather good agreement (Wüthrich, 1989).

At present, there are two main obstacles to extension of such methods to larger proteins, including membrane proteins solubilized in detergents. First, there is the homogeneous linebroadening associated with the relatively long rotational correlation time, which for spherical geometry depends on the radius cubed. This increased linebroadening corresponds to a more rapid decay of the nuclear transverse coherence in the time domain, i.e., in comparison to the indirect dipolar (spin-spin) couplings that are important for through-bond J-correlation experiments (COSY and TOCSY). Secondly, the nonhomogeneous linebroadening due to the increased molecular complexity yields spectral crowding, which is associated with chemical shift nonequivalence of the numerous protein functional groups, including both the polypeptide backbone and amino acid side chains. A number of innovative strategies have been developed for getting around these difficulties. These include: (i) reduction of the homogenous linebroadening by nonspecific substitution with ^{2}H; and (ii) the reduction of spectral overlap problems due to nonhomogeneous linebroadening by implementation of ^{13}C and ^{15}N isotope editing techniques, which involve three- and even four-dimensional NMR methods (Bax, 1989; Clore and Gronenborn, 1991). For proteins in solution, the practical limit at present for structural determination appears to be around 20 kD. Such high resolution NMR methods tend not to be applicable for structure determination in the case of relatively large membrane bound proteins solubilized in detergent micelles. However, the field is rapidly evolving, and one can anticipate that this may change in the future. Conventional multidimensional NMR techniques have been applied for some time to study small hydrophobic peptides in detergent micelles, including glucagon, melittin, gramicidin A, and viral coat proteins (Braun et al, 1983; Bystrov et al, 1986).

Application of Solid-State NMR Methods

An alternative is to employ solid-state NMR methods, in which the structural information is contained in the principal values and principal axes of the coupling tensors, due to the anisotropic quadrupolar, dipolar, or chemical shift interactions. In liquid state NMR experiments the above tensors are averaged to their traces, thus enabling the isotropic chemical shifts and relatively small indirect (through bond) dipolar interactions to be observed. However, in solid-state NMR methods the much larger quadrupolar, dipolar, and chemical shift interactions are studied directly. One pays a price for this conceptual simplification, of course, with increased instrumental demands, which include the need for rather high radiofrequency power to ensure adequate excitation of the broad spectral envelopes. Such experiments cannot typically be performed with standard high-resolution NMR spectrometers. Rapid digitization of the signals in the time domain is also required, together with solid echo techniques, and in some cases very high speed magic angle sample spinning (MAS) is used involving speeds in excess of 200,000 rev/min (Sefcik et al, 1983). The reward for implementing solid-state NMR techniques in membrane biophysics is that novel information is acquired regarding the conformations of membrane proteins and peptides, together with knowledge of the amplitudes and rates of their fluctuations.

One avenue involves the application of NMR methods developed originally for rotating solids to isotopically labeled membrane constituents. Most work has involved membrane bound proteins and peptides labeled with ^2H, ^{13}C, and ^{15}N. Here the quadrupolar, dipolar, or chemical shift tensors are modulated or effectively averaged to their traces by high-speed mechanical rotation about an axis inclined at the magic-angle (54.7°) with respect to the main magnetic field. The secular part of the second-rank spin interactions is eliminated in such magic-angle spinning methods. However, because random powder-type samples are employed, a drawback is that information regarding the orientation of the principal axes of the coupling tensor within the frame of the membrane is lost. The principal values of the chemical shift tensor, or more often the trace or isotropic value, can then be compared to data for appropriate model compounds, and moreover the chemical shielding can be further interpreted using Karplus-Pople theory. Such methods have been applied to studies of retinal proteins including bacteriorhodopsin (de Groot et al, 1990) and rhodopsin (Smith et al, 1991, 1990, 1987) with the retinal chromophore labeled with ^{13}C at various positions. Moreover, ^{15}N NMR studies of the retinal Schiff base nitrogen of the retinal prosthetic group of bacteriorhodopsin have been carried out (de Groot et al, 1990). These studies of bacteriorhodopsin with the retinylidene group isotopically labeled have led to the conclusion that the retinal Schiff base is protonated in the dark-adapted ground state, comprising a mixture of 13-cis, C=N syn and all-trans, C=N anti forms, in which the conformation of the cyclohexene ring is 6-s-trans.

Other approaches based on magic-angle spinning techniques are capable of providing novel information on average internuclear distances, which in turn can enable the relative orientations of functional groups within membrane proteins to be investigated. Information on the average relative distances of the nuclei is acquired, rather than knowledge of orientational fluctuations. One method is referred to as rotational resonance (RR or R^2), and relies upon the fact that two chemically nonequivalent nuclei undergo a rapid exchange of their magnetization (energy transfer) when the rate of sample rotation matches the difference in their isotropic chemical shifts (Creuzet et al, 1991). The rate of this magnetization transfer depends sensitively on the homonuclear dipolar coupling constant, and enables average distances to be determined between pairs of dipolar coupled nuclei separated by less than about 5–6 Å. As a rule, isotopic labeling is required, such as ^{13}C double labeling. Rotational resonance measurements have allowed the investigation of inter- and intramolecular distances in bilayers of 1,2-dipalmitoyl-sn-glycero-3-phosphocholine (DPPC) (Smith et al, 1994a), as well as the structure and orientation of the transmembrane domain of glycophorin A in lipid bilayers (Smith et al, 1994b). In addition, these RR methods have been applied fruitfully to study internuclear distances and the conformation of the retinal chromophore of bacteriorhodopsin doubly labeled with ^{13}C (McDermott et al, 1994). Another method, termed Rotational-Echo Double-Resonance (REDOR) (Gullion and Schaefer, 1989) provides similar knowledge regarding average internuclear distances in high-speed rotating samples, e.g., having ^{13}C–^{15}N isotope pairs. This method has been used successfully to investigate the glutamine membrane transport system of *Escherichia coli* (Hing et al, 1994), and clearly offers future promise for the study of membrane proteins.

In addition to the above techniques employing magic-angle spinning, one can apply solid-state NMR methods to hydrophobic peptides and membrane proteins incorporated within lipid bilayers oriented on planar supports (Ketchem et al, 1993; Prosser et al, 1994). Here the formalism and methodology described earlier is used to obtain information on the average orientations of the isotopically labeled groups, e.g., labeled with ^2H, ^{13}C, or ^{15}N. Such methods have been used to determine the average structure of gramicidin A in a lipid bilayer, using the orientational constraints provided by the quadrupolar, dipolar, and chemical shift tensors involving ^1H, ^2H, ^{13}C, and ^{15}N nuclei (Ketchem et al, 1993). Similar studies have also been carried out of membrane-bound peptides such as mellitin, gramicidin A, and the magainins (Hing et al, 1990; Koeppe et al, 1994; Separovic et al, 1994; Smith et al, 1989). Recently, bacteriorhodopsin having the retinylidene chromophore labeled with ^2H has been investigated in membrane bilayers oriented on planar supports by Heyn and coworkers (Ulrich et al, 1994, 1992). A representative series of ^2H NMR spectra as a function of bilayer orientation (tilt series) for bacteriorhodopsin labeled with ^2H at the retinylidene C_{18} and C_{20} methyl groups is shown in Figure 13 (Ulrich et al, 1994), together

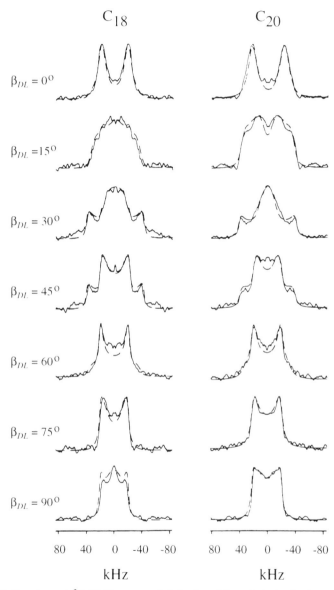

Figure 13. Experimental ^{2}H NMR spectra of dark-adapted bacteriorhodopsin deuterated at the C_{18} methyl group (left) and the C_{20} methyl group (right) of the retinylidene chromophore as a function of the bilayer tilt angle relative to the main magnetic field ($\nu_0 = 61.4$ MHz; T $= -60°$C). Theoretical spectra calculated using Equation (38) of the text for a uniaxial immobilized sample are indicated by the dashed lines. Comparison of the experimental and simulated ^{2}H NMR spectra enables determination of the orientation of the methyl 3-fold rotor axis relative to the frame of the membrane bilayer. (Ulrich et al, 1994. Used by permission.)

with spectral simulations using Equation (38). Comparison of the theoretical spectra to the experimental spectra enables the orientation of the methyl rotor axes to be determined relative to the membrane frame with relatively high precision. Similar studies of bacteriorhodopsin labeled with ^{2}H at the C_{18}, C_{19}, and C_{20} methyl groups have provided interesting new information regarding the intramolecular conformation and orientation of the retinylidene moiety within the dark-adapted ground state of the bacteriorhodopsin molecule. It appears that the polyene chain is not straight, but rather has an in-plane curvature and possibly an out-of-plane twist, together with a 6-s-trans conformation of the cyclohexene ring as mentioned above. Furthermore, ^{2}H NMR studies of bR labeled at the C_{19} methyl group of the retinal chromophore trapped in the M state suggest that a slight upward tilting of the chromophore long-axis occurs in the deprotonated 13-cis isomer of the M intermediate versus the dark-adapted state (Ulrich et al, 1995). It can be concluded from these and other studies that the application of solid-state NMR methods to membrane bound proteins and peptides is a very promising new area of research, one that can be expected to grow with time, and hopefully yield new insights into membrane structure and function.

Concluding Remarks and Future Perspectives

In this chapter we have attempted to summarize the application of NMR methods to membranes including both lineshape and relaxation studies. A brief overview is given of the use of NMR spectroscopy for investigations of membrane constituents, encompassing lipid bilayers in the fluid (L_α) phase, as well as cholesterol and membrane bound peptides and proteins. Solid-state NMR methods are described mainly in relation to membrane lipid bilayers, which are illustrative of liquid-crystalline materials in general. Here the NMR lineshapes yield information pertinent to motion averaging of the quadrupolar, dipolar, or chemical shift tensors. The corresponding nuclear spin relaxation rates provide knowledge of the amplitudes and rates of the fluctuations about the residual or time-averaged values of the coupling tensors. Deuterium NMR lineshape studies of lipid bilayers in the liquid-crystalline phase are described; NMR relaxation methods involving ^{2}H and ^{13}C are reviewed as a means of obtaining information regarding dynamical properties; and brief mention is made of NMR studies of integral membrane proteins and membrane bound peptides.

A number of intriguing questions at the cutting edge of membrane structural biology can be fruitfully approached at present. Many of the necessary experimental and theoretical tools have been developed as part of a sustained effort of basic research by many workers over the years. Knowledge of the equilibrium and dynamical properties from NMR spectroscopy is in principle contained in the correlation functions describing the orientational fluctuations of the membrane constituents. The latter encompass the flexible phospholipids, the rigid cholesterol moiety, and membrane bound peptides or proteins such as bacteriorhodopsin or rhodopsin. However, NMR spectroscopy differs from other experimental techniques such as fluorescence depolarization in that the orientational correlation functions are not observed in real time (Brown, 1979). Rather, the power spectrum of the stochastic fluctuations, that is to say the spectral densities of motion $J_m(\omega)$, governs the observable nuclear spin relaxation rates by a weak collisional process. Motions with correlation times as short as picoseconds to nanoseconds yield relaxation times in the range of milliseconds to seconds, which can be measured with an ordinary clock. The apparently bewildering variety of NMR techniques can be considered by the uninitiated to entail measurement of various spectral densities, corresponding to magnetic or electrical interactions of the nuclei.

Among the major conclusions to date from NMR studies of membrane constituents are the following. First, because NMR is a nonperturbing intrinsic probe technique, it is capable of providing knowledge regarding the entire system of interest. With regard to membrane lipids, information pertinent to both the polar head groups and the nonpolar acyl chains is obtained. Investigations of the various phospholipid head groups using ^{2}H and ^{31}P NMR spectroscopy have yielded knowledge of their average conformations and orientations relative to the membrane surface, as well as conformational changes due to ion-binding associated with the electrostatic properties of membranes (Seelig et al, 1987). In addition, segmental order parameters for the acyl chains have been determined in the fluid phase using ^{2}H and ^{13}C NMR spectroscopy, which in effect yields atomic level information in the highly disordered liquid-crystalline state (Brown and Chan, 1995). Such order profiles are amenable to interpretation at the microscopic level in terms of average properties, including the mean projected acyl length and cross sectional area per molecule, and are relevant to other biophysical studies of membrane bilayers, including small angle X-ray scattering (SAXS) and neutron scattering (SANS). Likewise, knowledge of the structural properties of lipids in nonlamellar (H_I and H_{II}) phases can be obtained (Thurmond et al, 1993), which is relevant to the force balance involving the polar head group and acyl chain regions. Recently we have described how this force balance can govern the conformational energetics of integral membrane proteins such as rhodopsin (Brown, 1996, 1994). Multinuclear NMR studies of membrane bound polypeptides and proteins employing solid-state NMR techniques have yielded new information about the average conformations and dynamics of the polypeptide backbones and amino acid side chains, as well as the conformation and orientation of prosthetic groups such as retinal in bacteriorhodopsin and rhodopsin.

Besides such knowledge of the equilibrium properties of membrane constituents, measurement and interpretation of the NMR relaxation rates provide a novel means of studying experimentally their dynamics. The NMR relaxation depends on the rates and amplitudes of various motions as well as the degree of their coupling. Such investigations indicate that the motion-averaged NMR spectral lineshapes of membrane lipids in the fluid phase originate from a broad distribution of motions, including rapid dihedral angle isomerizations, molecular motions, and collective bilayer disturbances. Interestingly, the results of theoretical calculations in closed form are born out by quantitative experimental relaxation studies. In the liquid-crystalline (L_α) state a square-law dependence of the NMR relaxation rate (R_{1Z}) and order profiles (S_{CD}) is obtained, which represents a characteristic signature of the influences of slow motions. Such fluctuations in the local ordering set up by faster segmental motions may arise from molecular motions and/or collective disturbances of the bilayer. The frequency dependence of the ^{2}H and ^{13}C spin-lattice (R_{1Z}) relaxation rates of membrane lipid bilayers yields evidence for collective fluctuations, and is most indicative of a broad distribution of motion components in the MHz regime, described by a $\omega^{-1/2}$ dispersion law. The presence of a distribution of motion components may need to be considered with regard to statistical mechanical analyses of the equilibrium properties of fluid bilayers, which mainly consider local isomerizations of the acyl chains.

One can also ask: Is the proposed fluctuation spectrum in lipid bilayers comprising segmental, molecular, and collective motions biologically significant? There are reasons why this may be the case, some of which are briefly enumerated below. Local segmental motions of the acyl chains are related to the microscopic viscosity of the bilayer interior, indicating that small molecules such as quinone diffuse in a medium comprising essentially liquid hydrocarbon. On account of the rapid dihedral angle isomerizations and other internal motions of the lipid acyl chains, the main barrier to diffusion of small nonpolar solutes probably involves the lipid/water interface. Perhaps surprisingly, the question of the microviscosity of the hydrocarbon interior of the bilayer has remained a somewhat open question (Brown, 1982; Brown et al, 1983; Venable et al, 1993). The low microviscosity of the bilayer interior is quantitatively in accord with the earlier classic oil drop model of Overton. Moreover, it calls attention to the importance of the force balance, involving both the membrane lipid polar head groups as well as the acyl chains, with regard to the bilayer physicochemical properties. By analogy to microemulsions, this force balance can be usefully described in terms of a flexible surface model (Olsson and Wennerström, 1994). We have recently discussed how the energetics of membrane deformation may be related to conformational transitions of membrane proteins such as rhodopsin (Brown, 1994). One can expect in general that rotational and translational diffusion of integral membrane proteins are strongly modulated by structural and dynamical features of the aqueous interfacial region. Formation of transient complexes of membrane proteins may thus be dependent on the membrane lipid mobility. Finally, the presence of collective bilayer excitations may be associated with undulatory and other repulsive forces involving the membrane surface (due to entropic confinement), which in turn may be linked to biological phenomena such as membrane fusion, including endo- or exocytosis.

Clearly there is a need for further experimental NMR and other measurements in combination with theoretical developments as a means of testing the above concepts. Computer simulations of the dynamics of lipid bilayers employing molecular mechanics (Venable et al, 1993) have tended to substantiate these earlier conclusions (derived with paper and pencil) from NMR spectroscopy. Simulation methods represent a powerful new means of incorporating the results of various experimental studies into a comprehensive force field whose parameters account for the membrane properties. At present such molecular dynamics computer simulations are limited to relatively short time scales, comprising ≈ 200 ps or less. Although the trajectories can in principle be extended into the nanosecond regime by Langevin or Brownian dynamics simulations, here there is a drawback in that the collision frequency γ is an adjustable parameter, leaving the time-scale undefined. In addition, Monte Carlo and molecular dynamics computer simulations are sensitive to periodic boundary conditions, and thus to the size of the system, as well as to the functional form of the force field. It follows that there is an element of circularity to the approach in that the conclusions tend to substantiate the premises. Clearly NMR relaxation rate studies comprise one of the main experimental benchmarks for detailed comparison. The interplay between experimental NMR relaxation studies and theoretical dynamics simulations constitutes a very promising area for future research (Pastor et al, 1991; Venable et al, 1993). With the availability of faster computers, together with further development and refinement of theoretical methods, one can investigate substantially longer time scales, and thus address more specifically the motion distribution in bilayer membranes proposed on the basis of NMR spectroscopy (Brown, 1982; Brown et al, 1983).

ACKNOWLEDGMENT

This basic research is sponsored in part by grants from the US National Science Foundation, the US National Institutes of Health, and the Swedish Natural Science Research Council. The author wishes to extend his sincere appreciation to all current and previous members of the laboratory for their dedication, hard work, and many contributions to this research. Slutligen vill jag även tacka alla mina vänner på Fysikalisk Kemi 1 vid Lunds Universitet för frikostig gästfrihet och insiktsfulla diskussioner under så många trevliga sommrar!

Appendix: A Primer of Irreducible Tensor Methods in Membrane Biophysics

In what follows, the application of irreducible tensor methods is described briefly with regard to NMR spectroscopy of membranes. The formalism is equally applicable to other branches of spectroscopy including fluorescence and infrared spectroscopy (Brown, 1979). Irreducible tensors and their transformations under rotations are central to the application of angular momentum theory in NMR spectroscopy. The sooner one is able to convert from a Cartesian (x, y, z) basis to an irreducible or spherical basis the better off one is! This is in contrast to the use of direction cosines, which become quite cumbersome if more than a single rotation of coordinates is considered. As a rule, the coupling Hamiltonians in NMR spectroscopy can be formulated as the scalar products of two irreducible tensors, one of which corresponds to the spin angular momentum operators, and the other to the physical coupling involving the nuclear spins, e.g., due to electric quadrupolar or magnetic dipolar interactions, or the anisotropic chemical shift. First we shall summarize the contruction of irreducible tensor operators from the more familiar vector operators within a Cartesian basis. Then we describe briefly the relation between irreducible and Cartesian tensors, together with some of their useful properties. We shall state without proof the following results from group theory which are covered in greater detail elsewhere (Brink and Satchler, 1968; Häberlen, 1976; Rose, 1957).

Consider two vector operators $\hat{A}$ and $\hat{B}$ in a Cartesian frame, for example corresponding to the nuclear spin angular momentum. The irreducible (or spherical) components of rank $j = 1$ are formed from linear combinations of the Cartesian components as follows: $\hat{T}_0^{(1)} \equiv \hat{A}_0 = \hat{A}_Z$, $\hat{T}_{\pm 1}^{(1)} \equiv \hat{A}_{\pm 1} = \mp \sqrt{\frac{1}{2}}(\hat{A}_X \pm i \hat{A}_Y)$, and analogously for $\hat{B}_0$ and $\hat{B}_{\pm 1}$. Taking the direct product of the above irreducible tensor operators, one can construct irreducible tensor operators $\hat{T}_m^{(j)}$ of higher rank j, where the projection quantum number m runs from $-j$ to $+j$ in integral steps. This is described in standard texts on group theory (Rose, 1957). For the quadrupolar or dipolar interactions one has that $j = 2$, leading to the result:

$$\hat{T}_0^{(2)} = \frac{1}{\sqrt{6}}(3\hat{A}_0\hat{B}_0 - \hat{A} \cdot \hat{B}), \tag{A1a}$$

$$\hat{T}_{\pm 1}^{(2)} = \frac{1}{\sqrt{2}}(\hat{A}_0\hat{B}_{\pm 1} + \hat{A}_{\pm 1}\hat{B}_0), \tag{A1b}$$

$$\hat{T}_{\pm 2}^{(2)} = \hat{A}_{\pm 1}\hat{B}_{\pm 1}. \tag{A1c}$$

Substituting $\hat{A} = \hat{I}$ and $\hat{B} = \hat{S}$ gives Equations (2) of the text in terms of the angular momentum operators.

Moreover, it is necessary to consider the nature of the coupling interactions involving the angular momentum in NMR spectroscopy. These are described most generally by second-rank tensors V in terms of the principal values and principal axis system, rather than by construction from known vector operators. Second-rank Cartesian tensors can be decomposed most generally into a sum of irreducible tensors of ranks $j = 0, 1$, and 2 with respect to the full three-dimensional rotation group, that is to say $V = V^{(0)} + V^{(1)} + V^{(2)}$. For the case of quadrupolar or dipolar interactions, it is the traceless symmetric part $V^{(2)}$ that is of interest. The irreducible components $V_m^{(2)}$ of the coupling tensors are related to their Cartesian components V_{ii} ($i = x, y, z$) within the principal axis system (PAS) by (Häberlen, 1976):

$$V_0^{(2)} = \sqrt{\frac{3}{2}}\delta, \tag{A2a}$$

$$V_{\pm 1}^{(2)} = 0, \tag{A2b}$$

$$V_{\pm 2}^{(2)} = -\frac{1}{2}\delta\eta. \tag{A2c}$$

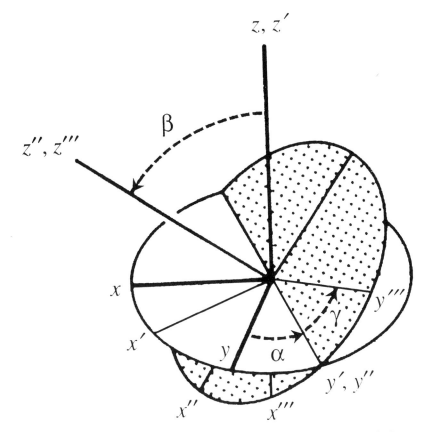

Figure A1. Illustration of Euler angles employed in transformation of irreducible (spherical) tensors from an initial (x, y, z) coordinate frame to a final (x''', y''', z''') system. The first rotation is about the initial z-axis by the Euler angle α. The second rotation is about the new y'-axis by the Euler angle β. The third rotation is about the penultimate z''-axis by the Euler angle γ. Consideration of the Euler angles allows one to formulate explicitly various models for the rotational dynamics of membrane constituents, including both lipids and proteins (see text).

In the above expressions, $\delta \equiv V_{zz}$ is the largest principal value and $\eta \equiv (V_{yy} - V_{xx})/V_{zz}$ is the asymmetry parameter. For a given coupling interaction (label λ) the above results correspond to Equations (3) of the text.

The usefulness of spherical tensor calculus in various branches of spectroscopy is that the irreducible tensor operators transform under rotations in a simple, well-defined manner. Consequently, the treatment of multiple rotations is relatively straightforward, once a few elementary concepts have been learned. Rotations are described most generally in terms of three parameters, the most useful for our purposes being the three Euler angles α, β, and γ. Figure A1 indicates how a physical system within an initial coordinate frame is transformed via the three Euler angles α, β, and γ to a final coordinate frame (Rose, 1957). First the system within the (x, y, z) frame is rotated about the z-axis by the angle α

to the (x', y', z') frame. Next the primed system is rotated about the new y'-axis by the angle β to the (x'', y'', z'') system. Finally the double-primed system is rotated about the z''-axis by the angle γ to yield the system transformed to the (x''', y''', z''') frame, where γ is the angle between the y'''-axis and the so-called line of nodes. The matrix that rotates the components of an irreducible tensor of rank j from the initial to the final coordinate frame is called the Wigner rotation matrix $\boldsymbol{D}^{(j)}$, and the elements for the case of $j = 2$ are summarized in Table A1.

Now the elements of the Wigner rotation matrix constitute generalized spherical harmonics, just as the spherical harmonics are a generalization of the Legendre polynomials. They describe the effect of rotating an axially symmetric system as a rigid body, and their dependence on the three Euler angles $\Omega \equiv (\alpha, \beta, \gamma)$ is given by:

$$D_{m'm}^{(j)}(\alpha, \beta, \gamma) = e^{-im'\alpha} d_{m'm}^{(j)}(\beta) e^{-im\gamma}. \tag{A3}$$

Here $d_{m'm}^{(j)}(\beta)$ indicates the elements of the reduced Wigner rotation matrix. The Wigner rotation matrix elements have a number of useful symmetry properties, including the following,

$$D_{m'm}^{(j)}(\Omega) = (-1)^{m'-m} D_{-m'-m}^{(j)*}(\Omega), \tag{A4}$$

where the asterisk indicates complex conjugation. Setting $m = 0$ yields the correspondence to the more familiar spherical harmonics; relabeling the indices yields:

$$D_{m0}^{(j)}(\alpha, \beta, 0) = \left(\frac{4\pi}{2j+1}\right)^{1/2} Y_m^{(j)*}(\beta, \alpha), \tag{A5}$$

where the connection to the spherical polar angles is given by $(\theta, \phi) = (\beta, \alpha)$. Finally, taking $m = 0$ in Equation (A5) one obtains the more familiar Legendre polynomials, given by:

$$D_{00}^{(j)}(\Omega) = d_{00}^{(j)}(\beta) = P_j(\cos \beta), \tag{A6}$$

in which $P_j(x)$ is a Legendre polynomial of rank j, and $x \equiv \cos \beta$. For convenience the Legendre polynomials up to rank $j = 4$ are summarized in Table A2.

We are now in a position to indicate how the elements of an irreducible (spherical) tensor rotated by the Euler angles $\Omega \equiv (\alpha, \beta, \gamma)$ correspond to the elements of the irreducible tensor in the original frame. Under such a unitary transformation, the irreducible tensor elements become:

$$\hat{T}_n^{(j)} = \sum_p \hat{T}_p^{(j)} D_{pn}^{(j)}(\Omega), \tag{A7}$$

where the projection indices $n, p = -j, \ldots, j$. The above transformation under rotation in fact defines an irreducible tensor operator of rank j. Note that the elements of the transformed irreducible tensor operator $\hat{T}_n^{(j)}$ correspond to a linear combination of the components of the same tensor operator $\hat{T}_p^{(j)}$ where $D_{pn}^{(j)}(\Omega)$ indicates the elements of the Wigner rotation matrix.

Table A1. Elements $D^{(2)}_{m'm}(\alpha, \beta, \gamma)$ of the Second-Rank Wigner Rotation Matrix

m' \ m	-2	-1	0	1	2
-2	$e^{+2i\alpha}\left(\dfrac{1+\cos\beta}{2}\right)^2 e^{+2i\gamma}$	$e^{+2i\alpha}\sin\beta\left(\dfrac{1+\cos\beta}{2}\right)e^{+i\gamma}$	$e^{+2i\alpha}\sqrt{\dfrac{3}{8}}\sin^2\beta$	$-e^{+2i\alpha}\sin\beta\left(\dfrac{\cos\beta-1}{2}\right)e^{-i\gamma}$	$e^{+2i\alpha}\left(\dfrac{\cos\beta-1}{2}\right)^2 e^{-2i\gamma}$
-1	$-e^{+i\alpha}\sin\beta\left(\dfrac{1+\cos\beta}{2}\right)e^{+2i\gamma}$	$e^{+i\alpha}(2\cos\beta-1)\left(\dfrac{\cos\beta+1}{2}\right)e^{+i\gamma}$	$e^{+i\alpha}\sqrt{\dfrac{3}{2}}\sin\beta\cos\beta$	$e^{+i\alpha}(2\cos\beta+1)\left(\dfrac{1-\cos\beta}{2}\right)e^{-i\gamma}$	$-e^{+i\alpha}\sin\beta\left(\dfrac{\cos\beta-1}{2}\right)e^{-2i\gamma}$
0	$\sqrt{\dfrac{3}{8}}\sin^2\beta\, e^{+2i\gamma}$	$-\sqrt{\dfrac{3}{2}}\sin\beta\cos\beta\, e^{+i\gamma}$	$\dfrac{1}{2}(3\cos^2\beta-1)$	$\sqrt{\dfrac{3}{2}}\sin\beta\cos\beta\, e^{-i\gamma}$	$\sqrt{\dfrac{3}{8}}\sin^2\beta\, e^{-2i\gamma}$
1	$e^{-i\alpha}\sin\beta\left(\dfrac{\cos\beta-1}{2}\right)e^{+2i\gamma}$	$e^{-i\alpha}(2\cos\beta+1)\left(\dfrac{1-\cos\beta}{2}\right)e^{+i\gamma}$	$-e^{-i\alpha}\sqrt{\dfrac{3}{2}}\sin\beta\cos\beta$	$e^{-i\alpha}(2\cos\beta-1)\left(\dfrac{\cos\beta+1}{2}\right)e^{-i\gamma}$	$e^{-i\alpha}\sin\beta\left(\dfrac{1+\cos\beta}{2}\right)e^{-2i\gamma}$
2	$e^{-2i\alpha}\left(\dfrac{1-\cos\beta}{2}\right)^2 e^{+2i\gamma}$	$e^{-2i\alpha}\sin\beta\left(\dfrac{\cos\beta-1}{2}\right)e^{+i\gamma}$	$e^{-2i\alpha}\sqrt{\dfrac{3}{8}}\sin^2\beta$	$-e^{-2i\alpha}\sin\beta\left(\dfrac{1+\cos\beta}{2}\right)e^{-i\gamma}$	$e^{-2i\alpha}\left(\dfrac{1+\cos\beta}{2}\right)^2 e^{-2i\gamma}$

Table A2. Legendre Polynomials $P_j(x)$ up to Fourth Order

$$P_0(x) = 1$$
$$P_1(x) = x$$
$$P_2(x) = \tfrac{1}{2}(3x^2 - 1)$$
$$P_3(x) = \tfrac{1}{2}(5x^3 - 3x)$$
$$P_4(x) = \tfrac{1}{8}(35x^4 - 30x^2 + 3)$$

Let us next consider the effect of consecutive transformations of the irreducible tensor to a succession of different coordinate frames. Application of Equation (A7) then involves the Euler angles Ω_1, and a second application gives:

$$\hat{T}_m^{(j)} = \sum_n \hat{T}_n^{(j)} D_{nm}^{(j)}(\Omega_2), \tag{A8a}$$

$$= \sum_n \sum_p \hat{T}_p^{(j)} D_{pn}^{(j)}(\Omega_1) D_{nm}^{(j)}(\Omega_2), \tag{A8b}$$

$$= \sum_p \hat{T}_p^{(j)} \left[\sum_n D_{pn}^{(j)}(\Omega_1) D_{nm}^{(j)}(\Omega_2) \right]. \tag{A8c}$$

But one can also consider the overall rotation from the initial coordinate frame to the final frame, with Euler angles Ω_3, which yields:

$$\hat{T}_m^{(j)} = \sum_p \hat{T}_p^{(j)} D_{pm}^{(j)}(\Omega_3). \tag{A9}$$

Hence one arrives at the closure property of the group of rotations:

$$D_{pm}^{(j)}(\Omega_3) = \sum_n D_{pn}^{(j)}(\Omega_1) D_{nm}^{(j)}(\Omega_2). \tag{A10}$$

The above formula is a generalization of the spherical harmonic addition theorem to the three Euler angles $\Omega \equiv (\alpha, \beta, \gamma)$ involving the Wigner rotation matrices. It describes how an overall rotation through Euler angles Ω_3 is obtained by first rotating the system by the Euler angles Ω_1, and then by the Euler angles Ω_2. By induction, one can use closure to decompose the Wigner rotation matrix elements for any overall rotation into those for an arbitrary number of intermediate rotations. Alternatively, closure can be invoked to collapse the summations involving various intermediate rotations.

Equation (A10) indicates that the Wigner rotation matrix elements themselves transform as irreducible tensors. Thus by forming their direct products, one can construct rotation matrices of higher rank j in the manner discussed above. In NMR spectroscopy, tensors up to second-rank are combined, yielding Wigner rotation matrices up to rank $j = 4$ (Brown, 1979). The products of the

Wigner rotation matrix elements can be expanded in a Clebsch-Gordan series (Brink and Satchler, 1968), giving for the squared moduli:

$$|D^{(2)}_{m'm}(\Omega)|^2 = (-1)^{m-m'} \sum_{j} (2j+1) \begin{pmatrix} 2 & 2 & j \\ m' & -m' & 0 \end{pmatrix} \begin{pmatrix} 2 & 2 & j \\ m & -m & 0 \end{pmatrix} D^{(j)}_{00}(\Omega).$$

(A11)

Here (m', m) and Ω are generic projection indices and Euler angles, respectively, and $j = 0, \ldots, 4$ in integral steps, corresponding to the triangle condition in quantum mechanics. The quantities in the large parentheses denote Wigner 3-j symbols, which are related to the Clebsch-Gordan or vector coupling coefficients. The 3-j symbols have a number of useful symmetry properties; extensive tables exist (Rotenberg et al, 1959). This simplifies evaluation of the mean-squared values of the rotation matrix elements and related quantities. The above formalism is directly pertinent to studies of the equilibrium and dynamical properties of membrane constituents, including both lipids and proteins as described in the text.

REFERENCES

Abragam A (1961): *The Principles of Nuclear Magnetism*. London: Oxford University Press

Akutsu H, Seelig J (1981): Interaction of metal ions with phosphatidylcholine bilayer membranes. *Biochemistry* 20:7366–7373

Altbach MI, Mattingly M, Brown MF, Gmitro AF (1991): Magnetic resonance imaging of lipid deposits in human atheroma via a stimulated-echo diffusion technique. *Magn Reson Med* 20:319–326

Atkins PW (1990): *Physical Chemistry*, 4th Ed. San Francisco: Freeman

Auger M, Carrier D, Smith ICP, Jarrell HC (1990): Elucidation of motional modes in glycoglycerolipid bilayers. A ^{2}H NMR relaxation and line-shape study. *J Am Chem Soc* 112:1373–1381

Barry JA, Trouard TP, Salmon A, Brown MF (1991): Low temperature ^{2}H NMR spectroscopy of phospholipid bilayers containing docosahexaenoyl (22:6ω3) chains. *Biochemistry* 30:8386–8394

Bax A (1989): Two-dimensional NMR and protein structure. *Ann Rev Biochem* 58:223–256

Blinc R, Luzar M, Vilfan M, Burgar M (1975): Proton spin-lattice relaxation in smectic TBBA. *J Chem Phys* 63:3445–3451

Bloom M, Morrison C, Sternin E, Thewalt JL (1992): Spin echoes and the dynamic properties of membranes. In: *Pulsed Magnetic Resonance: NMR, ESR, Optics*, Bagguley DMS, ed. Oxford: Clarendon Press

Bonmatin J-M, Smith ICP, Jarrell HC, Siminovitch DJ (1990): Use of a comprehensive approach to molecular dynamics in ordered lipid systems: cholesterol reorientation in ordered lipid bilayers. A ^{2}H NMR relaxation case study. *J Am Chem Soc* 112:1697–1704

Braun W, Wider G, Lee KH, Wüthrich K (1983): Conformation of glucagon in a lipid-water interphase by ^{1}H nuclear magnetic resonance. *J Mol Biol* 169:921–948

Brink DM, Satchler GR (1968): *Angular Momentum.* London: Oxford University Press

Brown MF (1996): Influences of membrane lipids on the photochemical function of rhodopsin. In: *Structure and Biological Roles of Lipids Forming Non-Lamellar Structures,* Epand RM, ed. Greenwich: JAI Press

Brown MF (1994): Modulation of rhodopsin function by properties of the membrane bilayer. *Chem Phys Lipids* 73:159–180

Brown MF (1990): Anisotropic nuclear spin relaxation of cholesterol in phospholipid bilayers. *Mol Phys* 71:903–908

Brown MF (1984a): Theory of spin-lattice relaxation in lipid bilayers and biological membranes: dipolar relaxation. *J Chem Phys* 80:2808–2831

Brown MF (1984b): Unified picture for spin-lattice relaxation of lipid bilayers and biomembranes. *J Chem Phys* 80:2832–2836

Brown MF (1982): Theory of spin-lattice relaxation in lipid bilayers and biological membranes. ^{2}H and ^{14}N quadrupolar relaxation. *J Chem Phys* 77:1576–1599

Brown MF (1979): Deuterium relaxation and molecular dynamics in lipid bilayers. *J Magn Reson* 35:203–215

Brown MF, Chan SI (1995): Bilayer membranes: deuterium & carbon-13 NMR. In: *Encylcopedia of Nuclear Magnetic Resonance,* Grant DM, Harris RK, eds. New York: Wiley

Brown MF, Davis JH (1981): Orientation and frequency dependence of the deuterium spin-lattice relaxation in multilamellar phospholipid dispersions: implications for dynamic models of membrane structure. *Chem Phys Lett* 79:431–435

Brown MF, Seelig J (1978): Influences of cholesterol on the polar region of phosphatidylcholine and phosphatidylethanolamine bilayers. *Biochemistry* 17:381–384

Brown MF, Seelig J (1977): Ion-induced changes in head group conformation of lecithin bilayers. *Nature* 269:721–723

Brown MF, Söderman O (1990): Orientational anisotropy of nuclear spin relaxation in phospholipid membranes. *Chem Phys Lett* 167:158–164

Brown MF, Deese AJ, Dratz EA (1982): Proton, carbon-13, and phosphorus-13 NMR methods for the investigation of rhodopsin-lipid interactions in retinal rod outer segment membranes. *Methods Enzymol* 81:709–728

Brown MF, Ellena JF, Trindle C, Williams GD (1986): Frequency dependence of spin-lattice relaxation times of lipid bilayers. *J Chem Phys* 84:465–470

Brown MF, Miljanich GP, Dratz EA (1977): Interpretation of 100- and 360-MHz proton magnetic resonance spectra of retinal rod outer segment disk membranes. *Biochemistry* 16:2640–2648

Brown MF, Ribeiro AA, Williams GD (1983): New view of lipid bilayer dynamics from ^{2}H and ^{13}C NMR relaxation time measurements. *Proc Natl Acad Sci USA* 80:4325–4329

Brown MF, Salmon A, Henriksson U, Söderman O (1990): Frequency dependent ^{2}H NMR relaxation rates of small unilamellar phospholipid vesicles. *Mol Phys* 69:379–383

Brown MF, Seelig J, Häberlen U (1979): Structural dynamics in phospholipid bilayers from deuterium spin-lattice relaxation time measurements. *J Chem Phys* 70:5045–5053

Bystrov VF, Arseniev AS, Barsukov IL, Lomize AL (1986): 2D NMR of single and double stranded helices of gramicidin A in micelles and solutions. *Bull Magn Reson* 8:84–94

Clore GM, Gronenborn AM (1991): Two-, three-, and four-dimensional NMR methods for obtaining larger and more precise three-dimensional structures of proteins in solution. *Ann Rev Biophys Biophys Chem* 20:29–63

Creuzet F, McDermott A, Gebhard R, van der Hoef K, Spijker-Assink MB, Herzfeld J, Lugtenburg J, Levitt MH, Griffin RG (1991): Determination of membrane protein structure by rotational resonance NMR: bacteriorhodopsin. *Science* 251:783–786

Davies MA, Schuster HF, Brauner JW, Mendelsohn R (1990): Effects of cholesterol on conformational disorder in dipalmitoylphosphatidylcholine bilayers. A quantitative IR study of the depth dependence. *Biochemistry* 29:4368–4373

Davis JH (1983): The description of membrane lipid conformation, order and dynamics by ^{2}H-NMR. *Biochim Biophys Acta* 737:117–171

de Groot H, Smith SO, Courtin J, van den Berg E, Winkel C, Lugtenberg J, Griffin RG, Herzfeld J (1990): Solid-state ^{13}C and ^{15}N NMR study of the low pH forms of bacteriorhodopsin. *Biochemistry* 29:6873–6883

Deese AJ, Dratz EA, Brown MF (1981a): Retinal ROS lipids form bilayers in the presence and absence of rhodopsin: a ^{31}P NMR study. *FEBS Lett* 124:93–99

Deese AJ, Dratz EA, Dahlquist FW, Paddy MR (1981b): Interaction of rhodopsin with two unsaturated phosphatidylcholines: a deuterium NMR study. *Biochemistry* 20:6420–6427

Deisenhofer J, Michel H (1989): The photosynthetic reaction center from the purple bacterium *Rhodopseudomonas viridis*. *Science* 245:1463–1473

Dill KA, Naghizadeh J, Marqusee JA (1988): Chain molecules at high densities at interfaces. *Ann Rev Phys Chem* 39:425–461

Dodd SW, Brown MF (1989): Disaturated phosphatidylcholines in the liquid-crystalline state studied by deuterium NMR spectroscopy. *Biophys J* 55:102a

Ellena JF, Pates RD, Brown MF (1986): ^{31}P NMR spectra of rod outer segment and sarcoplasmic reticulum membranes show no evidence of immobilized components due to lipid-protein interactions. *Biochemistry* 25:3742–3748

Epand RM (1990): Relationship of phospholipid hexagonal phases to biological phenomena. *Biochem Cell Biol* 68:17–23

Fong TM, McNamee MG (1987): Stabilization of acetylcholine receptor secondary structure by cholesterol and negatively charged phospholipids in membranes. *Biochemistry* 26:3871–3880

Gally HU, Seelig A, Seelig J (1976): Cholesterol-induced rod-like motion of fatty acyl chains in lipid bilayers. A deuterium magnetic resonance study. *Hoppe-Seyler's Z Physiol Chem* 357:1447–1450

Gibson NJ, Brown MF (1993): Lipid headgroup and acyl chain composition modulate the MI-MII equilibrium of rhodopsin in recombinant membranes. *Biochemistry* 32:2438–2454

Gruner SM (1989): Stability of lyotropic phases with curved interfaces. *J Phys Chem* 93:7562–7570

Gullion T, Schaefer J (1989): Detection of weak heteronuclear dipolar coupling by rotational-echo double-resonance nuclear magnetic resonance. *Adv Magn Reson* 13:57–83

Häberlen U (1976): *High Resolution NMR in Solids. Selective Averaging.* New York: Academic Press

Hall JE, Vodyanoy I, Balasubramanian TM, Marshall GR (1984): Alamethicin: a rich model for channel behaviour. *Biophys J* 45:223–247

Halle B (1991): ^{2}H NMR relaxation in phospholipid bilayers. Toward a consistent molecular interpretation. *J Phys Chem* 95:6724–6733

Hing AW, Adams SP, Silbert DF, Norberg RE (1990): Deuterium NMR of Val^1...(2-^{2}H)Ala^3...gramicidin A in oriented DMPC bilayers. *Biochemistry* 29:4144–4156

Hing AW, Tjandra N, Cottam PF, Schaefer J, Ho C (1994): An investigation of the ligand-binding site of the glutamine-binding protein of *escherichia coli* using rotational-echo double-resonance NMR. *Biochemistry* 33:8651–8661

Huang TH, Skarjune RP, Wittebort RJ, Griffin RG, Oldfield E (1980): Restricted rotational isomerization in polymethylene chains. *J Am Chem Soc* 102:7377–7379

Ipsen JH, Karlström G, Mouritsen OG, Wennerström HW, Zuckermann MJ (1987): Phase equilibria in the phosphatidylcholine-cholesterol system. *Biochim Biophys Acta.* 905:162–172

Israelachvili J (1992): *Intermolecular and Surface Forces*, 2nd Ed. New York: Academic Press

Israelachvili JN, Wennerström H (1992): Entropic forces between amphiphilic surfaces in liquids. *J Phys Chem* 96:520–531

Jansson M, Thurmond RL, Barry JA, Brown MF (1992): Deuterium NMR study of inter-molecular interactions in lamellar phases containing palmitoyllysophosphatidylcholine. *J Phys Chem* 96:9532–9544

Jeffrey KR, Wong TC, Burnell EE, Thompson MJ, Higgs TP, Chapman NR (1979): Molecular motion in the lyotropic liquid crystal system containing potassium palmitate: a study of proton spin-lattice relaxation times. *J Magn Reson* 36:151–171

Jensen JW, Schutzbach JS (1984): Activation of mannosyltransferase II by nonbilayer phospholipids. *Biochemistry* 23:1115–1119

Kelusky EC, Smith ICP (1983): Characterization of the binding of the local anesthetics procaine and tetracaine to model membranes of phosphatidylethanolamine: a deuterium nuclear magnetic resonance study. *Biochemistry* 22:6011–6017

Ketchem RR, Hu W, Cross TA (1993): High-resolution conformation of gramicidin A in a lipid bilayer by solid-state NMR. *Science* 261:1457–1460

Kinnunen PKJ, Kõiv A, Lehtonen JYA, Rytömaa M, Mustonen P (1994): Lipid dynamics and peripheral interactions of proteins with membrane surfaces. *Chem Phys Lipids* 73:181–207

Koeppe RE II, Killian JA, Greathouse DV (1994): Orientations of the tryptophan 9 and 11 side chains of the gramicidin channel based on deuterium nuclear magnetic resonance spectroscopy. *Biophys J* 66:14–24

König S, Pfeiffer W, Bayerl T, Richter D, Sackmann E (1992): Molecular dynamics of lipid bilayers studied by incoherent quasi-elastic neutron scattering. *J Phys II* 2:1589–1615

Kroon PA, Kainosho M, Chan SI (1976): Proton magnetic resonance studies of lipid bilayer membranes. Experimental determination of inter- and intramolecular nuclear relaxation rates in sonicated phosphatidylcholine bilayer vesicles. *Biochim Biophys Acta* 433:282–293

Lafleur M, Cullis P, Fine B, Bloom M (1990): Comparison of the orientational order of lipid chains in the L_α and H_{II} phases. *Biochemistry* 29:8325–8333

Lindblom G, Orädd G (1994): NMR studies of translational diffusion in lyotropic liquid crystals and lipid membranes. *Prog NMR Spectrosc* 26:483–515

Lindblom G, Rilfors L (1989): Cubic phases and isotropic structures formed by membrane lipids—possible biological relevance. *Biochim Biophys Acta* 988:221–256

Lipari G, Szabo A (1982): Model-free approach to the interpretation of nuclear magnetic resonance relaxation in macromolecules. I. Theory and range of validity. *J Am Chem Soc* 104:4546–4559

Lundbæk JA, Andersen OS (1994): Lysophospholipids modulate channel function by altering the mechanical properties of lipid bilayers. *J Gen Physiol* 104:645–673

Marqusee JA, Warner M, Dill KA (1984): Frequency dependence of NMR spin lattice relaxation in bilayer membranes. *J Chem Phys* 81:6404–6405

McConnell HM (1976): Molecular motion in biological membranes. In: *Spin Labeling Theory and Applications*, Berliner LJ, ed. New York: Academic Press

McDermott AE, Creuzet F, Gebhard R, van der Hoef K, Levitt MH, Herzfeld J, Lugtenberg J, Griffin RG (1994): Determination of internuclear distances and the orientation of functional groups by solid-state NMR: rotational resonance study of the conformation of retinal in bacteriorhodopsin. *Biochemistry* 33:6129–6136

McIntosh TJ, Simon SA (1986): Area per molecule and distribution of water in fully hydrated dilauroylphosphatidylethanolamine bilayers. *Biochemistry* 25:4948–4952

McLaughlin S (1989): The electrostatic properties of membranes. *Ann Rev Biophys Biophys Chem* 18:113–136

Michelangeli F, Grimes EA, East JM, Lee AG (1991): Effects of phospholipids on the function of $(Ca^{2+}-Mg^{2+})$-ATPase. *Biochemistry* 30:342–351

Morrison C, Bloom M (1994): Orientation dependence of ^2H nuclear magnetic resonance spin-lattice relaxation in phospholipid and phospholipid:cholesterol systems. *J Chem Phys* 101:749–763

Nagle JF (1993): Area/lipid of bilayers from NMR. *Biophys J* 64:1476–1481

Nagle JF, Wiener MC (1988): Structure of fully hydrated bilayer dispersions. *Biochim Biophys Acta* 942:1–10

Navarro J, Toivio-Kinnucan M, Racker E (1984): Effect of lipid composition on the calcium/adenosine 5′-triphosphate coupling ratio of the Ca^{2+}-ATPase of sarcoplasmic reticulum. *Biochemistry* 23:130–135

Newton AC (1993): Interactions of proteins with lipid headgroups: lessons from protein kinase C. *Ann Rev Biophys Biomol Struct* 22:1–25

Nordio PL, Segre U (1979): Rotational dynamics. In: *The Molecular Physics of Liquid Crystals*, Luckhurst GR, Gray GW, eds. New York: Academic Press

Olsson U, Wennerström H (1994): Globular and bicontinuous phases of nonionic surfactant films. *Adv Coll Interf Sci* 49:113–146

Pastor RW, Venable RM, Karplus M (1991): Model for the structure of the lipid bilayer. *Proc Natl Acad Sci USA* 88:892–896

Pastor RW, Venable RM, Karplus M (1988): Brownian dynamics simulation of a lipid chain in a membrane bilayer. *J Chem Phys* 89:1112–1127

Pearlman JD, Zajicek J, Merickel MB, Carman CS, Ayers CR, Brookeman JR, Brown MF (1988): High-resolution ^1H NMR spectral signature from human atheroma. *Magn Reson Med* 7:262–279

Peters GH, Toxvaerd S, Larsen NB, Bjørnholm T, Schaumburg K, Kjaer K (1995): Structure and dynamics of lipid monolayers: implications for enzyme catalysed lipolysis. *Struct Biol* 2:395–401

Petersen NO, Chan SI (1977): More on the motional state of lipid bilayer membranes: interpretation of order parameters obtained from nuclear magnetic resonance experiments. *Biochemistry* 16:2657–2667

Pfeiffer W, König S, Legrand JF, Bayerl T, Richter D, Sackmann E (1993): Neutron spin echo study of membrane undulations in lipid multibilayers. *Europhys Lett* 23: 457–462

Prosser RS, Daleman SI, Davis JH (1994): The structure of an integral membrane peptide: a deuterium NMR study of gramicidin. *Biophys J* 66:1415–1428

Rajamoorthi K, Brown MF (1991): Bilayers of arachidonic acid containing phospholipids studied by ^{2}H and ^{31}P NMR spectroscopy. *Biochemistry* 30:4204–4212

Recktenwald DJ, McConnell HM (1981): Phase equilibria in binary mixtures of phosphatidylcholine and cholesterol. *Biochemistry* 20:4505–4510

Rommel E, Noack F, Meier P, Kothe G (1988): Proton spin relaxation dispersion studies of phospholipid membranes. *J Phys Chem* 92:2981–2987

Rose ME (1957): *Elementary Theory of Angular Momentum*. New York: Wiley

Rotenberg M, Bivins R, Metropolis N, Wooten JK Jr (1959): *The 3-j and 6-j Symbols*. Cambridge, MA: The Technology Press of the Massachusetts Institute of Technology

Roth M, Lewit-Bentley A, Michel H, Deisenhofer J, Huber R, Oesterhelt D (1989): Detergent structure in crystals of bacterial photosynthetic reaction centre. *Nature* 340:659–662

Salmon A, Dodd SW, Williams GD, Beach JM, Brown MF (1987): Configurational statistics of acyl chains in polyunsaturated lipid bilayers from ^{2}H NMR. *J Am Chem Soc* 109:2600–2609

Saupe A (1964): Kernresonanzen in kristallinen Flüssigheiten und kristallin-flüssigen Lösungen. *Z Naturforsch A* 19:161–171

Schindler H, Seelig J (1975): Deuterium order parameters in relation to thermodynamic properties of a phospholipid bilayer. A statistical mechanical interpretation. *Biochemistry* 14:2283–2287

Scott HL (1986): Monte Carlo calculations of order parameter profiles in models of lipid-protein interactions in bilayers. *Biochemistry* 25:6122–6126

Scott HL, Kalaskar S (1989): Lipid chains and cholesterol in model membranes: a Monte Carlo study. *Biochemistry* 28:3687–3691

Seddon JM (1990): Structure of the inverted hexagonal (H_{II}) phase, and non-lamellar phase transitions of lipids. *Biochim Biophys Acta* 1031:1–69

Seelig J (1977): Deuterium magnetic resonance: theory and application to lipid membranes. *Quart Rev Biophys* 10:353–418

Seelig J, Seelig A (1980): Lipid conformation in model membranes and biological membranes. *Quart Rev Biophys* 13:19–61

Seelig J, Macdonald PM, Scherer PG (1987): Phospholipid headgroups as sensors of electric charge in membranes. *Biochemistry* 26:7535–7541

Sefcik MD, Schaefer J, Stejskal EO, McKay RA, Ellena JF, Dodd SW, Brown MF (1983): Lipid bilayer dynamics and rhodopsin-lipid interactions: new approach using high-resolution solid-state ^{13}C NMR. *Biochem Biophys Res Commun* 114:1048–1055

Separovic F, Gehrmann J, Milne T, Cornell BA, Lin SY, Smith R (1994): Sodium ion binding in the gramicidin A channel. Solid-state NMR studies of the tryptophan residues. *Biophys J* 67:1495–1500

Shinitzky M, Barenholz Y (1978): Fluidity parameters of lipid regions determined by fluorescence polarization. *Biochim Biophys Acta* 515:367–394

Siegel DP, Green WJ, Talmon Y (1994): The mechanism of lamellar-to-inverted hexagonal phase transitions: a study using temperature-jump cryo-electron microscopy. *Biophys J* 66:402–414

Siminovitch DJ, Ruocco MJ, Olejniczak ET, Das Gupta SK, Griffin RG (1988a): Anisotropic ^{2}H-nuclear magnetic resonance spin-lattice relaxation in cerebroside- and phospholipid-cholesterol bilayer membranes. *Biophys J* 54:373–381

Siminovitch DJ, Wong PTT, Berchtold R, Mantsch HH (1988b): A comparison of the effect of one and two mono-unsaturated acyl chains on the structure of phospholipid bilayers: a high pressure infrared spectroscopic study. *Chem Phys Lipids* 46:79–87

Slichter CP (1990): *Principles of Magnetic Resonance*, 3rd Ed. Heidelberg: Springer-Verlag

Smith R, Thomas DE, Separovic F, Atkins AR, Cornell BA (1989): Determination of the structure of a membrane-incorporated ion channel. Solid-state nuclear magnetic resonance studies of gramicidin A. *Biophys J* 56:307–314

Smith SO, Griffin RG (1988): High resolution solid-state NMR of proteins. *Ann Rev Phys Chem* 39:511–535

Smith SO, Courtin J, de Groot H, Gebhard R, Lugtenberg J (1991): ^{13}C Magic-angle spinning NMR studies of bathorhodopsin, the primary photoproduct of rhodopsin. *Biochemistry* 30:7409–7415

Smith SO, Hamilton J, Salmon A, Bormann BJ (1994a): Rotational resonance NMR determination of intra- and intermolecular distance constraints in dipalmitoylphosphatidylcholine bilayers. *Biochemistry* 33:6327–6333

Smith SO, Jonas R, Braiman M, Bormann BJ (1994b): Structure and orientation of the transmembrane domain of glycophorin in lipid bilayers. *Biochemistry* 33:6334–6341

Smith SO, Palings I, Miley ME, Courtin J, de Groot H, Lugtenberg J, Mathies RA, Griffin RG (1990): Solid-state NMR Studies of the mechanism of the opsin shift in the visual pigment rhodopsin. *Biochemistry* 29:8158–8164

Smith SO, Palings I, Copié V, Raleigh DP, Courtin J, Pardoen JA, Lugtenberg J, Mathies RA, Griffin RG (1987): Low-temperature solid-state ^{13}C studies of the retinal chromohore in rhodopsin. *Biochemistry* 26:1606–1611

Söderman O (1986): The interaction constants in ^{13}C and ^{2}H nuclear magnetic resonance relaxation studies. *J Magn Reson* 68:296–302

Stohrer J, Gröbner G, Reimer D, Weisz K, Mayer C, Kothe G (1991): Collective lipid motions in bilayer membranes studied by transverse deuteron spin relaxation. *J Chem Phys* 95:672–678

Szabo A (1984): Theory of fluorescence depolarization in macromolecules and membranes. *J Chem Phys* 81:150–167

Tanford C (1980): *The Hydrophobic Effect,* 2nd Ed. New York: John Wiley

Thurmond RL, Dodd SW, Brown MF (1991): Molecular areas of phospholipids as determined by ^{2}H NMR spectroscopy: comparison of phosphatidylethanolamines and phosphatidylcholines. *Biophys J* 59:108–113

Thurmond RL, Lindblom G, Brown MF (1993): Curvature, order, and dynamics of lipid hexagonal phases studied by deuterium NMR spectroscopy. *Biochemistry* 32:5394–5410

Thurmond RL, Lindblom G, Brown MF (1990): Influences of membrane curvature in lipid hexagonal phases studied by deuterium NMR spectroscopy. *Biochem Biophys Res Commun* 173:1231–1238

Thurmond RL, Otten D, Brown MF, Beyer K (1994): Structure and packing of phosphatidylcholines in lamellar and hexagonal liquid-crystalline mixtures with a nonionic detergent: a wide-line deuterium and phosphorus-31 NMR study. *J Phys Chem* 98:972–983

Torchia DA, Szabo A (1982): Spin-lattice relaxation in solids. *J Magn Reson* 49:107–121

Trouard TP, Alam TM, Brown MF (1994): Angular dependence of deuterium spin-lattice relaxation rates of macroscopically oriented dilaurylphosphatidylcholine in the liquid-crystalline state. *J Chem Phys* 101:5229–5261

Trouard TP, Alam TM, Zajicek J, Brown MF (1992): Angular anisotropy of ^{2}H NMR spectral densities in phospholipid bilayers containing cholesterol. *Chem Phys Lett* 189:67–75

Ulrich AS, Heyn MP, Watts A (1992): Structure determination of the cyclohexene ring of retinal in bacteriorhodopsin by solid-state deuterium NMR. *Biochemistry* 31:10390–10399

Ulrich AS, Wallat I, Heyn MP, Watts A (1995): Re-orientation of retinal in the M-photointermediate of bacteriorhodopsin. *Struct Biol* 2:190–192

Ulrich AS, Watts A, Wallat I, Heyn MP (1994): Distorted structure of the retinal chromophore in bacteriorhodopsin resolved by ^{2}H-NMR. *Biochemistry* 33:5370–5375

van der Ploeg P, Berendsen HJC (1982): Molecular dynamics simulation of a bilayer membrane. *J Chem Phys* 76:3271–3276

Venable RM, Zhang Y, Hardy BJ, Pastor RW (1993): Molecular dynamics simulations of a lipid bilayer and of hexadecane: an investigation of membrane fluidity. *Science* 262:223–226

Vist MR, Davis JH (1990): Phase equilibria of cholesterol/dipalmitoylphosphatidylcholine mixtures: ^{2}H nuclear magnetic resonance and differential scanning calorimetry. *Biochemistry* 29:451–464

Vold RR, Vold RL (1991): Deuterium relaxation in molecular solids. *Adv Magn Opt Reson* 16:85–171

Wennerström H, Lindman B, Söderman O, Drakenberg T, Rosenholm JB (1979): ^{13}C magnetic relaxation in micellar solutions. Influences of aggregate motion on T_1. *J Am Chem Soc* 101:6860–6864

Wiedmann TS, Pates RD, Beach JM, Salmon A, Brown MF (1988): Lipid-protein interactions mediate photochemical function of rhodopsin. *Biochemistry* 27:6469–6474

Wiener MC, White SH (1992): Structure of a fluid dioleoylphosphatidycholine bilayer determined by joint refinement of x-ray and neutron diffraction data. III. Complete structure. *Biophys J* 61:434–447

Williams GD, Beach JM, Dodd SW, Brown MF (1985): Dependence of deuterium spin-lattice relaxation rates of multilamellar phospholipid dispersions on orientational order. *J Am Chem Soc* 107:6868–6873

Wittebort RJ, Olejniczak ET, Griffin RG (1987): Analysis of deuterium magnetic resonance line shapes in anisotropic media. *J Chem Phys* 86:5411–5420

Wüthrich K (1989): Protein structure determination in solution by nuclear magnetic resonance spectroscopy. *Science* 243:45–50

Yeagle PL, Kelsey D (1989): Phosphorus NMR studies of lipid-protein interactions: human erythrocyte glycophorin and phospholipids. *Biochemistry* 28:2210–2215

Zaccai G, Büldt G, Seelig A, Seelig J (1979): Neutron diffraction studies on phosphatidylcholine model membranes. II. Chain conformation and segmental disorder. *J Mol Biol* 134:693–706

Part III

Small Molecules and Peptides in Biomembranes

The first two sections complete the description of "neat" biomembrane structure and dynamics, while the next two sections present insights into how small molecules, peptides, and proteins interact with lipid assemblies. This section explores this research area from both the experimental and theoretical point of view. Studying peptide/lipid interactions using molecular simulations is limited by the relaxation time scales of biomembranes; thus, this area benefits greatly from a combined experimental/theoretical effort. The first chapter (Chapter 8) of Section III describes the diffusion of small molecules in bilayers. This chapter by Terry Stouch and Donna Bassolino, emphasizing the role that concerted movements can play in allowing the rapid transit of small molecules within a bilayer structure, provides a good example of how computer simulation can be used to gain information about aspects of lipid membranes that are not easily accessible experimentally. In Chapter 9, Huey Huang reviews the problem of membrane active peptides, with a particular attention to the biological function of small amphiphilic helix-forming peptides. In Chapter 10, Randy Ketchem, Benoît Roux, and Tim Cross describe the combined use of solid-state NMR and theoretical techniques to elucidate the three-dimensional structure of an integral membrane peptide. Energy restraints based on observed solid-state NMR properties are introduced in Monte Carlo simulations to refine the structure of the gramicidin channel in a lipid bilayer. In Chapter 11, K. V. Damodaran and Ken Merz examine the influence of small peptides on the structure of a bilayer. In particular the affect peptides have on alkyl chain order parameters and head-group hydration is emphasized.

8

Movement of Small Molecules in Lipid Bilayers: Molecular Dynamics Simulation Studies

TERRY R. STOUCH AND DONNA BASSOLINO

Abstract

To assist in understanding the mechanism of membrane permeation, the movement of four different molecules within hydrated lipid bilayer membranes were studied via over 15 nanoseconds of atomic-level molecular dynamics simulation. In particular, the simulations were used to explain the anomolously high rate of permeation seen for small molecules. These simulations support the hypothesis that the rate of diffusion of small solutes is enhanced because they can move rapidly within and jump between spontaneously arising voids. The enhanced diffusion rate is greatest in the bilayer center where the voids are most frequently found and of the largest size. Molecules the volume of benzene or smaller experience this enhanced movement, however larger molecules, those the size of adamantane or larger, do not. The details of the diffusional mechanisms of these molecules are discussed. The role of hydrogen bonding for the interactions between drugs (a nifedipine analog) and membranes is discussed.

Introduction

A key role of biomembranes is to limit the passage of small molecules between aqueous compartments. A key problem in drug design is bioavailability which is largely dependent on membrane permeability. Experiments in this area can be difficult to perform and interpret. Additionally, such experiments, of course, require the availability of the compounds in question. Yet, from a drug design perspective, it is most desirable to predict a molecule's properties prior to synthesis. Hence, a computational method to predict membrane permeation rates of small molecules would be of value. Additionally, an understanding of the

Biological Membranes
K. Merz, Jr. and B. Roux, Editors
© Birkhäuser Boston 1996

interactions between small solutes and membranes is directly relevant to understanding the interactions between protein side chains with membranes. Although the movement of many ions or small molecules, such as sugars, is mediated by proteins, the passage of many other molecules such as water, fatty acids, and bile acids, is known to occur passively (Anel et al, 1993; Kamp and Hamilton, 1993). Also, the passive partitioning of drugs into membranes and their subsequent location and movement within membranes is thought to be important to their interaction with membrane-bound receptors (Mason et al, 1991; Xiang and Anderson, 1994).

Despite extensive research, the process, energetics, and kinetics of membrane permeation is still unclear. Permeation rate is often estimated through correlations with the partition coefficients between water and an organic phase (Overton, 1899). However, this approach suffers from several weaknesses. For example, although the field of medicinal chemistry has adopted n-octanol as a standard, there is considerable unresolved debate about just which is the best organic phase to use (Diamond and Katz, 1974; Dix et al, 1974; Lieb and Stein, 1971). It is likely that this will be dependent on the nature of the solutes under investigation and on the nature of the particular membrane in question. In those instances where these correlations can be determined, they extend only to the total free energy of partitioning; often the entropic and enthalpic contributions to partitioning from water are quite different between membrane and a neat organic phase (Diamond and Katz, 1974), indicating that the process of partitioning is different between the systems. These correlations also fail to account for the internal structure of membranes and the details of the process of diffusion of small molecules within membranes. Even within homologous series of molecules, there is a size dependence to permeation which ruins the correlations with partition coefficients by altering the rate of diffusion within the membrane for small molecules (Bassolino et al, 1993; Leib, 1986). The enthalpic and entropic contributions to the free energy of partitioning are also affected by the curvature of the lipid layer. The partitioning of hydrophobic solutes into bilayers is entropically driven for planar bilayers, as expected; however it becomes enthalpically driven for highly curved vesicles (Beschiaschvili and Seelig, 1992).

It is important to understand that even when correlations are seen, these are more a convenience than they are physically informative. It is quite clear, both from experiment and simulation, that the membrane exhibits a structure that would not be expected at an interface between an organic solvent and water. The gross features of a bilayer are clear: the polar regions of the lipids that interface with water and the hydrocarbon interior. However, the structure of the bilayer has a much finer detail than this and has many more regions. Although experiment shows much of this, it is not always fully appreciated. As will be shown here, simulations have additionally helped to define these regions. Perhaps the best known experimental observable that clearly shows this is the NMR derived order parameter which indicates the degree of order of the lipid hydrocarbon chains at transverse positions within the bilayer (Seelig and Seelig, 1974). This makes it clear that lipid bilayers are more ordered near the water/lipid interface and become progressively less ordered as the bilayer center is approached. Even in the center, however, the order parameters do not suggest the complete disorder expected in fluid hydrocarbons. Also, X-ray scattering clearly shows an electron density profile that is substantially nonhomogeneous; it is high near the headgroups, drops substantially in the upper regions of the hydrocarbon chains, and exhibits a characteristic drop at the very center of the bilayer (Franks, 1976; Nelander and Blaurock, 1978). Further, it is clear that bilayer properties are in part derived from water molecules at the water/bilayer interface, which are oriented by the lipid headgroups (Alper et al, 1993b; Gawrisch et al, 1992). As will be shown here, these features of bilayer structure are important to the passage of small molecules.

Some have tried to predict diffusion rates within membranes through calculations involving measured permeation rates and partition coefficients (Miller, 1986). These suffer from the fact that, as a heterogeneous construct, different regions of the bilayer might have very different affinities for a solute. Attempts have been made to account for this by employing the concept of membrane compartments. However, although regions (polar headgroup, hydrocarbon interior) exist in membranes, the boundary between them is gradual, and estimates of the thickness of these compartments is difficult to determine and will vary from membrane to membrane as well as with different physical states of the membrane (Miller, 1986). Also, the heterogeneity of the membrane means that identical partition coefficients for two solutes does not mean that they interact with the membrane in the same way. Interactions between the solutes and the different regions of the bilayer can be quite different. Additionally, the application of simple relationships to calculate solute diffusion from experimentally determined permeation rates also suffers from the fact that permeation and partitioning experiments are not easy to perform.

A simpler and more direct approach to including some aspect of the details of membrane structure into predictions of permeation has been to try to correct the correlations between permeation and partition coefficients (Walter and Gutknecht, 1986). As previously mentioned, it is known from experiment that small molecules exhibit anomalously high rates of permeation relative to those of larger molecules (Lieb and Stein, 1969; Leib, 1986; Walter and Gutknecht, 1986) and so deviate from predictions made by Overton's rule. Leib and Stein proposed that, as in soft polymers, diffusion might be accelerated by the presence of dynamical voids on the order of the size of the solute, or larger (Lieb and Stein, 1971; Lieb and Stein, 1969; Trauble, 1971). The permeation rates of these small solutes would be enhanced by their ability to move within and between these voids. Support for this was provided by Walter and Guknecht who found that the permeability of these small solutes was inversely proportional to their size (Walter and Gutknecht, 1986). If the hole-jumping hypothesis was correct, one would expect this; the probability of occurrence of a void might also be expected to be inversely proportional to its size. It is additionally interesting that ^{13}C NMR relaxation experiments suggested that diffusion within the bilayer center is faster than nearer the headgroups (Dix et al, 1978). Coupled with density profiles and the NMR and theoretical work of Dill (DeYoung and Dill, 1988; Marqussee and Dill, 1986), this suggested that the hole-jumping mechanism, if in fact it existed, might be position dependent.

The anomalous rate of permeation of small molecules, the genesis of this phenomenon, and the hole-jumping hypothesis, will be the focus of this monograph. We will discuss the results of atomic-level molecular dynamics simulations of solutes in lipid bilayers that help to verify this hypothesis, explain the experimental results and the atomic-level phenomena that are responsible for them, and provide additional details of the process.

In short, we will review results that clearly support the hypothesis that molecules of a small size move through the bilayer by a mechanism unavailable to larger molecules. More specifically, free voids exist in the hydrocarbon interior of the bilayer that are commonly of a size large enough to engulf benzene. These are most common in the bilayer center. Molecules this size and smaller move more rapidly in the bilayer center than closer to the headgroups by a factor of about 3 to 4. They do so in part because they can travel within the voids, but also because they experience fairly large movements (jumps) within a short period of time (about 6–8 Å in 2.5 ps). Molecules too large to fit within these voids show consistent rates of diffusion regardless of position.

Molecular Dynamics (MD) Simulations of Lipid Bilayers

Clearly, there is much that is unknown about solute permeation and a number of computational approaches have been applied to understand it. In addition to the analytical and correlational work mentioned above, several attempts have been made to apply more detailed, simulation-based approaches to understanding this phenomenon including lattice-based approaches (Jorgensen et al, 1991a; Jorgensen et al, 1991b; Marqussee and Dill, 1986). In order to verify the theory and hypothesis suggested by experiment, as well as understand the details of the mechanism of movement of solutes at an atomic level, we employed molecular dynamics simulations of atomic level resolution models of hydrated lipid bilayers. We were studying a range of solutes of different sizes in order to understand the effects of size, shape, and chemical function on permeation (Alper and Stouch, 1995; Bassolino et al, 1995; Bassolino et al, 1993). The method of molecular dynamics was employed because the properties of membranes—more so than those of any other biological construct—are dependent on membrane fluidity, a property that is tightly regulated by cells. This fluidity and motion governs the passage of solutes through lipid bilayer membranes, a fact of central importance to our topic.

This was the first work to report atomic-level computational attempts to understand membrane permeation. The only previously-reported closely related studies were MD simulations of the diffusion and solvation of monatomic or diatomic gases in simple, homogeneous polymers or zeolites (Müller-Plathe et al, 1993; Nicholas et al, 1993; Takeuchi and Okazaki, 1990), but nothing had been reported on heterogeneous biological membranes. We applied similar MD methods to studies of the diffusion of benzene (Bassolino et al, 1995; Bassolino et al, 1993), nifedipine, a calcium channel blocker (Alper and Stouch, 1995), adamantane, and methane within phospholipid bilayers. We found that movement within the membrane is a complex event that is dependent on the size and complexity of the transiting molecule and on its position within the membrane or its aqueous environment. In all cases, we find that the headgroup interfacial region presents a barrier to movement between water and the membrane.

Previous studies in our own lab (Stouch, 1993; Stouch et al, 1994) and those of others (Damodaran et al, 1992; Edholm and Johansson, 1987; Egberts and Berendsen, 1988; Jonsson et al, 1986; Pastor et al, 1991; Scott and Kalaskar, 1989; Van der Ploeg and Berendsen, 1983; Watanabe et al, 1988), have demonstrated that classical MD simulations employing empirical atomic/molecular force fields are able to duplicate a wide range of experimentally observable, structural and dynamical properties for neat lipid bilayers. Electron density, the ordering of hydrocarbon chains, bilayer thickness, distribution of atomic positions throughout the bilayer, lateral diffusion rates of lipid molecules, rates of hydrocarbon chain conformational intercoversion, and residence times of water about lipid headgroups are only some of the properties that can be duplicated by simulation. Even subtle aspects of structure, such as the ordering of water at the membrane/water interface, have been reproduced (Alper et al, 1993b).

It is appropriate to mention something about the requirements of these simulations. It goes without saying that the quality of the empirical force field that determines interatomic interactions is a central component in obtaining such good results (Stouch et al, 1991; Williams and Stouch, 1993). However, the details of the simulation itself are of equal importance. Of particular importance to the quality of the reproduction of experimental observables are the length of the simulation and the treatment of long-range interactions. If a simulation is true to the physics of the problem, the length of the simulation must match the timescale of the phenomena under investigation or the phenomena influencing those processes. For example, as we will show, the rate of the isomerization of the torsions in the hydrocarbon chains is in some cases hundreds of picoseconds. Because of this, simulations must be conducted for lengths of time several times greater than this in order to sample alternate configurations. In accord with the varying hydrocarbon torsion angles and the consequent hydrocarbon chain movement, we previously showed that duplication of NMR derived order parameters of the chains requires as much as one nanosecond to reproduce (Stouch et al, 1994). The simulations discussed here were conducted for a minimum of a nanosecond. Particularly sobering in light of the effort required for even multinanosecond simulations are studies by De Loof who has shown that tens of nanoseconds of simulation are required to duplicate properties dependent on whole molecule rotation of the lipids, a physically reasonable result (De Loof et al, 1991). These highly charged and pressure-dependent lipid assembles are particularly sensitive to the treatment of atomic and molecular interactions that occur over large distances. It has been shown that to obtain converged properties and duplicate properties of the membrane/water interfaces, the electrostatic forces are important to a distance of 30 Å (Alper et al, 1993a). Note that even the dispersive part of the Lennard-Jones function can be important beyond 15 Å for duplicating properties of pressure and some dynamical properties. In the studies reported here, interactions of the polar solutes were calculated to 30 Å.

The agreement with experiment seen for pure bilayers provided confidence in the physical validity of the bilayer simulations, and we extended them to study the movement of solutes within the bilayer and the partitioning of solutes between water and lipid bilayers. Benzene, a molecule whose size is of particular interest, has been most extensively studied. The movement within the bilayer of nifedipine, a calcium channel blocker, will also be discussed at some length. Additionally, this molecule's rich functionality provides it with properties and interactions lacking in the featureless benzene. In order to complete an understanding of the effects of size on membrane permeation, the movement of methane and adamantane will also be discussed.

Important Bilayer Features

Before the movement of the solutes themselves are discussed, we will note the features of the bilayer salient to this mechanism of movement: (1) the density distribution in the bilayer; (2) the size and distribution of empty voids in the bilayer, which are reflected by the density distribution; (3) the rate of torsional isomerization of the lipid hydrocarbon chains; and (4) and the position of hydrogen bonding groups of the lipids and water.

First, as mentioned above, the bilayer structure has a distinctly lower density at its center than it does in the intermediate regions of the hydrocarbon chains and much less than in the headgroup regions, where polar headgroups and water pack tightly (Figure 1). This lower density is due to the orientation of the lipid molecules (which is imposed by the headgroup/water interactions), the thermal energy of the bilayer, and the considerable conformational freedom of the hydrocarbon chains available due to this energy. This lowered density is a manifestation of a substantial amount of free volume. As we have shown (Stouch et al, 1994), at any one instant in time, this free volume is not evenly distributed, but is localized as voids, which are frequent and large in the bilayer center (Figure 2). These voids are frequently many tens and perhaps hundreds of cubic angstroms, large enough to easily accommodate small molecules. Their distribution and size varies somewhat with temperature, decreasing in size and becoming more probable in regions nearer the headgroups as the temperature increases.

The rate of the interconversion of conformers by rotation about single bonds in the hydrocarbon chains is another factor that contributes to the diffusion of small molecules. It is well established, by both experiment and simulation (Stouch et al, 1994), that the rates of interconversion are on the order of several hundreds of picoseconds in the more ordered region high in the chains but

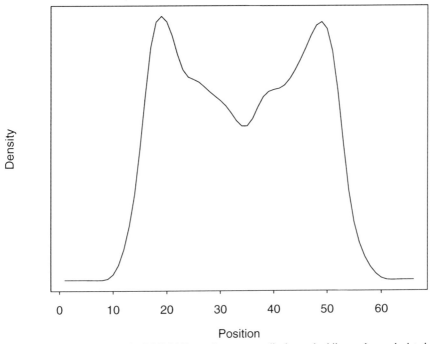

Figure 1. Electron density of a DMPC bilayer shown perpendicular to the bilayer plane calculated from an MD simulation. Only the lipid atoms were included (water was excluded). The peaks represent the two phosphate regions.

decrease to several tens of picoseconds close to the terminal methyl groups of the chains.

The last bilayer feature that needs to be addressed is the positioning of water and hydrogen bonding groups. The carbonyl groups of acylglycerophospholipids are those groups of the lipid molecules closest to the hydrocarbon region that are able to make hydrogen bonds. These, however, make hydrogen bonds to water molecules which must be considered as integral parts of lipid bilayer structures. These water molecules extend hydrogen bonding capability deeper into the bilayer. Additionally, in contrast to the carboxyl groups, they serve not only as hydrogen bond acceptors, but also as donors (Figure 3).

MD Simulations of Small Solutes

The effects of these bilayer features on the movement of benzene will first be examined. Benzene is of particular interest because it is just of a size to fit within the larger of the more frequent voids. Also, it is compact, without any conformational degrees of freedom, and featureless, with no long-range electrostatic interactions or hydrogen-bonds. Although most in the biomolecular

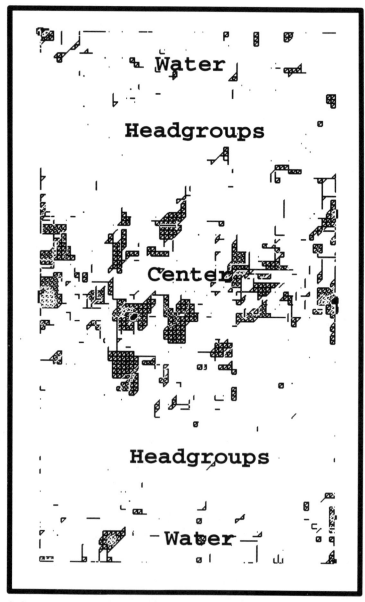

Figure 2. Cross section of a DMPC bilayer showing free voids. This is a time composite plot, five different times (separated by 50 ps) are represented by shades of gray.

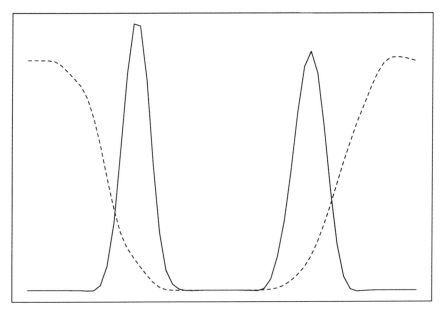

Figure 3. Transverse probability density through a DMPC bilayer for fatty acyl carbonyl carbon atoms (solid) and water (dotted) showing substantial overlap of the distributions.

simulation field would consider it to be a relatively small molecule, it should be noted that at the time it was done, it was the largest solute whose diffusion had been studied in an MD simulation in the condensed phase. The next largest was propane in a zeolite (Nicholas et al, 1993); others were monoatomic or diatomic gases in polymers or biological molecules.

A number of MD simulations were conducted with benzene as a solute in hydrated DMPC bilayers with ratios of 1:36 to 1:9 benzene: DMPC. Over four nanoseconds of simulation were conducted at 320 K (about 23 degrees above the gel to L-α phase transition of hydrated DMPC) and over a nanosecond of simulation was done at each of 310 K, 330 K, and 340 K of a 1:9 benzene: DMPC system in order to study the temperature dependence of benzene diffusion (Bassolino et al, 1993; Bassolino et al, 1995).

The most important aspect of any simulation, prior to its use to understand new phenomena, is its ability to predict known phenomena. We mentioned above the ability of simulations to reproduce many properties of pure lipid bilayers. In addition, the calculated diffusion constants and rotational correlation times for benzene were in good agreement with ESR studies of solutes of similar size and shape.

In addition to showing agreement with experiment, the Einsteinian diffusion coefficients were different in different regions of the bilayer, in agreement with NMR experiments (Dix et al, 1978). The rate of diffusion in the bilayer center, where the size and occurrence of voids are both greatest, was 3–4 times the rate high in the hydrocarbon chains (nearer the headgroups) where voids of any size seldom occur.

The difference between these values is significant; however, the Einstein diffusion coefficient (which has become almost a standard for analyzing molecular movement) is not always applicable. Also, it is a macroscopic quantity derived from a line-fit to averages of the movements over time and so removes all detail from the observations. Further, although these values show that the rates of benzene diffusion are different in different regions of the bilayer, they do not tell us why and do not, in themselves, confirm or refute the hole-jumping hypothesis.

Luckily, atomic-level simulation allows us to examine these motions at whatever levels of detail we desire. Unfortunately, the complexity of the bilayer makes the movements complex and difficult to monitor. The solutes can move in any direction, and movements can be of a wide range of magnitudes. The kinetic energy of the solutes can change due to the thermal motion and elastic collisions with the lipid molecules. Also, the lipids themselves can move and, as will be discussed, this motion can give rise to new components in the movement of the solutes. A range of measures is used to understand and characterize some of the features of benzene movement, some borrowed from studies of gas diffusion in polymers.

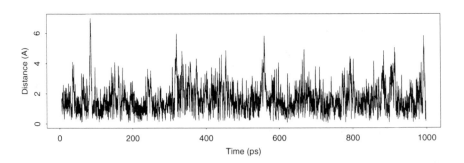

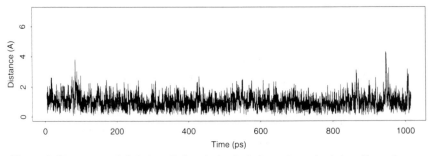

Figure 4. Distance traveled during all 5 ps intervals during a 1 ns simulation. Top, a benzene molecule localized near the bilayer center . Bottom a benzene molecule localized high in the hydrocarbon chains, near the headgroup.

Particularly illuminating are the magnitudes of the individual movements of the solutes over increasing intervals of time. At short times, 100 fs, the size of all movements are of very similar magnitude and appear to be dependent only on the system temperature and not on position within the bilayer. At 500 fs, however, some individual movements are much greater than others and, on average, the movements near to the center are greater than those near the headgroups (Figures 4 and 5). The differences in the magnitudes of individual movements as well as the difference between the average movements in the center in comparison to those in higher regions of the bilayer increases further with longer times. Substantial increases occur only up to about 5 ps, however, and the rate of increase in the magnitudes of individual movements slows considerably at longer times. The largest contiguous movements are on the order of 6–8 Å in about 2.5 ps.

Figure 4 shows that these jumps occur frequently in the bilayer center; however, they essentially never occur close to the headgroups. In addition to quantifying the average magnitude of solute movement, Figure 5 shows an interesting consequence of bilayer structure, the inflection at the midway point. The change in magnitude of the distance moved by the solutes is not gradual from the center to the headgroups, but seems to exhibit two distinct states, with a narrow region of variation in between. It should be noted that the order pa-

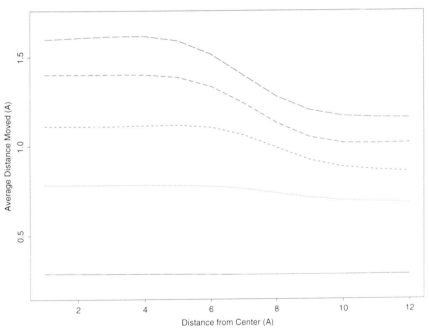

Figure 5. Average distance traveled by benzene vs. distance from the bilayer center. This shows the evolution of the differential in rates between the bilayer center and region near the headgroups. Different lines are for successively increasing times. Bottom to top: 0.1, 0.5, 1.0, 2.5, 5.0 ps.

rameters of the hydrocarbon chains, after a plateau in values in the upper 1/2 to 2/3 of the hydrocarbon chain, undergo a distinct, rapid reduction in magnitude close to the bilayer center. It seems likely that the causes of these phenomena are related.

Figure 6 (see color insert at the end of this chapter) shows a three-dimensional trace of one benzene molecule's trajectory over 1 nanosecond. Most of the motion occurs over short distances; the solute moves within a particular region or slowly works its way into closely adjacent regions. However, several distinct jumps can be seen between different parts of the trace, for example, between the orange region in the center and that to the right, and between the light green in the center and that on the left of the plot.

If the distributions of these motions are plotted (Figure 7), a bimodal distribution is seen for solutes in the center but not for those nearer the headgroups. This shows a small population of motions whose mean magnitude is over three times larger than that of the larger population common to all molecules regardless of position. It has been difficult to quantitate the exact contribution of this distribution to the overall rate of diffusion, but it seems likely that the enhancement in the rates of diffusion in the center are due largely to these additional types of motion, the jumps.

The jumps seem to be of several kinds. As seen in other systems, some of the jumps are coincident with an increase in kinetic energy of the solute, indicating an activated process. In other words, some energy barrier must be overcome in order for the jump to take place. This is not true for all jumps, however, and some jumps are essentially isoenergetic. Many, but not all, jumps are preceded by a conformational change in the surrounding lipid molecules. These changes move the positions of the lipids' hydrocarbon chains and create a pathway between voids in the bilayer or alter the shape and size of a void, allowing the solute to either move between voids or make larger unimpeded motions within the larger void, respectively. These motions could well be isoenergetic or could be preceded by an energy increase. Radial distribution functions of the chain carbon atoms have peaks at about 5–7 Å and 9–10 Å. Movement of a neighboring chain then could account for the observed 6–8 Å magnitude of the jumps.

The term jump suggests a continuous, unimpeded movement. From the trajectories, however, it is difficult to separate one continuous movement from several contiguous rapid movements, such as might occur if a solute bounced along the walls of a void in generally the same direction. It is worth noting, however, that our estimate of the length and time scale of a jump of about 6–8 Å in about 2.5 ps is close to the average velocities of the solutes. In other words, a solute travelling at its average velocity would have to move in one continuous movement in order to move a jump's distance in a jump's timescale.

The mechanism of diffusion for benzene has been fairly well characterized. For even smaller solutes, such as methane, a similar mechanism of motion is seen accompanied by the same 3 to 4-fold increase in diffusion rate in the center compared to that near the headgroups. At first, this appears substantially different from the ten times increase in rate seen by Marrink and Berendsen for

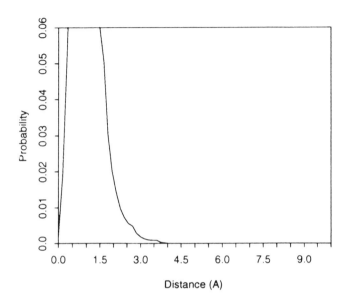

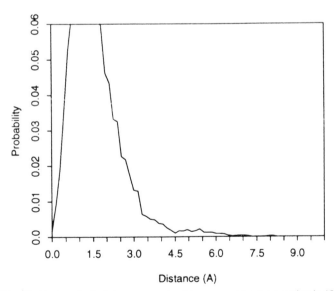

Figure 7. The distribution of all of the movements of benzene molecules occuring in 10 ps over a 1 ns trajectory. Bottom, the center of the bilayer. Note the additional distribution at about 6 Å. Top, near to the headgroups, no second distribution is seen.

water (Marrink and Berendsen, 1994), which is comparable in size to methane. It seems likely, however, that these results are not in disagreement since water's diffusion rate in the headgroups might be slowed by its hydrogen bonding with the headgroups and ester carbonyl groups. The effects of hydrogen bonding on solute movement will be discussed presently.

For small molecules, then, it seems clear that the hole-jumping occurs and the rate of diffusion of small molecules is accelerated in the center of the bilayer due to an additional component of motion, i.e., large, rapid movements within and between the voids known to occur there. These motions are often moderated by movements of the lipid molecules themselves, which are also faster in the center.

However, two questions still remain. First, is the hole jumping unique to small molecules? Second, provided that this is true, for what absolute size of a solute does this mechanism of diffusion become important?

MD Simulations of Large Solutes

An analog of nifedipine (Figure 8), which is a drug in wide-spread use because of its function as a calcium channel blocker, was simulated within the bilayer for over 4 ns. Study of this drug is of particular interest because a number of experimental studies have been done to localize it, or related molecules, in the bilayer (Bauerle and Seelig, 1991; Mason et al, 1990). This molecule is over three times more massive than benzene, and more chemically interesting due to the ability of its amine and ester groups to hydrogen bond. Along with its greater mass, it has a substantially greater volume than benzene. Of particular significance is that its volume is substantially larger than that routinely obtained by the voids in the bilayer. This is reflected by the rate of diffusion of the

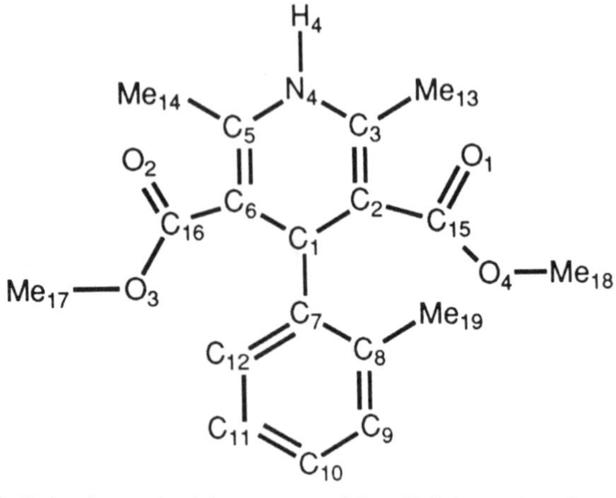

Figure 8. Molecular graph of the structure of the nifedipine analog discussed here.

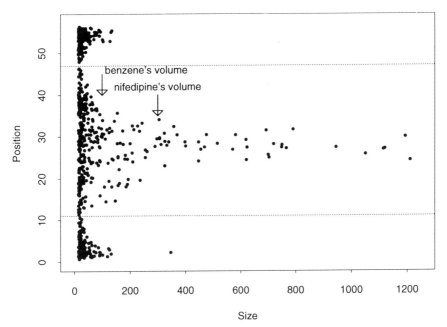

Figure 9. Transverse position through the bilayer (Å) vs. size (Å^3). Dotted lines note average positions of the lipid phosphate groups. The approximate sizes of benzene and nifedipine are indicated. This is a time composite plot and voids from 23 configurations are plotted.

nifedipine analog, which, unlike those of benzene and methane, is constant regardless of location. This larger molecule does not show the enhanced rate of diffusion in the bilayer center seen for the smaller molecules, presumably because it is too large to fit within the voids and add the process of jumps between these voids to its mechanism of diffusion (Figure 9).

This answers the first of the questions posed at the end of the last section. Small solutes do have an additional component of motion that is unavailable to larger solutes and that accelerates their rate of movement within the bilayer.

This drug analog shows other behavior also worth noting. In contrast to benzene, which appears to translate equally in all directions, the drug diffuses a factor of three to five times faster lateral to the bilayer than transverse to the bilayer. In some cases, particularly when the drug is near the headgroups, the translational movements correlate with those of neighboring lipid molecules. Although the diffusion coefficient for this molecule (about 1×10^{-6} cm^2/s) is, as expected, of the same magnitude but somewhat smaller than that of benzene (because of its increased mass), a substantial part of that diffusion is due to its movement parallel to the bilayer plane, whereas benzene seems to move isotropically. It appears that not only is the molecule too large to take advantage of the free voids in the bilayer, but also the molecule might be too large to freely move through and between the lipid molecules. Instead, its movement is dependent partly on the movement of the lipid molecules themselves. In that the lateral diffusion rates of the lipid molecules (approximately 10^{-8} to 10^{-7} cm^2/s) are at least one order of magnitude slower than that of the solute, the solute's mechanism of diffusion must also have a faster component. However, this does not appear to depend on its transverse position.

In our MD simulation, we found that the nifedipine analog rotated at different rates in different regions. Benzene, however, showed very similar rotational characteristics regardless of location. Presumably, once situated in a void of sufficient size, the symmetry of benzene allows it to tumble freely, at least within the plane of its ring. The drug analog, as expected, rotated more slowly than benzene when in the bilayer center. However, near the headgroups, it exhibited only small reorientations in position and could not be observed to rotate during 2 ns of residence at the lipid/water interface. A closer examination showed clearly that this slowed motion was due, at least in part, to hydrogen bonding between the drug and, not only the lipid molecules, but also the water molecules at the interface. The molecule took part in a hydrogen bonding network that propagated from the drug into and through the water and polar headgroups. In fact, much of the reorientation that occurred at the interface optimized the hydrogen bonding interactions. This molecule has one hydrogen bond donor and two carbonyl oxygens that can act as acceptors (two ester ether oxygens could hydrogen bond, but less frequently). The donor seemed to preferentially form hydrogen bonds with lipid fatty ester carbonyl oxygen atoms, although it transiently hydrogen bonded with phosphoether oxygens in the phosphate groups of the lipid molecules and was often surrounded by several water molecules. For a substantial part of the simulation, the drug's lateral movement was matched by that of a lipid molecule with which its amine function hydrogen bonded with one or both of the lipid's carbonyl oxygen atoms. The drug reoriented itself during the simulation to tilt in such a way that its carbonyl oxygens could make favorable hydrogen bonding contacts with the only other donor in the system, water. If the drug oriented perpendicular to the bilayer with its amine function toward the water, the carbonyl oxygen atoms would be too deep in the hydrocarbon region of the bilayer to make any hydrogen bonds. However, one carbonyl group can be made accessible to solvent if the molecule tilts. When exposed in this way to the water, the single bonds within the ester side chains of the drug rotated much more freely than when they interacted only with the lipid molecules (such as when the drug analog was in the bilayer center), suggesting that packing interactions with the lipid molecules were tight.

Although the molecule formed hydrogen bonds with the water molecules, both as donor and acceptor, and participated in hydrogen bonding networks, it is worth noting that individual water molecules moved in and out of this relationship and resided for only a few tens of picoseonds. When an individual molecule moved away from the drug, it was replaced by another molecule in such a way as to give a consistent average hydrogen bonding environment whose lifetime was longer than that of the participation of an individual water molecule.

This is in accord with other studies of water structure about the lipid headgroups. In particular, this is similar to the preferred orientation of water that we see at the membrane/water interface, which is due not to a static population of waters but to a dynamic interface wherein when one oriented water molecule leaves its position, it is replaced by another that takes on a similar orientation (Alper et al, 1993).

Refinement of the Size Estimate

Our studies show that benzene is of a size that exhibits enhanced rates of diffusion due to multiple mechanisms of diffusion, and in particular, because of its ability to make use of the peculiarities of bilayer structure. Nifedipine would appear to be too large to do this. However, nifedipine also is irregularly shaped and is capable of hydrogen bonding, both of which would alter its motions. In order to remove these latter features from consideration, and also to further refine the relationship between molecular volume and enhanced diffusion rates, we also conducted studies of adamantane (Figure 10) (see color insert at the end of this chapter). This essentially featureless, spherical molecule is substantially smaller than nifedipine and about twice benzene's volume. Four nanoseconds of simulation within the bilayer showed similar results to those of the nifedipine analog—the same rate of diffusion regardless of location. This suggests that the size beyond which a molecule can no longer show the enhanced diffusion rate is between that of benzene and that of adamantane. This is in agreement with our studies of void size, which show voids of the size of benzene reasonably frequently but few voids substantially larger than this.

Effects of Solutes on Lipid Bilayer Structure

It is important to note that the presence of the solutes had only small effects on the bilayer structure itself. Early theories of anesthesia stated that anesthetics elicited their effects by swelling biological membranes, thus increasing the capacitance of neural membranes and decreasing the effect of the transbilayer potential. That theory was disproven by X-ray scattering studies that showed no appreciable swelling by clinical concentrations of anesthetics (Franks and Lieb, 1979; Turner and Oldfield, 1979). White and coworkers subsequently showed that even very high concentrations of molecules as large as hexane have little effect on the gross morphology of lipid bilayers (Jacobs and White, 1984). In all of our studies of solutes, including a number not mentioned here, we see no change in bilayer thickness. This is despite the fact that this property is fairly responsive to simulation conditions and can vary significantly, by several angstroms (Stouch, 1993).

However, NMR studies have shown slight decreases in the order parameters of the hydrocarbon chains, which are paralleled by our simulations. The smaller solutes move quickly enough through the bilayer that specific effects on the order parameters are difficult to quantify. However, the drug analog moves slowly enough so that its effects on the order parameters can be divined. When residing at the interface, it is roughly pointed into the bilayer and its two rings penetrate a substantial distance into the hydrocarbon chain region. At this position, the drug acts much like other molecules that situate at the interface, such as cholesterol, and it causes an ordering of the regions of the chains coincident with the drug, i.e., high in the chains to about the center of the chains. When the drug is in the center of the bilayer, however, the opposite effect, a slight decrease in the order parameters in the center region of the chains, is seen. This is likely because lipid molecules that would normally occupy the space that the drug occupies must tilt or assume less ordered conformations in order to avoid its bulk. This difference in the effect on hydrocarbon order is another consequence of the unique structure of the bilayer as compared with that of bulk hydrocarbons.

Effects of Temperature and Variation in Bilayer Structure

The influence of bilayer structure on solute movements was further demonstrated by variation of the temperature of the simulations with benzene as a solute. Rather than simply modifying the kinetic energy of the benzene molecules and hence their diffusion rates, variation in temperature additionally altered the bilayer structure slightly. The voids became more fragmented and more evenly distributed in the hydrocarbon regions, and, as expected, the rates of torsional isomerization increased at all positions. The results were that, on average, as the temperature increased, the distribution of the solutes drifted toward the headgroup region. This agreed with experimental results, which suggested a similar phenomenon, as well as with the intuitive conclusions drawn from those experiments (Diamond and Katz, 1974; Dix et al, 1974). As the temperature of a bilayer is decreased, the ends of the hydrocarbon chains (near the center of the bilayer) will lose their fluidity last. Because of this, the presence of solutes in this region would have fewer entropic effects than the solutes would have in the more ordered and more tightly packed regions of the bilayer closer to the headgroups. These results also agreed with the lattice-theory-derived conclusions of Dill and coworkers, who noted the entropic effects of chain ordering on solute expulsion (Marqussee and Dill, 1986).

Discussion and Summary

In summary, MD simulations have shown that molecules of a size on the order of that of benzene and smaller diffuse faster in the bilayer center than elsewhere in the bilayer. The increased rate appears tied to the ability of the small solutes to make rapid, large motions, jumps, between and within the plentiful voids that exist in the bilayer center. This enhanced diffusion is unavailable for molecules such as adamantane and nifedipine whose volumes are, respectively, two and three times greater than that of benzene. Voids seldom appear that can accommodate the larger molecules. It seems likely that this increased rate of diffusion is at least partly responsible for the faster-than-expected permeation of small molecules.

An important aspect of any simulation is its trueness to reality. Wherever possible we have made comparisons to experiment of both the bilayer structure as well as the characteristics of the solutes. We find that bilayer structure and dynamics, as well as solute diffusion and rotation rates, are well reproduced as are the responses of the bilayer to the presence of solutes.

These studies were performed using bilayers of DMPC with a constant average surface area in the L-α phase. However, lipid bilayers can be made of many different lipids and combinations of lipids, and biological membranes are quite heterogeneous mixtures of lipids, sterols, and proteins. The gross properties of these more complex bilayers are expected to be similar to those of the DMPC bilayer (which is often used experimentally as a model of membranes, for just this reason), and the effects that we note here would be expected to be similar also. However, it is interesting to consider how the details of the process of diffusion might change as lipids with unsaturation, lipids with varying hydrocarbon chain lengths, lipids with varying surface area, sterols, and proteins are added to this system. The studies of the temperature dependence of membrane structure and solute diffusional mechanism help us to begin to understand this, and additional studies to probe the effects of membrane diversity are underway.

The solutes discussed here were uncharged. Bilayers and membranes pass charged molecules orders of magnitude slower than they pass uncharged molecules. However, passage is measurable, particularly for hydrophobic ions. For ions or other solutes with prominent electrostatic features, an additional factor becomes important, the membrane dipole potential. It has been known for some time that positive hydrophobic ions (such as tetraphenylphosphonium) cross lipid bilayers much more slowly than do negative hydrophobic ions (such as tripheynylborate), leading to the conclusion that even uncharged, symmetrical lipid bilayers must have a positive electrostatic potential in their interior (Gennis, 1989). The genesis of this potential has been debated for some time and is now known to have at least two causes. The first cause is the orientation of the lipid fatty acyl carbonyl groups, which, on average, orient with their oxygen atoms pointing toward water. This means that the positive end of the carbonyl dipole is closer to the bilayer center than is the negative end. The other, less obvious, cause is the orientation of water molecules at the membrane/water interface. It has been shown both by experiment and by simulation that these water molecules are oriented by the membrane so that the positive end of their dipoles are, on average, oriented slightly closer to the bilayer center than is the negative end of their dipole (Alper et al, 1993b; Gawrisch et al, 1992). The summed contributions of the large number of water molecules in this region accounts for 100–200 mvolts of potential. For charged solutes, then, the structure of another component of the bilayer, the hydrating water, becomes important.

Based on a number of studies, some of which are discussed here, it seems that the carbonyl groups of glycerol-ester linkages play an important role in acyl-glycerophospholipids. They have the polar and hydrogen bonding function closest to the bilayer center. Additionally, they serve to attract water molecules close to the hydrophobic hydrocarbon region of the bilayer.

In the interest of space and time, we have discussed here only studies of movement within the bilayer. Permeation is dependent not only on movement in the bilayer, but partitioning into the bilayer. In concept, this property is also accessible to simulation. Preliminary studies have been reported for water partitioning into membranes (Marrink and Berendsen, 1994). Determination of this property for a number of other solutes is underway in our laboratory.

Although the size of the voids was discussed, we said nothing of the shape of these voids. This might be important for the movement of oblong solutes. Conceivably, it might change with a different mix of lipid molecules. One attempt has been made to quantitate these shapes (Xiang and Anderson, 1994) that indicated that the voids could be quite long. However that study was done without benefit of solvating waters that might serve to decrease the length of the observed voids. Although we have not make a detailed study of this, our method of examining voids shows a range of sizes and shapes.

It is satisfying that the atomic level simulations can enhance our knowledge of biochemistry by verifying hypotheses such as the void-jumping hypothesis of Lieb and Stein and by providing additional detail to less detailed theoretical methods such as those of Dill. Simulations of the size and scope of ours have only become possible recently due to the increased availability of very fast, high capacity computers. The success of these preliminary studies promises that as both theory and computational hardware advance, in the future we will be able to calculate permeation rates for a range of molecules through simulation with an ever more complex membrane model.

ACKNOWLEDGMENTS

The authors would like to thank Malcolm Davis for his assistance in preparing the figures for this manuscript. The authors are also indebted to R. Shaginaw, J. Stringer, R. Gopstein, and G. Burnham (BMS High Performing Computer Center) and S. Samuels (BMS Department of Macromolecular Modeling) for orchestrating our Cray Y-MP and Silicon Graphics computing network and providing essential computer support. We would also like to thank J. Novotny and J. Villafranca for encouraging these studies.

REFERENCES

Alper HE, Stouch TR (1995): Orientation and diffusion of a drug analog in biomembranes: Molecular dynamics simulations. *J Phys Chem* 99:5724–5731

Alper HE, Bassolino DA, Stouch TR (1993a): Computer simulations of a phospholipid monolayer/water system: The effect of long range forces on water structure and dynamics. *J Chem Phys* 98:9798–9807

Alper HE, Bassolino D, Stouch TR (1993b): The limiting behavior of water hydrating a phospholipid monolayer: A computer simulation study. *J Chem Phys* 99:5547–5559

Anel A, Richieri GV, Kleinfeld AM (1993): Membrane partition of fatty acids and inhibition of T-cell function. *Biochemistry* 32:530–536

Bassolino D, Alper HE, Stouch TR (1995): Mechanism of solute diffusion through lipid bilayer membranes by molecular dynamics simulation. *J Amer Chem Soc* 117:4118–4129

Bassolino D, Alper HE, Stouch TR (1993): Solute diffusion in lipid bilayer membranes: An anatomic level study by molecular dynamics stimulation. *Biochemistry* 32:12624–12637

Bauerle H-D, Seelig J (1991): Interaction of charged and uncharged calcium channel antagonists with phospholipid membranes. Binding equilibrium, binding enthalpy and membrane location. *Biochemistry* 30:7023

Beschiaschvili G, Seelig J (1992): Peptide binding to lipid bilayers. Nonclassical hydrophobic effect and membrane-induced pK shifts. *Biochemistry* 31:10044–10053

Damodaran KV, Merz KM Jr, Gaber BP (1992): Structure and dynamics of the dilauroylphosphatidylethanolamine (DLPE) lipid bilayer. *Biochemistry* 31:1–20

De Loof H, Harvey SC, Segrest JP, Pastor RW (1991): Mean field stochastic boundary molecular dynamics simulation of a phospholipid in a membrane. *Biochemistry* 30:2099–2113

DeYoung LR, Dill KA (1988): Solute partitioning into lipid bilayer membranes. *Biochemistry* 27:5281–5289

Diamond JM, Katz Y (1974): Interpretation of non-electrolyte partition coefficients between dimyristoyl lecithin and water. *J Membr Biol* 17:121–154

Dix JA, Diamond JM, Kivelson D (1974): Translational difusion coefficient and partition coefficient of a spin-labeled solute in lecithin bilayer membranes. *Proc Nat Acad Sci USA* 71:474–478

Dix JA, Kivelson D, Diamond JM (1978): Molecular motions of small molecules in lecithin bilayers. *J Membr Biol* 40:315–342

Edholm, O. and Johansson, J. (1987): Lipid bilayer polypeptide interactions studied by molecular dynamics simulation. *Eur Biophys J* 14:203–209

Egberts E, Berendsen HJC (1988b): Molecular dynamics simulation of a smectic liquid crystal with atomic detail. *J Chem Phys* 89:3718–3732

Franks NP (1976): Structural analysis of hydrated egg lecithin and cholesterol bilayers. *J Mol Biol* 100:345–358

Franks NP, Lieb WR (1979): The structure of lipid bilayers and the effects of general anesthetics. *J Mol Biol* 133:469–500

Gawrisch K, Ruston D, Zimmerberg J, Parsegian VA, Rand RP, Fuller N (1992): Membrane dipole potentials, hydration forces, and ordering of water at membrane surfaces. *Biophys J* 61:1213–1223

Gennis RB (1989): Biomembranes, molecular structure and function. In: *Springer Advanced Texts in Chemistry* Cantor CR, ed. New York: Springer-Verlag

Jacobs RE, White SH (1984): Behavior of hexane dissolved in DMPC bilayers: An NMR and calorimetric study. *J Am Chem Soc* 106:915–920

Jonsson B, Edholm O, Teleman O (1986): Molecular dynamics simulations of a sodium octanoate micelle in aqueous solution. *J Chem Phys* 85:22663–2271

Jorgensen K, Ipsen JH, Mouritsen OG, Bennett D, Zuckermann MJ (1991a): The effects of density fluctuations on the partitioning of foreign molecules into lipid bilayers: Application to anasthetics and insecticides. *Biophys Acta* 1067:241–253

Jorgensen K, Ipsen JH, Mouritsen OG, Bennett D, Zuckermann MJ (1991b): A general model for the interaction of foreign molecules with lipid membranes: Drug and anaesthetics. *Biophys Acta* 1062:227–238

Kamp F, Hamilton JA (1993): Movement of fatty acids, fatty acid analogues and bile acids across phospholipid bilayers. *Biochemistry* 32:1074–11086

Lieb WR (1986): *J Membr Biol* 92:111–119

Lieb WR, Stein WD (1971): The molecular basis of simple diffusion within biological membranes. *Curr Top Membr Transp* 2:1–39

Lieb WR, Stein WD (1969): Biological membranes behave as non-porous polymer sheets with respect to the diffusion of non-electrolytes. *Nature* 224:240–243

Marqussee JA, Dill KA (1986): Solute Partitioning into chain molecule interphases: Monolayers, bilayer membranes, and micelles. *J Chem Phys* 85:434–444

Marrink S-J, Berendsen HJC (1994): Simulation of water transport through a lipid membrane. *J Phys Chem* 98:4155

Mason PR, Moring J, Herbette LG (1990): A molecular model involving the membrane bilayer in the binding of lipid soluble drugs to their receptors in the heart and brain. *Nucl Med Biol* 17:13

Mason RP, Rhodes DG, Herbette LG (1991): Reevaluating equalibrium and kinetic binding parameters for the lipophilic drugs based on a structural model for drug interaction with biological membranes. *J Med Chem* 34:869–877

Miller DM (1986): *Biochim Biophys Acta* 856:27–35

Müller-Plathe F, Rogers SC, van Gunsteren WF (1993): Gas sorption and transport in polyisobutylene. *J Chem Phys* 98:9895–9904

Nelander JC, Blaurock AE (1978): Disorder in nerve myelin: Phasing the higher order reflections by means of the diffuse scatter. *J Mol Biol* 118:497–532

Nicholas JB, Trouw FR, Mertz JE, Iton LE, Hopfinger AJ (1993): MD simulations of propane and methane in silicate. *J Phys Chem* 97:4149–4163

Overton E (1899): *Vierteljahrsschr Naturforsch Ges Zuerich* 44:88–135

Pastor R, Venable RM, Karplus M (1991): Model for the structure of the lipid bilayer. *Proc Nat Acad Sci USA* 88:892–896

Scott HL, Kalaskar S (1989): Lipid chains and cholesterol in model membranes: A Monte Carlo study. *Biochemistry* 28:3687–3691

Seelig A, Seelig J (1974): The dynamics structure of phopholipids measured by deuterium NMR. *Biochemistry* 13:4839

Stouch TR (1993): Lipid membrane structure and dynamics studied by all-atom molecular dynamics simulations of hydrated phospholipid bilayers. *Mol Sim* 10:317–345

Stouch TR, Alper HE, Bassolino D (1994): Supercomputing studies of biomembranes *Internat J Supercomp App* 8:6–23

Stouch TR, Ward KB, Altieri A, Hagler AT (1991): Simulations of lipid crystals: Characterization of the potential energy functions and parameters for lecithin molecules. *J Comp Chem* 12:1033–1046

Takeuchi H, Okazaki K (1990): Molecular dynamics simulation of diffusion of simple gas molecules in a short chain polymer. *J Chem Phys* 92:5643–5652

Trauble H (1971): The movement of molecules across lipid membranes: A molecular theory. *J Membr Biol* 4:193–208

Turner GL, Oldfield E (1979): Effect of a local anesthetic on hydrocarbon chain order in membranes. *Nature* 277:669–670

Van der Ploeg P, Berendsen HJC (1983): Molecular dynamics of a bilayer membrane. *Mol Phys* 49:233–248

Walter A, Gutknecht J (1986): Permeability of small molecules though lipid bilayer membranes. *J Membr Biol* 90:207–217

Watanabe K, Ferrario M, Klein M (1988): *J Chem Phys* 92:818–821

Williams DE, Stouch TR (1993): Characterization of force fields for lipid molecules: Applications to crystal structures. *J Comp Chem* 14:1066–1076

Xiang T, Anderson B (1994): Molecular distributions in interphase: Statistical mechanical theory combined with molecular dynamics simulation of a model lipid bilayer. *Biophys J* 66:561

Figure 6. The trajectory of the movement of one benzene molecule colored according to time (red, early; idigo, late). See color insert, opposite (top).

Figure 10. Snapshot of the simulation of four molecules of adamantane in the lipid bilayer. Water, cyan; phosphocholine headgroups, yellow; hydrocarbon chains, green; ester oxygens, red; adamantane molecules, CPK representations. See color insert, opposite (bottom).

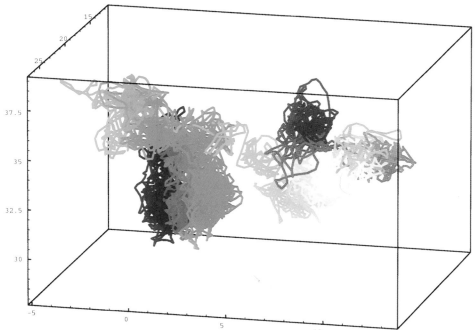

Figure 6.

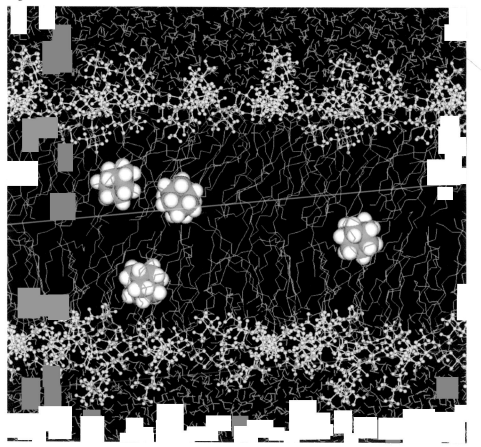

Figure 10.

9

Structural Basis and Energetics of Peptide Membrane Interactions

Huey W. Huang

Introduction

Over the past twenty-five years, tremendous progress has been made in understanding the physical properties of lipid bilayer membranes (Israelachvili, 1992; Lipowsky et al, 1991; Nelson et al, 1989). It seems appropriate to ask how the properties of lipid bilayers are related to the interaction of membranes with proteins. In this chapter, I will review our own work on the problem of membrane active peptides, in particular on the biological function of small amphiphilic helix-forming peptides.

Small amphiphilic helix-forming peptides (20–40 amino acids) are produced by bacteria and fungi as well as animals. Examples are: lantibiotics from bacteria (Sahl, 1993); alamethicin, suzukacillin, and trichotoxin from fungi (Latorre and Alvarez, 1981); magainins (Zasloff, 1987), bombinin, and bambinin-like peptides (Gibson et al, 1991) from amphibians; melittins from bees (Habermann, 1972); cecropins from moths (Steiner et al, 1981); and cecropin P1 from pigs (Lee et al, 1989). These peptides are characterized by their tendency to form amphiphilic helices when associated with a membrane. Their primary function is to rupture the membranes, which causes cytolysis. (In very low concentrations, they also form discrete ion channels and affect the cellular potentials, but whether that is the intended function is not known.) Indeed all of them are known as antibiotics or antimicrobials (except for melittin which is a hemolytic toxin). The evidence that these peptides exert their activity directly on the lipid bilayer rather than on some protein targets is threefold: (1) the synthetic enantiomers made of D-amino acids exhibit the same activities as the natural L enantiomers (Bessale et al, 1990; Wade et al, 1990); (2) the peptides lyse bacteria rapidly, often within minutes, consistent with a direct

Biological Membranes
K. Merz, Jr. and B. Roux, Editors
© Birkhäuser Boston 1996

attack on membrane (Boman et al, 1994); and (3) the antibacterial activity of these peptides correlate with the lysis of liposomes (Steiner et al, 1988; Matsuzaki et al, 1995).

A helical peptide of 20 amino acids has a length ($\approx$30 Å) comparable to the thickness of the hydrocarbon region of a phospholipid bilayer. The amino acids of an amphiphilic helical peptide are distributed in such a way that on average one side along the helix is hydrophilic and the other side hydrophobic. For the peptides found in nature (that is, in contrast to de novo designed synthetics), these descriptions are at best qualitative. The hydrophilic side often contains a number of hydrophobic amino acids and vice versa. Figure 1 shows the helical wheel diagrams of some examples. The helicities of these peptides in their membrane associted forms are mostly determined by circular dichroism (CD). Some of these peptides contain one or two prolines. How much disruption on the helicity is caused by these prolines is not clear. Such imperfections in amphiphilicity and helicity may be important properties related to their biological functions. We should keep these possible complications in mind, when we, for the convenience of discussion, describe a peptide as an amphiphilic rod. (Another point to keep in mind is that the physical length may not be equal to the hydrophobic length, the part compatible with the hydrocarbon chains.)

We imagine that such molecules can associate with a lipid bilayer in two ways: the peptide either adsorbs with helical axis parallel to the bilayer surface or inserts perpendicularly into the bilayer. In the surface state, the peptide is presumably adsorbed at the hydrophilic-hydrophobic interface between the polar head group region and the hydrocarbon region of the bilayer. In the inserted state, the peptide molecules presumably form aggregates with their hydrophobic sides in contact with the hydrocarbon chains of the lipid and their hydrophilic sides in contact with each other or with incorporated water. Thus we designed our experiment first to detect the orientation of the peptide. Then we investigated the high-order structures formed by these peptides in membranes. We also investigated by X-ray diffraction how the structure of bilayer is affected by peptides. Due to its availability at low cost, alamethicin has been used extensively in our pilot experiments. Recently we have extended our experiments to other peptides, particularly magainins.

Our experiments showed that amphiphilic helical peptides insert into a membrane when their concentration (specifically the amount of membrane bound peptide per lipid) exceeds a critical value. Below the critical concentration, the peptides mostly adsorb on the membrane surface. In the insertion state, the peptide forms pores, thereby killing the cell. The dependence of the critical concentration on the lipid composition of the membrane provides a plausible explanation for the cell-type selectivity exhibited by the host-defense peptides, because cell membranes do have specific lipid compositions. Finally we will show that the existence of the critical concentrations can be understood from the elastic property of lipid bilayers.

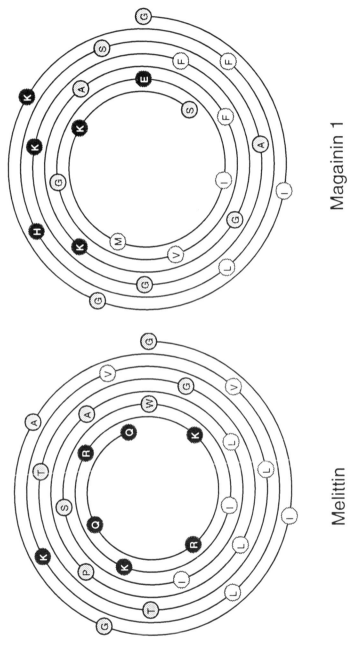

Figure 1. Helical wheel diagrams of two natural peptides, melittin and magainin 1. The black circles represent hydrophilic amino acids, the white hydrophobic, and the gray ambivalent.

Experimental Approach

Because we are interested in the structural aspect of the peptide-membrane interaction, we take an approach similar to that of protein crystallographers. We believe that it is easiest to extract the structural information if there is order in the samples. However, we cannot crystallize the lipid-peptide system because it is important that the lipid bilayers are maintained in the liquid state as in cell membranes. Thus we prepared our peptide-lipid samples in the smectic liquid-crystalline state, which is often called an oriented multilayer sample, either on one substrate surface or between two parallel surfaces. This sample preparation has the following advantages. (1) Because smectic liquid crystals exhibit unique defect features, one can use polarized microscopy to identify the thermodynamic phase and examine the degree of alignment (Asher and Pershan, 1979; Huang and Olah, 1987; Power and Pershan, 1977). (2) If our simple-minded picture of peptide-membrane interactions is correct, the helices are either oriented parallel or perpendicular to the substrate surface. These orientations are detectable by CD. (3) The samples are ideal for diffraction studies. Compared with a powder sample, which is a collection of multilamellar vesicles, an oriented sample produces diffraction patterns of higher resolutions. This is because the Lorentz factors and the effects of thermal undulations (Nallet et al, 1993) both favor the oriented sample. (4) In-plane X-ray and neutron scattering can be used to study the high-order structures formed by these peptides in membranes (He et al, 1993a).

Oriented Circular Dichroism

CD of oriented samples is called oriented CD or OCD. Figure 2 shows OCD of one alamethicin sample in two different hydration conditions: spectra A, B, and C in a high humidity condition, and D, E, and F in a low humidity condition. These spectra were measured with light incident at an angle α with respect to the normal to the multilayers, so $\alpha = 0°$ corresponds to normal incidence. Spectrum A represents the normal incident OCD of helices oriented parallel to the incident light, and spectrum F for helices perpendicular to the incident light. To prove the orientation, the sample was tilted at angles oblique to the incident light. According to the Moffitt theory (Moffit, 1956), a helix's exciton band at 205 nm is polarized parallel to the helical axis. The spectra A, B, and C indicate that the helix is oriented perpendicular to the plane of the bilayers in high humidity. As predicted by the theory, the magnitude of the amplitude at 205 nm increases with $\sin^2 \alpha$. On the other hand, the spectra D, E, and F indicate that the helix is oriented parallel to the plane of the bilayers in low humidities. Here the magnitude of the amplitude at 205 nm decreases as $[1 - (\sin^2 \alpha)/2]$ as predicted by the theory. Another theoretical constraint is the requirement that (1/3)(perpendicular spectrum A) + (2/3)(parallel spectrum F) reproduces the familiar solution CD spectrum of helices (Wu et al, 1990). A

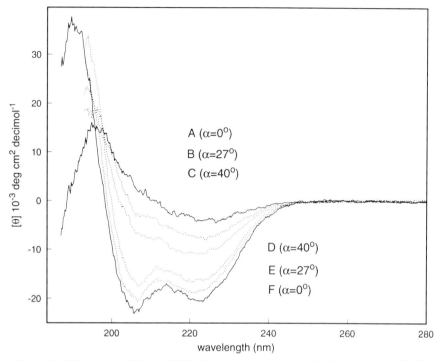

Figure 2. OCD of alamethicin in DPhPC bilayers at the peptide to lipid molar ratio 1/80 (Wu et al, 1990). When the sample was in high humidities, alamethicin was oriented perpendicular to the bilayers, exhibiting the spectra A, B, and C. α is the angle between the incident light and the normal to the multilayers. When the sample was equilibrated at low humidities (RH < 90%), alamethicin was oriented parallel to the plane of the bilayers, exhibiting the spectra D, E, and F. The lipid used in this experiment was made with phytol isolated from pumpkin seeds (Avanti Polar Lipids). Its purity was not clear. More recent experiment using DPhPC made with chemically synthesized phytol (purity > 99%) showed that alamethicin in DPhPC at 1/80 stays parallel to the bilayers at all hydration levels (Wu et al, 1995; see Figure 3). This is an example of how the critical concentration for insertion depends on the lipid composition: P/L* ∼ 1/120 in pumpkin seeds DPhPC, P/L* ∼ 1/40 in synthetic DPhPC. To show that this change was indeed due to the altered composition in DPhPC, not due to any other factors, we have reproduced the phase diagram of alamethicin in DOPC that was published in Huang and Wu (1991).

complete spectral analysis and comparison with the Moffitt theory were given in Wu et al (1990). (In fact the OCD measurement of alamethicin provided the first unambiguous experimental proof to the Moffitt theory) (Olah and Huang, 1988). OCD measurement at an oblique angle requires a special sample chamber, otherwise the oblique dielectric interfaces give rise to linear dichroism which in turn produces artifactual CD (Wu et al, 1990). Fortunately spectra A and F are clearly distinguishable. Thus, for the purpose of determining the peptide orientation, normal incident OCD is sufficient. Measurement of normal incident OCD is as simple as ordinary solution CD. It can be performed in any standard CD spectropolarimeter.

We have used OCD to study a variety of peptides in different lipids. In all cases, the peptide helices are either parallel or perpendicular to the bilayers. Most interestingly, the parallel state and the perpendicular state seem to represent two distinct phases. An example of phase diagram is shown in Figure 3. In general, at low peptide-to-lipid ratios (P/L), the peptides are in the surface state (the parallel state). Above a critical concentration P/L* (we will call this the critical concentration for insertion or CCI), there is a coexistence region in which a fraction of the peptide molecules are inserted and the rest remain on the surface. The inserted fraction increases (from zero at P/L*) with P/L until the peptide is completely inserted. The CCI for a given peptide varies with the lipid composition of the bilayer. Indeed in most cases, CCI is either very high or very low, so that within the concentration limitation of measurement, only one orientation of the peptide, either parallel or perpendicular to the membrane, was detected. (This is not surprising considering that the concentration limitations of measurement, either due to the difficulties for sample prepara-

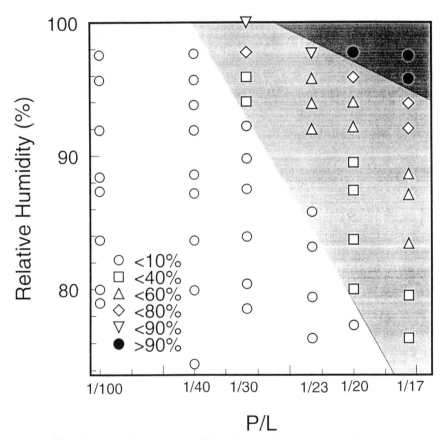

Figure 3. Phase diagram of alamethicin orientation in (synthetic) DPhPC bilayers.

tion or due to the sensitivity of the measuring instrument, are rather narrow ($\sim$1/10 > P:L > $\sim$1/200).

In activity assays, these peptides show critical concentrations for lysis (Cruciani et al, 1991; Jen et al, 1987; Juretic et al, 1989; Westerhoff et al, 1989). We hypothesized that the existence of CCI is the reason why a peptide kills some cells without harming the others (Ludtke et al, 1994). At a given solution concentration of peptide, different cell membranes have different affinities for binding the peptide. Different cell membranes are also likely to have different CCI. Since lysis occurs only if the concentration of the bound peptide exceeds CCI, some degree of cell type selectivity can be explained. (Huang and Wu, 1991; Ludtke et al, 1994; Wu et al, 1990)

In-Plane Scattering

Freeze-fracture electron microscopy has been used to study the protein organization on membranes. However the resolution of this technique is not fine enough to detect small peptides. Also the possibility of artifacts in the freeze-fracture process is difficult to assess (Pearson et al, 1984). Atomic force microscopy or its modified version is potentially a powerful tool for imaging membrane proteins. But so far the fluidity of the lipid bilayer has made the imaging impossible (Mou et al, 1995). Thus the greatest advantage of the scattering method is its applicability to the liquid crystalline (L_α) state of membranes. With suitably labeled samples, in-plane X-ray or neutron scattering provides a direct measurement of the lateral particle distribution in membranes.

With aligned multilayer samples, in-plane scattering can be performed in the transmission mode, which is much simpler than the reflection method (Pershan et al, 1987). For example, we performed in-plane scattering on a 1.5 KW X-ray machine to study the gramicidin channels in both the conducting and nonconducting states (He et al, 1993a; 1994). The interpretation of in-plane scattering is also relatively simple. In particular, with today's computer power, simulations in two dimensions can be efficiently carried out on a PC.

Neutron scattering has the advantage of utilizing the different scattering power of hydrogen and deuterium to selectively reveal particular aspects of complex biological assemblies. In order to investigate the high-order structures formed by the antibiotic peptides in membranes, we performed neutron in-plane scattering of alamethicin in the inserted phase, hydrated with D_2O (Figure 4). We interpret the scattering curve with a model schematically shown in Figure 5. The inserted peptide forms discrete pores in the barrel-stave fashion. With the momentum transfer $\mathbf{Q}$ in the plane of the bilayer, the primary contrast against the lipid background comes from the D_2O within the pores. As expected, when the D_2O was replaced by H_2O (by exposing the sample to H_2O vapor) the signal disappeared. This proves that water is part of the high-order structure of the inserted alamethicin.

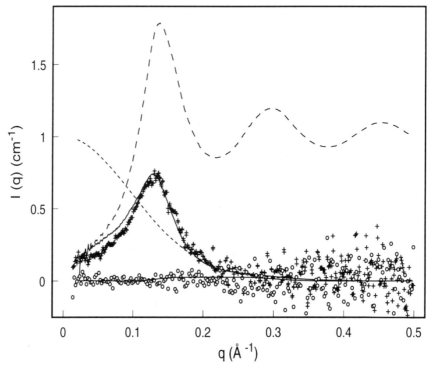

Figure 4. Neutron in-plane scattering curve of alamethicin pores in DLPC bilayers hydrated with D_2O (data +). After the sample was exposed to H_2O vapor for 48 hours, the neutron scattering was indistinguishable from the constant incoherent background (data o). Long-dashed line is the simulated structure factor $S(Q)$ for 8-monomer pores, assuming that all of the peptide at P:L = 1:10 forms such pores. Short-dashed line is the square of the corresponding form factor $|F(Q)|^2$ for D_2O hydration. The product $|F(Q)|^2S(Q)$ (solid line) agrees with data +. The solid line near the background is the theoretical curve $|F(Q)|^2S(Q)$ with $|F(Q)|^2$ now calculated for H^2O hydration. (Reprinted from He et al, 1995a; 1995b with permission.)

The basic law of scattering is well-known (Bacon, 1975). For in-plane scattering, the scattering intensity $I(Q)$ as a function of Q is given by (He et al, 1993a)

$$I(Q)/N = |F(Q)|^2S(Q), \tag{1}$$

with

$$S(Q) = 1 + \int [n_c(r) - \bar{n}]J_0(Qr)2\pi r dr \tag{2}$$

where $S(Q)$ is the structure factor; N the number of pores; $F(Q)$ the scattering amplitude by individual pore, called the form factor; $n_c(r)2\pi r dr$ the average number of pores within the ring of radius r and width dr, centered at an arbitrarily chosen pore; $\bar{n}$ the mean number density of the pores; $J_0(Qr)$ the zeroth order Bessel function of Qr. The presence of a scattering peak tells us that we

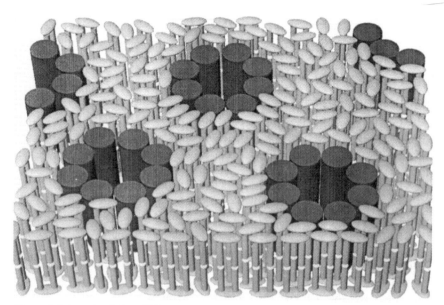

Figure 5. Schematic of alamethicin pores in DLPC bilayer. The pore is composed of 8 alamethicin monomers arranged in the barrel-stave fashion. Each monomer is represented by a cylinder of ~11 Å in diameter. The outer diameter of the pore is ~40 Å. The diameter of the water pathway is ~18 Å. Lipid molecules (two-legged objects) are drawn approximately to scale. The surface density of pores roughly corresponds to peptide to lipid molar ratio 1:10. (Reprinted from He et al, 1995a with permission.)

have well-defined scattering objects. And the peak position (Q_{max}) alone gives us a rough measure of their size. Using this estimate, the diameter of the pores is about $2\pi/Q_{max} \sim 49$ Å. This can be seen in the following way. Suppose that in a hyothetical model the pores can pass through each other without interference when they collide. Then, relative to a given pore the distribution of the other pores is uniform, in other words, n_c is just equal to $\bar{n}$ or $I(Q)/N = |F(Q)|^2$; there would be no interference between pores to produce a maximum in the scattering curve. The difference between this hypothetical case and the real case is that, in the latter, there is volume exclusion between pores. A maximum in scattering usually indicates there is a well-defined distance betweem scattering objects at $\sim 2\pi/Q_{max}$. This distance has to be the result of volume exclusion, so it is roughly two times the pore radius, but it also weakly depends on the pore concentration. $2\pi/Q_{max}$ slowly increases with decreasing concentration (He et al, 1993b). The estimate 49 Å is 20% larger than the correct diameter of the pore.

For quantitative analysis, we compute the structure factor $S(Q)$ by simulation. We assume that the pore is lined by n alamethicin helices in the barrel-stave fashion, leaving a cylindrical water pathway in the middle. From the molecular dimensions of an alamethicin monomer (Fox and Richards, 1982), the outside diameter, 2R and inside diameter, 2r can be estimated. Since the molecular cross sections of lipids are also known (Wu et al, 1995), the total area of the membrane containing N cylinders can be estimated from the peptide to lipid ratio. We let 1000 pores diffuse randomly within the area with the constraint that no two pores can overlap with each other. After the system reached equilibrium, the structure factor $S(Q) = |\sum_j \exp(i\mathbf{Q} \cdot \mathbf{r}_j)|^2$ where $\mathbf{r}_j$ is the position of the center of the jth pore, was computed and averaged over time. In Figure 4 we show that the S(Q) for pores of 8 monomers multiplied by the square of the corresponding form factor F(Q) agrees well with the data. In this example, the peptide's contrast against the lipid background is small, so F(Q) is most sensitive to the radius of the aqueous pore, r. However the peptide contributes to the scattering by limiting the distance of closest approach. Indeed S(Q) is most sensitive to the outside radius R (He et al, 1993b). As a result, both the radii of the aqueous pore and the channel are determined quite accurately (± 1 Å). In a previous study on in-plane scattering (He et al, 1993b), we have shown that the peak width and peak amplitude of the structure factor are sensitive to the density of the scattering object. The neutron data are consistent with the assumption that all of the alamethicin at P:L = 1:10 is involved in pore formation. More extensive analyses showed that alamethicin forms pores in a narrow range of size (Ke et al, 1995b). Most of them (>70%) are made of n and $n \pm 1$ monomers. In DLPC at P/L = 1/10, the mean size n is 8, that has an outer diameter about 40 Å and a water pathway of 18 Å in diameter, as schematically shown in Figure 5. The size of the pores appears to vary somewhat with the water content. It also varies with lipid. In DPhPC, the mean size n is about 11, that has an outer diameter about 50 Å and a water pathway of 26 Å in diameter. Thus pore formation appears to be the antimicrobial mechanism of alamethicin. (He et al, 1995a; 1995b)

X-Ray Lamellar Diffraction

It is well known that cell membranes are fluid structures. The lipid molecules are in the L_α phase. Membranes are studied in this phase, because in other phases proteins and peptides may pack differently with the lipid. (For example, peptides, including alamethicin, melittin, and magainin, inserted in the liquid phase are pushed out to the surface if the membranes are cooled or dehydrated to a gel phase; Vogel, 1987; Huang and Wu, 1991; Ludtke et al, 1994.) Consequently, membrane structure is not described by the atomic positions as for a crystal. Nonetheless a membrane has a fairly rigid z-dimensional structure. There are well defined structural characteristics that play important roles in interaction with peptides as we will demonstrate below. One of these characteristics is the membrane thickness. In particular, in a phospholipid bilayer, the mean phosphorus-phophorus distance across the bilayer can be measured to high precision. (Note that this does not violate the often stated rule of thumb: resolution of atomic position $\sim 2\pi/Q$ of the highest Bragg order. For example, the lattice constant can be determined to high precision from the lowest Bragg peak.) Other well defined structural characteristics are the degree of chain disorder and the average tilt angle of the headgroup. Our main concern here is the membrane thickness.

Determination of the structural characteristics is complicated by the static defects often present in the sample and by the long-range (small wave vector) thermal undulations of lipid bilayers. What we want to determine is the local membrane structure which is the average of the local thermal motions of atoms and molecules. The long-range undulations of the bilayer do not affect the local structure but do affect the measurement, particularly the lamellar diffraction (Caillé, 1972). Our goal is to determine the local structure with the minimum influence of the static defects and the long range undulations.

We have experimented with many diffraction methods on membrane samples, including the often used powder diffraction method and the method of multilayers supported on a curved surface (Frank and Lieb, 1979). We found that the method using aligned multilayers and $\omega - 2\theta$ scan produced diffraction patterns of highest resolution.

To study how the structure of the lipid bilayer changes with peptide concentration, we have performed X-ray lamellar diffraction of DPhPC bilayers containing alamethicin at various concentrations. The scans have been repeated through the hydration-dehydration cycles for the purpose of phase determination (the so-called swelling method). This experimental procedure has other benefits. The hydration-dehydration cycles tend to align the multilayers and help the sample reach equilibrium. It is also reassuring when the data are reproduced through a cycle. As the humidity approaches 100% RH, the long-range thermal undulations of lipid bilayers intensify (Caillé, 1972). Consequently the high order Bragg peaks are suppressed and the line shapes of the remaining low order peaks broaden. It is impossible to obtain high-resolution structural information directly from membranes in excessive water, because there is no diffraction in the high Q range. However in slightly dehydrated conditions, high resolution diffraction patterns are obtainable. Thus our strategy is to investigate the membrane structures in the slightly dehydrated states and extrapolate the results to the fully hydrated states.

Typical diffraction data were shown in Figure 6. The data reduction is straightforward because of the well-defined diffraction geometry in $\omega - 2\theta$ scan (Warren, 1960). With the phases determined by the swelling method (Blaurock, 1971), the relative scattering amplitudes were Fourier-transformed to obtained the scattering density profile ρ_{sc}. This profile is related to the true electron density ρ by $\rho = c\rho_{sc} + b$, with constants b and c. The constants b and c arise from the fact that we did not measure the absolute intensities and that the zeroth order diffraction peak was not measurable. These two constants can be determined if we know the composition of the sample, the molecular areas of the components in the plane of the bilayer, and the value of ρ at one point.

We define the bilayer thickness by the peak-to-peak distance of the profile, t, which corresponds to the phosphate-to-phosphate distance in the bilayer. This is a well-defined length from the diffraction data, because the peak-to-peak dis-

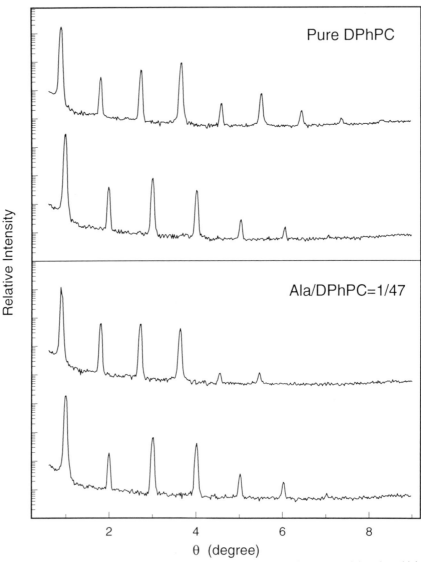

Figure 6. X-ray diffraction patterns of DPhPC bilayers and DPhPC bilayers containing alamethicin at the peptide to lipid ratio 1/47. The top curves of both samples were 95–98% relative humidity, the bottom curves near 60% relative humidity. (Reprinted from Wu et al, 1995 with permission.)

tance is independent of the normalization for the electron density profile, or the choices of b and c. (The peak positions are defined by $d\rho(z)/dz = 0$ which is equivalent to $d\rho_{sc}(z)/dz = 0$.) In Figure 7 we show t vs. D for the samples of pure DPhPC, P/L = 1/150, 1/80, and 1/47. OCD measurement indicated that in all these cases, alamethicin is in the surface state. It is clear, therefore,

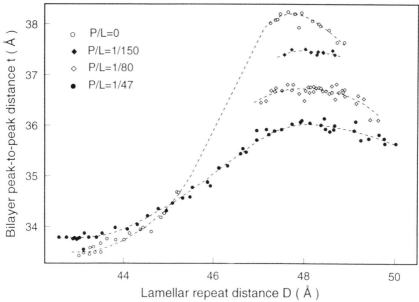

Figure 7. Bilayer thickness t vs. lamellar spacing D for pure DPhPC (P/L = 0), and for P/L = 1/150, 1/80, and 1/47. In all these cases, the peptide is on the bilayer surface. (Reprinted from Wu et al, 1995 with permission.)

that peptide in the surface state reduces the bilayer thickness. In fact the decrease of the bilayer thickness is proportional to the peptide concentration P/L (Figure 8).

We have interpreted this result with a simple physical picture. In a planar bilayer of pure lipid, the polar region (the head group and the associated water molecules) and the hydrocarbon chain region must maintain the same cross-sectional area. Suppose now that some peptide molecules are adsorbed in the polar region. Then the added cross-sectional area in the polar region due to the adsorbed peptide molecules must be matched by a corresponding areal increase in the chain region. In general, the cross-sectional area of a lipid is larger if its chains are more disordered. Since the volume of the chains is, to the first order, constant during an order-disorder transition (e.g., the volume change at the gel to L_α phase transition is 4% for DPPC [Nagle and Wilkinson, 1978]), the fractional increase in the cross section $\Delta A/A_O$ (per lipid) equals the fractional decrease in the thickness $-\Delta t/t_O$. $\Delta S = \Delta A(L/P)$ is then the expanded area due to each adsorbed peptide molecule. The experimental values of ΔS match the crystallographic value of alamethicin (Fox and Richards, 1982). This is consistent with the assumption that the peptide is adsorbed at the interface, and the adsorbing peptide pushes the lipid head groups laterally to create an additional area ΔS in the polar region. (Wu et al, 1995 and Ludtke et al, 1995)

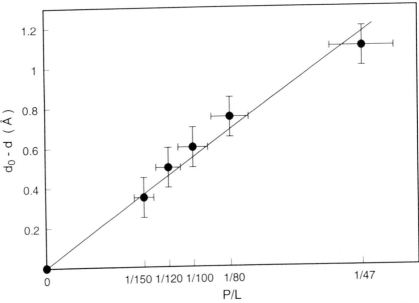

Figure 8. The decrease in the thickness of the chain region $(d_0 - d) = (t_0 - t)/2$ is proportional to the peptide concentration P/L. The straight line is a line of constant proportion. (Reprinted from Wu et al, 1995 with permission.)

Deformation Free Energy of Lipid Bilayer

We now examine the energetics of membrane thinning. Consider a tensionless lipid bilayer, consisting of two identical monolayers, parallel to the x–y plane before peptide adsorption. Let "u ± (x, y)" be the displacement of the top and bottom interfaces from their equilibrium positions and "a" the equilibrium thickness of each monolayer. Then $D(x, y) = u_+ - u_-$ is the change in the bilayer thickness (or more precisely the thickness of the hydrocarbon chain region) from the equilibrium value 2a and $M(x, y) = (u_+ + u_-)/2$ is the displacement of the mid-plane from its equilibrium position. The deformation free energy of the bilayer, per unit area of the unperturbed system, is given by (Huang, 1986; 1995)

$$f = aB \left(\frac{D}{2a}\right)^2 + \frac{K_c}{8}(\Delta D)^2 + \frac{K_c}{2}[\Delta M - C_O(x, y)]^2. \qquad (3)$$

B is the compressibility modulus of the bilayer (to be distinguished from the bulk modulus for layer compression of a multilayer stack [Caillé, 1972; De Gennes, 1969]). K_C is Helfrich's bending rigidity for a bilayer (Helfrich, 1973). $C_O(x, y)$ is the local spontaneous curvature (Helfrich, 1973) induced by peptide adsorption. At the moment, there is no reliable way of computing $C_O(x, y)$. However, in this model the D-mode and the M-mode are independent on each other. And we will be mainly concerned with the D-mode deformation. The deformation of the bilayer thickness is determined by the Euler equation

$$(B/a)D + (K_c/2)\Delta^2 D = 0, \tag{4}$$

obtained from the minimization of the total free energy. For mathematical simplicity, let us assume that the x–y cross section of a peptide adsorbed on the interface is a circular disk of radius r_O. The deformation induced by an adsorbed peptide is then described by $D = c\,kei(r/\lambda) + d\,ker(r/\lambda)$ (Watson, 1966), with $\lambda = (aK_C/2B)^{1/4}$, and the total deformation energy

$$F = (c^2 + d^2)\pi(BK_c/8a)^{1/2}I(r_O/\lambda) \tag{5}$$

with

$$I(z) = z(kei(z)ker'(z) - ker(z)kei'(z)). \tag{6}$$

The constants of integration c and d are determined by the boundary conditions at $r = r_O$. The thickness change at the boundary $D(r_O) = D_O$ is determined by the area expansion due to the peptide adsorption as discussed in the last section. The derivative at the boundary should be slightly negative (see Figure 9 below), since we expect the boundary lipid molecules to tilt in such directions to fill the void created by the peptide. For simplicity we use the approximation $D'(r_O) = 0$. The profile of the thickness change is shown in Figure 9 with a magnified D_O.

The most important parameter determining the deformation is λ. For a numerical estimate, we use $K_C = 50k_BT = 2 \times 10^{-12}$ erg (Chiruvolu et al, 1994), $B = 5 \times 10^8$ ergcm^{-3} (Hladky and Gruen, 1982; Huang, 1986), and $a = 15$ Å (Wu et al, 1995) to obtain $\lambda = 13$ Å and we let $r_O = 10$ Å so the cross section of an absorbed peptide, Γ, is about 300 Å^2 (Wu et al, 1995). As noted above, conservation of the chain volume implies

$$\Gamma = (1/a) \int D\,dxdy, \tag{7}$$

-80	-60	-40	-20	0	20	40	60	80

Å

Figure 9. The profile of bilayer thickness change $D(r) = u_+(r) - u_-(r)$ when a peptide molecule (shaded object) is adsorbed on it. The amplitude is magnified; the actual D_O is about -1.9 Å if the peptide cross section Γ is 300 Å^2. (Reprinted from Huang, 1995 with permission.)

from which D_O is determined to be -1.9 Å. The total deformation energy $F = 1.9 k_B T$. We note that the area of deformation by one adsorbed peptide is quite large, as much as 100 Å in diameter. This, to some extent, justifies the use of continuum theory for discussing peptide-membrane interactions.

We can further show that the membrane-mediated potential between two adsorbed peptide molecules is repulsive if the distance between them is $\lesssim 2\sqrt{2}\lambda$ (~ 37 Å) and essentially constant if $\gtrsim 2\sqrt{2}\lambda$. And for high peptide concentrations, the deformation energy of the membrane is proportional to the square of the peptide concentration. Thus the critical concentration for insertion can now be understood as follows. The free energy of adsorption (per peptide molecule) consists of two parts, the energy of binding to the interface $-\varepsilon_B$ and the energy of membrane deformation f_M. At low P/L, f_M is independent of concentration and OCD experiment implies $-\varepsilon_B + f_M \ll -\varepsilon_I$, the free energy of insertion. However, at high concentrations, f_M increases linearly with the peptide concentration. Therefore at a sufficiently high peptide concentration, the energy of adsorption can exceed the energy of insertion. The critical concentration for insertion is reached when $-\varepsilon_B + f_M = -\varepsilon_I$. (Huang, 1995)

ACKNOWLEDGMENT

This research was supported in part by the National Institutes of Health grant AI34367; by the Department of Energy grant DE-FG03-93ER61565; and by the Robert A. Welch Foundation.

REFERENCES

Asher SA, Pershan PS (1979): Alignment and defect structures in oriented phosphatidyl-choline multilayers. *Biophys J* 27:137–152

Bacon GE (1975): *Neutron Diffraction,* 3rd Ed. Oxford: Clarendon Press

Bessalle R, Kapitkovsky A, Gorea A, Shalit I, Fridkin M (1990): All-D-magainin: chirality, antimicrobial activity and proteolytic resistance. *FEBS Lett* 274:151–155

Blaurock AE (1971): Structure of the nerve myelin membrane: proof of the low resolution profile *J Mol Biol* 56:35–52

Boman HG, Marsh J, Goode JA (1994): *Antimicrobial Peptides.* Chichester, England: John Wiley and Sons

Caillé A (1972): Remarques sur la diffusion des rayons X dans les smectiques. *C R Acad Sci Serie B* 274:891–893

Chiruvolu S, Warriner HE, Naranjo E, Idziak S, Radler JO, Plano RJ, Zasadzinski JA, Safinya CR (1994): A phase of liposomes with entangled tubular vesicles. *Science* 266:1222–1225

Cruciani RA, Barker JL, Zasloff M, Chen H-C, Colamonici O (1991): Antibiotic maga-inins exert cytolytic activity against transformed cell lines through channel formation. *Proc Natl Acad Sci USA* 88:3792–3796

De Gennes PG (1969): Conjectures sur l'état smectique. *J Phys France* 30:65–71

Fox RO, Richards FM (1982): A voltage-gated ion channel model inferred from the crystal structure of alamethicin at 1.5-Å resolution. *Nature* 300:325–330

Franks NP, Lieb WR (1979): The structure of lipid bilayers and the effects of general anaesthetics: an x-ray and neutron diffraction study. *J Mol Biol* 133:469–500

Gibson BW, Tang D, Mandrell R, Kelly M, Spindel ER (1991): Bombinin-like peptides with antimicrobial activity from skin secretions of the asian toad, *Bombina orientalis. J Biol Chem* 266:23103–23111

Habermann E (1972): Bee and wasp venoms *Science* 177:314–322

He K, Ludtke SJ, Huang HW, Worcester DL (1995a): Antimicrobial peptide pores in membranes detected by neutron in-plane scattering. *Biochemistry* 34:15614–15618

He K, Ludtke SJ, Huang HW, Worcester DL (1995b): Neutron Scattering in the Plane of Membrane: Structure of Alamethicin Pores. *Biophys J:* submitted

He K, Ludtke SJ, Wu Y, Huang HW, Andersen OS, Greathouse D, Koeppe RE (1994): Closed State of Gramicidin Channel Detected by X-ray In-Plane Scattering *Biophys Chem* 49:83–89

He K, Ludtke SJ, Wu Y, Huang HW (1993a): X-ray scattering with momentum transfer in the plane of membrane: application to gramicidin organization. *Biophys J* 64:157–162

He K, Ludtke SJ, Wu Y, Huang HW (1993b): X-ray scattering in the plane of membrane. *J Phys France IV* 3:265–270

Helfrich W (1973): Elastic properties of lipid bilayers: theory and possible experiment. *Z Naturforsch* 28C:693–703

Hladky SB, Gruen DWR (1982): Thickness fluctuations in black lipid membranes. *Biophys J* 38:251–158

Huang HW (1995): Elasticity of lipid bilayer interacting with amphiphilic helical peptide. *J Phys II France* 5:1427–1431

Huang HW (1986): Deformation free energy of bilayer membrane and its effect on gramicidin channel lifetime. *Biophys J* 50:1061–1070

Huang HW, Olah GA (1987): Uniformly Oriented Gramicidin Channels Embedded in Thick Monodomain Lecithin Multilayers. *Biophys J* 51:989–992

Huang HW, Wu Y (1991): Lipid-alamethicin interactions influence alamethicin orientation. *Biophys J* 60:1079–1087

Israelachvili J (1992): *Intermolecular and Surface Forces.* London: Academic Press

Jen W-C, Jones GA, Brewer D, Parkinson VO, Taylor A (1987): The antibacterial activity of alamethicin and zervamicins. *J Applied Bacteriology* 63:293–298

Juretic D, Chen H-C, Brown JH, Morell JL, Hendler RW, Westerhoff HV (1989): Magainin 2 amide and analogues. *FEBS Lett* 249:219–223

Latorre R, Alvarez S (1981): Voltage-dependent channels in planar lipid bilayer membranes. *Physiol Rev* 61:77–150

Lee JY, Boman A, Chuanxin S, Andersson M, Jornvall H, Mutt V Boman HG (1989): Antibacterial peptides form pig intestine: Isolation of a mammalian cecropin. *Proc Natl Acad Sci USA* 86:9159–9162

Lipowsky R, Richter D, Kremer K (1991): *The Structure and Conformation of Amphiphilic Membranes.* Berlin: Springer-Verlag

Ludtke SJ, He K, Huang HW (1995): Membrane thinning caused by magainin 2. *Biochemistry* 34:16764–16769

Ludtke SJ, He K, Wu Y, Huang HW (1994): Cooperative membrane insertion of magainin correlated woth its cytolytic activity. *Biochim Biophys Acta* 1190:181–184

Matsuzaki K, Sugishita K, Fujii N, Miyajima K (1995): Molecular basis for membrane selectivity of an antimicrobial peptide, magainin 2. *Biochemistry* 34:3423–3429.

Moffitt W (1956): Optical rotatory dispersion of helical polymers. *J Chem Phys* 25:467–478

Mou J, Czajkowsky DM, Shao Z (1995): Gramicidin A aggregation in supported gel-state phosphatidylcholine bilayers. *Biochemistry:* in press

Nagle JF, Wilkinson DA (1978): Lecithin bilayers: density measurements and molecular interactions. *Biophys J* 23:159–175

Nallet F, Laversanne R, Roux D (1993): Modelling x-ray or neutron scattering spectra of lyotropic lamellar phases: interplay between form and structure factors. *J Phys II France* 3:487–502

Nelson D, Piran T, Weinberg S (1989): *Statistical Mechanics of Membranes ans Surfaces.* Singapore: World Scientific

Olah GA, Huang HW (1988): Circular dichroism of oriented α helices. I. Proof of the exciton theory. *J Chem Phys* 89:2531–2537

Pearson LT, Edelman J, Chan SI (1984): Statistical mechanics of lipid membranes. Protein correlation functions and lipid ordering. *Biophys J* 45:863–871

Pershan PS, Braslau A, Weiss AH, Als-Nielsen J (1987): Smectic layering at the free surface of liquid crystals in the nematic phase: x-ray reflectivity. *Phys Rev A* 35:4800–4813

Powers L, Pershan PS (1977): Monodomain samples dipalmitoyl-phosphatidylcholine with varying concentrations of water and other ingredients. *Biophys J* 20:137–152

Sahl HG (1991): Pore formation in bacterial membranes by cationic lantibiotics. In: *Nisin and novel lantibiotics,* Jung G and Sahl HG, eds. Leiden, Netherlands: Escom

Steiner H, Hultmark D, Engström A, Bennich H, Boman HG (1981): Sequence and specificity of two antibacterial proteins involved in insect immunity *Nature* 292 246–248

Vogel H (1987): Comparison of the conformation and orientation of alamethicin and melittin in lipid membranes. *Biochemistry* 26:4562–4572

Wade D, Boman A, Wåhlin B, Drain CM, Andreu D, Boman HG, Merrifield RB (1990): All-D amino acid-containing channel-forming antibiotic peptides. *Proc Natl Acad Sci USA* 87:4761–4765

Warren BE (1990): *X-ray Diffraction.* New York: Dover

Watson GN (1966): *Theory of Bessel Functions.* Cambridge, England: Cambridge University Press

Westerhoff HV, Juretic D, Hendler RW, Zasloff M (1989): Magainins and the disruption of membrane-linked free-energy transduction. *Proc Natl Acad Sci USA* 85:910–913

Wu Y, Huang HW, Olah GA (1990): Method of Oriented Circular Dichroism *Biophys J* 57:797–806

Wu Y, He K, Ludtke SJ, Huang HW (1995): X-ray diffraction study of lipid bilayer membrane interacting with amphiphilic helical peptides: diphytanoyl phosphatidylcholine with alamethicin at low concentrations. *Biophys J* 68:2361–2369

Zasloff M (1987): Magainins, a class of antimicrobial peptides form *Xenopus* skin: Isolation, characterization of two active forms and partial cDNA sequence of a precursor. *Proc Natl Acad Sci USA* 84:5449–5453

10

Computational Refinement Through Solid State NMR and Energy Constraints of a Membrane Bound Polypeptide

RANDAL R. KETCHEM, BENOÎT ROUX AND TIMOTHY A. CROSS

Introduction

The determination of macromolecular structures in anisotropic environments such as membranes is vital to the field of structural biology. While solid state nuclear magnetic resonance spectroscopy (SSNMR) methods have been demonstrated for obtaining three dimensional structures of membrane bound polypeptides (Cross and Opella, 1983; Ketchem et al, 1993; Opella et al, 1987), computational refinement methods are needed for optimally utilizing these constraints in such a molecular environment. Methods for structural determination and refinement of macromolecules in solution have fully evolved (Brünger et al, 1986; Clore et al, 1985; Havel and Wüthrich, 1985), but the nature of the constraints obtained for membrane proteins are such that a new refinement procedure must be developed. Described here is such a technique that has the ability to optimize the structure of a membrane protein in order to best represent the experimental data and determine its high-resolution structure.

The cation channel formed by dimers of gramicidin A (gA) is used here as a model membrane bound polypeptide for the description of this refinement procedure. While the structure of gA has been determined in organic solvents (Langs, 1988; Pascal and Cross, 1992), in SDS micelles (Bystrov et al, 1987; Lomize et al, 1992) and in lipid bilayers (Ketchem et al, 1993), this refinement procedure applied to the latter environment leads to a level of structural resolution never before obtained for a membrane polypeptide. The gA sequence, VGALAVVVWLWLWLW, consists of alternating L and D amino acids with the N-terminus blocked by a formyl group and the C-terminus blocked by ethanolamine. The alternating stereochemistry causes the side chains to be on the same side of a β-sheet structure, forcing it to curl into a β-helix (Urry,

Biological Membranes
K. Merz, Jr. and B. Roux, Editors
© Birkhäuser Boston 1996

1971). This structure was shown to be right-handed in lipid bilayers (Nicholson and Cross, 1989) and the initial structure of both the backbone (Ketchem et al, 1993; Teng et al, 1989) and side chains (Hu et al, 1993; Lee and Cross, 1994; Lee et al, 1995) has been determined. Computational refinement of this membrane polypeptide is necessary in order to join the initial backbone and side chain structures while incorporating all of the structural data and minimizing the structural energy.

Experimental Methods

The data used in this refinement procedure are orientational constraints obtained via SSNMR. These structural constraints define the orientation of an interaction axis or tensor with respect to an external vector, the applied magnetic field. SSNMR is the spectroscopy of samples that lack isotropic motions faster than the time scale of the nuclear spin interaction, typically on the kHz range. A sample prepared so that all possible orientations of the observed interaction are present exhibits a broad spectral line in which the frequency range represents

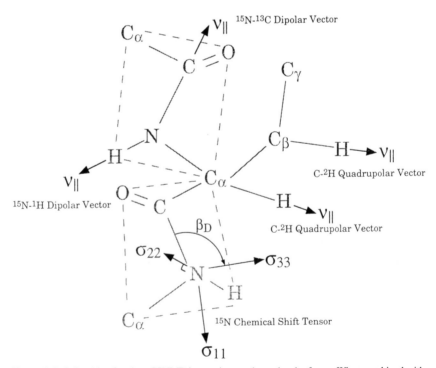

Figure 1. Relationship of various SSNMR interactions to the molecular frame. When combined with the experimentally observed orientations of these interactions with respect to the external magnetic field, the orientation of the molecular frame with respect to the external magnetic field is determined.

the orientational range. A sample prepared so that the molecules are uniformly oriented with respect to the magnetic field exhibits a sharp spectral line representing a single orientation for the interaction tensor. The resonance frequency of the observed interaction is dependent upon the orientation of the interaction tensor to the magnetic field. Since the interaction tensor has a fixed orientation to the molecular frame and since the sample alignment axis is fixed in the molecular frame, the observed interaction frequency constrains the molecular structure with respect to the magnetic field and alignment axis. The different types of experimentally observed interactions are illustrated in Figure 1.

Oriented samples are prepared by first codissolving isotopically labeled gramicidin with dimyristoyl phosphatidylcholine (DMPC) in a 95% benzene/5% ethanol solution. This solution is then spread on twenty-five 5 mm × 20 mm glass slips, and the glass slips are vacuum dried. These slips are then stacked into a square glass tube with one end sealed, and the sample is hydrated to approximately 50% by weight. The glass tube is then fully sealed and allowed to incubate at 40°C for a minimum of two weeks. This results in a sample containing gramicidin oriented in fully hydrated lipid bilayers. The orientation is such that the channel axis of the gramicidin helix, as well as the normal to the lipid bilayer, is parallel to the normal of the glass slips (Fields et al, 1988). This provides an oriented sample from which the molecular structure can be determined.

General Strategy

The goal is to determine the three-dimensional structure of the channel that encompasses all the available data. If a reduced set of experimental data is used and it is assumed that the geometry of the peptide units takes only the accepted values for an ideal polypeptide (i.e., bond lengths and bond angles are kept constant and peptide linkages are perfectly planar with a torsion angle ω equal to $180°$), it is possible to determine the channel structure through analytical calculations (Teng et al, 1989; see below). However, when all data are incorporated into the structure, the situation corresponds to a complex multidimensional optimization problem. A powerful strategy for determining the best structure consists in setting up a generalized global penalty function, incorporating all available information extracted from SSNMR and other chemical information about the local geometries and energies of polypeptides. The best structure is obtained from a geometry optimization by minimizing this global penalty function. To perform the minimization in such a high-dimensional configurational space, a very effective computational technique called simulated annealing is used (Kirkpatrick et al, 1983; Metropolis et al, 1953). In the early stages, the channel backbone conformational space is searched by introducing torsion moves while the standard peptide geometry is maintained. Such reductions in the dimensionality of the optimization problem is useful to accelerate the convergence of the minimization procedure. At this early stage the backbone SSNMR data and the backbone hydrogen bonds with no other energy considerations were included in the penalty function. In the final stage, the complete geometry of the polypeptide was allowed to vary. During this last stage, the penalty function incorporates all available experimental data about the backbone and side chains, the hydrogen bond distances, and the full CHARMM empirical energy function.

Determination of the Initial Structure

The first step in the determination of the structure of gramicidin is the analytical calculation of the initial backbone structure from experimentally determined $^{15}N-^{1}H$ and $^{15}N-^{13}C$ dipolar splittings for each peptide plane. The dipolar splittings are used to calculate the orientations of the unique axes of the dipole interactions (the N–H and N–C internuclear vectors) with respect to the external magnetic field. The orientations of the dipole interactions, in turn, serve to orient the peptide planes with respect to the magnetic field. By taking advantage of these peptide plane orientations and the known geometry of the C_α carbon that joins adjacent planes, the relative orientations of adjacent planes can be determined. In so doing the ϕ and ψ torsion angles can be calculated (Teng et al, 1989). The paired peptide planes, or diplanes, about each alpha carbon are determined for each residue, and the initial backbone structure is built by overlapping the shared peptide planes between alpha carbons as shown in Figure 2.

The resulting initial backbone structure has many positive attributes. The use of local orientational constraints in the determination of the initial backbone structure leads to a structure whose local and intermediate range interactions are well defined. Even though the nature of the data is such that the constraints are local to the individual peptide planes, the initial backbone structure displays a high level of long range conformational consistency. This is evident in the fact that the secondary structure of the molecule is shown as a right handed β-helix and that the intramolecular hydrogen bonding pattern is clearly identified. The initial backbone structure is therefore in an appropriate starting position from which to consider further experimental constraints and a computational refinement protocol.

The next step in the determination of the structure of gramicidin is the calculation of the side chain torsion angles through experimentally determined $C-^{2}H$ quadrupolar splittings from uniformly aligned samples. These splittings define the orientation of the $C-^{2}H$ bonds with respect to the magnetic field and can therefore be used as a means to orient the side chains with respect to the peptide backbone and to define the side chain torsion angles. The β_2 polar angle (Lee et al, 1995) describes the angle between the $C_\alpha-C_\beta$ bond vector and the magnetic field. The side chain torsion angles are systematically varied along with a narrow range of β_2 consistent with the determination of β_2 from the backbone structure and with the error bar from this determination. The rmsd between the observed and calculated quadrupolar splittings is used

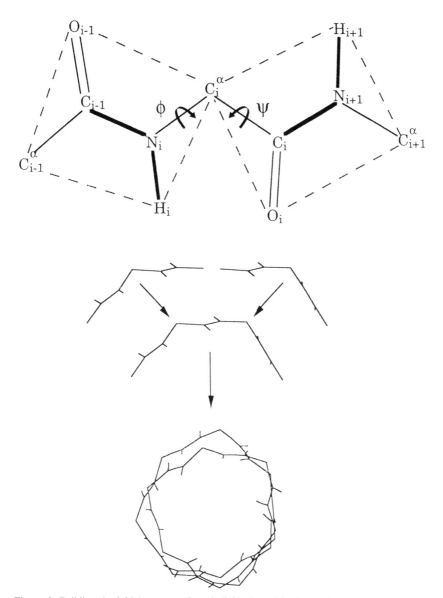

Figure 2. Building the initial structure from individual peptide planes. Once the orientations of the peptide planes are independently determined with respect to the external magnetic field, their orientations with respect to each other can be determined. The resulting diplanes can then be overlapped to build the initial backbone structure.

to determine the acceptable conformational states of the side chain. If multiple conformational states are shown at a single β_2 angle, the solution that falls within a rotameric state is accepted. The Val_1 and Val_7 side chain data represent fast exchange between rotameric states (Lee et al, 1995). Their orientations are therefore represented by modified data such that only the dominant rotameric state is considered. By this method, the torsion angles for each individual side chain are determined.

The final step in the initial structure determination is the joining of the backbone and side chain structures. Since the backbone and the side chains share C_α–C_β bond vectors, the side chains can be placed directly on the backbone while maintaining their relative orientations with respect to the magnetic field axis. Attaching the side chains to the backbone leads to the global initial structure.

Characteristics of the Initial Structure

While the initial backbone structure defines the peptide fold, identifies the intramolecular hydrogen bonds, and defines the side chain orientations relative to the backbone, it does so by using a subset of the SSNMR data that has been acquired. The torsion angles for the backbone structure are calculated solely from the ^{15}N–^1H and ^{15}N–^{13}C dipolar splittings, without consideration of the ^{15}N and ^{13}C chemical shifts or the C_α–^2H quadrupolar splittings. Since the backbone structure is built using the dipolar splitting data, the structure meets the requirements of this data. The calculation of all the acquired data from the initial structure shows that minor modifications in the initial structure are necessary in order to move the structure to within the experimental error of the observed data.

The side chain structures are determined by making full use of the observed C–^2H data for each individual side chain. However, side chain torsion angles are determined relative to the backbone structure, and, therefore, changes in the backbone structure will cause changes in the side chain torsion angles to achieve the same C–^2H bond orientations with respect to the magnetic field. Also, the tryptophan indole ^{15}N chemical shifts and ^{15}N–^1H dipolar splittings have been observed (Hu et al, 1993) and need to be included in the determination of the tryptophan orientations.

Further aspects of the initial structure indicate that structural modifications are necessary in order to obtain the global structure. The intramolecular backbone hydrogen bonds are clearly identified in the initial structure, but do not exhibit optimal hydrogen bond characteristics (Figure 3). Also, the side chains are initially considered as independent local structures, so when they are included as extensions on the backbone structure, close van der Waals (VDW) contacts become evident between some of the side chains (Figure 3). Another aspect of the initial structure is the assumption that the peptide linkage is planar (ω torsion angle $= 180°$). Preliminary experimental results suggest that a signif-

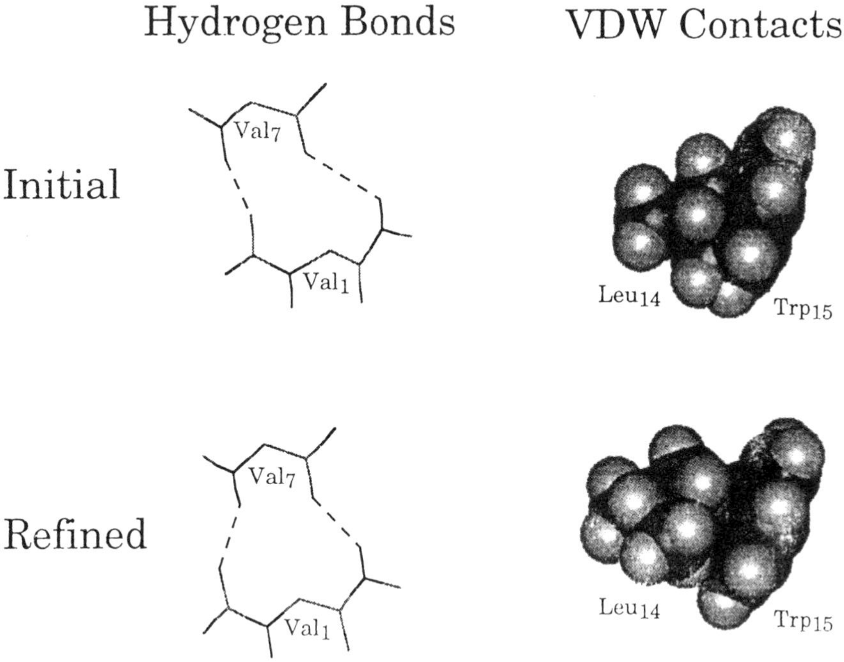

Hydrogen Bonds VDW Contacts

Initial

Refined

Figure 3. Hydrogen bonding geometry and VDW interactions before and after computational refinement. The initial backbone structure identifies the intramolecular hydrogen bonds and the side chain orientations. Refinement is necessary in order to optimize these interactions.

icant deviation from planarity exists for many of the peptide linkages. Analysis of the data calculated from the initial structure also indicates that deviations from planarity in the peptide linkages must be introduced in order to bring the structure within experimental error of all the observed SSNMR data. Another assumption in the determination of the initial structure is a static, identical geometry for the bond lengths and bond angles across all amino acid types. Analysis of various high-resolution structures (Engh and Huber, 1991) indicates that the atomic geometry must be relaxed.

The characteristics of the initial structure are such that it is evident that a suitable starting structure has been found, but that computational refinement is necessary in order to meet the wealth of experimental data, to optimize the hydrogen bonds and VDW interactions, and to remove some of the assumptions made in calculating the initial structure.

Structural Refinement Using Simulated Annealing

The application of simulated annealing to this global optimization problem requires a definition of the system configuration, a method by which the configuration is varied, and a penalty function by which the structural variations are controlled. In this case, the system configuration is defined as the atomic coordinates of the peptide, from which a description of the structure can be obtained. Many methods for introducing structural variations exist, such as changes in the torsion angles or variations in the individual atomic coordinates themselves. A penalty function for controlling the structural variations is dependent upon the method by which the structural variations are induced. If modifications of the torsion angles are used to alter the peptide configuration, then the experimental data and hydrogen bond distances are all that is required to define the penalty function. If direct modification of the individual atomic coordinates is the method used, then functions that describe the atomic interactions are needed and must be included in the penalty function along with the experimental data. The computational refinement of the initial gramicidin structure requires both torsion and atomic structural modifications in a two step procedure and therefore requires two different penalty functions.

Refinement in a two step procedure is required since atom moves alone are not sufficient to introduce the structural modifications necessary to obtain structural agreement with the experimental data. Even with a high starting temperature, atom moves are inhibited by the high energy contribution to the penalty resulting from VDW interactions. Refinement by torsion moves as a first step allows for structural changes without consideration of the atomic interactions and thus allows larger structural modifications.

As stated earlier, the initial structure exhibits steric VDW contacts to a degree requiring structural modification. The VDW interactions for the leucine side chains cause a substantial increase in the structural energy. In order to alleviate the worst of the bad contacts, the Leu_{10}, Leu_{12}, and Leu_{14} side chains and the ethanolamine have been energy minimized. The orientations of the resulting minimized side chains have been observed and found to be in an acceptable range from which to begin the refinement procedure. Since the refinement procedure progresses directly from torsion moves to atom moves, this step in the refinement procedure is performed before the introduction of torsion moves.

The first step in the computational refinement is the modification of the gramicidin backbone structure through torsion moves. Randomly chosen backbone torsion angles are changed a random amount between defined limits, $\pm 3.0°$ for ϕ and ψ and $\pm 0.1°$ for ω. Changes in the torsion angles are implemented symmetrically about the chosen bond which has the effect of modifying the entire structure instead of just the region after the chosen bond. The total change in an individual torsion angle is not constrained to fall within the defined limit since a single torsion angle may be chosen multiple times during a refinement. As a result, the introduction of structural modifications through torsion moves allows for an adequate search of local conformational space. While this is desired as a means for moving out of a local minimum, torsion moves are used primarily as a means for obtaining an approximate structural agreement to the backbone SSNMR data and a reasonable hydrogen bond geometry.

The final step in the refinement procedure is the introduction of atom moves to the torsion refined structure. The Cartesian coordinates are altered by introducing random displacements between ± 0.001 Å for each dimension. By using relatively small displacements, atom moves do not search a large conformation space but are used to generate changes in the structural geometry and to introduce minor atomic modifications in order to find a better match to the experimental data than the torsion refined structure. A beneficial result of the use of atom moves in the final refinement is that a precise structural agreement to the experimental data can be found while at the same time satisfying the constraints on the peptide geometry, without deviating significantly from the torsion refined structure.

The penalty function used to describe the structural deviation from the defined constraints is the sum of the structural penalties plus the energy.

$$\text{Total Penalty} = \sum (\text{Structural Penalties}) + \text{Energy}$$

where the individual structural penalties are calculated as:

$$\text{Penalty} = \begin{cases} \sum_{i=1}^{N} \dfrac{1}{2} \left(\dfrac{\text{limit}}{\text{Error}} \right)^2 & \text{if limit} \geq 0 \\ 0 & \text{if limit} < 0 \end{cases}$$

$$\text{limit} = |\text{Calculated} - \text{Observed}| - \text{Error}$$

where N is the number of measurements of a specific data type. For the hydrogen bond penalties, the observed value is the accepted hydrogen bond distance, and the error is the allowed distance range for the hydrogen bond. The use of the experimental error in the definition of the penalty serves several purposes. One use is to equate the various data types used in the penalty. The data originates from several very different observations, such as the chemical shift frequency and quadrupolar splittings. These different data types are associated with different observed magnitudes and are therefore difficult to equate directly. Each experimental error is of a magnitude relative to the observed interaction size and, therefore, division by the error has the result both of scaling the different data types so that they contribute equally to the total penalty and of making the penalty for the individual data types dimensionless. The penalty functions resulting from various data types indicate that the different penalties are of equal magnitude as a function of the error. Another use of the error is to define a region of the interaction space within which the penalty is zero. Moreover, it is important to have the ability to define separate error values within a particular data type, since experimental error may vary from site to site. Incorporating the error into the penalty function allows for the appropriate structural flexibility during the refinement procedure at the various measured sites.

The contributions to the penalty function differ depending upon the method of structural modification, either torsion or atom moves. Modifications of the backbone torsion angles do not require the use of side chain data, since the purpose of the torsion moves is to generate a better backbone structure from which to continue the global refinement. The constraints imposed on the structure

Table 1. Contributions to the Penalty Function for the Refinements
by Both Torsion and Atom Moves[a]

Torsion moves	λ	Initial penalty	Refined penalty
^{15}N Chemical Shift	1.0	1.0	0.3
^{13}C Chemical Shift	1.0	0.0	0.0
^{15}N–^{13}C Dipolar Splitting	1.0	0.0	0.9
^{15}N–^{1}H Dipolar Splitting	1.0	0.0	0.0
Distance	1.0	10.8	0.0
C_α–^{2}H Quadrupolar Splitting	1.0	55.6	0.3
Total Penalty		67.4	1.5

Atom moves	λ	Initial penalty	Refined penalty
^{15}N Chemical Shift	10.0	3.3	0.0
^{13}C Chemical Shift	10.0	0.0	0.0
^{15}N Indole Chemical Shift	10.0	4.5	0.0
^{15}N–^{13}C Dipolar Splitting	10.0	8.6	0.6
^{15}N–^{1}H Dipolar Splitting	10.0	0.0	0.0
^{15}N–^{1}H Indole Dipolar Splitting	10.0	31.6	0.0
Distance	10.0	0.0	0.4
$C_{\alpha,\beta,\ldots}$–^{2}H Quadrupolar Splitting	10.0	16199.1	1.5
CHARMM Energy	1.0	22659.0	255.0
Total Penalty		38906.1	257.5

[a] The penalty values shown have been multiplied by the λ values indicated and therefore reflect the final penalty used during refinement and not the calculated penalty of the individual constraints.

during torsion refinement are fifteen ^{15}N chemical shifts, two ^{13}C chemical shifts, fourteen ^{15}N–^{13}C dipolar splittings, fifteen ^{15}N–^{1}H dipolar splittings, ten N–O and ten H–O hydrogen bond distances, and twelve C_α–^{2}H quadrupolar splittings, for a total of seventy-eight constraints. For the refinement by atom moves, all data is used since all atom positions are modified. In addition to the structural constraints used in the torsion refinement, the atom refinement uses four indole ^{15}N chemical shifts, four indole ^{15}N–^{1}H dipolar splittings, an additional fifty-four C–^{2}H quadrupolar splittings, and the energy, for a total of one hundred and forty-one constraints. The values of the individual contributions to the penalty function for both the torsion and atom refinements are shown in Table 1.

The ^{15}N and ^{13}C chemical shift tensors are characterized from unoriented samples so that the magnitudes of the tensor elements and the orientation of the tensor with respect to the molecular frame can be determined (Mai et al, 1993). The chemical shifts observed from oriented samples are compared to chemical shifts calculated using the molecular coordinates and the tensor characteristics.

A change in the orientation of the atomic coordinates leads to a change in the calculated chemical shifts and a resultant change in the penalty.

The $^{15}N-^{1}H$ and $^{15}N-^{13}C$ dipolar splittings and the ^{2}H quadrupolar splittings are observed in oriented samples and reflect the orientation of the unique interaction tensor element, the internuclear vector, with respect to the external magnetic field. The dipolar splittings are described by the equation $\Delta\nu = \nu_{\parallel}(3\cos^2\theta - 1)$, where $\Delta\nu$ is the observed dipolar splitting, $\nu_{\parallel}$ is the magnitude of the dipolar interaction, and θ is the angle of the dipolar interaction with respect to the external magnetic field. The magnitude of the dipolar interaction is proportional to the product of the gyromagnetic ratios of the two nuclei divided by the cube of the internuclear distance. The quadrupolar splittings are described by $\Delta\nu = (3/4)QCC(3\cos^2\theta - 1)$, where $\Delta\nu$ is the observed quadrupolar splitting, QCC is the quadrupolar coupling constant, and θ is the angle of the unique quadrupolar interaction tensor element with respect to the external magnetic field. QCC is experimentally defined from model compound studies. Unlike the dipolar interaction, often the quadrupolar interaction is slightly asymmetric. Because the asymmetry is small ($\eta = 0.05\text{-}0.03$) it has been ignored, and the unique tensor element has been assumed to lie on the internuclear vector. The dipolar and quadrupolar splittings can be calculated from the atomic coordinates using the above equations and compared to the observed splittings during refinement.

The initial structure identifies the intramolecular hydrogen bonds. The hydrogen bonds are therefore included as direct contributions to the penalty function as is routinely done in structural refinements from solution NMR (Case and Wright, 1993; Logan et al, 1994). During the refinement the internuclear distances associated with the intramolecular hydrogen bonds are calculated and compared to accepted H–O and N–O internuclear distances for β-sheet structures with values of 1.96 ± 0.3 Å and 2.91 ± 0.3 Å, respectively (Jeffrey and Saenger, 1994).

The simulated annealing refinement procedure has been performed according to the Metropolis Monte Carlo algorithm (Metropolis et al, 1953; see also the chapter by H. L. Scott in this volume). Acceptance of an attempted move is controlled by both the temperature and the difference in the penalty before and after the attempted move. A move that causes a decrease in the penalty is always accepted. A move that increases the penalty, however, is only sometimes accepted. The choice to accept an uphill move is made by first choosing a random number between 0 and 1. If this random number is less than $\exp(-\Delta\text{penalty}/T)$, the uphill move is accepted. This is to say that $\exp(-\Delta\text{penalty}/T)$ defines the probability of the uphill move. The higher this probability, the greater chance of choosing a random number that is less than the calculated probability and therefore accepting the move.

The simulated annealing refinement procedure is controlled by a temperature parameter and an annealing schedule. The Monte Carlo algorithm generates configurations corresponding to the Boltzmann distribution of a canonical ensemble at a given temperature. The global minimization is controlled by an annealing schedule, i.e., the rate at which the temperature is lowered during the course of the refinement. The focus of this refinement strategy is to introduce minor structural modifications to the initial structure. Large changes would lead to conformational space that has already been shown to be excluded through the development of the initial structure. Therefore, the initial value of the temperature is set at 300 K for refinement both by torsion and atom moves so that large structural changes are not possible. The system configuration undergoes 2000 modifications or 200 successful modifications, whichever is first, before the temperature is lowered by 10%. As a result, in the beginning of the refinement the temperature is dropped relatively fast as a function of attempted structural modifications since many successful moves are initially found. As the refinement continues, the temperature is dropped less often since fewer accepted moves are found. The refinement is terminated when either the penalty is zero or no successful structural modifications are found at a particular temperature.

For the refinement by torsion moves, a zero penalty indicates that the structure has reached a point at which the calculated data matches the observed data to within the experimental error and that the imposed hydrogen bond distances are met to within an accepted range. For the refinement by atom moves, a zero penalty will not be found due to the inclusion of the energy.

Multiple parameters are used in the refinement by atom moves, such as the starting temperature, the diffusion parameter for the atom moves, and the λ values used to assign relative weights to the individual constituents of the penalty function. Combinations of various values for these parameters are used to determine the optimal conditions for refinement, taking into consideration the final fit to the experimental data, the final energy of the system, and the computational time required. The values for the optimized parameters are a starting temperature of 300 K and a diffusion parameter of 0.0005 Å. The λ values for the refinement by torsion moves are all 1.0 so that all of the contributions to the penalty are weighted equally. For the refinement by atom moves, the energy contribution tends to overwhelm the penalty calculated from the experimental constraints, so the λ values for all constituents of the penalty function other than the energy, for which λ is set to 1.0, are set to 10.0. Maintaining a λ value of at least 1.0 for the energy contribution is important since a λ value less than one corresponds to an effectively higher temperature and would lead to a nonrealistic value for the energy of the refined structure, thus compromising the structural integrity.

The refinement procedure discussed here has been implemented in a program called TORC (TOtal Refinement of Constraints), which has been written entirely in C. Though it would not be feasible to include the more than 5000 lines of code here, the general refinement procedure is shown in Figure 4. The code has been incorporated into CHARMM (Brooks et al, 1983) in order to take advantage of its ability to calculate the structural molecular mechanical energy for use in the refinement by atom moves. The all-atoms PARAM22 (Mackerell et al, 1992) version of the force field of CHARMM has been used to describe

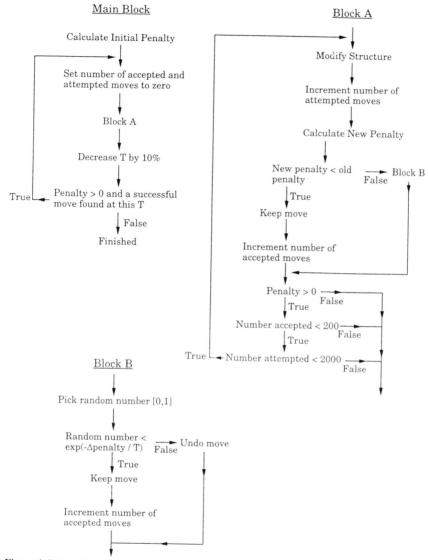

Figure 4. Schematic representation of the algorithm used to drive the refinement procedure for both torsion and atom moves. This figure demonstrates the general means by which the refinement is executed.

the internal energy (bonds, angles, dihedrals), as well as the nonbonded interactions (VDW and electrostatics). In contrast with other biomolecular force fields, the hydrogen bonding is described solely in terms of nonbonded interactions, and no special function is introduced. The IMAGE facility of CHARMM has been used to impose the dimer symmetry of the gramicidin channel. The VDW and electrostatic nonbonded interactions have been calculated on the basis of a group-based pair list. The interactions have been smoothly truncated at a distance of 10 Å using a 2 Å switching function. A dielectric constant of 5 has been used.

Refinement Results

A typical single refinement, starting from an initial structure and refining first with torsion moves and then with atom moves, requires approximately 12 CPU hours on a Silicon Graphics 4xR8000 Power Challenge running on a single CPU. During the course of the torsion refinement some 1,000,000 attempted moves are made with approximately 30,000 of the moves being accepted. Atom refinement attempts approximately 625,000 moves and accepts some 30,000 moves. The majority of the accepted moves occur in the initial stages of the refinement when the temperature is high and the structure is in a high penalty state.

The torsion and atom move refinements are represented structurally in Figure 5. As the figure indicates, the torsion move refinement introduces substantial structural changes within local conformational space as evident in the structural regularity of the helix, the corrected hydrogen bonding geometry, and relaxed ω torsion angles. While the average helical pitch for the initial to torsion refined structure only changes from 4.9 Å to 5.0 Å, the value of the residues per turn is heavily affected. The initial structure has a value of 6.8 residues per turn, and the torsion refined structure has a value of 6.4 residues.

The characteristic most easily observed in the structural representation of the atom refinement is the alleviation of undesired VDW contacts as seen in the altered side chain orientations. The backbone is barely modified, but it still shows a change in helical pitch from 5.0 Å to 5.1 Å and a change in residues per turn from 6.4 to 6.5.

During the refinement procedure the calculated data for an individual amino acid and the values for the contributions to the penalty function are saved as a function of attempted moves. These trajectories serve to illustrate the refinement procedure since they show the fluctuations in the calculated data and the behavior of the energy and the data penalties during the course of the refinement. Trajectories for selected data types, the penalties associated with the different data types, and the penalties resulting from the data for both the torsion and atom move refinements are shown in Figure 6.

Plots of the data values as a function of attempted moves, the left hand column in Figure 6, show the initial modification of the data, illustrating the level of the search of conformational space. The plots indicate that at the beginning of the refinement, the structure can be modified to an extent that the calculated data moves well outside of the experimental error range for the observed data, thus allowing for structural deviation from a local minimum. As the refinement progresses, the structural modifications decrease in both amplitude

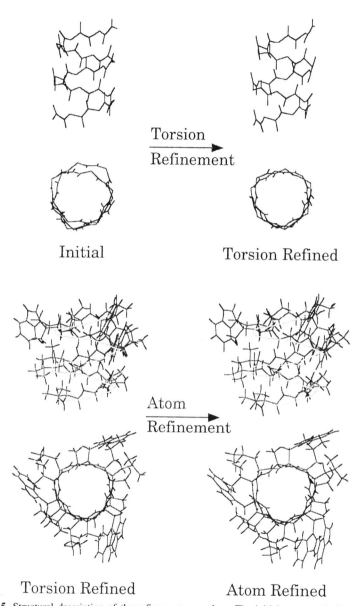

Figure 5. Structural description of the refinement procedure. The initial structure is first refined through modification of the backbone torsion angles without consideration of the side chains. The resulting torsion refined structure is then further refined by moving all atoms, backbone and side chain atoms included. The atom refined structure is the end of the refinement procedure and fully represents the constraints used in both stages of the refinement.

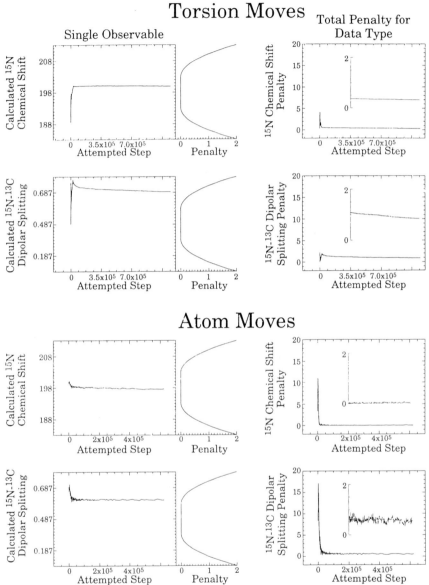

Figure 6. Refinement trajectories for selected data types. The left hand column illustrates the variability of the indicated SSNMR observable for a single site over the length of the refinement. The y-axis is labeled with the value for the experimentally observed interaction in the center and twice the experimental error above and below the observed value. The center column shows the penalty curve for the separate observables and indicates that though the relative values of the observables are quite different, the associated penalties are identical with respect to deviation from experimental error. The right hand column is the summed penalty associated with the individual constraints. A magnification of the penalty trajectory tails are inset and follow the same value of the x-axis as the full trajectory.

and frequency and the calculated data moves toward an agreement with the observed data to within experimental error. The torsion move trajectories initially deviate much more than the atom move trajectories since refinement by torsion moves introduces larger structural modifications.

The data trajectories further indicate that the individual calculated data values are initially within or very close to the range of experimental error. Even for those sites that remain within experimental error during the entire refinement, structural modifications are apparent at the beginning of the refinement trajectory. The trajectories for the sites that lie closer to the edge of experimental error also show structural fluctuations at the beginning of the refinement and later settle close to or within the experimental error.

Plots of the individual experimental data contributions to the penalty, the right hand column in Figure 6, show the ability of the refinement procedure to introduce structural modifications that significantly lower the penalty. At the beginning of the refinement for both torsion and atom moves, fluctuations in the calculated penalties illustrate the acceptance of both downhill and uphill structural changes. These structural changes are clearly directed toward a penalty minimum. Near the end of the refinement, the stabilization of the penalty at or near zero for the individual contributors indicates that the refinement has succeeded in finding a structure that meets the experimental data. This is further shown in the details of the initial and final penalties for both the refinement by torsion moves and atom moves as seen in Table 1.

The energy trajectory for the refinement using atom moves indicates that the structure resulting from the torsion refinement has a relatively high energy. The major contribution to the high energy, 22659 kcal/mol, is undesired VDW contacts. The contacts are easily relieved early in the refinement, and the energy contribution quickly reaches a state in which the driving force to minimize the penalty is the relaxation of the static geometry. Upon refinement by atom moves, the energy moves to a reasonable 255 kcal/mol. The final energy is not due to a single contribution but is spread over all constraints contributing to the energy as shown in Table 2. This indicates that no single interaction is being negatively influenced by the simultaneous use of energy and SSNMR data in the refinement procedure.

The backbone hydrogen bond distances are shown in Figure 7. As the figure indicates, the hydrogen bonds in the initial structure, though close to the accepted values, indicate that structural refinement is necessary. Out of ten hydrogen bonds, only two are within the accepted H–O and N–O limits. Including the hydrogen bonds in the penalty function during refinement by torsion moves induces structural changes that bring the hydrogen bonds within the accepted range except for two hydrogen bonds that lie very close to the edge of the range. Subsequent refinement by atom moves does not significantly alter the hydrogen bonding geometry.

The refinement procedure has a substantial affect on the geometry of the peptide linkages. The torsion move refinement, although choosing modifications for ω within the small range of $\pm 0.1°$, introduces significant deviations from

Table 2. Energy Distribution for the Structure Before and After Refinement by Atom Moves[a]

	Torsion refined	Atom refined
Bonds	40.2	15.5
VDW	20833.3	8.5
VDW (Image)	1572.8	−1.2
Electrostatics	17.0	21.4
Electrostatics (Image)	−0.6	−0.8
Angles	109.9	114.4
Urey-Bradley	8.9	14.5
Dihedrals	77.5	80.2
Impropers	0.1	2.5
Total	22659.1	255.0

[a] The full CHARMM energy is used as a structural constraint during refinement by atom moves. The structure resulting from torsion moves is at a high starting energy due mainly to steric VDW contacts. Refinement including the energy successfully alleviates these contacts.

planarity in several of the ω torsion angles. The largest deviation is 13°, while the average deviation is 6.4° with four deviations over 10°. The atom move refinement includes the energy and, therefore, the CHARMM force field is imposed on the ω torsion angles. As a result, the average deviation from planarity for the ω torsion angles is reduced to 4.8° with the largest deviation being 18° and only two deviations over 10°. These results show that even in the presence of the CHARMM force field, the structural constraints force large nonplanar peptide linkages.

The incorporation of the CHARMM energy into the penalty function for the refinement by atom moves has multiple effects on the final structure. As well as providing a means by which undesired VDW interactions are removed and maintaining the covalent geometry, the energy imposes subtle changes in the covalent geometry to match the definition set by the CHARMM force field. The initial and torsion refined structures maintain a static geometry using accepted values for bond lengths and angles (Teng et al, 1989). These values are used throughout the initial structure without consideration of the amino acid type or external interactions. The geometry of the atom refined structure, however, has been relaxed and therefore contains bond lengths and bond angles that more accurately correspond to their local environment.

The relaxation of the covalent geometry is evident in the consideration of the N–C bond lengths and the N–C$_\alpha$–C bond angles. The N–C bond lengths are all set to an initial value of 1.340 Å. Refinement with atom moves causes lengthening and shortening of these bonds for different amino acids with changes distributed from −0.015 Å to 0.024 Å, the average final N–C bond length being 1.35 Å. The N–C$_\alpha$–C bond angle is changed an average of 2.8° from 110.0° to

Hydrogen Bond Distances

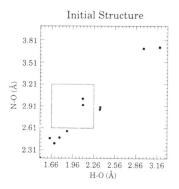

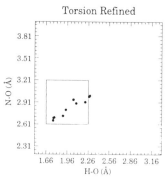

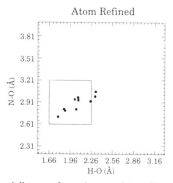

Figure 7. The hydrogen bond distances for each stage of the refinement procedure. The box indicates the range of allowed H–O and N–O distances and is centered on the accepted H–O and N–O values. The initial structure defines the intramolecular hydrogen bonds, but refinement is necessary in order to bring the hydrogen bond distance within the range of accepted values. Refinement by torsion moves succeeds in refining the hydrogen bonds and refinement by atom moves makes only minor modifications to the hydrogen bonds.

107.2°, and ranges from 102.5° to 111.9° for the different amino acids. These changes in the covalent geometry indicate that the local amino acid environments are taken into consideration during the structural refinement.

Although the trajectories show that the refinement procedure is searching conformational space, the all atom rmsd of 1.74 Å between the initial and final structure indicates that only a small region of conformational space is searched. Since the initial structure is calculated analytically, this is all that is desired. The all atom rmsd of 1.72 Å between the initial and torsion refined structures is an indication that the majority of the structural modifications occur during the refinement by torsion moves. The all atom rmsd of 0.32 Å between the torsion and atom refined structures shows that refinement by atom moves serves primarily to introduce small structural adjustments to the approximate structure obtained via torsion refinement.

The final structure obtained through this refinement procedure satisfies well the constraints imposed on the structure. The final penalty is very low and reflects the ability of the refinement to introduce structural modifications in such a way as to satisfy imposed constraints, such as energy and hydrogen bonds, while simultaneously satisfying the experimentally derived constraints.

Discussion

The refinement procedure described here has been demonstrated to introduce minor structural modifications leading to a structure that encompasses the experimental data and the calculated structural energy. The simultaneous use of the experimental data and the energy as contributors to the penalty function produces a refined structure that satisfies well all imposed constraints without being biased toward either the data or the energy.

The characteristics of the initial structure that prompted the development of a computational refinement procedure have been successfully addressed. The refinement provides the means to obtain a global structural solution that meets the imposed constraints. The refined backbone structure is defined not only by the $^{15}N-^{1}H$ and $^{15}N-^{13}C$ dipolar interaction vectors used to analytically calculate the initial structure, but by all of the experimental backbone data. The refined global structure includes both the backbone and side chains and therefore conforms to the complete set of experimental data used. The imposed hydrogen bond distance constraints are also met in the refined structure and result in a well-defined helix. The inclusion of the energy in the refinement by atom moves served to remove undesired VDW contacts in the structure and to relax the covalent geometry, without adversely affecting the structural match to the experimental data.

The use of SSNMR as the experimental method by which the structural constraints are obtained has the advantage of defining the global peptide structure in terms of local structural units. The orientations of these units are individually determined, and the initial structure can then be analytically determined from these quantitative constraints without the need for a large number of qualitative constraints as required for other methods, such as solution-state NMR. All that is then required of the computational refinement is minor modifications within the local conformational space in order to best fit additional constraints imposed upon the structure.

Apparent in the refined structure is the fact that the final penalty from the experimental data is not zero, indicating that while most of the refined structure lies within the experimental error of the observed data, some sites do not. Many reasons for a nonzero penalty exist, of which some can be addressed in the future and others are inherent in the method by which the data is obtained.

One contribution to a nonzero penalty that can be addressed is the parameters used in the calculation of the experimental data from the atomic coordinates. For the calculation of the chemical shifts as a function of the tensor orientation with respect to the external magnetic field, the tensor elements and the tensor orientation with respect to the molecular frame are characterized from dry powders while the observed oriented chemical shifts are determined in fully hydrated lipid bilayers. In order to obtain more accurate calculated chemical shifts the tensors should be characterized in fully hydrated lipid bilayers to better represent the channel conformation of this peptide (Lazo, et al, 1995; Lazo, et al, 1993). Two other parameters are the magnitudes of the dipole interactions, $\nu_\parallel$, used in the calculation of the $^{15}N-^{1}H$ and $^{15}N-^{13}C$ dipolar splittings from the atomic coordinates. These dipolar splittings, which serve to orient the N–H and N–C bond vectors, are directly proportional to the values of $\nu_\parallel$. As stated earlier, $\nu_\parallel$ is directly proportional to the product of the gyromagnetic ratios for the two nuclei and is inversely proportional to the cube of the internuclear distance. The internuclear distance is set to a static value for calculating $\nu_\parallel$. This value for $\nu_\parallel$ is then used without modification throughout the refinement procedure, even though atom moves vary the internuclear distance. This results in the calculated dipolar splitting being dependent upon the orientation of the bond vector alone and not upon the internuclear distance.

Furthermore, molecular dynamics are present in the sample (Lazo et al, 1995; Lazo et al, 1993; Nicholson et al, 1991; North, 1993; North and Cross, 1993) and motionally average the observed experimental data. The amplitude and axis of the local backbone motions have been determined for many sites and these motions will affect $\nu_\parallel$. Incorporating motional averaging into the value of $\nu_\parallel$ will lead to more realistic N–H and N–C bond orientations.

Molecular fluctuations occurring in the sample may be such that it is not possible to accurately represent all of the structural data in a single, static structure. An accurate simulation of the molecular dynamics based on the experimentally observed backbone and side chain dynamics would allow for the calculation of the time averaged observables and could be used to determine a structure that meets the experimental structural constraints over the period of the molecular dynamics simulation. This would give not a single, static structure that best represents the experimental data but would give a series of structures of which the time average meets the time averaged experimental data.

The inclusion of the structural energy, while maintaining the covalent geometry and removing undesired VDW contacts, influences the structure in a way that may not correspond to the influence due to the experimental data. The energy is calculated for the structure in a vacuum and will therefore bias the structure toward a point that does not fall completely within the bounds of the experimental error for all sites. As a result, the penalty due to the experimental data is reduced to a point that is very close to zero but cannot be further reduced. At the same time, the calculated energy is significantly reduced during refinement but not to the extent possible in the absence of experimental constraints. This indicates that although the penalties from the two different types of structural constraints are both necessary and are reduced during refinement, they compete for control of the structural modifications.

The general applications for this refinement procedure include both the determination of high-resolution structures of membrane bound polypeptides and the determination of structural details of specific regions of interest in macromolecules. The ability of this procedure to refine a structure by taking full advantage of the complete spectrum of orientational, distance and energy constraints is an important development in the field of structural biology.

ACKNOWLEDGMENTS

This work was supported by the National Science Foundation in a grant to TAC, MCB 9317111.

REFERENCES

Brünger AT, Clore GM, Gronenborn AM, Karplus M (1986): Three-dimensional structure of proteins determined by molecular dynamics with interproton distance restraints: Application to crambin. *Proc Natl Acad Sci USA* 83:3801–3805

Bystrov VF, Arseniev AS, Barsukov IL, Lomize AL (1987): 2D NMR of single and double stranded helices of gramicidin A in micelles and solutions. *Bull Magn Reson* 8:84–94

Brooks BR, Bruccoleri RE, Olafson BD, States DJ, Swaminathan S, Karplus M (1983): CHARMM: A program for macromolecular energy minimization and dynamics calculations. *J Comput Chem* 4:187–217

Case DA, Wright PE (1993): Determination of high-resolution NMR structures of proteins. In: *NMR of Proteins* Clore GM, Gronenborn AM, eds. London: The Macmillan Press Ltd

Clore GM, Gronenborn AM, Brünger AT, Karplus M (1985): Solution conformation of a heptadecapeptide comprising the DNA binding helix F of the cyclic AMP receptor protein of *Escherichia coli*. Combined use of ^{1}H nuclear magnetic resonance and restrained molecular dynamics. *J Mol Biol* 186:435–55

Cross TA, Opella SJ (1983): Protein structure by solid state NMR. *J Am Chem Soc* 105:306–308

Engh RA, Huber R (1991): Accurate bond and angle parameters for X-ray protein structure refinement. *Acta Cryst* A47:392–400

Fields CG, Fields GB, Petefish J, Van Wart HE, Cross TA (1988): Solid phase peptide synthesis and solid state NMR spectroscopy of [Ala3–^{15}N][Val1] gramicidin A *Proc Nat Acad Sci USA* 85:1384–1388

Havel TF, Wüthrich K (1985): An evaluation of the combined use of nuclear magnetic resonance and distance geometry for the determination of protein conformations in solution. *J Mol Biol* 182:281–294

Hu W, Lee K-C, Cross TA (1993): Tryptophans in membrane proteins: indole ring orientations and functional implications in the gramicidin channel. *Biochemistry* 32:7035–7047

Jeffrey GA, Saenger W (1994): *Hydrogen Bonding in Biological Structures.* Berlin, Germany: Springer-Verlag

Ketchem RR, Hu W, Cross TA (1993): High-resolution conformation of gramicidin A in a lipid bilayer by solid-state NMR. *Science* 261:1457–1460

Kirkpatrick S, Gelatt Jr CD, Vecchi MP (1983): Optimization by simulated annealing. *Science* 220:671–680

Langs DA (1988): Three-dimensional structure at 0.86 Å of the uncomplexed form of the transmembrane ion channel peptide gramicidin A. *Science* 241:188–191

Lazo ND, Hu W, Cross TA (1995): Low-temperature solid-state ^{15}N NMR characterization of polypeptide backbone librations. *J Mag Res Ser B* 106:43–50

Lazo ND, Hu W, Lee K-C, Cross TA (1993): Rapidly-frozen polypeptide samples for characterization of high definition dynamics by solid-state NMR spectroscopy. *Biochem Biophys Res Commun* 197:904–909

Lee K-C, Cross TA (1994): Side-chain structure and dynamics at the lipid-protein interface: Val₁ of the gramicidin A channel. *Biophys J* 66:1380–1387

Lee K-C, Huo S, Cross TA (1995): Lipid-peptide interface: Valine conformation and dynamics in the gramicidin channel. *Biochemistry* 34:857–867

Logan TM, Zhou M-M, Nettesheim DG, Meadows RP, Van Etten RL, Fesik SW (1994): Solution structure of a low molecular weight protein tyrosine phosphatase. *Biochemistry* 33:11087–11096

Lomize AL, Orechov VY, Arseniev AS (1992): Refinement of the spatial structure of the gramicidin A transmembrane ion-channel. *Bioorg Khim* 18:182–200

Mackerell AD Jr, Bashford D, Bellot M, Dunbrack RL, Field MJ, Fischer S, Gao J, Guo H, Ha S, Joseph D, Kuchnir L, Kuczera K, Lau FTK, Mattos C, Michnick S, Nguyen DT, Ngo T, Prodhom B, Roux B, Schlenkrich B, Smith J, Stote R, Straub J, Wiorkiewicz-Kuczera J, Karplus M (1992): Self-consistent parameterization of biomolecules for molecular modeling and condensed phase simulations. *Biophys J* 61:A143

Mai W, Hu W, Wang C, Cross TA (1993): Orientational constraints as three-dimensional structural constraints from chemical shift anisotropy: The polypeptide backbone of gramicidin A in a lipid bilayer. *Protein Sci* 2:532–542

Metropolis N, Rosenbluth AW, Rosenbluth MN, Teller AH, Teller E (1953): Equation of state calculations by fast computing machines. *J Chem Phys* 21:1087–1092

Nicholson LK, Cross TA (1989): Gramicidin cation channel: An experimental determination of the right-handed helix sense and verification of the β-type hydrogen bonding. *Biochemistry* 28:9379–9385

Nicholson LK, Teng Q, Cross TA (1991): Solid-state nuclear magnetic resonance derived model for dynamics in the polypeptide backbone of the gramicidin A channel. *J Mol Biol* 218:621–637

North CL (1993): Peptide backbone librations of the gramicidin A transmembrane channel as measured by solid state nuclear magnetic resonance. Implications for proposed mechanisms of ion transport (dissertation). Tallahasee, FL: Florida State University

North CL, Cross TA (1993): Analysis of polypeptide backbone T_1 relaxation data using an experimentally derived model. *J Mag Res Ser B* 101:35–43

Opella SJ, Stewart PL, Valentine KG (1987): Protein structure by solid state NMR spectroscopy. *Q Rev Biophys* 19:7–49

Pascal SM, Cross TA (1992) Structure of an isolated gramicidin A double helical species by high-resolution nuclear magnetic resonance. *J Mol Biol* 226:1101–1109

Teng Q, Nicholson LK, Cross TA (1989): Experimental determination of torsion angles in the polypeptide backbone of the gramicidin A channel by solid state nuclear magnetic resonance. *J Mol Biol* 218:607–619

Urry DW (1971): The gramicidin A transmembrane channel: A proposed $\Pi_{(L,D)}$ helix. *Proc Natl Acad Sci USA* 68:672–676

11

Bilayer-Peptide Interactions

K.V. Damodaran and Kenneth M. Merz, Jr.

Introduction

Biomembranes have a variety of important functions in living systems. Besides forming an envelope to the cell, they also have a regulatory function, serving as a barrier to the transport of matter between the cell and the outside world (Gennis, 1989). These latter processes involve interaction of the membrane lipids with membrane proteins and a variety of small molecules such as water, hormones, etc. Moreover, a detailed understanding of the interactions between lipid bilayers and small molecules and peptides can give useful insights into a variety of membrane phenomena such as passive and active permeation across cell membranes (Deamer and Bramhall, 1986), function of drugs and anesthetics (Lichtenberger et al, 1995), and membrane fusion and fusion inhibition (Bentz, 1993; Burger and Verkleij, 1990; White, 1992). Furthermore, bilayer models involving peptides can also be used as prototypes for understanding membrane protein interactions.

Membrane Partitioning: Structure and Thermodynamics

The thermodynamic and structural aspects of membrane partitioning by different kinds of guests have been widely investigated (Beschiaschvili and Seelig, 1992; Davis et al, 1983; Huschilt et al, 1985; Jacobs and White, 1989; 1987; 1986; Seelig and Ganz, 1991; Wimley and White, 1993). Seelig and coworkers have studied the thermodynamics of a number of amphiphilic and hydrophobic molecules partitioning into lipid bilayers (Bauerle and Seelig, 1991; Beschiaschvili and Seelig, 1992; Seelig and Ganz, 1991). A unique feature of these molecules is that they exhibit the so called nonclassical hydrophobic effect in

Biological Membranes
K. Merz, Jr. and B. Roux, Editors
© Birkhäuser Boston 1996

which the enthalpy of transfer makes the significant contribution to the free
energy of transfer, as opposed to the classical hydrophobic effect in which the
entropic contribution is the dominant part. However, such factors as the vesicle
radii have a strong influence on the relative contributions from enthalphy and entropy of transfer (Beschiaschvili and Seelig, 1992). The studies by Wimley and
White on a series of amphiphilic indole compounds have shown that hydrophobic and bilayer contributions are important (Wimley and White, 1993). Further,
there have been other systems in which the entropic contribution is the significant part. The hydrophobic tripeptides of the series (Ala-X-Ala-O-*tert*-butyl;
X = Trp, Phe, Leu, Ala, Gly) investigated by Jacobs and White show the classical hydrophobic-type behavior during the partitioning into phosphatidylcholine
(PC) based bilayers (Jacobs and White, 1989; 1987; 1986). The orientation of
these peptides in the membrane bound form has been investigated by Brown and
Huestis using nuclear Overhauser enhancement (NOE) measurements (Brown
and Huestis, 1993). These studies have shown that Ala-Phe-Ala-O-*tert*-butyl
adopts a preferred conformation with the side chain of the central residue (Phe)
and the *tert*-butyl group remaining in close proximity to each other. The Phe
and *tert*-butyl groups then have been found to be buried below the lipid head
groups. The availability of a large amount of data on the structure and energetics of this system makes it a particularly attractive one to investigate using
theoretical techniques.

Fusion Peptides and Fusion Inhibiting Peptides

Fusion peptides and fusion inhibiting peptides in the membrane bound form
have drawn significant attention in recent years (Blobel et al, 1992; Burger and
Verkleij, 1990; Kelsey et al, 1991; 1990; Muga et al, 1994; Richardson and
Choppin, 1983; Richardson et al, 1980). Membrane fusion is an important step
in many biological processes such as endocytocis, fertilization, and infection of
host cells by enveloped viruses. In the case of viral fusion, the initial attachment
of the virus to the target membrane and the subsequent fusion are mediated by
spike glycoproteins of the viral membrane. The region of these surface proteins
identified as responsible for inducing fusion is called fusion peptides. For example, in the case of the glycoprotein hemagglutinin (HA) of the influenza virus,
there are two subunits (HA1 and HA2) in which HA1 contains the sialic acid
binding site and HA2 contains the fusion peptide (Bentz et al, 1992; White,
1992). Fusion peptides contain sequences that are relatively hydrophobic and
can be modeled as an α-helix with the hydrophobic side chains located on one
side of the helix, suitable for partitioning into the hydrocarbon region of the
bilayer (Blobel et al, 1992; White, 1992), although this aspect has been ques-
tioned in the case of PH-30, the fusion peptide involved in sperm-egg fusion
(Muga et al, 1994).
    Fusion inhibiting peptides are small (2–3 residues) hydrophobic peptides
with sequences similar to the N-termini of fusion peptides (Kelsey et al, 1991;
1990; Richardson and Choppin, 1983; Richardson et al, 1980; Yeagle et al,

1992). Yeagle and coworkers have demonstrated that the fusion inhibiting peptides are effective in certain types of virus-vesicle as well as vesicle-vesicle fusion involving large unilamellar vesicles of N-Methyl dioleoylphosphatidylethanolamine (N-Me-DOPE) (Kelsey et al, 1991; 1990). These workers have further investigated the structural characteristics of these peptides in relation to their ability for fusion inhibition. Epand et al have shown that the blocking carbobenzoxy group at the N-terminus and the partial negative charge on these peptides are crucial for the fusion inhibiting activity (Epand et al, 1993). Their experiments have shown that by having an unblocked N-terminus (which could be protonated) and a blocked C-terminus these peptides could act as fusion peptides enhancing nonbilayer phase formation and membrane leakage, rather than inhibiting these events (Epand et al, 1993). However, the mechanism of inhibition by these peptides is not of universal nature. For example, Stegmann has shown that the fusion inhibiting peptides are not effective in Ca^{2+} induced fusion or virus-vesicle fusion involving the influenza virus mediated by hemagglutinin (HA) (Stegmann, 1993a,b).

The molecular mechanisms involved in the action of fusion peptides and fusion inhibiting peptides are far from clear, although structural characteristics of the bilayers participating in fusion have been considered an important factor, particularly in model systems (Bentz et al, 1992). For example, fusion activity has been observed in N-methyldioleoylphosphatidylethanolamine (N-Me-DOPE) and N-Me-DOPE/DOPC bilayers in a temperature range below the lamellar to inverted hexagonal (H_{II}) phase transition temperature (T_H) where an isotropic ^{31}P resonance is observed (Ellens et al, 1989; 1986). These data have given rise to the point of view that fusion peptides may be functioning by promoting the formation of nonlamellar fusion intermediate structures, and ability of the lipids to form such structures may enhance fusion activity since a fusion pathway can be conceived based on such nonlamellar intermediate structures. Similarly the activity of the fusion inhibiting peptides has also been linked to their ability to suppress the formation of these fusion intermediates. Electron microscopic studies by Yeagle et al have shown that these peptides inhibit the formation of highly curved phospholipid structures and have suggested a mechanism of fusion inhibition based on this (Yeagle et al, 1992). However, these fusion mechanisms based on the formation of nonlamellar fusion intermediates do have exceptions because the polypeptide melittin has been found to induce fusion in vesicles of phosphatidylcholine (PC) based lipids, which have little tendency to form inverted hexagonal-like structures (Murata et al, 1987). A similar observation has also been made in the case of the influenza virus (HA) in PC based vesicles in the $L_{\beta'}$ phase (Stegmann, 1993).

Computer Simulations

Computer simulations (Allen and Tildesley, 1987) employ both quantum and classical mechanical Hamiltonians to investigate the structural and dynamical properties at the atomic and molecular levels. With the recent availability of massively parallel computational methods on a variety of platforms, higher computational speed has become more commonly accessible. This has made computer simulation an important tool in the study of biomolecular systems in particular (McCammon and Harvey, 1987; Wipff, 1994). A variety of simulation techniques are available to the computational chemist which vary in the degree of completeness of the representation of the system and the level of theoretical treatment of the model. Ideally, one would like to have the entire system to be treated using quantum mechanical (QM) techniques. However, in spite of the tremendous improvements in computational resources, this is possible only for rather small systems (of the order of a few hundred heavy atoms). Long time scale, large size simulations have to use classical representations using molecular mechanical (MM) interaction potentials. Coupled QM-MM approaches can also be used where a more rigorous QM (ab initio, semi-empirical) based representation is used for a small region of interest such as the active site of an enzyme, while the MM representation is used for the rest of the system (Field et al, 1990; Gao and Xia, 1992; Stanton et al, 1995). Even in the realm of molecular mechanical representation, one can adopt varying degrees of approximation such as using united atoms for (CH_n) groups or by using frictional terms for the solvent environment as in Brownian dynamics (BD) (Dickinson, 1985) or stochastic boundary molecular dynamics (SBMD) (Brooks et al, 1985). For a detailed description of these techniques the reader is referred to reviews and monographs listed in the references (Brooks et al, 1988; McCammon and Harvey, 1987). It is only natural that as one strives to expand on a particular aspect (longer simulation, for example), one has to sacrifice on another front (use a simplified representation). Thus, as we have stated earlier (Damodaran and Merz, 1994a), different simulation techniques complement, rather than function as alternatives for, one another. Once the reliability of a model is established through the agreement of experimentally observable properties with calculated ones, it can be used with confidence to investigate the system further, at the molecular level. Moreover, one can also make use of unphysical but computationally convenient techniques like free energy perturbation and potential of mean force calculations (Mezei and Beveridge, 1986) to investigate processes at the microscopic level not possible experimentally, while correlating the results at the macroscopic level with experimentall observable quantities.

Computer simulation methods have been used extensively in recent years to study the structure and dynamics of bilayer models (Damodaran and Merz, 1994a,b; Damodaran et al, 1992; Egberts and Berendsen, 1988; Huang et al, 1994; Marrink et al, 1993; Stouch, 1993). Apart from investigating the structural details of the liquid-crystalline state, these methods have been used to study bilayer fluidity (Venable et al, 1993), solute diffusion (Alper and Stouch, 1995; Bassolino-Klimas et al, 1995; 1993) and interaction of polypeptides with the bilayers (Scott, 1991; Woolf and Roux, 1994; Xing and Scott, 1989). In this review, we discuss some of the recent developments in the investigation of host-guest type phenomena involving lipid bilayers using computer simulation methods.

Modeling of Bilayer-Peptide Systems

Modeling Strategies: Lipid and Solvent Environments

Various kinds of approximations can be adopted to represent the lipid and/or solvent environments when the interest is solely on the guest molecule. Phenomenological energy terms can be used to represent the effect of the lipid-solvent environments on the guest molecule and to account for the hydrophobic and hydrogen bonding interactions as well as the ordering effect due to the lipids. Long simulations can be done using this representation since only a small number of atoms are involved. Milik and Skolnick have used this scheme to investigate the insertion of a number of polypeptides forming α-helices into membranes using Monte Carlo simulations (Milik and Skolnick, 1993; 1992; see also the chapter by Skolnick and Milik in this volume). Starting from a random conformation for the peptide outside the membrane region, they have obtained the peptide position and orientation in the membrane bound state for systems such as Magainin 2, M2δ, and melittin in good accord with NMR results. Similarly, Edholm and Jahnig have studied the membrane spanning segment of glycophorin using molecular dynamics using a similar representation (Edholm and Jahnig, 1988). However, these models must be used with caution to study dynamical properties. For example, the large scale diffusive (Milik and Skolnick, 1993) and other types of motions (Edholm and Jahnig, 1988) observed for the membrane spanning helices in these simulations could have been amplified. Nevertheless, the structural insights from these simulations along with NMR and X-ray results could be used in more detailed simulations.

Another system studied in a similar fashion is the gramicidin channel. A number of studies have been performed on this ion channel using different kinds of representations for the lipid and solvent environments (for a review, see Roux and Karplus, 1994). These include restraining potentials applied directly on the channel atoms (Chiu et al, 1993) and hydrophobic environment created by Lennard-Jones (LJ) spheres (Roux and Karplus, 1993). For example, Chiu et al (1991; 1989) have used restraints with different characteristic time constants on the channel to mimic the effect of the lipids to investigate the dynamics of the water molecules in the channel and the role of the flexibility of the peptide in the transport process. They have found that channel flexibility has significant influence on the diffusion of the center of mass of the channel water chain. Roux and Karplus (1993) have used LJ spheres to mimic the lipid environment of the channel in their simulations to estimate the free energy profile of a sodium ion along the axis of the channel.

MD Simulations of Bilayer-Peptide Systems

The advantage of using explicit lipid and solvent environments is that one can also study the behavior of these environments. Insights into the nature of the lipid-water interface and the interaction of the protein with this region are important (Woolf and Roux, 1994). The perturbations on the static (average) and short time dynamical behavior of the lipid and solvent environments can be elucidated by comparison to a neat (unperturbed) system. The most important limitation in using a detailed representation, however, is the system size itself. Since the system typically consists of several thousands of atoms, the simulation times have to be limited to several hundreds of picoseconds to a few nanoseconds. However, one can still get important insights into the behavior of guest molecules in lipid bilayers. For example, Bassolino-Klimas et al have conducted very long simulations of benzene diffusion in dimyristoylphosphatidylcholine (DMPC) bilayer (Bassolino-Klimas et al, 1995; 1993). The rate of diffusion has been found to be different in different regions of the bilayer. Large, hopping motion of the benzene molecules moderated by the torsional motion of alkyl chains has been observed at the bilayer center. The presence of the benzene molecules has resulted in a small decrease in the alkyl chain order, particularly near the lipid carbonyls. Further, Alper and Stouch have studied a still larger molecule, a nifedipine analogue interacting with DMPC bilayers through nanosecond scale MD simulations (Alper and Stouch, 1995). Although this molecule has shown slightly different diffusional characteristics from Benzene, the bilayer properties are not significantly affected. Woolf et al have investigated the ion channel gramicidin in the presence of explicit lipid and water environment. They have studied the energetics and life times of various lipid-protein interactions in the case of the ion channel, particularly the importance of the Trp side chains near the bilayer-water interface (Woolf and Roux, 1994; Woolf et al, 1994).

Bilayer models in general require long times to equilibrate due to their low molecular mobility. Hence it is important that the starting conformation be set up accurately with as much structural data available as possible. Information regarding the physical parameters, such as the area per lipid and bilayer thickness are vital, since these parameters affect the order parameters. It is very difficult to achieve structural homogeneity through lipid rotations within the usual equilibration times employed, typically ≈ 100 ps. Hence starting models based on crystal structures are of little use for simulating the physiologically important liquid-crystalline phase (L_α Phase) (Damodaran and Merz, 1994; Stouch, 1993). In the case of bilayer-peptide systems, it is also important to have structural information based on NMR diffraction experiments about the location and orientation of the peptide in the bilayer, since peptides placed in an incorrect orientation can get trapped in place and may not reorient even after long equilibration runs.

We have used molecular dynamics simulations to investigate small peptides interacting with phosphatidylcholine (PC) and phosphatidylethanolamine (PE) based lipid bilayers. Here we describe two examples from these studies, namely, a hydrophobic tripeptide (Ala-Phe-Ala-O-*tert*-butyl, AFAtBU) intercalated into a DMPC bilayer and the interaction of fusion inhibiting peptides (FIPs; Z-D-Phe-L-Phe-Gly) on N-Methyl dioleoylphosphatidylethanolamine (N-Me-DOPE) bilayers. Details of the simulation methodology will be published elsewhere (Damodaran and Merz, 1995; Damodaran et al, 1995). Briefly, we have used the AMBER force field (Weiner et al, 1984) along with partial charges obtained by electrostatic potential (ESP) fitting procedure performed on lipid and peptide molecules (Besler et al, 1990; Merz, 1992). All the simulations have been carried out using parallelized versions of MINMD (AMBER 4.0) and SANDER (AMBER 4.1) modules running on the IBM-SP1/SP2 at the Pennsylvania State University's Center for Academic Computing, the Cornell Theory Center, and the CRAY-T3D at the Pittsburgh Supercomputing Center.

Ala-Phe-Ala-O-*tert*-butyl (AFAtBu): DMPC

The interaction of a series of tripeptides (Ala-X-Ala-O-*tert*-butyl) with different PC based bilayers have been widely investigated using a number experimental techniques (Jacobs and White, 1989; 1987; 1986). Neutron diffraction studies have shown that the peptide is largely confined to the bilayer-water interface with only large hydrophobic side chains of the central residues (X = Trp, Phe) being exposed to the hydrocarbon region of the bilayer (Jacobs and White, 1989). However, more specific data regarding the orientation and dynamics became available from the nuclear Overhauser enhancement (NOE) measurements by Brown and Huestis (1993). According to their data, the Phe-side chain as well as the *tert*-butyl groups are located below the head group region. The initial structure for the MD simulation has been built by placing 32 DMPC lipids forming two leaflets (16 × 2) with the head groups facing each other with a surface area of 66 Å²/lipid (Lewis and Engelman, 1983). The lipids are randomly oriented in the plane of the leaflets, with particular care taken to avoid any bad van der Waals contacts. This has been considered a reasonable model since at the level of hydration and lipid-lipid spacings used in the model, the lipids can undergo rotational diffusion along their long axis and may not have a preference to the layered parallel chain arrangement found in the crystal structure. Moreover this procedure also eliminates the collective tilt observed when the starting structures built from crystal structures are used (Damodaran and Merz, 1994). The lipid neighbors of the peptide in the starting configuration are shown in Figure 1a.

Two 450 ps trajectories have been generated with the peptides bound in different orientations. Earlier neutron diffraction data (Jacobs and White, 1989) have suggested that the peptide backbone might be parallel to the bilayer surface with the central residue (Trp in their experiments) exposed to the hydrophobic region. Our initial model has been built according to this data (designated as Model-I). In this model the *tert*-butyl group has been exposed to the aqueous region and has remained so throughout the simulation. However, subsequent NMR investigations by Brown and Huestis (1993) have provided more specific information on the peptide. A second trajectory has been generated (Model-II) in which the peptide is initially minimized with the NOE restraints before it is docked into the bilayer. Since this model has been built based on more specific data, it is considered a better representation, and the trajectory has been analyzed in greater detail. In both cases there are two peptides docked to one of the leaflets

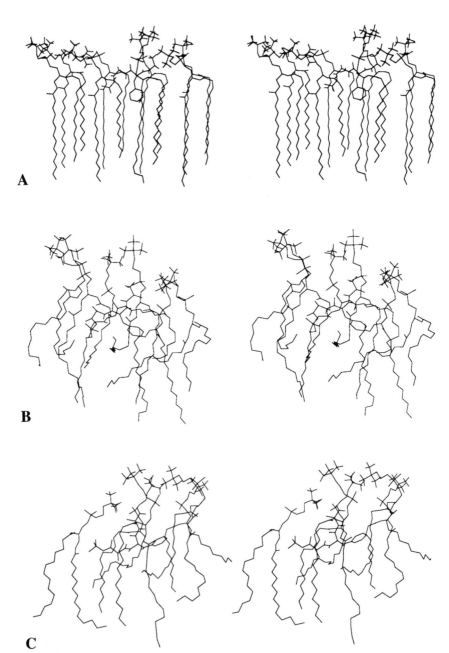

Figure 1. Stereo plots showing AFAtBu peptides and lipid neighbors in model-II described in the text. (a) The starting structure. Only peptide-1 is shown. Peptide-2 was an exact copy of peptide-1. (b) and (c) show snap shots from the end of the MD trajectory for peptide-1 (b) and peptide-2 (c).

of the model using the graphics program MIDASPlus (University of California at San Francisco). The bilayer-peptide system was solvated by adding about 25 SPC/E water molecules (Berendsen et al, 1987) per lipid in the interlamellar region. Two Cl⁻ counterions were placed in the aqueous region to render the total system neutral.

Density Profiles

A typical snapshot of the peptides and their neighboring lipids from the MD trajectory of model-II is shown in Figure 1b,c. The location of the peptides in the bilayer, as well as various lipid and solvent regions along the bilayer normal, are also shown using density profiles in Figure 2, calculated for Model-II. The peptide distribution, shown as shaded regions, overlaps partially with the alkyl chains region and the head groups. The density profiles shown in Figure 2 is in good agreement with neutron diffraction results (Jacobs and White, 1989). The height of the terminal methyl distribution and the minimum in the alkyl chain methylene distribution at the bilayer center indicate the degree of overall disorder in the hydrocarbon region. For example, in the all-trans state, the terminal methyl distribution is very narrow and intense while the minimum in the methylene distribution has a narrow deep minimum at the bilayer center. As the disorder increases, the width of the methyl distribution increases and the height of the minimum in the methylene distribution increases.

Lipid and Solvent Environments

Graphical visualization of the bilayer peptide system shows that the *tert*-butyl groups as well as the Phe-side chains remain buried in the hydrocarbon region while the amino termini anchor themselves at the lipid-water interface region, forming hydrogen bonds with water and the nonesterified oxygens of the phosphate groups. This is evident from the distinct hydration shells in the water radial distribution functions and the clear first peaks in the histogram for the phosphate nonesterified oxygens shown in Figure 3a,b. However, the peptide conformations are slightly different for the two peptides. While one of the peptides (referred to as peptide-1) has the *tert*-butyl groups in close proximity to the Phe-side chain, as suggested by NMR results (Brown and Huestis, 1993), the second peptide (peptide-2) has the *tert*-butyl group slightly more extended into the alkyl chain region. There does not seem to be any energy penalty for this, since the van der Waals interactions are satisfied in both conformations. In this context, it is also important to keep in mind the different time scales involved in NMR experiments and MD simulations.

The *tert*-butyl groups and the Phe-side chains are buried in the hydrocarbon region and have only very little solvent concentration around them, as shown by the radial distribution functions in Figure 3c,d. The histograms for the alkyl

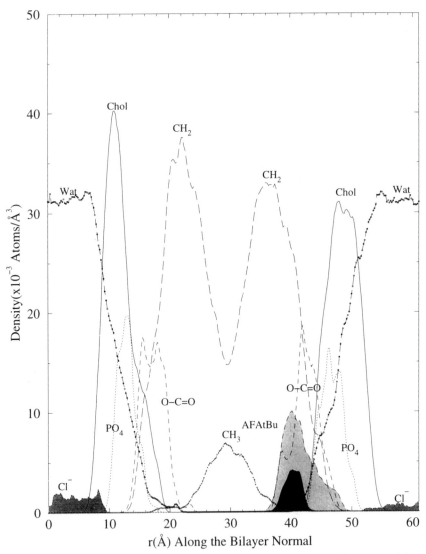

Figure 2. Density profiles of the peptides, lipids and water regions along the bilayer normal from the AFAtBu:DMPC simulations (model-II).

chains shown in Figure 3e,f are similar for both peptides and have a clear first neighbor peak. The flexible nature of the alkyl chains allows even the terminal methyl carbons to have close interactions with these peptide groups and to contribute to the first peak.

The location of the peptides and the orientation in the bilayer are in good agreement with the experimental results. However, the microenvironments for the two peptides in the simulation differ slightly. This is clearly visible in the

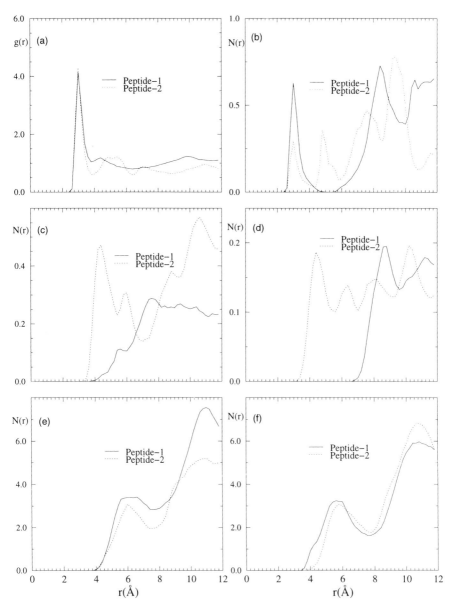

Figure 3. Water oxygen radial distribution functions (g(r)) and lipid histograms (N(r)) of different lipid regions from the AFAtBu peptides in model-II. (a) g(r) of water from the peptide N-termini, (b) histograms for the lipid phosphate group from the peptide N-termini, (c) g(r) of water from the *tert*-butyl group, (d) g(r) of water from the Phe-side chain, (e) histogram of alkyl chain carbon atoms from the *tert*-butyl group, and (f) histogram of alkyl chain carbon atoms from the Phe-side chain.

water histograms for the *tert*-butyl groups and the Phe-chains shown in Figure 3c,d and the histogram for the phosphate group oxygens in Figure 3b. Such differences can be related to the heterogeneous nature of the bilayer-water interface, the low mobility of atoms in this region and the finite length of the simulation. One way to alleviate this problem is to perform a number of relatively shorter simulations with different starting configurations and average the results over the individual trajectories. The system size also can be increased to include many guest molecules. Since we do not have such data currently, we feel it is appropriate to present the environments of each peptide independently.

Bilayer Perturbations

We have also examined the bilayer perturbations due to the peptides by calculating the order parameter profiles. The calculated order parameter profiles of the two leaflets in the system are shown in Figure 4 along with the profiles from a neat DMPC simulation (Damodaran and Merz, 1994) and an experimental profile for DPPC by Seelig and Seelig (1974). The experimental profile is at a lower reduced temperature (Seelig and Browning, 1978) ($T_r = 0.5$; $T_r = (T - T_c)/T_c$, where T_c is the gel liquid-crystalline phase transition temperature and T is the simulation temperature) than both neat DMPC ($Tr = 0.60$) and the peptide/DMPC simulations ($Tr = 0.67$) (Jacobs and White, 1986). Hence the experimental order parameters are higher than the calculated values. We have also shown the calculated standard deviations as error bars for the perturbed and unperturbed leaflets. Although standard deviations of the order of 0.03, 0.05, and 0.07 have been observed at the carbonyl, midchain, and methyl regions respectively, the average order parameters converge to a much better degree (typically within 0.01). In model-II (Figure 4b), the insertion of the peptide causes a slight increase in the order parameters at the carbonyl end of the alkyl chains while no clear difference has been observed from the neat system in the mid-chain and methyl terminal regions. In model-I (Figure 4a), however, a slight decrease has been observed at the carbonyl region while at the methyl end, order parameters are close to the neat DMPC values.

NMR measurements on DMPC bilayers with different Ala-X-Ala-O-*tert*-butyl tripeptides have shown that the peptide introduced disorder in the alkyl chains (Jacobs and White, 1987), the effect being more pronounced for residues with larger side chains such as Trp and Phe at the X position. However, this effect is not clearly borne out in our simulations. The overall perturbation caused by the peptide on the bilayer properties is minimal, particularly the dynamical properties. Spectral density plots calculated from velocity autocorrelation function for the alkyl chains and head groups are nearly identical for neat DMPC and the leaflet with the peptide (data not shown). This is probably due to a lower peptide:lipid ratio used in our simulation (1:8 compared to 1:5 used in NMR experiments).

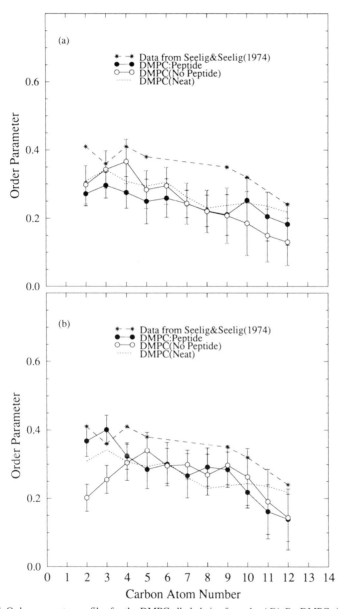

Figure 4. Order parameter profiles for the DMPC alkyl chains from the AFAtBu:DMPC simulations. Model-I (a); Model-II (b).

Peptide Dynamics

AFAtBu shows essentially a random conformation in solution while in the bilayer bound form, it adopts a preferred conformation with the Phe rings having close interaction with Ala-3 C_α and the *tert*-butyl group (Brown and Huestis, 1993). Although one of the two peptides in our MD simulations has remained in this conformation, the second peptide has its *tert*-butyl group somewhat extended into the alkyl chain region. We have also compared the peptide dynamics in the bilayer and in solution using RMS deviations for the heavy atoms. These

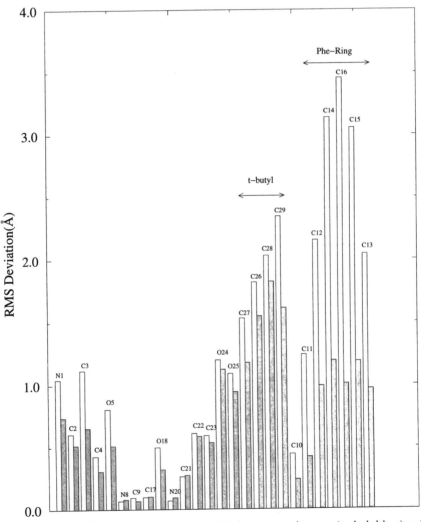

Figure 5. Calculated RMS deviations of the peptide heavy atoms in water (unshaded bars) and bilayer (shaded bars). Data shown for Model-II. The bilayer data were obtained as the average over both peptides in the model.

data also support this conclusion. The RMS deviations shown in Figure 5, not unexpectedly, are larger in the solution phase. More interesting is the significant increase in the RMS deviations for the Phe-ring (atoms C10–C13) in solution than in other regions, including the *tert*-butyl group (atoms C27–C29) of the peptide. Figure 5 also suggests that the *tert*-butyl group is more mobile in the bilayer, while in solution the Phe-ring has the highest motional freedom.

Fusion Inhibiting Peptide (Z-D-Phe-L-Phe-Gly): N-Methyl DOPE

Unlike the previous case, no structural data regarding the location of the FIPs or other parameters, such as the change in the area per lipid, have been available for building a starting structure for the FIP:N-Methyl DOPE system. In order to get insights into the orientation of these peptides in lipid bilayers, several molecular dynamics simulations have been carried out in different solvent environments. We have simulated the peptide solvated in water, hexane, and a mixed solvent environment of hexane and water using the SPC/E water model (Berendsen et al, 1987) and a united atom representation for hexane (Jorgensen et al, 1984). The starting geometry for the mixed solvent simulation has been created by first solvating the peptide in a shell of water molecules and subsequently solvating the peptide-water configuration in hexane. These simulations have been run with a 60 ps of equilibration phase followed by 120 ps of accumulation of coordinates and velocities. Graphical visualization of the snapshots from the mixed solvent runs has shown that the water molecules have accumulated on one side of the peptide such that peptide/water hydrogen bonds could be retained, while minimizing water-hexane contacts. A typical snapshot obtained from ZfFG⁻ in water:hexane medium is shown in Figure 6. The phenylalanyl

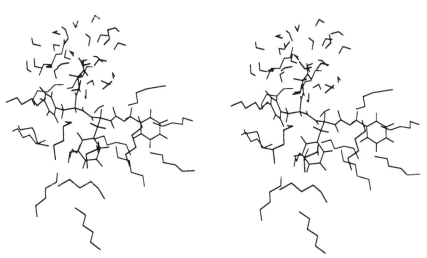

Figure 6. Stereo view of ZfFG⁻ in water-hexane environment.

side chains and the carbobenzoxy ring also have developed favorable van der Waals contacts with hexane molecules. Thus, from these simulations, we consider an orientation of the peptide in the bilayer-water interface with the plane of the peptide backbone parallel to the bilayer surface (as opposed to an orientation perpendicular to the bilayer surface) as a favorable starting structure. The peptides placed in this manner have the hydrophobic phenyl rings intercalated into the membrane interior (roughly at the carbonyl region) while the C-terminal region of the peptide is in the interfacial region of the bilayer as shown in Figure 7. This is facilitated by the fact that the C-terminal residue is a (relatively small) glycine residue. Thus the structural characteristics of the peptide are designed for such an orientiation in the bilayer bound form. We have

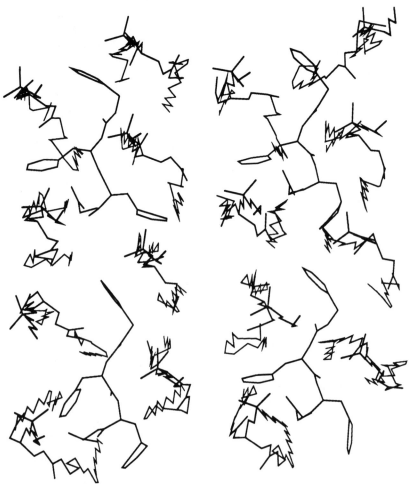

Figure 7. A snap shot of the starting structure for ZfFG0:N-Me-DOPE simulations showing the placement of the Phe side chains between lipid alkyl chains. Hydrogen atoms on the lipids and peptides are not shown for clarity. However, they were included in the calculations.

built our bilayer-peptide model by docking four peptides to one of the leaflets of 16 N-Methyl DOPE lipids in this geometry. Since the intercalation of the peptides may be expected to increase the surface area, we have used an area per lipid 10% larger than the neat N-Me-DOPE (72.6 Å^2 versus 66 Å^2). The bilayer model has been solvated by adding about 25 SPC/E waters (Berendsen et al, 1987) in the head group region. All the simulations are run at 310 K, well above the gel to liquid-crystalline phase transition temperature of neat N-Me-DOPE (258–263K). We have carried out two simulations; with the peptide C-terminus both in the neutral (ZfFG0) and negatively charged (ZfFG$^-$) states. The final configuration from the run with neutral peptides (ZfFG0:N-Me-DOPE) has been used to generate the starting configuration for the run with negatively charged peptides (ZfFG$^-$:N-Me-DOPE). Two other related simulations carried out are: (1) a neat N-Me-DOPE bilayer containing 32 (16 × 2) lipids at a surface area of 66 Å^2 per lipid; (2) an expanded neat N-Me-DOPE bilayer with a surface area of 72.6 Å^2, the same value used in the bilayer-peptide model; and (3) a relatively shorter run ($\approx$100 ps equilibration + 80 ps sampling) of (ZfFG0:N-Me-DOPE) in which four ZfFG0 peptides are intercalated on both monolayers forming the bilayer. Coordinates and velocities have been collected using constant volume periodic boundary conditions. One more improvement over the previous example is that all the hydrogen atoms are also included in the simulations.

Lipid Perturbations

The most interesting property from the point of view of fusion inhibition is the behavior of the order parameters. The order parameter profiles for the leaflet containing the peptides (both neutral and charged models) are shown in Figure 8. We have also included in this figure the profile from the neat N-Me-DOPE simulations at the two surface areas. The intercalation of the FIPs gives rise to higher order parameters near the carbonyl end (the upper part) of the alkyl chains. This is due to the additional van der Waals interactions between the alkyl chains of the lipids and the phenyl rings of the FIPs. The order parameters are higher in the charged model due to the insertion of a larger fraction of the phenyl rings. For carbons 10–17, the order parameters are lower for the bilayer-peptide system compared to the original neat N-Me-DOPE. This decrease in order is mainly due to the larger area per lipid used in the bilayer-peptide system. We have verified this by calculating the order parameter profile from the expanded neat N-Me-DOPE simulation. Expanding the surface area by 10% results in a dramatic decrease in the order parameters from the original neat simulation. Comparison of the profile from the bilayer-peptide and the expanded neat N-Me-DOPE simulations also suggests that the peptides indeed do have an ordering effect even in the methyl terminal region, while the effect is much stronger near the carbonyls. We also note that, while the order parameters for the monolayer without the peptides are very similar to those from the expanded N-Me-DOPE simulation, the values are somewhat lower in the region of carbon

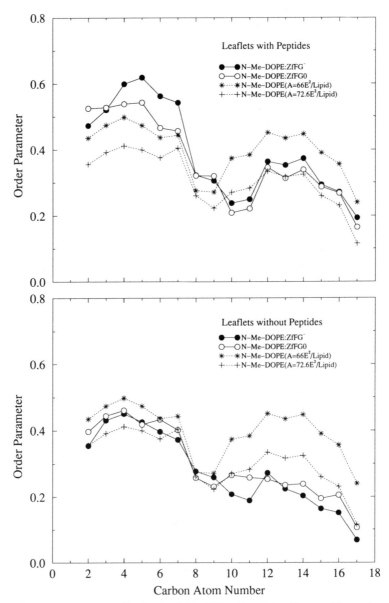

Figure 8. Order parameter profiles for the N-Me-DOPE alkyl chains. The neat bilayer data were averaged over both leaflets constituting the bilayer while the perturbed bilayer data were averaged only over one monolayer with or without the peptides as the case may be.

atoms 10–17. The is probably due to the transmission of a perturbation from the leaflet containing the peptides to the one without the peptides resulting in the loss of interleaflet alkyl contacts. For example, the leaflet containing the peptides responds to their presence by forming pockets around the peptides which results in reduced interleaflet contacts. However, the order parameters calculated from the model with peptides intercalated on both monolayers show similar (increased) values for both monolayers (data not shown).

We would like to consider the influence of the FIPs on the alkyl chain order as the manifestation of a plausible molecular mechanism for fusion inhibition. The increased packing density due to the presence of the phenylalanyl side chains in the alkyl chain region gives rise to larger order parameters and renders the bilayer gel-like instead of driving it towards the H_{II} phase, giving rise to the formation of fusion intermediates. The effect is primarily due to van der Waals interactions as evidenced by the fact that both neutral and negatively charged peptides give rise to larger order parameters. However, the partial negative charge helps the peptide C-terminus to be better anchored at the head group region, while the side chains arranged in a propeller-like structure can penetrate between the alkyl chains without causing bilayer disruption.

Density Profiles

The density profiles for the lipid molecules, FIPs, and water molecules along the bilayer normal from the ZfFG⁻:N-Me-DOPE simulation are shown in Figures 9 and 10. The bilayer thickness, as estimated by the peak-to-peak distance between N-methyl ethanolamine density distributions, is ≈48.2 Å for the neat system, ≈43.8 Å for the neutral FIP model, and 45.4 Å for the charged FIP model. X-ray diffraction studies by Gruner et al (1988) have determined the neat bilayer thickness to be 39 ± 5 Å. Thus, compared to the experimental data, the bilayer thickness for the neat system is somewhat larger and, as a result, slightly more ordered. The smaller dimension for the bilayer with the peptides could be due partly to increased disorder in the alkyl chains caused by the larger area per lipid used in the model.

The FIPs' distribution overlaps partially with those of the lipid head groups, glycerol backbone and alkyl chains. The penetration of the side chains is further illustrated by the distribution of the rings shown separately in Figure 10. The side chains of the charged peptides have been partitioned significantly better than the neutral peptide. Of the four peptides included in the model, three have all their rings intercalated while a fourth one has two rings in the alkyl chain region and a third ring remaining in the head group region. However, in the neutral peptide case, only one peptide has all of its side chains intercalated into the lipid hydrocarbon region. Two peptides have 2 rings in the alkyl chain region while the third ring remains in the head group region. The fourth peptide has only one of the rings in the alkyl chain region and a third ring remaining in the head group region.

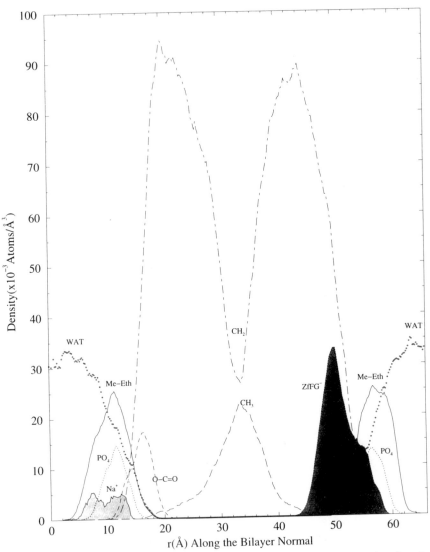

Figure 9. Density profiles along the bilayer normal for the peptide, lipid, and water regions from the N-Me-DOPE:ZfFG⁻ simulation. In this figure the counterion (Na⁺) distribution has been amplified by a factor of 10 to have visible intensity.

The water distribution in the density profiles suggests a reduction in the extent of water penetration in the monolayer with the peptides. Considering that these peptides have large hydrophobic side chains, this observation is not unexpected. This can be best seen from the pair distribution functions (pdfs) for water-carbonyl interactions in neat bilayer and bilayer-peptide simulations given

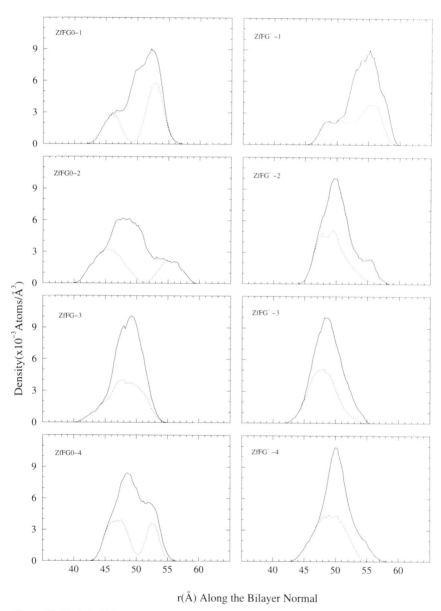

r(Å) Along the Bilayer Normal

Figure 10. Total (solid lines) and side chain (dotted lines) probability distributions for the peptides along the bilayer normal. The bilayer center is on the low r side (below ≈45 Å) and solvent/head group region is on the high r side (≈50 Å and above) in these plots.

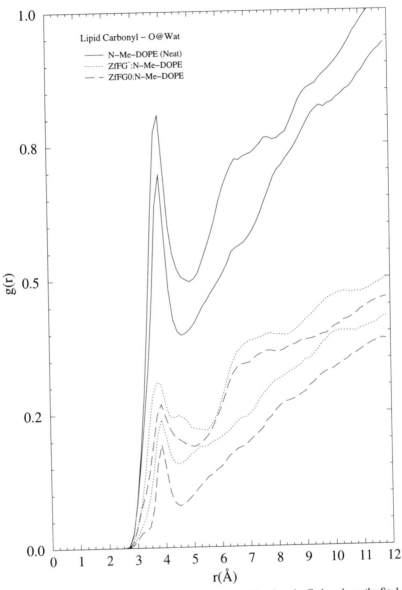

Figure 11. Pair distribution functions for water from the lipid carbonyls. Carbonyls on the Sn-1 and Sn-2 positions are shown independently.

in Figure 11, which shows that the neat pdfs are greater in intensity than the pdfs for the leaflets containing peptides. This observation, combined with the order parameter data, strongly suggests that the FIPs force the leaflet containing the peptides to become more gel-like even at a temperature at which the liquid-crystalline phase is normally observed.

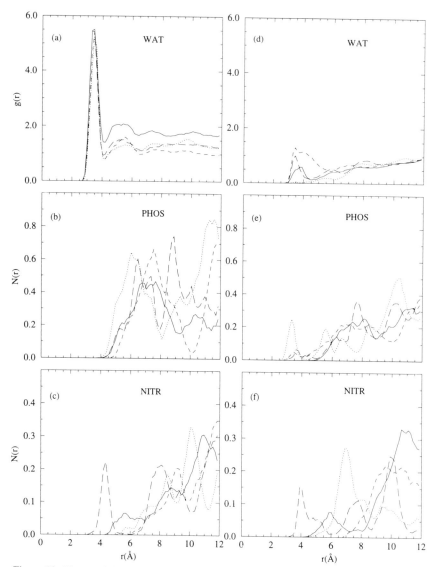

Figure 12. Water and lipid head group environments of the C-termini of the charged (a–c) and neutral (d–f) peptides in the bilayer. Abbreviations: WAT: water; PHOS: nonesterified oxygens of the lipid phosphate group and NITR: nitrogen atom of the N-Methyl Ethanolamine head group. The environments from the four peptides are shown independently.

The partial negative charge on the C-terminal carboxylate group of the peptides helps to anchor them in the head group region of the bilayer. The peptide-solvent-lipid interactions involved in this process have been analyzed using the water pair distributions and lipid histograms given in Figure 12. In general the carboxyl groups have been found to develop indirect interactions with the phosphate groups through their hydration shells (Figure 12a,b), rather than forming direct hydrogen bonds with the R-NH$_2$Me$^+$ group of ethanolamine (Figure 12c). However, one peptide does not follow this trend and this can be seen as a clear first shell in the carboxyl-nitrogen histogram in Figure 12c for one peptide and an absence of a peak for the remaining peptides. This ob-

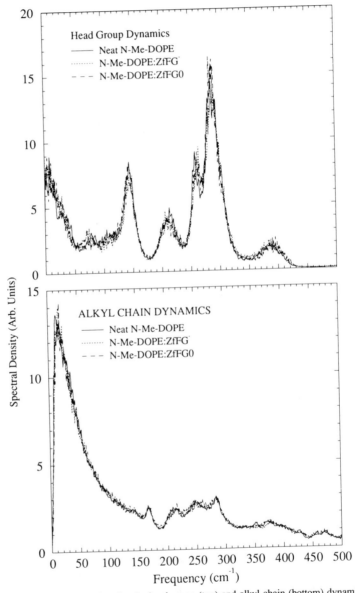

Figure 13. Spectral density plots for the head group (top) and alkyl chain (bottom) dynamics from the neat N-Me-DOPE-Me and N-Me-DOPE:FIP simulations.

servation is reasonable, since the R-NH$_2$Me$^+$ groups are located farther from the peptide C-termini than the phosphate groups. As for the neutral peptides, the pdfs/histograms for these interactions are more broadened and less defined, resulting in a lack of the anchoring effect (see Figures 12d–f). This data is also consistent with the fact that the model with charged peptides shows more uniform intercalation of the rings and has higher order parameters in the alkyl chains.

Lipid Dynamics

It is interesting that the presence of the FIPs has not affected the short time lipid dynamical properties significantly in spite of their strong influence on the order parameters. The spectral densities obtained for the head group and alkyl chain motions for the perturbed lipids are almost identical to those from the neat N-Me-DOPE simulations as shown in Figure 13. In both the charged and uncharged states of the peptides the predominant mode of interaction with the lipids is van der Waals, while the main electrostatic interactions are mostly limited to the phosphate groups mediated by intervening water molecules. The phosphate groups are relatively immobile and do not contribute to the head group dynamics significantly and, hence, we would expect that a perturbant might not have a strong influence on the short timescale dynamics associated with this region of the lipid molecules. Even in alkyl chain dynamics it appears that a similar effect is present, i.e., the contribution from carbon atoms away from the FIPs are masking that from the carbon atoms at the carbonyl end which have stronger interactions with the peptide. Thus it appears that the dynamics has to be studied on a much longer time scale to detect any effect associated with the change in order parameters. However, the spectral density peaks of the peptides show a noticeable broadening in the bilayer bound form compared to the free peptide in the solution (data not shown).

Conclusions, Future Outlook

We have presented two cases of the application of molecular dynamics simulation techniques for studying lipid-peptide interactions using an explicit representation. We feel this approach can be very insightful for such problems, as shown by these examples. The first example shows the feasibility of modeling membrane bound states of simple peptides in comparison with the experimental results, while in the case of fusion inhibiting peptides we have applied this approach to a more complex problem.

Many improvements are possible both in the methodology as well as in the models. For example, advanced pressure and temperature coupling techniques have been developed for MD simulations, and incorporation of long range electrostatic interactions based on Ewald sum will give a better representation of the model. These endeavors are currently underway in our laboratory. As far as the models themselves are concerned, the most important aspect to consider is the starting geometry, including the sampling of more regions of conformational space. As we have mentioned earlier, several relatively shorter (typically about 100 ps of equilibration, 150–200 ps of sampling) simulations starting from different initial configurations may be required to get very good averages, particularly in the interface regions where the molecular mobilities are rather small. Although this would mean an increase in the simulation length, with more accessibility of computational speed this approach seems not only feasible but can be adopted to investigate such problems as studying relative binding affinities of mutants of a peptide using free energy perturbation techniques.

We have not been able to observe any significant perturbation on the lipid dynamics induced by the guest molecules in the two examples we have studied, although the peptides' dynamics were clearly affected by intercalating into the bilayer. While this was somewhat surprising, we believe this is due to the relatively smaller nature of the guests, location of the guests near the interface, and finally, due to the shorter time scale on which the dynamics were investigated. It is likely that dynamics have to be studied over a much longer time scale (several hundreds of picoseconds) to detect any difference corresponding to the increased order parameters we have observed in the case of FIPs interacting with N-Me-DOPE bilayers, while we have been able to analyze them typically only for $\approx$60 ps. It may be possible to investigate the perturbation of the lipid dynamics on a much longer time scale either by using stochastic boundary molecular dynamics or long time step molecular dynamics simulations currently being developed based on normal-mode and implicit integration techniques (Zhang and Schlick, 1994).

REFERENCES

Allen MP, Tildesley DJ (1987): *Computer Simulation of Liquids*. Oxford: Clarendon Press

Alper HE, Stouch TR (1995): Orientation and diffusion of a drug analogue in biomembranes: Molecular dynamics simulations. *J Phys Chem* 99:5724–5731

Bassolino-Klimas D, Alper HE, Stouch TR (1995): Mechanism of solute diffusion through lipid bilayer membranes by molecular dynamics simulation. *J Am Chem Soc*: in press

Bassolino-Klimas D, Alper HE, Stouch TR (1993): Solute diffusion in lipid bilayer membranes: An atomic level study by molecular dynamics simulation. *Biochemistry* 32:12624–12637

Bauerle HD, Seelig J (1991): Interaction of charged and uncharged calcium channel antagonists with phospholipid membranes. Binding equilibrium, binding enthalpy and membrane location. *Biochemistry* 30:7203–7211

Bentz J, ed. (1993): *Viral fusion mechanisms*. Boca Raton: CRC Press

Bentz J, Alford D, Ellens H (1992): Liposomes, membrane fusion and cytoplasmic delivery. In: *The Structure of Biological Membranes*, Yeagle PL, ed. Boca Raton: CRC Press

Berendsen HJC, Grigera, JR, Straatsma TP (1987): The missing term in effective pair potentials. *J Phys Chem* 91:6289–6271

Beschiaschvili G, Seelig J (1992): Peptide binding to lipid bilayers. nonclassical hydrophobic effect and membrane-induced pK shifts. *Biochemistry* 31:10044–10053

Besler BH, Merz KMJ, Kollman PA (1990): Atomic charges derived from semiempirical methods. *J Comput Chem* 11:431–439

Blobel CP, Wolfsberg TG, Turck CW, Myles DG, Primakoff P, White JM (1992): A potential fusion peptide and an integrin ligand domain in a protein active in sperm-egg fusion. *Nature* 356:248–252

Brooks CL, Karplus M, Pettit BM (1988): *Proteins: A Theoretical Perspective of Dynamics, Structure, and Thermodynamics.* New York: John Wiley and Sons

Brooks CLI, Brunger A, Karplus M (1985): Active site dynamics in protein molecules: A stochastic boundary molecular dynamics approach. *Biopolymers* 24:843

Brown JW, Huestis WH (1993): Structure and orientation of a bilayer-bound model tripeptide. A ^1H NMR study. *J Phys Chem* 97:2967–2973

Burger KNJ, Verkleij AJ (1990): Membrane fusion. *Experientia* 46:631–644

Chiu S-W, Jakobsson E, Subramaniam S, McCammon, JA (1991): Time correlation analysis of simulated water motion in flexible and rigid gramicidin channels. *Biophys J* 60:273

Chiu SW, Novotny JA, Jakobsson E (1993): The nature of ion and water barrier crossings in a simulated ion channel. *Biophys J* 64:98–108

Chiu SW, Subramaniam S, Jakobsson E, McCammon JA (1989): Water and polypeptide conformations in the gramicidin channel. A molecular dynamics study. *Biophys J* 56:253–261

Damodaran K, Merz KM, Jr. (1995): Interaction of the fusion inhibiting peptide carbobenzoxy-D-Phe-L-Phe-Gly with N-methyldioleoylphosphatidylethanolamine lipid bilayers. *J Amer Chem Soc* 117:6561–6571

Damodaran KV, Merz KM, Jr. (1994a): Computer simulation of lipid systems. In: *Reviews in Computational Chemistry,* Lipkowitz KB, Boyd DB, eds. New York: VCH Publishers

Damodaran KV, Merz KM, Jr. (1994b): A comparison between DMPC and DLPE based lipid bilayers. *Biophys J* 66:1076–1087

Damodaran KV, Merz KM, Jr., Gaber BP (1995). Interaction of small peptides with lipid bilayers. *Biophys J* 69:1299–1308

Damodaran KV, Merz KM, Jr., Gaber BP (1992): Structure and dynamics of the dilauroylphosphatidylethanolamine lipid bilayer. *Biochemistry* 31:7656–7664

Davis JH, Clare DM, Hodges RS, Bloom M (1983): Interaction of a synthetic amphiphilic polypeptide and lipids in a bilayer structure. *Biochemistry* 22:5298–5305

Deamer DW, Bramhall J (1986): Permeability of lipid bilayers to water and ionic solutes. *Chem Phys Lipids* 40:167–188

Dickinson E (1985): Brownian dynamics with hydrodynamic interactions: The application to protein diffusional problems. *Chem Soc Rev* 14:421

Edholm O, Jahnig F (1988): The structure of a membrane-spanning polypeptide studied by molecular dynamics. *Biophys Chem* 30:279–292

Egberts E, Berendsen HJC (1988): Molecular dynamics simulation of a smectic liquid crystal with atomic detail. *J Chem Phys* 89:3718–3732

Ellens H, Bentz J, Szoka FC (1986): Fusion of phosphatidylethanolamine-containing liposomes and mechanism of the $L_a - H_{II}$ phase transition. *Biochemistry* 25:4141–4147

Ellens H, Seigel DP, Alford D, Yeagle PL, Boni L, Lis LJ, Bentz QPJ (1989): Membrane fusion and inverted phases. *Biochemistry* 28:3692–3703

Epand RM, Epand RF, Richardson CD, Yeagle PL (1993): Structural requirements for the inhibition of membrane fusion by carbobenzoxy-D-Phe-Phe-Gly. *Biochim Biophys Acta* 1152:128–134

Field MJ, Bash PA, Karplus MJ (1990): A combined quantum mechanical and molecular mechanical potential for molecular dynamic simulations. *J Comput Chem* 11:700–733

Gao J, Xia X (1992): A priori evaluation of aqueous polarization effects through Monte-Carlo QM-MM simulations. *Science* 258:631–635

Gennis RB (1989): *Biomembranes: Molecular Structure and Function* New York: Springer-Verlag

Gruner SM, Tate MW, Kirk GL, So PTC, Turner DC, Keane DT, Tilcock CPS, Cullis PR (1988): X-ray diffraction study of the polymorphic behavior of N-methylated dioleoylphosphatidylethanolamine. *Biochemistry* 27:2853–2866

Huang P, Perez JJ, Loew GH (1994): Molecular dynamics simulations of phospholipid bilayers. *J Biomol Str Dyn* 11:927–956

Huschilt JC, Hodges RS, Davis JH (1985): Phase equilibria in an amphiphilic peptide-phospholipid model membrane by deuterium nuclear magnetic resonance difference spectroscopy. *Biochemistry* 24:1377–1386

Jacobs RE, White SH (1989): The nature of the hydrophobic binding of small peptides at the bilayer interface: Implications for the insertion of transbilayer helices. *Biochemistry* 28:3421–3437

Jacobs RE, White SH (1987): Lipid bilayer pertubations induced by simple hydrophobic peptides. *Biochemistry* 26:6127–6134

Jacobs RE, White SH (1986): Mixtures of a series of homologous hydrophobic peptides with lipid bilayers: A simple model system for examining the protein-lipid interface. *Biochemistry* 25:2605

Jorgensen WL, Madura JD, Swenson CJ (1984): Optimized intermolecular potential functions for liquid hydrocarbons. *J Am Chem Soc* 106:6638–6646

Kelsey DR, Flanagan TD, Young J, Yeagle PL (1991): Inhibition of sendai virus fusion with phospholipid vesicles and human erythrocyte membranes by hydrophobic peptides. *Virology* 182:690–702

Kelsey DR, Flanagan TD, Young J, Yeagle PL (1990): Peptide inhibitors of enveloped virus infection inhibit phospholipid fusion and sendai virus fusion with phospholipid vesicles. *J Biol Chem* 265:12178–12183

Lewis BA, Engelman DM (1983): Lipid bilayer thickness varies linearly with acyl chain length in fluid phosphatidylcholine vesicles. *J Mol Biol* 166:211–217

Lichtenberger LM, Wang ZM, Romero JJ, Ulloa C, Perez JC, Giraud MN, Barreto JC (1995): Non-steroidal anti-inflammatory drugs (NSAIDs) associate with zwitterionic phospholipids: Insight into the mechanism and reversal of NSAID-induced gastrointestinal injury. *Nature Med* 1:154–158

Marrink SJ, Berkowitz M, Berendsen HJC (1993): Molecular dynamics simulation of a membrane/water interface: The ordering of water and its relation to the hydration force. *Langmuir* 9:3122–3131

McCammon JA, Harvey SC (1987): *Dynamics of Proteins and Nucleic Acids.* New York: Cambridge University Press

Merz KM, Jr (1992): Analysis of a large database of electrostatic potential derived atomic point charges. *J Comput Chem* 13:749–767

Mezei M, Beveridge DL (1986): Free energy simulations. *Ann NY Acad Sci* 482:1–23

Milik M, Skolnick J (1993): Insertion of peptide chains into lipid membranes: An off-lattice Monte Carlo dynamics model. *Proteins: Struc Func Genet* 15:10–25

Milik M, Skolnick J (1992): Spontaneous insertion of polypeptide chains into membranes: A Monte Carlo study. *Proc Natl Acad Sci USA* 89:9391–9395

Muga A, Neugebauer W, Hirama T, Surewicz WK (1994): Membrane interaction and conformational properties of the putative fusion peptide of PH-30, a protein active in sperm-egg fusion. *Biochemistry* 33:4444–4448

Murata M, Nagayama K, Ohnishi S (1987): Membrane fusion activity of succinylated melittin is triggered by protonation of its carboxyl groups. *Biochemistry* 26:4056–4062

Richardson CD, Choppin PW (1983): Oligopeptides that specifically inhibit membrane fusion by paramyxoviruses: Studies on the site of action. *Virology* 131:518–532

Richardson CD, Scheid A, Choppin PW (1980): Specific inhibition of paramyxovirus and myxovirus replication by oligopeptides with amino acid sequences similar to those at the N-termini of the F_1 or AH_2 viral polupeptides. *Virology* 105:205

Roux B, Karplus M (1994): Molecular dynamics simulations of the gramicidin channel. *Annu Rev Biophys Biomol Struct* 23:731–761

Roux B, Karplus M (1993): Ion transport in the gramicidin channel: Free energy of the solvated right-handed dimer in a model membrane. *J Am Chem Soc* 115:3250–3262

Scott HL (1991): Lipid-cholesterol interactions (Monte Carlo simulations and theory). *Biophys J* 59:445–455

Seelig A, Seelig J (1974): The dynamic structure of fatty acyl chains in a phospholipid bilayer measured by deuterium magnetic resonance. *Biochemistry* 13:4839–4845

Seelig J, Browning JL (1978): General features of phospholipid conformation in membranes. *FEBS Lett* 92:41

Seelig J, Ganz P (1991): Nonclassical hydrophobic effect in membrane binding equilibria. *Biochemistry* 30:9354–9359

Stanton RV, Little LR, Merz KM Jr (1995): Quantum free energy perturbation study within a PM3/MM coupled potential. *J Phys Chem* 99:483–486

Stegmann T (1993)a: Influenza hemagglutinin-mediated membrane fusion does not involve inverted phase lipid intermediates. *J Biol Chem* 268:1716–1722

Stegmann T (1993)b: Membrane fusion-inhibiting peptides do not inhibit influenza virus fusion or the Ca^+-induced fusion of negatively charged vesicles. *J Biol Chem* 268:26886–26892

Stouch TR (1993): Lipid membrane structure and dynamics studied by all-atom molecular dynamics simulations of hydrated phospholipid bilayers. *Mol Simulation* 102–106:335–362

Venable RM, Zhang Y, Hardy BJ, Pastor RW (1993): Molecular dynamics simulations of a lipid bilayer and of hexadecane: An investigation of membrane fluidity. *Science* 262:223–226

Weiner SJ, Kollman PA, Case DA, Singh UC, Ghio C, Alagona G, Profeta S, Weiner P (1984): A new force field for molecular mechanical simulation of nucleic acids and proteins. *J Am Chem Soc* 106:765–784

White JM (1992): Membrane fusion. *Science* 258:917–924

Wimley WC, White SH (1993): Membrane paritioning: Distinguishing bilayer effects from the hydrophobic effect. *Biochemistry* 32:6307–6312

Wipff G, ed. (1994): *Computational Approaches in Supramolecular Chemistry,* NATO ASI Series. Dordrecht: Kluwer Academic Publishers

Woolf TB, Desharnais J, Roux B (1994): Structure and dynamics of the sidechains of gramicidin in a DMPC bilayer. In: *Computational Approaches to Supramolecular Chemistry,* Wipff G, ed. Dordrecht: Kluwer Academic Publishers

Woolf TB, Roux B (1994): Molecular dynamics simulation of the gramicidin channel in a phospholipid bilayer. *Proc Natl Acad Sci USA* 91:11631–11635

Xing J, Scott HL (1989): Monte Carlo studies of lipid chains and gramicidin A in a model membrane. *Biochem Biophys Res Comm* 165:1–6

Yeagle PL, Young J, Hui SW, Epand RM (1992): On the mechanism of inhibition of viral and vesicle membrane fusion by carbobenzoxy-D-phenylalanyl-L-phenylalanylglycine. *Biochemistry* 31:3177–3183

Zhang G, Schlick T (1994): The Langevin/implicit-Euler/normal-mode scheme for molecular dynamics at large time steps. *J Chem Phys* 101:4995

Part IV

MEMBRANE PROTEINS

This final section begins with a chapter by Barbara Seaton and Mary Roberts (Chapter 12) wherein they describe peripheral membrane proteins. In contrast to integral membrane proteins, in which a large portion of the protein is embedded in the lipid bilayer, subtle factors affecting lipid-protein interactions can play a key role in modulating the association of peripheral membrane proteins with the membrane surface, and consequently, their function. In Chapter 13, Thomas Heimburg and Derek Marsh describe what is known about the thermodynamics of protein/lipid interactions and how this information can be used to better understand the protein/lipid interface. In Chapter 14, Ole Mouritsen and Paavo Kinnunen use simplified molecular models to describe the important role played by lipid and protein composition in modulating the functional characteristics of membranes. The properties of "neat" lipid membranes are contrasted with those in which proteins are present.

The next two chapters deal with a very important problem in biology: the prediction of protein structure from sequence (the so-called protein folding problem). Much attention has been paid to predicting the structure of soluble proteins, but in Chapter 15, Xiche Hu, Dong Xu, Ken Hamer, Klaus Schulten, Juergen Koepke, and Hartmut Michel describe an effort to predict the structure of an integral membrane protein. Protein folding in a non-polar environment is an interesting problem since it involves the exposure of non-polar surfaces (as opposed to polar surfaces in soluble proteins) and the burial of polar faces into the protein interior (as opposed to non-polar surfaces in soluble proteins). Jeff Skolnick and Mariusz Milik (Chapter 16) use stochastic simulations based on simplified mean field models to examine how peptides and small proteins insert into a membrane to adopt spontaneously their folded configuration. Small peptides, magainin and M2d, as well as filamentous bacteriophage proteins, fd and Pf1, are considered. Their membrane-bound structure has been determined using NMR techniques; therefore, they represent ideal systems to test the validity of theoretical methods.

The structure and dynamics of the Pf1 coat protein interacting with a phospholipid membrane is also investigated in Chapter 17 by Benoît Roux and Thomas Woolf. In this final chapter, MD is used to generate a trajectory of Pf1 in a fully hydrated phospholipid membrane with all atomic details. It is particularly interesting to compare the results obtained from this detailed simulation with those obtained from the simplified mean field model described in Chapter 16. In addition, the chapter describes a step-by-step protocol for performing a detailed simulation of an intrinsic membrane protein in a membrane, starting from the construction and the equilibration of the initial configuration of the system.

12

Peripheral Membrane Proteins

BARBARA A. SEATON AND MARY F. ROBERTS

The founding principle behind structural biology is that form equals function. Nowhere is this relation more apparent than in proteins that operate in environments in which aqueous and lipophilic phases meet. This biphasic environment strongly influences both the structure and function of these proteins. Membrane proteins can be classified as integral or peripheral, depending upon the nature of the protein-membrane association. For proteins in the former category, the membrane is an integral part of their structures, which are greatly perturbed by disruption of the membrane by detergents. The extent of interaction with the lipid bilayer is usually obvious by simple inspection of the protein structure because of what is known about lipid physical chemistry, i.e., that there is a large energy cost to burying uncompensated polar or charged protein residues in a strongly hydrophobic milieu. For example, the membrane-embedded portions of transmembrane proteins, such as bacterial reaction centers and porins, are localized by the extensive hydrophobic regions, which are frequently bordered by aromatic side-chains, found on the surfaces of these proteins. For these transmembrane proteins, whose functional roles are to transfer energy or substances from one aqueous pool to another across a nonaqueous barrier, the membrane serves to organize the protein structure. In contrast to integral membrane proteins, those classified as peripheral can be released from their membrane attachment through gentler means, without disruption of either membrane or protein structure. Reversible association with membranes is frequently required for the biological functions of these proteins. Fewer structure-function generalizations have been drawn about their interactions with the lipid phase because their complex and dynamic behavior is difficult to classify on the basis of simple chemical principles. However given an adequate data base, it may be possible to recognize common types of protein-membrane interactions in peripheral as well as integral membrane proteins.

Biological Membranes
K. Merz, Jr. and B. Roux, Editors
© Birkhäuser Boston 1996

This chapter specifically addresses the need for a comparative look at peripheral membrane proteins that operate at the oil-water interfaces of membrane surfaces and other lipid aggregates. The proteins included for comparison are considered peripheral membrane proteins in that they may interact reversibly with membranes and generally do not require membrane disruption to break the protein-membrane association. For simplicity, the term membrane is used here to refer to the lipid phase, but the broader and more correct term is oil-water interface. Many interfacial proteins act on lipid aggregates (e.g., micelles, fatty globules) other than bilayers though the nature of the association is similar in both cases. In the following sections, structural and functional features of peripheral membrane (or interfacial) proteins are reviewed with the goal of identifying common themes. The proteins described in this chapter were chosen both because their molecular structures are known at high resolution (through X-ray crystallography or nuclear magnetic resonance, NMR) and because structural data are available regarding the mechanisms of protein-membrane association. That there are as yet relatively few well-studied examples of this protein class (five types of proteins are described in this chapter) reflects the newness of this area of structural biology. Yet several common structure/function themes do emerge from this preliminary comparison.

Phospholipases, Kinetic Models for Interfacial Activation

Phospholipases are a diverse series of enzymes designed for phospholipid catabolism. An important feature of many of the phospholipases is that while they are, in general, water-soluble, their substrates are insoluble and organized in a two-dimensional matrix. Hence, these phospholipases can be considered peripheral proteins since they are not embedded in the membrane. Since phospholipases often require Ca^{2+} (or other metal ions) for either activity and/or membrane association, they can be compared to other peripheral membrane proteins that associate with membranes in a Ca^{2+} and phospholipid dependent manner. These catabolic esterases are classified according to the phospholipid ester bond hydrolyzed (Figure 1). Phospholipase A_2 (PLA$_2$) catalyzes the hydrolysis of the sn-2 fatty acyl bond to liberate a fatty acid and a lysophospholipid (1-acyl-glycerophospholipid); phospholipase A_1 (PLA$_1$) activities hydrolyze the sn-1 fatty acyl chain and also exhibit activity toward the sn-2 chain. Phospholipase C (PLC) promotes hydrolysis of the glycerophosphate ester bond to liberate a phosphorylated alcohol and diacylglycerol (DAG). Phospholipase D (PLD) works on the headgroup phosphodiester bond to produce phosphatidic acid and an alcohol; it also has a transphosphorylation activity. These enzymes have been implicated as key controlling enzymes in a wide variety of biological processes including production of arachidonic acid for eicosanoid biosynthesis and regulation of intracellular Ca^{2+} levels, and they also produce lipid (e.g., DAG, phosphatidic acid and lysophosphatidic acid) and water-

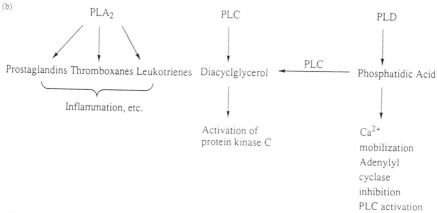

Figure 1. a) The sites of phospholipase cleavage on a phosphatidylcholine molecule. b) Second messenger molecules, formed as a result of phospholipid breakdown by phospholipases, and the processes they mediate.

soluble (e.g., inositol phosphates) second messenger molecules, all of which play important regulatory roles in cellular activation.

A common feature of lipolytic enzymes (lipases as well as phospholipases) is interfacial activation which is the kinetic preference they display toward substrates in aggregated as opposed to monomeric forms (Verheij et al, 1981). This kinetic feature was first noted by Roholt and Schlamowitz (1961) and later investigated in detail by de Haas and coworkers with pancreatic PLA$_2$ (De Haas et al, 1971; Pieterson et al, 1974). The mechanism by which substrate aggregation leads to catalytic rate increases may be quite different depending on the enzyme and its substrate specificity. For phospholipases, the degree of activation observed toward phospholipid aggregates varies among sources of the enzyme (e.g., pancreatic versus cobra venom PLA$_2$), specificity of the phospholipase (PLA$_2$ exhibits a 20–100-fold activation while nonspecific and PI-specific bacterial PLC enzymes exhibit only a 2–3-fold and 5–6-fold activation, respectively, upon substrate micellization), and type of aggregate employed (micelle versus bilayer) (De Haas, 1971; Lewis et al, 1993; Roberts, 1991; Roberts and Dennis,

1982; Wells, 1974). For PC, the most commonly used phospholipid substrate, packing the phospholipid in a micelle rather than a bilayer array leads to a higher apparent rate of hydrolysis (Roberts, 1991; Roberts and Dennis, 1989). It has been suggested that the tighter headgroup packing in bilayers reduces enzyme access to the substrate.

A complete understanding of interfacial catalysis and other kinetic hallmarks of phospholipases requires structural analysis of these enzymes both alone and in complexes with substrate analogs. The following sections provide an overview of phospholipase molecular structures, key interactions of the protein with substrate analogs, and enzyme modifications (e.g., phosphorylation) and/or cofactors (e.g., divalent cations) that modulate phospholipase activity.

Phospholipase A_2 (PLA$_2$)

Classified as A_2 by the location of the scissile bond (Figure 1), this phospholipase exists in two forms, secreted (sPLA$_2$) and cytosolic (cPLA$_2$). They are structurally and mechanistically distinct, though both are classified as PLA$_2$ enzymes by virtue of their common hydrolysis reaction. Considerably more data have been obtained on sPLA$_2$, which is one of the best-characterized peripheral membrane proteins. Due to the more recent discovery of cPLA$_2$, this form of the enzyme lacks the extensive structural data base of its secretory counterpart.

SECRETED PLA$_2$, (sPLA$_2$)

Extracellular forms of PLA$_2$ have been detected and isolated from a wide variety of sources including snake, bee and wasp venoms, pancreas, kidney, bacteria, and virtually every mammalian tissue that has been studied. Typically, these enzymes have molecular weights between 13 and 15 kDa. There is substantial homology among the different types of secreted PLA$_2$ enzymes. Best studied are the enzymes from snake venoms and from pancreatic tissues. The physiological role of pancreatic PLA$_2$ is the initial digestion of phospholipid components in dietary fat, whereas the toxins in snake venoms serve to immobilize prey promoting cell lysis and membrane disruption. The pancreatic sPLA$_2$ is produced as a proenzyme requiring cleavage of the terminus for activation (Verheij et al, 1981). sPLA$_2$ enzymes have an absolute requirement for Ca^{2+}, typically needing concentrations ≈ 1 mM for maximum activity. This divalent metal ion is a key component of catalysis. In some cases it may also promote binding of the enzyme to its substrate, although many sPLA$_2$ enzymes can bind to phospholipid aggregates without Ca^{2+} present.

Regardless of the role and source of the enzyme, the overall folded structure is similar for all secreted PLA$_2$ enzymes examined. Crystal structures have been determined for the unliganded bovine and porcine pancreatic enzymes (Dijkstra et al, 1989; 1983; 1981), both of which are monomeric. A crystal structure of a mutant porcine pancreatic PLA$_2$ with residues 62–66 deleted and several other residues altered has also been obtained (Thunnissen et al, 1990). Venom enzymes have also been crystallized, and these proteins tend to aggregate in dimer or trimer units (Brunie et al, 1985). While the oligomeric state for some of the venom enzymes may have no functional significance in the catalytic process, it may occur in vivo to maintain the protein in an inactive state until activity is required (Fremont et al, 1993). The protein consists of three major and two minor α-helical segments, a double-stranded antiparallel β-sheet, a primary Ca^{2+} binding loop (and in some structures a second Ca^{2+} site), and seven disulfide bonds. The structure of the bovine pancreatic enzyme with a covalent inhibitor bound to the active site His48 at Nδ has also been solved (Renetseder et al, 1988). These X-ray structures all show a similar cleft lined with hydrophobic residues, with His48, in the deepest part of the cleft, hydrogen bonded to a water molecule and an aspartic acid residue (Figure 2) in an orientation suggestive of what is observed in serine proteases.

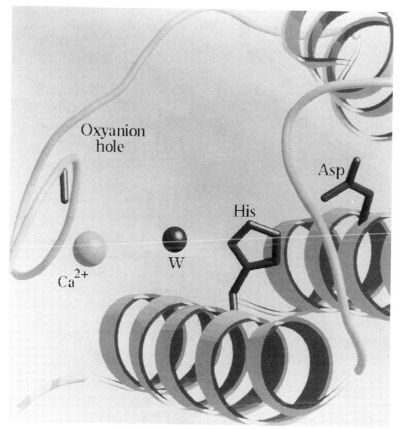

Figure 2. Active site in sPLA$_2$ (*Naja naja atra*) crystal structures. a) unliganded state, b) with phosphonate transition state analog bound. Overall folds shown for backbone; water molecule (W) and Ca^{2+} ion shown as spheres; side chains shown for active site histidine and aspartic acid.

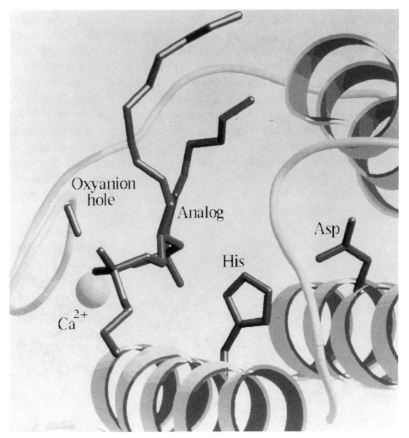

Figure 2b.

The crystal structure of PLA_2 (*Naja naja atra*) has been solved in an un-liganded state (White et al, 1990) as well as with a phosphonate transition state analog, *L*-1-O-octyl-2-heptylphosphonyl-*sn*-glycero-3-phosphoethanolamine (White et al, 1990), bound to the enzyme. The latter provides clues as to both the catalytic mechanism and the cause of interfacial activation. The phosphonate mimics the tetrahedral carbon produced as water is added to the *sn*-2 carbonyl. There are two Ca^{2+} sites. The active site Ca^{2+} is hepta-coordinated in a pentagonal bipyramidal cage formed by Asp49, the backbone carbonyls of the Ca^{2+} binding loop, and two oxygen atoms of the inhibitor. The secondary Ca^{2+} site is 6.6 Å from the catalytic Ca^{2+}, pentacoordinated and more loosely bound. The ethanolamine amino group is hydrogen bonded to Asn53; however, this interaction cannot occur with a choline headgroup. The *sn*-1 chain is less firmly fixed than the *sn*-2 chain; methylene and methyl groups furthest from the glycerol are least well-ordered. The bound lipid is bent sharply at the tetrahedral phosphorus yielding a lipid conformation similar to that of crystalline phospholipids. Both chains occupy the hydrophobic channel (Figure 3), con-

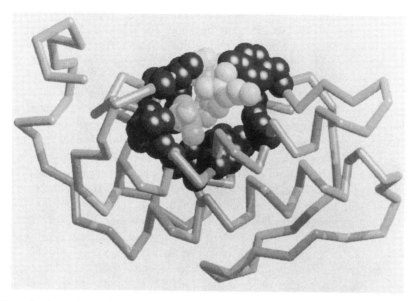

Figure 3. Ligand situated in hydrophobic cleft of sPLA$_2$ (*Naja naja atra*). Overall folds shown for backbone; analog shown in space-filling representation (light); hydrophobic side chains lining cleft shown in space-filling representation (dark).

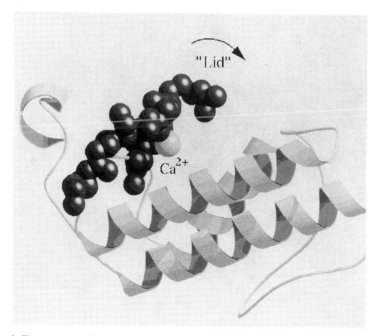

Figure 4. Flap movement in sPLA$_2$ (*Naja naja atra*). a) unliganded state, b) with phosphonate transition state analog bound. Backbone, mostly α-helix, shown in ribbon representation; Ca^{2+} ion shown as light sphere; lid region in space-filling representation (dark).

taining invariant Ile9 and Phe5 and nearly invariant Leu2 and Trp19 residues, which spans 14 Å from the active site histidine. The left wall of the channel is formed from a potentially mobile segment that acts as a flap, which is held in place only when substrate is bound (Figure 4). In the *N. naja atra* complex this is Tyr69, whose hydroxyl group is hydrogen bonded to the *sn*-3 phosphate of the inhibitor, which is also anchored to the primary Ca^{2+} ion. It is the existence of this hydrophobic channel that has led to a proposal that explains the interfacial activation of the enzymes. A substrate molecule is extracted from the aggregate matrix to some degree and diffuses into the hydrophobic channel leading to the active site (Scott et al, 1990). It is thought that polar groups can easily diffuse in since the left wall of the channel is not completely formed until the substrate is bound and anchored to the Ca^{2+}. The seal provided by the binding of the enzyme to the interface allows the transfer of substrate into the active site without solvation of the hydrophobic alkyl substituents. This is suggested as a key component of the observed interfacial activation of this enzyme towards activated substrate, (Figure 5) and, as will be discussed, it is similar to the view of lipases deduced from crystal structures. Experiments testing this concept using crosslinkable phospholipids are consistent with the view that substrate extraction is a crucial kinetic determinant (Soltys et al, 1993).

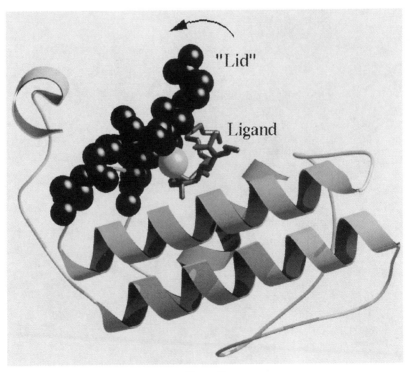

Figure 4b.

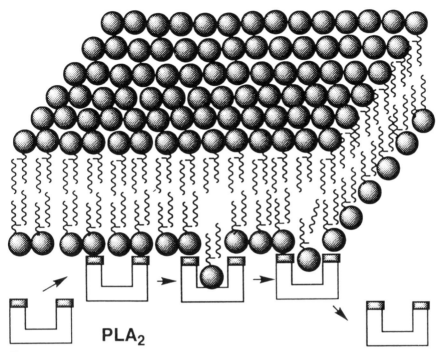

Figure 5. Proposed model for PLA$_2$ interfacial binding and phospholipid hydrolysis (figure kindly provided by W. Cho).

Secreted PLA$_2$ activities also occur in mammalian tissues and several of these have been well-studied because of their potential role in the inflammatory response (Kudo et al, 1993). However, most mammalian cells, for example, human platelets, have at least two types of PLA$_2$: a secreted form (sPLA$_2$) and a cytosolic form (cPLA$_2$). Intracellular Ca^{2+}-independent PLA$_2$ activities have also been isolated and characterized (Ackermann et al, 1994; Hazen and Gross, 1993). Physiological agonists activate PLA$_2$ to liberate arachidonic acid, which goes on to form prostaglandins and thromboxanes, and lysophospholipid, which can be used to generate platelet-activating factor (PAF). An acid-stable, Ca^{2+}-dependent (requiring mM Ca^{2+} levels) sPLA$_2$ (14 kDa) is released from secretory vesicles when a variety of cells are stimulated in a dose-dependent fashion. sPLA$_2$ enzymes have extensive homology with the well-studied group II (snake venom) and I (pancreatic) enzymes. Three crystal structures exist for sPLA$_2$ from human rheumatoid arthritic synovial fluid: (1) without Ca^{2+} (Wery et al, 1991), (2) with Ca^{2+} and (3) with the same transition state analog used previously (Scott et al, 1991). Both are quite similar to the structure of bovine pancreatic or snake venom enzymes. The active site His48 is hydrogen bonded to a water molecule, and the hydrophobic cleft is quite distinct. There is considerable debate over whether or not the sPLA$_2$ enzymes are the causative agents of inflammation, although they must have some role in mediating the inflammatory response (Dennis, 1994). Details of how the secretory or venom PLA$_2$ enzymes affect cellular activities are sparse, although receptors for extracellular PLA$_2$s (including type I, proPLA$_2$-I, and synovial PLA$_2$-II) have been identified (Ishizaki et al, 1994).)

CYTOSOLIC PLA$_2$, (cPLA$_2$)

The intracellular Ca^{2+}-dependent PLA$_2$ enzymes are considerably larger than the secreted enzymes and have been purified only recently. The 749 residue sequence of cPLA$_2$ has no homology to sPLA$_2$ (Kramer et al, 1991). cPLA$_2$ enzymes also exhibit lysophospholipase and transacylase activities (Reynolds et al, 1993), activities not associated with sPLA$_2$ enzymes. This strongly suggests that cPLA$_2$ catalyzed hydrolysis of phospholipids occurs by a completely different mechanism than that for the sPLA$_2$ enzymes. In unstimulated cells cPLA$_2$ is found in the cytosol, and at the membrane to some extent. Ca^{2+} binding and phosphorylation, which occur upon stimulation, enhance translocation of this enzyme to the membrane surface for interaction with its substrate (Clark et al, 1991). The cPLA$_2$ enzyme sequences so far deduced from analysis of genes all have a Ca^{2+}-binding domain responsible for binding the protein to the phospholipid interface. The bound Ca^{2+} may form complexes with anionic phospholipids that anchor cPLA$_2$ to the membrane in a fashion similar to the way Ca^{2+}-anionic phospholipids interact with the annexins (see later section of this chapter) or protein kinase C. Thus, the activity of this water-soluble enzyme is controlled by modifying its affinity for the membrane surface via Ca^{2+} binding and phosphorylation. Ca^{2+} also appears to be a catalytic cofactor for phospholipid hydrolysis, although, since no structure for cPLA$_2$ is available, the exact role of the catalytic Ca^{2+} is unclear. The enyzme isolated from platelets has a sequence reminiscent of the active site of lipases and is consistent with hydrolysis occurring via a covalent fatty acyl enzyme intermediate. Recently, a serine in this sequence (Ser-228) has been shown by mutagenesis to be absolutely required for catalysis (Sharp et al, 1994). Thus the catalytic mechanism of cPLA$_2$ may include nucleophilic attack by an activated serine, a feature that is shared with lipases and other serine hydrolases but not with sPLA$_2$.

Phospholipase C (PLC)

The action of PLC enzyme on phospholipids generates the hydrophobic product diacylglycerol as well as a water-soluble phosphate ester. There are two major classes of this type of phospholipase based on substrate specificity. PLC enzymes with specificity towards PI are a key component of phosphatidylinositol-mediated signaling pathways, a general mode of intracellular signal transduction in eukaryotic cells. Phosphatidylinositol-specific phospholipase C (PI-PLC) is responsible for the cleavage of membrane PIs (often phosphorylated) into diacylglycerol (DAG) and inositol phosphate(s) (Figure 6). PI-PLC enzymes exist both intracellularly and extracellularly in a wide variety of tissues and organisms.

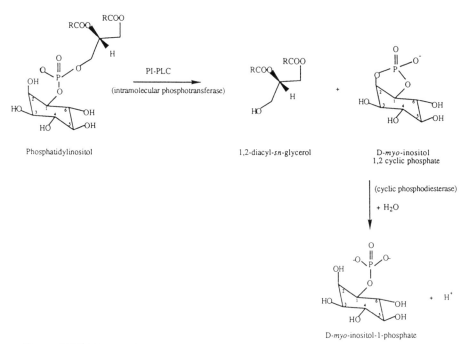

Figure 6. PI-PLC a) reaction. b) schematic of inositol binding in active site of PI-PLC (Reprinted from Heinz et al, 1995 with permission).

Extracellular PI-PLCs have been isolated from the culture media of several microorganisms, while intracellular PI-PLC isozymes are prevalent in mammalian cells. Both products of intracellular PI-PLC activities are second messengers: DAG is a membrane-soluble activator of protein kinase C, and inositol trisphosphate (IP_3) is a soluble agent responsible for Ca^{2+} mobilization. However, there are also nonspecific PLC enzymes, both secreted as well as intracellular. In mammalian systems, it is thought that the DAG generated by these nonspecific PLC enzymes is partially responsible for the sustained activation of protein kinase C.

PI-PLC

Extracellular PI-PLC enzymes are water-soluble and relatively specific for non-phosphorylated PI. PI-PLC activity is found in the culture media of several bacteria including the human pathogens *Staphylococcus aureus* (Low, 1981), *Listeria monocytogenes* (Camilli et al, 1991; Leimeister et al, 1991; Menguad et al, 1991), and *Clostridium novyi* (Taguchi and Ikezawa, 1978), as well as species not normally pathogenic (Artursson et al, 1992; Griffith et al, 1991; Ikezawa et al, 1981; Jager et al, 1991). Purified PI-PLC enzymes from two *Bacillus* species are nearly the same, with only 8 amino acids different and

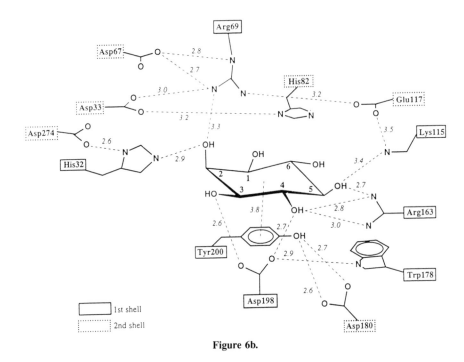

Figure 6b.

little homology with mammalian PI-PLC enzymes. The PI-PLC enzymes from *B. cereus* and *B. thuringiensis* catalyze the cleavage of the glycerophosphate linkage of PI to yield DAG and D-myo-inositol-1,2-cyclic monophosphate. This same enzyme can also slowly catalyze the hydrolysis of the latter product to D-myo-inositol-1-monophosphate (Griffith et al, 1991; Volwerk et al, 1990). The catalytic reaction is independent of Ca^{2+} and occurs in two steps (Figure 6): (1) an intramolecular phosphotransfer reaction to form inositol cyclic 1,2-monophosphate; and (2) hydrolysis of the water-soluble cyclic phosphodiester to produce inositol-1-phosphate. The second step has a much higher K_m and a much lower V_{max}. Tighter binding of the enzyme to the phospholipid substrate suggests that there are specific interactions of the enzyme with the interfacial substrate PI. The enzyme exhibits significant interfacial activation (5–6-fold) towards short-chain PI substrates (Lewis et al, 1993). The crystal structure of PI-PLC from *B. cereus* has recently been solved at 2.6 Å resolution (Heinz et al, 1995). This PI-PLC molecule consists of an imperfect 8-stranded α/β barrel; as such it bears no resemblance to other phospholipases, including the nonspecific PLC from the same organism. Key histidines and other conserved residues in the active site were identified in an inositol-PI-PLC complex (Figure 6b). The inositol-protein interactions observed in the complex suggest that the catalytic mechanism proceeds with two histidines acting as a general acid and base in a manner analogous to ribonuclease.

Mammalian PI-PLC enzymes also exhibit both intrinsic phosphotransferase and cyclic phosphodiesterase activities suggesting that their catalytic mechanisms may be similar to the bacterial enzymes. However, unlike the bacterial enzymes, mammalian PI-PLCs require Ca^{2+} for activity. The Ca^{2+} complexed to the anionic substrates PI, PIP, and PIP_2 can certainly act as an anchor in attaching the enzyme to the membrane. Interactions of the cation with the inositol phosphate groups (not the phosphodiester linkage) may be particularly important for the higher specificity of these enzymes for phosphorylated PI. Whether or not Ca^{2+} is also a catalytic cofactor for the enzymes is not clear at this time.

PI-PLC isoforms and their regulation have been extensively characterized (Rhee and Choi, 1992a). There are at least five immunologically distinct PI-PLC enzymes that appear to be separate gene products (Rhee et al, 1989). On the basis of their primary sequence similarity, the most studied PI-PLCs are classified as PLCβ, PLCγ, and PLCδ. They have significantly different molecular weights but common regions responsible for the catalytic activity. Alignment of sequences from all groups reveals that highly conserved residues are clustered into two regions: one of $\approx$170 residues and the other of $\approx$260 residues designated X and Y regions (Majerus, 1992; Rhee and Choi, 1992b). Kinetic specificities vary among the forms: β and δ exhibit a 40–60-fold preference for PIP_2 over PI at 1 mM Ca^{2+}, while γ shows only a two- to threefold preference for the phosphorylated lipid. There are two other families, PLCα and PLCε, whose regulation is different and not completely understood.

None of the PLC isoforms contains a membrane-spanning sequence, and indeed most of the PLC is purified from the cytosolic fraction. The purified membrane-associated PLC has been shown to be the same as the cytosolic PLC (Katan et al, 1988; Lee et al, 1987). Since PLC must bind to membranes before hydrolyzing PI, there must be a specific mechanism that is responsible for the translocation of PLC between cytosol and membrane. The only case in which this has been investigated is PLCγ1, which is translocated from cytosol to the particulate membrane fraction in vivo after it is activated in response to growth-factors. Phosphorylation of this isozyme on specific tyrosine residues by activated tyrosine kinase growth factor receptors induces the relocation of PLCγ from the cytosol to the plasma membrane, where, presumably, it is better able to interact with its phospholipid substrates (Vega et al, 1992).

NONSPECIFIC PLC

Although the breakdown of PI by a PI-PLC generates the DAG that initially activates protein kinase C, its sustained activation may be provided by the DAG formed upon hydrolysis of PC by a general nonspecific PLC or by PLD and subsequent hydrolysis of PA by PA phosphatase (Berridge, 1986; Billah and Anthes, 1990; Exton, 1990; Pelech and Vance, 1989). The DAG released by this mechanisms does not elicit Ca^{2+} mobilization since no IP_3 is involved. Bacterial PLC is a useful model for this mammalian activity, since it can mimic mammalian PLC activity in enhancing prostaglandin synthesis (Levine et al, 1988). Antibodies to the enzyme purified from B. cereus also crossreact with some of the mammalian activity suggesting that the bacterial enzyme is antigenically similar to mammalian PLC (Clark et al, 1986).

Both from a structural and mechanistic view, the PLC from B. cereus is the best characterized. PLC is a monomeric exocellular Zn^{2+} metalloenzyme. The gene encoding B. cereus has been sequenced showing that the enzyme is translated as a 283 residue precursor with a 24 residue signal peptide and a 14 residue propeptide (Johansen et al, 1988). The secreted form contains 245 amino acids. PLC is extremely stable, even in the presence of 8M urea at 40°C and can be reversibly unfolded and refolded in guanidinium hydrochloride by removal and subsequent addition of Zn^{2+} (Little and Johansen, 1979). Enzyme activity is closely related to the metal ions. Under non-denaturing conditions, removal or substitution of the Zn^{2+} ions causes inactivation of the enzyme and a change in substrate specificity (Little, 1981).

The crystal structure of PLC (Hough et al, 1989) shows a molecule roughly ellipsoidal in shape with overall dimensions of 40 Å × 30 Å × 20 Å (Figure 7a). It consists of ten α-helical regions which are folded into a tightly packed single domain. These helices contain 66% of the residues, the remainder forming loops between them. The molecule surface is smooth except for a cleft 8 Å deep by 5 Å wide that includes the three Zn^{2+} binding sites. The metal ion cluster is remarkably similar to that observed in alkaline phosphatase from Escherichia coli. The active site is amphiphilic in nature with hydrophobic (Phe, Ile, Asn, Leu, Tyr) and hydrophilic (Ser, Thr, Asp, Glu) residues surrounding it. Structures for the enzyme complexed with phosphate (Hansen et al, 1992) and a nonhydrolyzable PC substrate analog (Hansen et al, 1993) have also been determined, and a summary of distances between residues and the lipid is shown in Figure 7b. As in the PLA_2 structure, metal ion interactions with the phosphate moiety are crucial; however, the bound substrate analog phosphodiester linkage is in a strained and unusual conformation. In both of these the phosphate is bound to the three Zn^{2+}; the enzyme shows little change upon analog binding. However, in the PLC-dihexanoylphosphonate structure, the phosphate is in an unusual high energy conformation with the choline moiety folded back and nearly parallel to the acyl chains. This is an unusual conformation for a phospholipid, not observed in any pure lipid aggregates, and it may provide insights into how the enzyme carries out the hydrolysis of the phosphodiester bond. The novel orientation could induce a strain in the phosphate linkage that facilitates hydrolysis. Specific interactions of the protein with the choline moiety are sparse: the methyl carbon atoms on the positively charged nitrogen of the choline moiety are located less than 4 Å from the phenolic oxygen of Tyr56, the aromatic ring of Phe 66 and the carboxyl group of Glu4. A water molecule, located approximately apical to the DAG leaving group, seems to be the most likely candidate for the attacking nucleophile that initiates the reaction. The cavity where

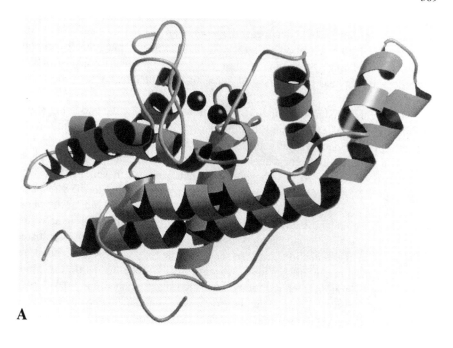

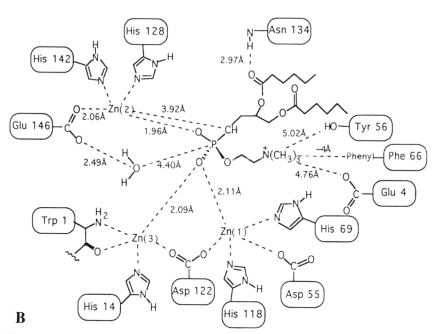

Figure 7. PLC from *Bacillus cereus*. a) crystal structure (overall folds shown for backbone; 3 Zn^{2+} ions shown as dark spheres). b) A schematic of the active site residues of PLC (*B. cereus*) shown in a complex with a nonhydrolyzable PC substrate analog. Note the nearly parallel relationship between the DAG moiety and the choline headgroup in the model (Hansen et al, 1992), as adapted by Martin et al (1994). Reprinted with permission.

the choline group is bound appears large enough for larger headgroups and has a pronounced hydrophobic character which suggests that more hydrophobic headgroups may be better substrates for the enzyme. In this structure, there does not appear to be any other area that might interact with the membrane interface.

More recent distance geometry calculations and molecular modeling (Sundell et al, 1994) have suggested that the phosphonate may not be bound in the conformation adopted by the actual substrate. The latter appears by modeling to be in a conformation consistent with the low energy state of normal phospholipids (i.e., similar to the phospholipid analog bound in venom PLA_2). In the phosphonate/PLC complex, all three Zn^{2+} are coordinated to the phosphate group; in the modeling only two of the Zn^{2+} are involved in binding the phosphate. The mechanism is thought to proceed by an in-line attack of an activated water molecule on the phosphodiester linkage to generate a pentacoordinated phosphorus in the transition state. In the substrate analog structure, the strain induced in the phosphodiester aids cleavage; in the modeling work, the phosphate group is thought to be reoriented after cleavage to stabilize the developing negative charge on the alkoxy leaving group. After bond cleavage, the phosphocholine (phosphomonoester) moiety is translated out of the active site leaving the DAG alkoxide behind for protonation and eventual release. Regardless of which of these mechanisms is correct, the X-ray analysis of the substrate analog/PLC complex has provided another very detailed example of how proteins can bind phospholipids.

Phospholipase D

A variety of membrane phospholipids can be converted to phosphatidic acid (PA), the simplest of phospholipids, via a hydrolase activity; the PA in turn can be degraded by a PLA_2 to lyso-PA (which has mitogenic properties) or it can be converted to a nonsignalling lipid via a PLD transferase activity and DAG to produce a bis-PA (Hilkmann et al, 1991; Van Blitterswijk and Hilkmann, 1993). In recent years, it has been shown that both PA and lyso-PA, added to cell medium, act as second messengers in a wide variety of cells (Ferguson and Hanley, 1992; Jalink et al, 1990; Moolenaar, 1994; Murayama and Ui, 1987; Salmon and Honeyman, 1980). While a mammalian PLD has not been purified, the enzyme from cabbage (Abousalham et al, 1993) has been purified and examined in more detail. The enzyme exhibits both hydrolase and transferase activities similar to the mammalian enzyme implied in signal transduction. All the PLD assayed thus far require Ca^{2+} for activity, although it is not known whether the ion is critical for binding of the enzyme to the phospholipid interface or for the actual catalytic event. Ca^{2+} will also interact strongly with the lipophilic PLD product PA when it is generated. It has been noted that PA appears to be a product inhibitor of PLD activity. This may simply reflect competition of the anionic PA with the enzyme for Ca^{2+}. A better understanding of what controls the way these enzymes interact with substrates in the membrane awaits the purification and molecular level characterization of at least one of these proteins.

Lipases, Flap Movements, and Interfacial Catalysis

Lipases constitute a family of triglyceride hydrolases that act at the lipid-water interface. They are widely distributed in nature and display a wide range of biological roles, structural features, and substrate specificities. Pancreatic lipases, secreted by the duodenum, play a central role in lipid metabolism by catalyzing the hydrolysis of dietary triglycerides into 2-monoglycerides and free fatty acids. Lipoprotein lipase, which is located on the endothelial cell walls of capillaries and adipose tissue cell surfaces, hydrolyzes plasma lipoprotein-bound lipids and helps transfer fatty acids to tissues. Microbial lipases hydrolyze many different substrates and several present potentially useful industrial applications. Pancreatic lipases require a protein cofactor, colipase, to avoid deactivation by bile salts, while lipases from simpler organisms have no such requirement. While the precise definition of a true lipase is challenged by some exceptions (e.g., guinea pig lipase; Hjorth et al, 1993), a major common property is that they become activated at oil-water interfaces (interfacial or surface activation). Thus lipases hydrolyze emulsified or aggregated substrates, while the closely related esterases act upon soluble substrates. The molecular mechanism of this interfacial activation has been a target for many structural, mutagenic, and kinetic studies.

Structurally, lipases are a diverse group, which has been studied extensively in recent years. The first crystal structures of a mammalian lipase, human pancreatic lipase (Winkler et al, 1990), and the first microbial lipase, RML (Brady et al, 1990) were published as back-to-back papers in *Nature*. The structural and evolutionary relationships in lipase mechanism and activation, based on the first few crystal structures, have been reviewed (Dodson et al, 1992). Since then, there has been an explosion in the number of lipase crystal structures reported. To date, enzymes from microbial (fungal, yeast, bacterial) and mammalian sources have been characterized crystallographically. They share a few characteristics, such as an α/β hydrolase fold (Ollis et al, 1992). In this motif, predominantly parallel β-sheets are flanked by α-helices. The enzymes vary considerably in size and complexity. The fungal lipase from *Rhizomucor meihei* (RML) is a roughly spherical, single-domain, 29 kDa molecule that folds into a simple 8-stranded sheet flanked by 6 helices. The fungal lipase from *Geotrichum candidum* (GCL) is also single-domain but larger, a 60 kDa molecule with an 11-strand core and 16 helices. The *Pseudomonas glumae* lipase (PGL) is smaller than RML but consists of 3 domains, only one of which contains the hydrolase fold. Mammalian pancreatic lipases from human or equine sources are 50 kDa molecules containing an 11-strand central sheet. Sequence homology is virtually absent among these enzymes except for a common motif, G-X-S-X-G, surrounding the active site, nucleophilic serine. They typically contain the Ser-His-Glu catalytic triad that reflects their hydrolytic properties, yet even the triad is not 100% conserved in all lipases.

Microbial Lipases

The simplest and best-studied lipase is RML (the earlier source name was *Mucor miehei*). From its study, several properties that are common to many lipases have been explained in structural terms. The first RML crystal structure (Brady et al, 1990) clearly shows the location of the catalytic Ser-His-Glu triad (Figure 8), which is stereochemically very similar to that found in serine proteases. The catalytic mechanism appears to closely follow the mechanism of other serine hydrolases: the Asp-His couple enhancing the reactivity of the nucleophilic serine; the hydroxyl oxygen attacks the acyl carbonyl carbon of the lipid to form a tetrahedral intermediate, a enzyme-substrate adduct; an oxyanion hole stabilizes the developing negative charge of the adduct; and the product is released with cleavage of the scissile bond. The active site itself is located fairly close to the protein surface.

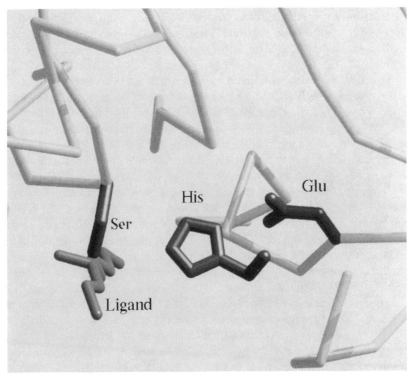

Figure 8. Catalytic triad in *Rhizomucor meihei* lipase.

Three distinct conformations that reflect different stages of activation have been observed for lipases. As observed in several crystal structures, the inactive or resting state has an active site that is buried by a short helical fragment, called the flap, within a long surface loop. The surface above this buried active site is predominantly hydrophilic. The active site channel is hydrophobic and has the nucleophilic serine situated at the bottom. This conformation has been observed in crystal structures of RML (Figure 9a) (Brady et al, 1990; Derewenda et al, 1992b), GCL (Figure 10a) (Schrag et al, 1991), PGL (Noble et al, 1993), and three other fungal lipases (Derewenda et al, 1994).

A different conformation is seen in complexes of microbial lipases with irreversible inhibitors. The active conformation has been observed for RML complexes with n-hexylphosphonate ethyl ester (Brzozowski et al, 1991) or di-ethylphosphate (Derewenda et al, 1992a), and *Humicola lanuginosa* complexes with diethylphosphate (DEP) or dodecylethylphosphate (C12) (Lawson et al, 1994). These structures mimic the tetrahedral transition state of lipolysis and freeze the enzyme in an interfacially activated state. In these complexes, the

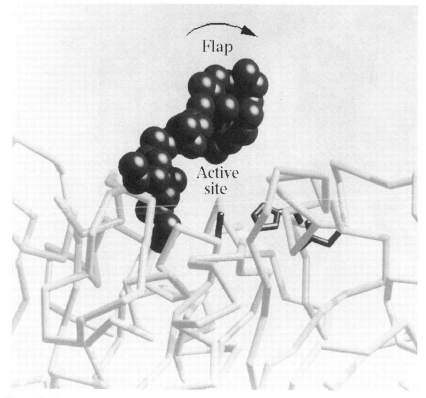

Figure 9. Flap movement in *Rhizomucor meihei* lipase. a) unliganded state, b) with bound ligand (light, space-filling). Shown are protein backbone (light sticks); catalytic triad residue side chains (dark sticks); flap region (dark, space-filling).

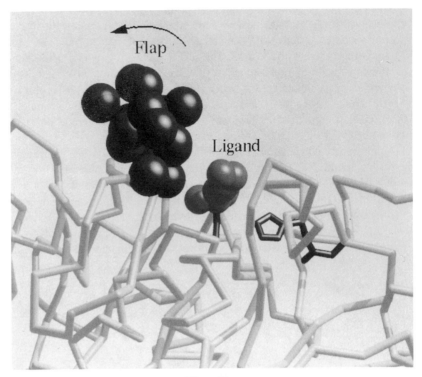

Figure 9b.

flap moves aside to unblock the active site channel and at the same time, exposes a large hydrophobic surface. In doing so, the flap undergoes a significant transition in which some atoms in the center move as much as 12 Å, in RML (Figure 9). The center, which is a short amphipathic helix (the lid), undergoes a rigid body movement from two hinges. Only the hinges undergo a change in their backbone dihedral angles. The loss of hydrophilic surface and creation of a more extensive hydrophobic surface presumably facilitates interactions with the apolar membrane phase. In RML, the native enzyme has a 10 Å-deep, hydrophilic surface depression that is filled with 18 water molecules. In the inhibited enzyme structure, the lid moves to fill this depression with its polar side, releasing nearly all of the bound water molecules. This depression thus serves as a space reserved for the lid in the active, open conformation of the enzyme.

Another substrate-free conformation, which may be intermediate between the resting and active states, has been observed that seems to explain features of the interfacial activation phenomenon. In this open form, the active site is accessible to solvent rather than buried. The crystal structure of a lipase from *Candida rugosa* (CRL; Grochulski et al, 1993), which is a close homologue to GCL, shows a flap that extends nearly perpendicular to the protein surface instead of capping the active site (Figure 10b). In this extended position, the

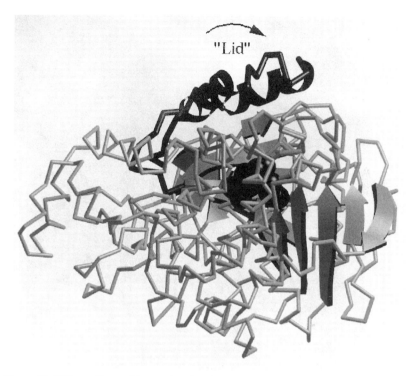

Figure 10. Different flap positions and active site accessibilities in unliganded, homologous microbial lipases. a) active site blocked by closed flap in *Geotrichum candidum* lipase, b) active site accessible due to open flap in *Candida rugosa* lipase. Shown are protein backbone (light sticks), with β-sheets in α/β hydrolase fold domain (light ribbons) and α-helical lid (dark ribbons); active site indicated by catalytic triad residue side chains (dark, space-filling).

flap forms one wall of a large depression around the active site. The wall facing the active site is especially hydrophobic around the nucleophilic serine. The opposite side of the flap is hydrophilic and interacts with polar atoms on the protein surface. Comparison between the closed GCL and open CRL structures reveals the nature of the conformational change. In these enzymes, which are larger than RML, three loops change position in the vicinity of the active site. The largest flap is 31 residues in CRL, and atoms in the tips of the flap are translocated by as much as 25 Å relative to the closed conformation observed in GCL (Figure 10). A short 8-residue loop partially blocks the active site in the open conformation, in which it is tucked under the larger flap. A third, 12-residue loop also shifts its orientation. A model for interfacial activation based on these structures (Grochulski et al, 1993) suggests that this transition involves unwinding and swinging of the flap around its hinged, disulfide-linked base. This movement exposes the hydrophobic face of the flap, which helps to form the substrate entry channel leading to the catalytic serine. Though it is not observed

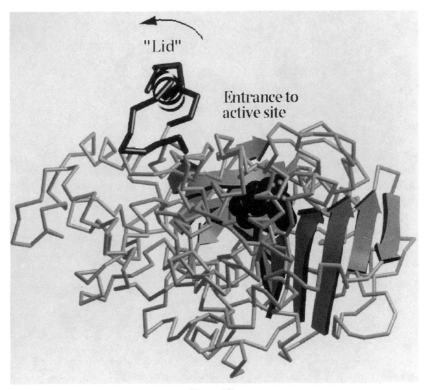

Figure 10b.

as such in the crystal structure, the open conformation is likely to be stabilized
by interactions with the apolar phase of the bilayer (Grochulski et al, 1993).
Two crystal forms of lipase B from *Candida antarctica* (CALB) also show the
enzyme in an open conformation (Uppenberg et al, 1994). The CALB active
site is accessible only through a narrow channel, 10 Å × 4 Å wide and 12 Å
deep, with hydrophobic walls, and the enzyme stereospecificity is attributed to
the channel shape. In all cases, hydrophobic crystal contacts appear to stabilize
the open conformation by mimicking the lipid bilayer. In addition, a monoclinic
crystal form of CALB contains bound detergent (β-octylglucoside) in the active
site which may help keep the lid open by mimicking substrate.

The enzyme-inhibitor complexes permit visualization of two new features
that are somewhat ambiguous in the unliganded structures. First, the lid closure over the inhibitor creates an oxyanion hole that is absent in the native
conformation, and details of the enzyme-substrate intermediate can be inferred.
Second, particularly with the longer-chain inhibitors, the disposition of the acyl
chains can be addressed in the structures of the complexes. The acyl chain of
the C12 inhibitor occupies a hydrophobic channel that extends from the active
site serine. This channel accommodates the first 7–8 acyl chain carbon atoms

of the inhibitor and is suggested to be the binding pocket for the departing fatty acid after hydrolysis. The sites for the other fatty acyl chains from the incoming triglyceride have also been modelled from these data (Lawson et al, 1994).

The hypothesis that the flap movements are critical for interfacial activation in lipases is supported by crystal structures of *Fusarium solain* cutinase (Martinez et al, 1992; 1994), another serine esterase. This enzyme hydrolyzes triglycerides and cutin, the insoluble lipid-polyester matrix covering plants, and it has been suggested that it bridges esterases and lipases (Martinez et al, 1992). Cutinase does not show kinetics consistent with interfacial activation, and it is active on soluble as well as emulsified substrates. From a structural standpoint, this may be because the active site serine is not buried by a flap but is solvent accessible. Instead, a small hydrophobic area surrounds the mouth of the active site channel. These mini-flaps in cutinase are more similar to features found in phospholipases than lipases to assist with interactions with the lipid interface. Moreover, the oxyanion hole is preformed in cutinase (Martinez et al, 1994) rather than formed as a result of conformational change, as in lipases.

While a common picture emerges from the structural studies of microbial lipases, inevitably there are exceptions. The bacterial enzyme, PGL, is unusual in several respects. First, it is the only 3-domain microbial lipase. Its largest domain, which contains the α/β hydrolase fold, is smaller than the homologous domain in other lipases with only a 6-stranded β-sheet surrounded by 5 α-helices. A second, 48-residue domain consist of three α-helices, one of which may be a lid to the active site. The third domain, with 58 residues, consists of four α-helices and a β-hairpin turn that extends out towards solvent. Another distinction of PGL is that the acidic residue of the catalytic triad is dispensable; it is possible that compensation is achieved through a neighboring Glu or water molecule. PGL also requires a Ca^{2+} ion, perhaps for structure stabilization near the active site.

Mammalian Lipases

Mammalian pancreatic lipases differ from microbial lipases in that binding to a helper protein, colipase, is required to overcome the inhibition effected by intestinal bile salts. These amphiphiles, along with phospholipids, interfere with the interfacial adsorption of lipases. The 10 kDa colipase forms a complex with lipase that is able to bind to the lipid-water interface and undergo interfacial activation. Both lipase and colipase are secreted by the exocrine pancreas, and their complex represents the biologically active form of the enzyme.

Human and horse pancreatic lipases are single-chain molecules of 449 residues each. These enzymes are close homologues, with 82% sequence identity, though the human lipase (HPL) is glycosylated whereas the horse enzyme (HoPL) is unmodified. The HPL and HoPL crystal structures of the native enzymes show the closed conformation, as described for microbial lipases. The most detailed structural data on these pancreatic lipases derives from the HoPL crystal structure refined at 2.3 Å resolution (Bourne et al,

1994). There are two well-defined domains in these lipases (Figure 11). The larger N-terminal domain, which represents about two-thirds of the molecular mass, exhibits catalytic activity, while the smaller C-terminal domain binds colipase. The catalytic domain contains the characteristic α/β hydrolase fold, formed around a central core of 11 β-strands. The colipase-binding domain possesses a β-sandwich type of topology consisting of two sheets of 4 antiparallel β-strands. This domain also exhibits some hydrolytic activity toward soluble substrates. A bound Ca^{2+} ion is observed in the refined HoPL crystal structure. This ion exhibits pentagonal bipyramidal coordination and is not located near the active site. Ca^{2+} has not been identified as a requirement for the activity of pancreatic lipases, so the role of this ion in the molecular structure remains unclear.

The structure of colipase resembles that of snake toxins in that three disulfide-linked fingers extend from one end of the molecule. The fingertips are hydrophobic and presumably participate in interfacial binding. Colipase binding produces no structural transition in its lipase partner. The lipase binding site, which is predominantly polar with two salt-bridges and several hydrogen-bonds, is on opposide side from the finger region. Therefore colipase acts as a bridge between lipase and the lipid-water interface. The complete interfacial binding site thus involves hydrophobic residues from the colipase fingertips and the lipase flap region.

Much has been learned about interfacial activation in pancreatic lipases from X-ray structures of HPL-procolipase complexes (van Tilbeurgh et al, 1992; 1993). Crystal structures are known for both the closed, inactive (van Tilbeurgh et al, 1992) and open, activated (van Tilbeurgh et al, 1993) enzyme conformations, as described for microbial lipases. To produce the latter structure, the complex is crystallized in the presence of mixed micelles of phospholipid (1,2-didodecanoyl-sn-3-glycerophosphorylcholine) and bile salt (sodium taurodeoxycholate). A phospholipid molecule occupies the active site, which reorganizes the lid as seen for the fungal lipase-inhibitor complexes. While the procolipase binding site on the lipase C-terminal domain is unchanged, new contacts are made between procolipase and the lipase catalytic domain. Most importantly, the open lid that exposes the active site is held in place by interactions with procolipase (Figure 12). As with the microbial lipases, the lid movement and changes in the active site region create an oxyanion hole. The two acyl chains are bound in a hydrophobic channel and groove, respectively, and completely cover the hydrophobic active-site channel. No density is seen for the choline head group, and the assignment of the two acyl chains as phospholipid rather than lysophosphatidylcholine and fatty acid could not be made unambiguously from the density.

As described by Hjorth et al (1993), the characteristic properties of pancreatic lipases are interfacial activation, reactivation by colipase, and an inability to hydrolyze phospholipids. Guinea pig pancreatic lipase (GPL) challenges this definition in all three respects. Neither interfacial activation nor bile salt inhibition (or colipase reactivation) is observed, and GPL also expresses phospholipase

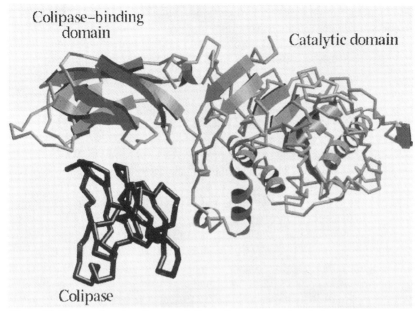

Figure 11. The two domains of pancreatic lipase (human), complexed with procolipase (porcine). Note that the lipase catalytic domain has the α-helices and β-sheet structure typical of the α/β hydrolase fold, found in other lipases.

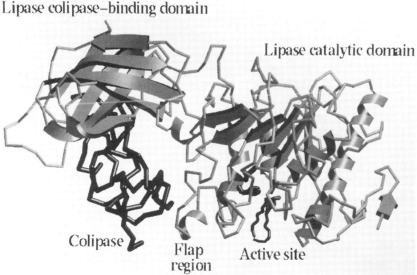

Figure 12. Pancreatic lipase (human), complexed with procolipase (porcine) and substrate analogs. The lipase active site (catalytic triad residues are darkened, along with bound substrate analog) is found exclusively in the lipase catalytic domain and is spatially removed from the colipase binding domain. Colipase, however, binds in an orientation where it interacts with the flap region adjacent to the active site.

activity. A model of GPL based on the crystal structure of HPL offers a molecular basis for these observations (Hjorth et al, 1993). The most important factor is that despite strong sequence homology with other mammalian pancreatic lipases, there is a significant deletion in the loop that corresponds to the flap region. Without the lid domain, the interfacial activation properties are lost. A mini-lid proposed in the GPL model would keep the active site freely accessible to monomeric short-chain substrates, even if the enzyme is not adsorbed to the interface. In this respect, GPL more closely resembles esterases than lipases. These authors suggest that lipases with lid domains also possess latent phospholipase activity that is depressed by interfacial adsorption, which favors lipase activity. Since GPL lacks this domain, the latent phospholipase activity of lipases is expressed.

The mammalian triglyceride lipase family includes not only pancreatic lipases, but also hepatic and lipoprotein lipases. A molecular model of lipoprotein lipase (LL) has been constructed (van Tilbeurgh et al, 1994), based the native HPL (Winkler et al, 1990) and HPL-procolipase (van Tilbeurgh et al, 1993) crystal structures. This enzyme displays about 30% sequence identity with the pancreatic lipases, but it has several functional differences. LL hydrolyzes plasma lipids in very low density lipoproteins (VLDLs) and causes fatty acid transfer to tissues. It localizes to the surface of adipocytes (fat-storing cells) and the epithelial cell walls of capillaries, where processes involving transfer from plasma to organ tissues takes place. This physiological localization is associated with a high affinity for heparin (a highly sulfated polysaccharide) and sulfate proteoglycans, which are components of cell surfaces. Unlike pancreatic lipases, LL is a noncovalent dimer that is activated by a 79-residue partner protein, apolipoprotein C-II. LL is modelled as a two-domain molecule similar to the pancreatic lipases, with a lid domain. The smaller, C-terminal domain is also predicted to form a β-sandwich in LL and is proposed, logically enough, to be the site for apolipoprotein C-II binding. Clusters of positively charged side-chains at an irregular surface loop at the junction between the two domains, and opposite the active site, are proposed to constitute some of the heparin binding determinants. The in vivo role of heparin binding is to keep LL fixed to the outer surface of epithelial cell walls, where hydrolysis of plasma lipids takes place. Dimerization, which is required for activity, may create a single, shared heparin binding site (van Tilbeurgh et al, 1994).

In summary, the prototypical lipase mechanism for interfacial adsorption and activation involves the opening of a flap to expose the active site to substrate molecules and to create a hydrophobic membrane binding site (Figure 13). Even atypical lipases, such as the mammalian GPL and cutinase, are exceptions that prove the rule: lack of a lid domain correlates well with the absence of interfacial activation. This hinged lid motion is not, however, unique for lipases but occurs in other proteins. In triose phosphate isomerase, a soluble metabolic enzyme, such a lid movement is associated with ligand binding, and an 11-residue surface loop moves more than 7 Å to cover the occupied ac-

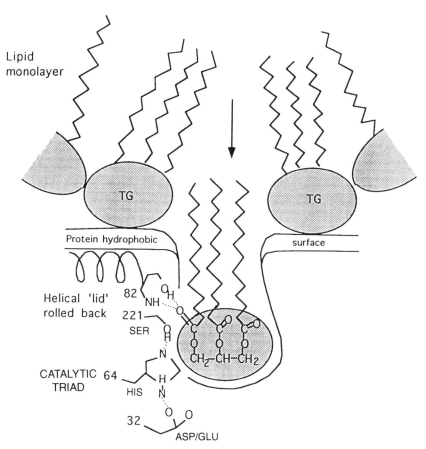

Figure 13. Proposed model for interfacial binding of lipase based on crystal structures (adapted from Blow, 1991 with permission).

tive site. Unlike lipases, the inactive, unliganded conformation is open, with the loop highly mobile and tending to disorder (Joseph et al, 1990). Annexin V, another peripheral membrane-binding protein described elsewhere in this chapter, also undergoes a large lid movement, which is thought to enhance interfacial adsorption (Concha et al, 1993; Meers and Mealy, 1993; Sopkova et al, 1993). It is interesting that in all three cases, the lids contain a conserved tryptophan residue. The great value of the tryptophan side-chain to such structural rearrangements, particularly as they involve the oil-water interface, may be its size and amphiphilicity. Indeed, in membrane-embedded proteins, tryptophans are often preferentially found at the interface rather than the interior of the bilayer.

Annexins, Models for Ca^{2+}-Bridging
in Adsorption to Membranes

The annexin family of Ca^{2+}-dependent interfacial proteins (previously known as lipocortins, endonexins, chromobindins, calpactins, calcimedins, calelectrins) offers an unusual opportunity to view protein-mediated processes occurring at the membrane surface, since they exist stably in water-soluble and membrane-bound forms (for reviews: Liemann and Lewit-Bentley, 1995; Moss, 1992; Raynal and Pollard, 1994; Seaton, 1996; Swairjo and Seaton, 1994). Annexins are widely distributed, appear to be ubiquitous in eukaryotes, and are found in very large amounts (1–2% total cell protein) in many tissues. Though definitive in vivo functions have yet to be established for annexins, their close associations with cell membranes implicate them in various Ca^{2+}-regulated membrane processes including membrane trafficking (exocytosis, endocytosis), cytoskeletal-membrane interactions, and ion movement across membranes. The term annexin (Geisow and Walker, 1986), for their common property of annexing phospholipid membranes, has been widely accepted as a basis for a common nomenclature (Crumpton and Dedman, 1990).

More than a dozen different annexins have been identified to date. Annexin family members differ in both tissue and subcellular localization, where they may colocalize with other annexins or be mutually exclusive. Immunocytochemical studies of annexins have shown them to be localized subadjacent to plasma membranes, near Ca^{2+}-sequestering intracellular organelles, or even in the nucleus. Even without signal sequences, some annexins have been found extracellularly, in lung lavage fluid, prostate fluid, and in the extracellular matrix. Levels of expression may vary with cellular growth state or hormonal influence. The coexistence of different annexins in a given tissue argues for distinct cellular roles among family members. The in vitro properties (e.g., promotion of vesicle aggregation and fusion) exhibited by many annexins appear to support their proposed functions (e.g., exocytosis; Creutz, 1992).

The common property underlying the in vitro properties of annexins, and probably in vivo functions, is their reversible Ca^{2+}-dependent binding of membranes containing phospholipids. In doing so, annexins interact directly with lipid molecules rather than with other proteins. Certain annexins (i.e., annexins II and XI) also form stable complexes with other proteins, but these protein-protein associations are believed to be distinct from membrane attachment. Annexins are considered peripheral, rather than integral membrane proteins; they can be stripped from the membrane simply by removal of Ca^{2+} rather than requiring disruption of the membrane. This property underscores the importance of the polar lipid backbone and headgroups in annexin-membrane interactions. Different family members exhibit some differences in Ca^{2+} affinity and phospholipid head group preference (Blackwood and Ernst, 1990). There is a strong preference for anionic phospholipids, though many annexins will also bind phosphatidylethanolamine (PE). Vesicles containing pure phosphatidylcholine (PC) are not bound by any annexins, and it is unclear whether PC is bound in mixed-phospholipid membranes. Binding of annexins to membranes in the absence of phospholipids or phospholipid analogues (Meers and Mealy, 1994) has not been observed.

Annexin V, A Structural Prototype for the Annexin Family

To date, annexin V is the best characterized member of this family and has served as a useful structural model. At least twelve annexins have been sequenced, and all show strong homology in the regions responsible for calcium and membrane binding. Annexin sequences and properties distinguish this family from the calmodulin (E-F hand), protein kinase C, and phospholipase A_2 families of Ca^{2+}-binding proteins. Most three-dimensional data has come from annexin V crystal structures, though annexin I has also been studied crystallographically. Electron microscopic images of two-dimensional crystals of annexins V and VI bound to lipid monolayers have provided views of the membrane-bound forms of these proteins.

Annexin sequences are characterized by a canonical motif in which a stretch of approximately 70 amino acids is repeated usually four times or, in the case of annexin VI, 8 times (Pepinsky et al, 1988). Within this common core region, annexins exhibit approximately 50% sequence identity, and most of the variations are conservative replacements. This region contains the familial annexin membrane-binding machinery. In addition to the conserved core, annexins also have a highly variable N-terminal region that distinguishes different family members. This region varies greatly in both sequence and length and confers individual annexin functions (Hoekstra et al, 1993). The N-terminal region varies from only a few residues (annexin V) or hundreds (annexins VII and XI). With the shortest N-terminal region in its family, annexin V can be considered a prototypical conserved core, and as such, its study has contributed much information about the common Ca^{2+}-dependent membrane-binding properties of annexins.

Annexin crystal structures have confirmed the hypothesis, inferred from sequence data, that the canonical primary structure typical of annexins corresponds to repeating homologous domains (Figure 14). Crystal structures have been determined from human (Huber et al, 1990, 1992), rat (Concha et al, 1993), and chicken (Bewley et al, 1993) annexin V and human [des 1-33] annexin I (Weng et al, 1993). The four-domain conserved core contains bound Ca^{2+} atoms along one molecular surface, consistent with its role in Ca^{2+} binding and membrane attachment (Figure 15). Less structural information is available about the variable N-terminal regions, which are either truncated (annexin I) or naturally short (annexin V) in the crystal structures.

As seen in the crystal structure of rat annexin V (Concha et al, 1993), the annexin core region contains mostly α-helical secondary structure (Figures 14, 15). The four core domains each exhibit a helical arrangement characterized as a four-helix bundle (helices designated A, B, D, and E) with a fifth capping helix (C) crossing over the top (Huber et al, 1990). Within the helix bundle are two

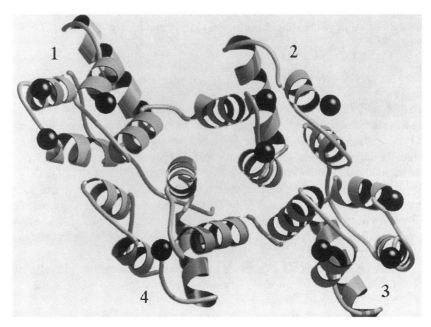

Figure 14. Annexin V (rat), showing highly α-helical backbone (ribbons) and Ca^{2+} (spheres) bound in all four domains. Top view, facing membrane-binding surface, shows symmetry of four canonical domains, numbered 1–4.

parallel helix-loop-helix substructures, held together by the capping helix and interhelical contacts. These differ from calmodulin-like helix-loop-helix units in that the loops are shorter and the helices are nearly antiparallel rather than angled across each other. The annexin conserved core structure appears to be unusually stable; annexin V, for example, is heat-stable and protease-resistant despite the absence of disulfide bonds. Overall, the molecular surface is hydrophilic and lacks extensive surface hydrophobic regions typical of membrane-embedded proteins.

The annexin molecule is slightly flattened and curved in the crystal structure, with opposing convex and concave faces (Huber et al, 1990). The convex surface includes the interhelical loops that link the A and B helices and the D and E helices, respectively (so designated in Weng et al, 1993). Considerable evidence supports the view that in the membrane-bound state, it is the convex surface of the annexin molecule that lies along the plane of the phospholipid membrane, with the loops in direct contact with membrane components. Electron microscopic evidence indicates that the molecule flattens out on the membrane surface, so that all Ca^{2+} binding loops (on the previously convex surface) are coplanar and can contact the membrane surface (Voges et al, 1994). The concave surface includes the capping helices, an extended, nonhelical connection between domains 2 and 3, and the N-terminus, which also has an extended con-

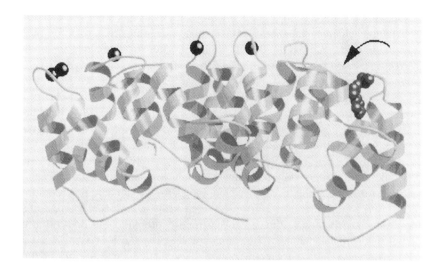

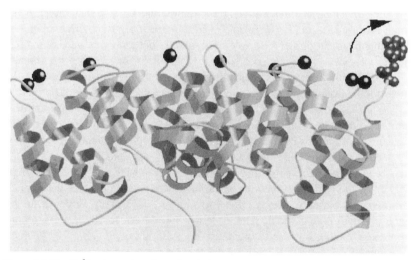

Figure 15. The Ca^{2+}-triggered conformational change in domain 3 of annexin V. Side view, rotated 90° around x axis from orientation in Figure 14, with membrane-binding surface facing up. a) If Ca^{2+} is locally absent from domain 3, the tryptophan side-chain is buried in the protein interior (top). b) If Ca^{2+} is locally present, that same tryptophan emerges from its buried position in the protein interior to interact with the lipid bilayer (bottom).

formation. The N-terminus thus faces the cytosol, where its variable sequence can be accommodated without perturbing Ca^{2+}-dependent membrane binding. In several annexins, this region is also the site of in vitro phosphorylation by tyrosine kinases or protein kinase C. A cytosol-facing orientation of the N-terminus may allow it to participate in interactions with other intracellular proteins or vesicular membranes.

Annexin V contains more Ca^{2+} binding sites than the other proteins discussed in this chapter, consistent with a membrane-anchoring role for these cations in annexin function. These cation-binding sites localize along the membrane-facing surface of annexin V. Each of the four domains contains up to three structurally heterogenous Ca^{2+} binding sites. In each domain, two Ca^{2+} sites occur in loops between the A and B helices (AB loops), and a third occurs in a loop between the D and E helices (DE loops). In the rat annexin V structure, ten of the twelve sites contain bound Ca^{2+} (Swairjo and Seaton, 1995), in good agreement with solution measurements of 10–12 bound Ca^{2+} atoms (Evans and Nelsesteun, 1994). The two sites in the AB loop comprise a double-Ca^{2+} site in that the metal ions are only separated by about 8 Å, and their proteinaceous ligands are mostly contributed by the same loop. In each domain, this loop protrudes from the molecular surface so it can contact the membrane. The DE loop differs from the AB loop in that it is even more hydrophilic and does not protrude but lies along the annexin molecular surface. The DE loop Ca^{2+} is about 11–12 Å from either Ca^{2+} in the AB loop in the same domain. Site-directed mutagenesis studies show that the AB sites are required for membrane attachment, while the DE sites are not sufficient for membrane binding but increase the binding affinity (Jost et al, 1994). The membrane-facing surface of annexin V is predominantly hydrophilic. The sole hydrophobic determinants on the annexin membrane-facing surface are located in the AB loops: one from each domain. In domain 3, this hydrophobic residue is the sole tryptophan in the molecule. This amino acid participates in a major conformational change in the protein.

Though all annexin crystal structures solved to date have been extremely similar, two distinct conformations have been seen (Figure 15). These concern alternate conformations in domain 3, where the tryptophan in annexin V resides. In one conformation, Ca^{2+} is bound in domains 1,2, and 4 but not 3 (e.g., Huber et al, 1992). The conformations of the AB and DE Ca^{2+} binding loops differs from the their counterparts in the other three domains, and the domain 3 tryptophan is buried in the hydrophobic core of that domain. However, in the open conformation (e.g., Concha et al, 1993; Sopkova et al, 1993), in which all four domains bind Ca^{2+}, the tryptophan side-chain is extended out into solvent, approximately 18 Å from its position in the closed conformation. The entire AB loop now protrudes into solvent, like the AB loops in the other three domains. The DE loop also has changed: instead of floating out into solvent with no contacts, it is now firmly anchored to the protein surface, where it binds a Ca^{2+} ion. Moreover, an acidic side-chain from the DE loop reaches over to conrtibute carboxylate oxygen ligands to the Ca^{2+} ion bound to the AB loop. This feature is seen in all four domains in the open conformation; however, the transition from closed to open appears to be confined to domain 3. This transition is also observed with fluorescence spectroscopy, and is associated with higher Ca^{2+} levels (Meers and Mealy, 1993). In the presence of liposomes, the fluorescence spectral changes suggest that the tryptophan goes from a buried position in the protein interior, at low Ca^{2+}, to embed itself in the membrane at higher Ca^{2+}. In the crystal structure corresponding to the open conformation of annexin V, this tryptophan is also involved in protein-protein contacts that may be important for the formation of the organized array seen on membrane surfaces, as described in a previous section. Thus, this conformational transition may play a central role in both Ca^{2+}-dependent attachment to phospholipid membranes.

The molecular details of the annexin V interaction with phospholipid head groups are recently revealed through the crystal structures of rat annexin V complexed, in the presence of Ca^{2+} and soluble phospholipid analogues (Swairjo et al, 1995). Binding of these compounds, which contain the polar heads but not acyl chains of phospholipids, is accomplished primarily through direct coordination of an annexin-bound Ca^{2+} ion and a phosphoryl oxygen (Figure 16). This mode of attachment is termed a calcium-bridge and has been hypothesized, but not yet observed, for other calcium-dependent peripheral membrane-binding proteins. ^{31}P-NMR measurements show that the phosphoryl resonance is in fast exchange between bound and unbound forms (Swairjo et al, 1994), though the attachment is very tight ($K_d \approx$ nanomolar; Tait et al, 1989). This bridging is

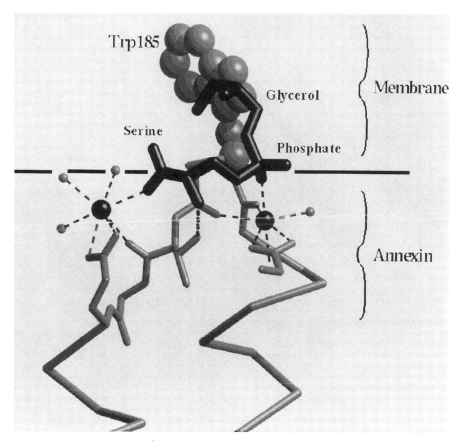

Figure 16. A Ca^{2+} bridge between annexin V and phospholipid analog.

readily reversible by the removal of calcium. The polar head groups are accommodated in hydrophilic surface binding sites on the protein, and the glycerol backbone extends away from the protein in the direction of the membrane.

Annexin V can self-assemble into trimers and higher oligomers on membrane surfaces (Brisson et al, 1991; Concha et al, 1992; Mosser et al, 1991; Pigault et al, 1994; Voges et al, 1994). The oligomerization is Ca^{2+}-dependent and requires a membrane surface. This oligomerization can take the form of ordered arrays, based on threefold symmetry. Extensive sheets of these annexin arrays should exert a noticeable effect on membrane structure. Cryo-electron micrographs of large unilamellar vesicles (LUVs) coated with annexin V have revealed the appearance of large, faceted edges on the LUVs (Andree et al, 1992). NMR studies provide evidence that annexin V also alters the curvature of smaller vesicles (Swairjo et al, 1994). Dynamic membrane properties such as bilayer permeability and lateral macromolecular diffusion may be influenced by these sheets of bound annexin. This coating of membrane surfaces has been invoked to explain the inhibition of membrane-bound proteins such as thrombin by annexin V (Andree et al, 1992) and may be related to the substrate sequestration mechanism invoked for inhibition of phospholipase A_2 (Russo-Marie, 1992). The ability to change local membrane morphology may itself constitute a biological role. In vivo, the structure of the lipid matrix undergoes considerable change, facilitated by diverse lipids, during events such as fusion. Annexin VI has been shown to regulate in vitro the budding of clathrin-coated pits, and this annexin also forms an organized array on membrane surfaces (Newman et al, 1989).

Can Annexin V Function Be Predicted from Its Structure?

Annexin V may be the most abundant and widely-distributed annexin. Endothelial cells of lung, kidney, liver, and placenta are particularly rich sources (Kaetzel et al, 1989; Pepinsky et al, 1988). These cells form the outer layer of organ tissues and are centrally involved in cellular processes such as solute transport and signal transduction between the circulation and tissues. Annexin V is primarily localized by immunofluorescence to the plasma membrane or cytosol, though it also occurs extracellularly in small amounts. The precise cellular role(s) of annexin V have not been established, but include roles in apoptosis (programmed cell death), anticoagulation, inhibition of other membrane proteins, membrane trafficking (though not membrane aggregation or fusion, which other annexins are associated with), and voltage-dependent Ca^{2+} ion channel activity (Raynal and Pollard, 1993; Seaton, 1996).

Whether annexin V, a peripheral membrane protein, really influences Ca^{2+} movement across membranes is currently controversial. Single-channel electrophysiological measurements support the assertion that annexin V is a voltage-dependent Ca^{2+} ion channel (Pollard et al, 1992). However, since known Ca^{2+} channels are transmembrane membrane proteins, and there is no evidence of annexin V spanning the membrane, annexin V must utilize a different mechanism to achieve Ca^{2+} channel activity. Huber and coworkers, in their first report of an annexin structure (Huber et al, 1990), described a hydrophilic central pore in the annexin V molecule. On the basis of the reported channel activity, they proposed that this pore corresponds to the ion channel site. As the crystallographic group further observed that the protein did not appear to embed itself in the membrane bilayer, they proposed a novel mechanism (electroporation effect) in which the annexin molecule, attached to the membrane surface, perturbs the membrane electrostatically and changes ion permeability properties (Karshikov, et al, 1992; Demange et al, 1994). More recently, Goossens et al (1995) have found that the apparent Ca^{2+} channel activity results from nonspecific binding that destabilizes the membrane, a phenomenon seen in other proteins, including bovine serum albumin. These researchers note that at higher Ca^{2+} concentrations, in which specific annexin-membrane binding is likely to take place, the membrane is actually stabilized and made less permeable to ion conductance across the bilayer. Voltage-dependent ion channel activity has been found even in protein-free model membrane systems, and it seems likely that the membrane-binding properties of annexin V may cause model membranes to behave in an atypical manner. Moreover, there are teleological difficulties with the specifics of the Ca^{2+} channel proposal. Annexin V levels are very high in many tissues, whereas very few ion channels are needed for channel activity to occur. Indeed, too much incoming Ca^{2+} would be harmful to the cell. The energy expended by the cell to produce so many annexin molecules does not appear to correlate well with the function of Ca^{2+} ion conductance. It seems more likely that, whatever its cellular function, annexin V influences membrane properties through a mechanism that is related to its high protein concentration, and tendency to form ordered arrays on cell membrane surfaces.

Blood Coagulation Proteins with Gla Domains

Blood coagulation consists of a series of physiological events that necessarily take place on cell membrane surfaces (Furie and Furie, 1988). The cascade that leads to the production of thrombin and the generation of fibrin clotting involves over a dozen extracellular proteins acting in concert. Formation of active complexes of clotting factors is made more efficient by localizing their individual components to membrane surfaces. For example, diffusional collision between two membrane-bound factors, VIIIa and IXa, in the plane of the bilayer results in a complex that enzymatically catalyzes conversion of membrane-bound factor X to Xa, a component in the prothrombinase complex. Ultimately, through similar steps involving other proteins, thrombin is produced from prothrombin and released from the cell surface. Thrombin cleaves fibrinogen to fibrin, which polymerizes to form an insoluble clot. The cell surfaces on which these cascade events take place in vivo belong to platelets, nonnucleated cells in the plasma. Platelets must be activated for initiation of the clotting cascade in order to avoid inappropriate formation of thrombi (clots). This activation is believed to involve exposure of anionic phospholipids, especially phosphatidylserine, on the platelet membrane surface. Like most mammalian cells, platelets do not normally present anionic phospholipids on the extracellular face of the cell membrane.

Many of the protein factors responsible for blood coagulation exhibit a modular organization. One of the modules is the so-called Gla domain, which contains γ-carboxyglutamic acid groups. The crystal structure of prothrombin fragment 1 (Figure 17), which contains the Gla domain (Soriano-Garcia et al, 1992), has established the basic fold as mostly α-helical, with an array of seven Ca^{2+} ions coordinated by Gla carboxylate side-chains and other ligands. Three of these ions are buried, with the remaining four located in surface exposed sites. There are little structural data that directly address the molecular basis of membrane binding. One model is that Ca^{2+} bridges connect the protein and polar phospholipid surface, as described for annexin V in another part of this chapter. It has also been suggested, for prothrombin, that Ca^{2+} binding may stabilize the creation of a phospholipid binding site (Borowski et al, 1986).

NMR solution structures of the apo form of the factor X Gla domain (Sunnerhagen et al, 1995) and the apo (Freedman et al, 1995a) and Ca^{2+}-bound forms of the factor IX Gla domain (Freedman et al, 1995b) are beginning to shed some light on Ca^{2+}-dependent changes in the structures of these modules. Sunnerhagen and coworkers observe that in the presence of Ca^{2+}, several hy-

Gla domain

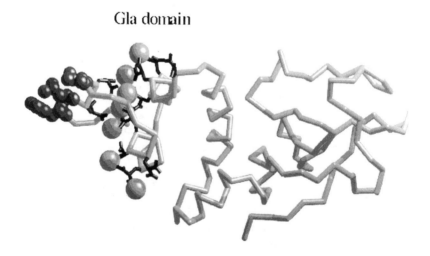

Figure 17. Prothrombin fragment 1, from crystal structure, showing EGF domain and Gla domain. The latter, thought to be responsible for membrane binding, contains a local cluster of hydrophobic side chains (dark, space-filling), and numerous gla side-chains (dark sticks) that bind 7 Ca^{2+} ions (light spheres) in a linear array.

drophobic residues (Phe 4, Leu 5, Val 8) change from a buried to a surface exposed orientation at the protein surface. They suggest that this exposure is primarily responsible for membrane attachment by the Gla domain. This transition is strongly reminiscent of that observed for annexin V (Concha et al, 1993; Meers and Mealy, 1993; Sopkova et al, 1993). For factor IX, it is established that the Gla domain and adjacent aromatic amino acids are responsible for phospholipid membrane binding (Jacobs et al, 1994). Freedman and coworkers (1995b) also note the existence of a hydrophobic surface patch (Leu 6, Phe 9, Val 10) that may play a role in binding factor IX to the membrane surface. Clearly, a common mode should operate in the Gla-domain proteins. However, this mode may include more than one mechanism, e.g., hydrophobic and hydrophilic, that cause the attachment of gla-containing proteins to cell membranes.

Recoverin, a Ca^{2+}-Myristoylation Switch Protein

Many proteins bind reversibly to membranes as a function of a post-translational modification, myristoylation. Recoverin, a 23 kDa homologue of calmodulin, the prototypical E-F hand Ca^{2+}-binding protein is one such protein. Recoverin occurs in retinal rod and cone cells, where it appears to function as a Ca^{2+} sensor for vision (Head, 1994). Ca^{2+} is known to play a role in light adaptation, and in the Ca^{2+}-bound form, recoverin prolongs the photoresponse (Gray-Keller et al, 1993).

The crystal structure of unmyristoylated recoverin (Flaherty et al, 1993) shows the presence of E-F hand domains, which are helix-loop-helix modules that may each bind a Ca^{2+} ion. Though there are four such domains in recoverin, as in calmodulin, only one binds Ca^{2+} in the recoverin crystal structure (Figure 18). Solution studies indicate that another, lower-affinity Ca^{2+} ion is bound in recoverin. These data suggest that the crystal structure is that of a half-occupied form, and the physiological consequences of this half-transition are not known. Recoverin differs from calmodulin, troponin C, and other known E-F hand molecular structures in that there are three extra helices, apart from the two contributed by each E-F hand module. The more compact arrangement of the recoverin domains also differs from that in calmodulin, which tends to adopt an extended, two-lobed dumbbell shape in its native state. Calmodulin is well known for undergoing a Ca^{2+}-induced conformational change that exposes two hydrophobic surface, one per lobe. A similar hydrophobic surface is observed in the crystal structure of the Ca^{2+}-bound recoverin. However in these two homologous proteins, the function of this exposed surfaces may differ. In calmodulin, the hydrophobic patches play a role in forming complexes with target peptides (Head, 1992). In recoverin, the hydrophobic site may be related to interfacial adsorption.

The solution structure of N-terminal myristoylated recoverin, as determined by three-dimensional heteronuclear NMR (Ames et al, 1994; Tanaka et al, 1995) shows that Ca^{2+}-free form of the protein is similar to that in the half-occupied

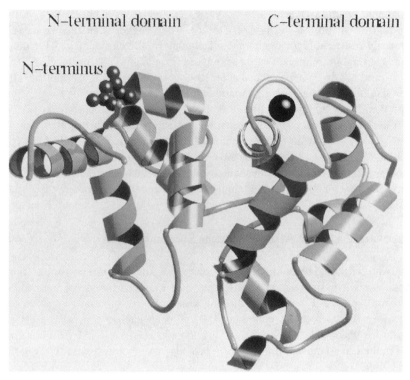

Figure 18. Crystal structure of unmyristoylated recoverin, with a single Ca^{2+} ion (dark sphere) bound in the C-terminal domain. This highly α-helical protein is a homolog of calmodulin, and contains two E-F hands in each half (N-terminal and C-terminal) of the molecule. Atoms at the N-terminus are indicated (dark, space-filling).

form seen in the crystal structure. In particular, the four E-F hand domains are present in both forms. However, in the myristoylated, Ca^{2+}-free form, the first helix (helix-loop-helix) in domain 2 is flexible whereas it is well-ordered in the unmyristoylated, Ca^{2+}-bound form. The ordering of this helix is suggested to reflect allosteric stabilization of the second domain when Ca^{2+} is bound to the site in the third domain, where the high-affinity Ca^{2+} site is located (Ames et al, 1994). Another difference between the two protein forms involves the N-terminal helix (residues 5–16). In the Ca^{2+}-free myristoylated form, it is much longer and is likely to be stabilized by the N-terminal myristoyl group. This helix is disordered in the crystal structure of the unmyristoylated protein, and myristoylation is known to protect the N-terminal region from proteolysis (Dizhoor et al, 1993).

A concerted allosteric model has been derived from results of solution studies of Ca^{2+} binding to myristoylated or unmodified recoverin (Ames et al, 1995) (Figure 19). In this model, Ca^{2+} affinities differ greatly in the T or R states. The R state has the greater Ca^{2+} affinity, by about 10,000-fold. In the T to R

Calcium-Myristoyl Switch

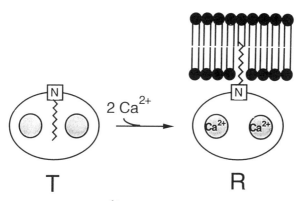

Figure 19. Allosteric model of the Ca^{2+}-myristoyl switch. The hydrophobic myristoyl group is sequestered in the T state, but extruded to interact with the lipid bilayer in the R state. The transition is triggered by Ca^{2+} binding. (Figure kindly provided by J. Ames, M. Ikura, and L. Stryer.)

transition, the myristoyl group moves from a buried location inside the protein to an extruded position. As seen in the three-dimensional solution structure of myristoylated recoverin (Tanaka et al, 1995), the myristoyl group in the T state extends down into a deep, hydrophobic pocket in the N-terminal domain of the protein. Several aromatic side-chains that line the pocket are invariant in recoverin and several of its N-terminal acylated homologues, suggesting common positions for the modified amino termini. In the extruded position, of course, the myristoyl group is buried in the bilayer where it anchors recoverin to the membrane.

The Ca^{2+}-induced membrane interaction proposed for recoverin is termed a Ca^{2+}-myristoylation switch. Such a switching mechanism is characteristic of many proteins involved in signal transduction processes. Calcineurin phosphatase and neuromodulin (GAP 43) are two examples where the (myristoylation or palmitoylation, respectively) switch mechanism may be activated by Ca^{2+}; other GTP-dependent switches have been described (Tanaka et al, 1995).

Conclusion

Despite considerable diversity in their structures and physiological roles, the proteins described in this chapter do appear to exhibit variations on several themes. The most striking general trend is that they each avail themselves of major components of their environment, notably Ca^{2+} ions, lipids, and the polarity and geometry of the membrane (or lipid aggregate). A second observation is the versatile use of conformational change, frequently driven by ligand-binding, to achieve reversibility in membrane associations. An overview of these common themes, as seen in proteins described in this chapter, is presented in Table 1.

Ca^{2+} ions are readily available at membrane surfaces and play functional roles in most of the proteins evaluated. Extracellular free Ca^{2+} concentrations are in the millimolar range, while intracellular levels (except within Ca^{2+}-storing organelles) are several orders of magnitude lower. However, inside the cell, free Ca^{2+} concentrations are somewhat higher near membranes, and cytosolic levels fluctuate significantly with cell stimulation. Both of these intracellular conditions enable certain proteins to use Ca^{2+} as a trigger mechanism, either in facilitating membrane interactions directly (e.g., Ca^{2+}-bridges), inducing conformational changes (e.g., recoverin), or recruiting proteins from cytosol to the membrane surface to become activated.

The use of the versatile Ca^{2+} ion is, of course, not exclusive to membrane-binding proteins, but the latter couple the properties of Ca^{2+} with specialized membrane-related activities. For these proteins, a particularly useful aspect of Ca^{2+} chemistry is its high affinity for phosphoryl groups. For the Ca^{2+}-binding proteins that bind or hydrolyze phospholipids, phosphoryl oxygens are Ca^{2+}-coordinating ligands, and this holds true for both intracellular and extracellular proteins. Based on observations of annexins and sPLA$_2$, a similar Ca^{2+}-phosphoryl interaction can be predicted for gla-domain blood coagulation proteins and protein kinase C, as well as for cPLA$_2$. Zn^{2+}, which has different coordination characteristics and does not substitute for Ca^{2+} as a dynamic signalling device and which also interacts well with the phosphoryl group. Thus it can play an important role in enzyme catalysis, as it does in bacterial PLC. The very different uses of Ca^{2+} in interfacial proteins are illustrated in the following examples. sPLA$_2$ uses the cation both as an intrinsic part of its catalytic machinery, one that specifically recognizes the phospholipid substrate. Recoverin (like its nonmembrane binding homolog, calmodulin) senses rising intracellular Ca^{2+} levels, and transduces the signal into biological function through a Ca^{2+}-induced conformational change. Finally, intracellular Ca^{2+} fluxes may play a role in recruiting cytosolic pools of annexins, cPLA$_2$, protein kinase C, and other proteins to the membrane, though this issue remains to be clarified.

The role of conformational changes at the oil-water interface is another common theme among these peripheral membrane-binding proteins. Phospholipases and lipases exhibit conformational transitions that enhance their hydrophobic interactions with the membrane. Aspects of the transitions can be strikingly similar, e.g., the open-close flap mechanisms in annexins and lipases, and, it is predicted, in gla-domain proteins. Tryptophan and other aromatic side-chains are involved more often than not in these interfacial structural transitions. Proteins use these transitions in different ways, depending upon their biological functions. Phospholipases and lipases use hydrophobic interfacial attachment mechanisms to solve the intrinsic problem of substrate accessibility: how to shuffle lipophilic substrates and products into and out of the active site of a water-soluble enzyme. Their mechanisms often require the existence of a hydrophobic active site cleft that, when introduced to the lipid phase, can form

Table 1. Characteristics of Peripheral Membrane Proteins

Protein	Divalent cation required (role)	Lipid-protein interaction	Hydrophobic exposure	Flap movement	Other
sPLA$_2$	Ca^{2+}, catalytic	phospholipid Ca^{2+}-PO, entire lipid	"seal" made over active site	Slight	Interfacial cataysis
PLC	Zn^{2+}, catalytic	phospholipid Zn^{2+}-PO, part of acyl chains	No	No	Interfacial catalysis
Lipases, microbial	Ca^{2+} bound (sometimes), not required	triglycerides, entire lipid	Yes	Yes	Interfacial catalysis requires flap
Lipases, pancreatic	Ca^{2+} bound, not required	triglycerides, entire lipid	Yes	Yes	Interfacial catalysis requires flap and colipase
Annexin (V)	Ca^{2+}; mb attachment	phospholipid Ca^{2+}-PO, polar only	Yes, trp, Ca^{2+}-triggered	Modest	Ca^{2+} bridges, self-assembly on mb surface
Gla domain	Ca^{2+}; mb attachment	phospholipid (esp PS), polar only?	Yes, aromatic "stack," Ca^{2+}-triggered	Modest (?)	Ca^{2+} bridges, protein assembly on mb surface
Recoverin	Ca^{2+}, signal	None	Yes, myristoyl, maybe more, Ca^{2+}-triggered	No (?)	Ca^{2+}-sensor, reversible mb attachment

a continuous hydrophobic phase for substrate diffusion. In constrast, proteins such as annexins that use the membrane surface as a platform on which to bind and self-assemble have little need for extensive hydrophobic interactions with the membrane. For such proteins, including gla-domain proteins, which associate with the membrane to facilitate formation of multiprotein complexes, the apolar determinants on the proteins are relatively modest. Finally, an important qualification for proteins that exhibit triggered membrane-binding characteristics is that the protein-interfacial interactions should not be so extensive as to be irreversible. For the membrane-bound form of the protein to be very much more stable than its water-soluble form would not be functionally advantageous. The myristoylation group in recoverin, for example, is sufficiently hydrophobic to assist membrane binding but its insertion into the membrane is readily reversible.

The existence of these common themes and motifs in otherwise unrelated proteins, to come full circle, is undoubtedly related to their unique biphasic environment. Simplicity suggests that the available and abundant materials would be used for diverse purposes, and that proteins sharing the same environment would use them in similar ways but to fit their own functional requirements. Thus lipids, Ca^{2+}, and the oil-water interface itself provide the means for peripheral membrane protein function.

REFERENCES

Abousalham A, Riviere M, Teissere M, Verger R (1993): Improved purification and biochemical characterization of phospholipase D from cabbage. *Biochim Biophys Acta* 1158:1–7

Ackermann EJ, Kempner ES, Dennis EA (1994): Ca^{2+}-independent cytosolic phospholipase A_2 from macrophage-like P388D1 cells. Isolation and characterization. *J Biol Chem* 269:9227–9233

Ames JB, Porumb T, Tanaka T, Ikura M, Stryer L (1995): Amino-terminal myristoylation induces cooperative calcium binding to recoverin. *J Biol Chem* 270:4526–4533

Ames JB, Tanaka T, Stryer L, Ikura M (1994): Secondary structure of myristoylated recoverin determined by three-dimensional heteronuclear NMR: implications for the calcium-myristoyl switch. *Biochemistry* 33:10743–10753

Andree HAM, Stuart MCA, Hermens W, Reutelingsperger CPM, Hemker HC, Frederik PM, Willems GM (1992): Clustering of lipid-bound annexin-V may explain its anticoagulant effect. *J Biol Chem* 267:17907–17912

Artursson E, Puu G (1992): A phosphatidylinositol-specific phospholipase C from *Cytophaga:* Production, purification, and properties. *Can J Microbiol* 38:1334–1337

Berridge MJ (1986): Phosphoinositides and receptor mechanisms. In: *Receptor Biochemistry and Methodology,* Putney JW, ed. New York: Alan Liss

Bewley MC, Boustead CM, Walker JH, Waller DA, Huber R (1993): Structure of chicken annexin V at 2.25-Å resolution. *Biochemistry* 32:3923–3929

Billah MM, Anthes JC (1990): The regulation and cellular functions of PC hydrolysis. *Biochem J* 269:281–291

Blackwood RA, Ernst JD (1990): Characterization of Ca^{2+}-dependent phospholipid binding, vesicle aggregation and membrane fusion by annexins. *Biochem J* 266:195–200

Blow D (1991): Lipases reach the surface. *Nature* 351:444–445

Borowski M, Furie BC, Bauminger S, Furie B (1986): Prothrombin requires two sequential metal-dependent conformational transition to bind phospholipid. *J Biol Chem* 261:14969–14975

Bourne Y, Martinez C, Kerfelec B, Lombardo D, Chapus C, Cambillau C (1994): Horse pancreatic lipase. The crystal structure refined at 2.3 Å resolution. *J Mol Biol* 238:709–732

Brady L, Brzozowksi AM, Derewenda ZS, Dodson E, Dodson G, Tolley S, Turkenberg JP, Christiansen L, Huge-Jensen B, Norskov L, Thim L, Menge U (1990): A serine protease forms the catalytic centre of a triacylglycerol lipase. *Nature* 343:767–770

Brisson A, Mosser G, Huber R (1991): Structure of soluble and membrane-bound human annexin V. *J Mol Biol* 220:199–203

Brunie S, Bolin J, Gewirth D, Sigler PB (1985): The refined crystal structure of dimeric phospholipase A$_2$ at 2.5 Å. Access to a shielded catalytic center. *J Biol Chem* 260:9742–9749

Brzozowski AM, Derewenda U, Derewenda ZS, Dodson GG, Lawson DM, Turkenberg JP, Bjorkling F, Huge-Jensen B, Patkar SA, Thim L (1991): A model for interfacial activation in lipases from the structure of a fungal lipase-inhibitor complex. *Nature* 351:491–494

Camilli A, Goldfine H, Portnoy DA (1991): *Listeria* monocytogenes mutants lacking PI-specific phospholipase C are equivalent. *J Exp Med* 173:751–754

Clark JD, Lin L, Kriz RW, Ramesha CS, Sultzman LA, Lin AY, Milona N, Knopf JL (1991) A novel arachidonic acid-selective cytosolic PLA$_2$ contains a Ca-dependent translocation domain with homology to PKC and GAP. *Cell* 65:1043–1051

Clark MA, Shorr RGL, Bomalski JS (1986): Antibodies prepared to *Bacillus cereus* phospholipase C crossreact with a phosphatidylcholine preferring phospholipase C in mammalian cells. *Biochem Biophys Res Commun* 140:114–119

Concha NO, Head JF, Kaetzel MA, Dedman JR, Seaton BA (1993): Rat annexin V crystal structure: Ca^{2+}-induced conformational changes. *Science* 261:1321–1324

Concha NO, Head JF, Kaetzel MA, Dedman JR, Seaton BA (1992): Annexin-V forms calcium dependent trimeric units on phospholipid vesicles. *FEBS Lett* 314:159–162

Creutz CE (1992): The annexins and exocytosis. *Science* 258:924–931

Crumpton MJ, Dedman JR (1990): Protein terminology tangle. *Nature* 345:212

De Haas GH, Bonsen PPM, Pieterson WA, Van Deenen LLM (1971): Studies on phospholipase A$_2$ and its zymogen from porcine pancreas. *Biochim Biophys Acta* 239:252–266

Demange P, Voges D, Benz J, Liemann S, Göttig P, Berendes R, Burger A, Huber R (1994): Annexin V: the key to understanding ion selectivity and voltage regulation? *Trends Biochem Sci* 19:272–276

Dennis EA (1994): Diversity of group types, regulation, and function of phospholipase A$_2$. *J Biol Chem* 269:13057–13060

Derewenda U, Brzozowski AM, Lawson DM, Derewenda ZS (1992a): Catalysis at the interface: The anatomy of a conformational change in a triglyceride lipase. *Biochemistry* 31:1532–1541

Derewenda ZS, Derewenda U, Dodson GG (1992b): The crystal and molecular structure of the *Rhizomucor miehei* triglyceride lipase at 1.9 Å resolution. *J Mol Biol* 227:818–839

Derewenda U, Swenson L, Green R, Wei Y, Dodson GG, Yamaguchi S, Haas MJ, Derewenda ZS (1994): An unusual buried polar cluster in a family of fungal lipases. *Nature Struct Biol* 1:36–47

Dijkstra BW, Kalk KH, Hol WGJ, Drenth J (1981): Structure of porcine pancreatic phospholipase A_2 at 1.7 Å resolution. *J Mol Biol* 147:97–123

Dijkstra BW, Renetseder R, Kalk KH, Hol WGJ, Drenth J (1983): Structure of porcine pancreatic phospholipase A_2 at 2.6 Å resolution and comparison with bovine phospholipase A_2. *J Mol Biol* 168:163–179

Dizhoor AM, Chen CK, Olshevskaya E, Sinelnikova VV, Phillipov P, Hurley JB (1993): Role of the acylated amino terminus of recoverin in Ca^{2+}-dependent membrane interaction. *Science* 259:829–832

Dodson GG, Lawson DM, Winkler FK (1992): Structural and evolutionary relationships in lipase mechanism and activation. *Faraday Discuss* 93:95–105

Evans TC Jr, Nelsestuen GL (1994): Calcium and membrane-binding properties of monomeric and multimeric annexin II. *Biochemistry* 33:13231–13238

Exton J (1990): Signaling through PC breakdown. *J Biol Chem* 265:1–4

Ferguson JE, Hanley MR (1992): Phosphatidic acid and lysophosphatidic acid stimulate receptor-regulated membrane currents. *Arch Biochem Biophys* 297:388–392

Flaherty KM, Zozulya S, Stryer L, McKay DB (1993): Three-dimensional structure of recoverin, a calcium sensor in vision. *Cell* 75:709–716

Freedman SJ, Furie BC, Furie B, Baleja JD (1995a): Structure of the metal-free τ-carboxyglutamic acid-rich membrane binding region of factor IX by two-dimensional NMR spectroscopy. *J Biol Chem* 270:7980–7987

Freedman SJ, Furie BC, Furie B, Baleja JD (1995b): Structure of the Ca^{2+}-bound τ-carboxyglutamic acid-rich factor IX. *Biochemistry*: in press

Fremont DH, Anderson DH, Wilsom IA, Dennis EA, Xuong NH (1993): Crystal structure of phospholipase A_2 from Indian cobra reveals a trimeric association. *Proc Natl Acad Sci USA* 90:342–346

Furie B, Furie BC (1988): The molecular basis of blood coagulation. *Cell* 53:505–518

Geisow MJ, Walker JH (1986): New proteins involved in cell regulation by Ca^{2+} and phospholipids. *Trends Biol Sci* 11:420–423

Goossens ELJ, Reutelingsperger CPM, Jongsma FHM, Kraayenhof R, Hermens W (1995): Annexin V perturbs or stabilises phospholipid membranes in a calcium-dependent manner. *FEBS Lett* 359:155–158

Gray-Kellor MP, Polans AS, Palczewski K, Detwiler PB (1993): The effect of recoverin-like calcium-binding proteins on the photo response of retinal rods. *Neuron* 10:523–531

Griffith OH, Volwerk JJ, Kuppe A (1991): Phosphatidylinositol-specific phospholipase C from *Bacillus cereus* and *Bacillus thuringiensis*. *Meth Enzymol* 197:493–502

Grochulski P, Li Y, Schrag JD, Bouthillier F, Smith P, Harrison D, Rubin B, Cygler M (1993): Insights into interfacial activation from an open structure of *Candida rugosa* lipase. *J Biol Chem* 268:12843–12847

Hansen S, Hansen LK, Hough E (1992): Crystal structures of phosphate, iodide, and iodate-inhibited phospholipase C from *Bacillus cereus* and structural investigations of the binding of reaction products and a substrate analogue. *J Mol Biol* 225:543–549

Hansen S, Hough E, Svensson LA, Wong Y-L, Martin SF (1993): Crystal structure of phospholipase C from *Bacillus cereus* complexed with a substrate analog. *J Mol Biol* 234:179–187

Hazen SL, Gross RW (1993): The specific association of a phosphofructokinase isoform with myocardial calcium-independent phospholipase A_2. *J Biol Chem* 268:9892–9900

Head JF (1994): Shedding light on recoverin. *Curr Biol* 4:64–66

Head JF (1992): A better grip on calmodulin. *Curr Biol* 2:609–611

Heinz DW, Ryan M, Bullock TL, Griffith OH (1995): Crystal structure of the phosphatidylinositol-specific phospholipase C from *Bacillus cereus* in complex with myoinositol. *EMBO J:* in press

Hilkman H, de Widt J, Vander Bend R (1991): Phospholipid metabolism in bradykinin-stimulated human fibroblasts. *J Biol Chem* 266:10344–10350

Hjorth A, Carrière F, Cudrey C, Wöoldike H, Boel E, Lawson DM, Ferrato F, Cambillau C, Dodson GG, Thim L, Verger R (1993): A structural domain (the lid) found in pancreatic lipases is absent in the guinea pig (phospho)lipase. *Biochemistry* 32:4702–4707

Hoekstra D, Buist-Arkema R, Klappe K, Reutelingsperger CPM (1993): Interaction of annexins with membranes: the N-terminus as a governing parameter as revealed with a chimeric annexin. *Biochemistry* 32:14194–14202

Hough E, Hansen LK, Birknes B, Jynge K, Hansen S, Hordvik A, Little C, Dodson E, Derewenda Z (1989): High-resolution (1.5 Å) crystal structure of phospholipase C from *Bacillus cereus*. *Nature* 338:357–360

Huber R, Berendes R, Burger A, Schneider M, Karshikov A, Luecke H, Römisch J, Pâques E (1992): Crystal and molecular structure of human annexin V after refinement. *J Mol Biol* 223:683–704

Huber R, Römisch J, Pâques E-P (1990): The crystal and molecular structure of human annexin V, an anticoagulant protein that binds to calcium and membranes. *EMBO J* 9:3867–3974

Ikezawa H, Taguchi T (1981): Phosphatidylinositol-specific phospholipase C from *Staphylococcus aureus*. *Meth Enzymol* 71:731–741

Ishizaki J, Hanasaki K, Higashino K, Kishino J, Kibuchi N, Ohara O, Arita H (1994): Molecular cloning of pancreatic group I phospholipase A_2 receptor. *J Biol Chem* 269:5897–5904

Jacobs M, Freedman SJ, Furie BC, Furie B (1994): Membrane binding properties of the factor IX γ-carboxyglutamic acid-rich domain prepared by chemical synthesis. *J Biol Chem* 269:25494–25501

Jager K, Stieger S, Brodbeck U (1991): Cholinesterase solubilizing factor from Cytophaga sp. is a PI-specific phospholipase C. *Biochem Biophys Acta* 1074:45–51

Jalink K, Van Corven EJ, Moolenaar WH (1990): Lysophosphatidic acid, but not phosphatidic acid, is a potent Ca^{2+}-mobilizing stimulus for fibroblasts. Evidence for an extracellular site of action. *J Biol Chem* 265:12232–12239

Johansen T, Holm T, Guddal PH, Sletten K, Haugli FB, Little C (1988): Cloning and sequencing of the gene encoding the phosphatidylcholine-preferring phospholipase C of *Bacillus cereus*. *Gene* 65:293–304

Joseph D, Petsko GA, Karplus M (1990): Anatomy of a conformation-al change: hinged "lid" motion of the triosephosphate isomerase loop. *Science* 249:1425–1428

Jost M, Weber K, Gerke V (1994): Annexin II contains two types of Ca^{2+}-binding sites. *Biochem J* 298:553–558

Kaetzel MA, Hazarika P, Dedman JR (1989): Differential tissue expression of three 35 kDa annexin calcium-dependent phospholipid-binding proteins. *J Biol Chem* 264:14463–14470

Karshikov A, Berendes R, Burger A, Cavalié, A, Lux HD, Huber R (1992): Annexin V membrane interaction: an electrostatic potential study. *Eur Biophys J* 20:337–344

Katan M, Kriz RW, Totty N, Meldrum E, Aldape RA, Knopf JL, Parker PJ (1988): Determination of the primary structure of PLC-154 demonstrates diversity of phosphoinositide-specific phospholipase C activities. *Cell* 54:171–177

Kramer RM, Roberts EF, Manetta J, Putnam JE (1991): The Ca^{2+}-sensitive cytosolic phospholipase A_2 is a 100 kDa protein in human monoblast U937. *J Biol Chem* 266:5268–5272

Kudo I, Murakami M, Hara S, Inoue K (1993): Mammalian non-pancreatic phospholipase A_2. *Biochim Biophys Acta* 1117:217–231

Lawson DM, Brzozowski AM, Rety S, Verma C, Dodson GG (1994): Probing the nature of the substrate binding site in *Humicola lanuginosa* lipase through X-ray crystallography and intuitive modelling. *Prot Engin* 7:543–550

Lee KY, Ryu SH, Suh PG, Choi WC, Rhee SG (1987): Phospholipase C associated with particulate fractions of bovine brain. *Proc Natl Acad Sci USA* 84:5540–5544

Leimeister-Wachter M, Domann E, Chakraborty T (1991): Detection of a gene-encoding Pi-specific phospholipase C that is coordinately expressed with listeriolysin in *Listeria monocytogenes*. *Mol Microbiol* 5:361–366

Levine L, Xiao DM, Little C (1987): Increased arachidonic acid metabolites from cells in culture after treatment with phospholi-pase C from *Bacillus cereus*. *Prostaglandins* 34:633–642

Lewis K, Garigapati V, Zhou C, Roberts MF (1993). Substrate requirements of bacterial phosphatidylinositol-specific phospholipase C. *Biochemistry* 32:8836–8841

Liemann S, Lewit-Bentley A (1995): Annexins: a novel family of calcium- and membrane-binding proteins in search of a function. *Structure* 3:233–237

Little C (1981): Effect of some divalent metal cations on phospholipase C from *Bacillus cereus*. *Acta Chem Scand* B35:39–44

Little C, Johansen S (1979): Unfolding and refolding of phospholipase C in solutions of guanidium chloride. *Biochem J* 179:509–514

Low, MG (1981): Phosphatidylinositol-specific phospholipase C from *Bacillus thuringiensis*. *Meth Enzymol* 71:741–746

Majerus PW (1992): Inositol phosphate biochemistry. *Annu Rev Biochem* 61:225–250

Martin SF, Wong Y-L, Wagman AS (1994): Design, synthesis and evolution of phospholipid analogues as inhibitors of the bacterial phospholipase C from *Bacillus cereus*. *J Org Chem* 59:4821–4831

Martinez C, De Geus P, Lauwereys M, Matthyssens G, Cambillau C (1992): *Fusarium solani* cutinase is a lipolytic enzyme with a catalytic serine accessible to solvent. *Nature* 356:615–618

Martinez C, Nicolas A, van Tilbeurgh H, Egloff M-P, Cudrey C, Verger R, Cambillau C (1994): Cutinase, a lipolytic enzyme with a preformed oxyanion hole. *Biochemistry* 33:83–89

Meers P, Mealy T (1994): Phospholipid determinants for annexin V binding sites and the role of tryptophan 187. *Biochemistry* 33:5829–5837

Meers P, Mealy T (1993): Relationship between annexin V tryptophan exposure, calcium, and phospholipid binding. *Biochemistry* 32:5411–5418

Mengaud J, Braun-Breton C, Cossart P (1991): Identification of phosphatidylinositol-specific phospholipase C activity in *Listeria monocytogenes:* A novel type of virulence factor? *Mol Microbiol* 5:367–372

Moolenaar WH (1994): LPA: a novel lipid mediator with diverse biological actions. *Trends Cell Biol* 4:213–219

Moss SE, ed. (1992): *The Annexins*. London: Portland

Mosser G, Ravanat C, Freyssinet J-M, Brisson A (1991): Sub-domain structure of lipid-bound annexin-V resolved by electron image analysis. *J Mol Biol* 217:241–245

Murayama T, Ui M (1987): Phosphatidic acid may stimulate membrane receptors mediating adenylate cyclase inhibition and phospholipid breakdown in 3T3 fibroblasts. *J Biol Chem* 262:5522–5529

Newman R, Tucker AD, Ferguson C, Tsernoglou D, Leonard K, Crumpton MJ (1989): Crystallization of p68 on lipid monolayers and as three-dimensional crystals. *J Mol Biol* 206:213–219

Noble MEM, Cleasby A, Johnson LN, Egmond MR, Frenken LGJ (1993): The crystal structure of a triacylglycerol lipase from *Pseudomonas glumae* reveals a partially redundant catalytic aspartate. *FEBS Lett* 331:123–128

Ollis DL, Cheah E, Cygler M, Dijkstra B, Frolow F, Franken SM, Sussman JL, Verschueren KHG, Goldman A (1992): The α/β hydrolase fold. *Prot Engin* 5:197–211

Pelech SL, Vance DE (1989): Signal transduction via phosphatidylcholine cycles. *Trends Biochem Sci* 14:28–30

Pepinsky RB, Tizard R, Mattaliano RJ, Sinclair LK, Miller GT, Browning JL, Chow EP, Burne C, Huang K-S, Pratt D, Wachter L, Hession C, Frey AZ, Wallner BP (1988): Five distinct calcium and phospholipid binding proteins share homology with lipocortin I. *J Biol Chem* 263:10799–10811

Pieterson JC, Vidal JC, Volwerk JJ, DeHaas GH (1974): Zymogen-catalyzed hydrolysis of monomeric substrates and the presence of a recognition site for lipid-water interfaces in phospholipase A_2. *Biochemistry* 13:1455–1460

Pigault C, Follenius-Wund A, Schmutz M, Freyssinet J-M, Brisson A (1994): Formation of two-dimensional arrays of annexin V on phosphatidylserine-containing liposomes. *J Mol Biol* 236:199–208

Pollard HB, Guy HR, Arispe N, de la Fuente M, Lee G, Rojas EM, Pollard JR, Srivastava M, Zhang-Keck Z-Y, Merezhinskaya N, Caohuy H, Burns AL, Rojas E (1992): Calcium channel and membrane fusion activity of synexin and other members of the annexin gene family. *Biophys J* 62:15–18

Raynal P, Pollard, HB (1994): Annexins: the problem of assessing the biological role for a gene family of multifunctional calcium- and phospholipid-binding proteins. *Biophys Biochim Acta* 1197:63–93

Renetseder R, Dijkstra BW, Huizing K, Kalk KH, Drenth J (1988): Crystal structure of bovine pancreatic phospholipase A_2 covalently inhibited by p-bromo-phenacyl-bromide. *J Mol Biol* 200:181–188

Reynolds LJ, Hughes LL, Louis AI, Kramer RM, Dennis EA (1993): Metal ion and salt effects on the phospholipase A_2, lysophospholipase, and transacylase activities of human cytosolic phospholipase A_2. *Biochim Biophys Acta* 1167:272–280

Rhee SG, Choi KD (1992a): Regulation of inositol phospholipid-specific phospholipase C isozymes. *J Biol Chem* 267:12393–12396

Rhee SG, Choi KD (1992b): Multiple forms of phospholipase C isozymes and their activation mechanisms. *Adv Second Messenger Phosphoprot Res* 26:35–61

Rhee SG, Suh PG, Ryu S-H, Lee KY (1989): Studies of inositol phospholipid-specific phospholipase C. *Science* 244:546–550

Roberts MF (1991): Using short-chain phospholipids to assay phospholipases. *Meth Enzymol* 197:95–112

Roberts MF, Dennis EA (1989): The role of phospholipases in phosphatidylcholine catabolism. In: *Phosphatidylcholine Metabolism,* Vance DE, ed. New York: CRC Press

Roholt O, Schlamowitz M (1961): L-α-(dicaproyl)lecithin, a soluble substrate for lecithinase A and D. *Arch Biochem Biophys* 94:364–379

Russo-Marie F (1992): Annexins, phospholipase A_2 and the glucocorticoids. In: *The Annexins*, Moss S, ed. New York: Portland Press

Salmon DM, Honeyman TW (1980): Proposed mechanism of cholinergic action in smooth muscle. *Nature* 284:344–345

Schrag JD, Li Y, Wu S, Cygler M (1991): Ser-His-Glu triad forms the catalytic site of the lipase from *Geotrichum candidum*. *Nature* 351:761–764

Scott D, White SP, Browning JL, Rosa JJ, Gelb MH, Sigler PB (1991): Structures of free and inhibited human secretory phospholipase A_2 from inflammatory exudate. *Science* 254:1007–1010

Scott D, White SP, Otwinowski Z, Yuan W, Gelb M, Sigler PB (1990): Interfacial catalysis: the mechanism of phospholipase A_2. *Science* 250:1541–1546

Seaton BA, ed. (1996): *Annexins: Molecular Structure to Cellular Function*, Austin, TX: R G Landes Company

Sharp JD, Pickard RT, Chiou XG, Manetta JV, Kovacevic S, Miller JR, Roberts EF, Strifler BA, Brems DN, Kramer RM (1994): Serine 228 is essential for catalytic activities of 85-kDa cytosolic phospholipase A_2. *J Biol Chem* 269:23250–23254

Soltys CE, Bian J, Roberts, MF (1993): Polymerizable phosphatidylcholines: Importance of phospholipid motions for optimum phospholipase A_2 and C activity. *Biochemistry* 32:9545–9552

Sopkova J, Renouard M, Lewit-Bentley A (1993): The crystal structure of a new high-calcium form of annexin V. *J Mol Biol* 234:816–825

Soriano-Garcia M, Padmanabhan K, de Vos A, Tulinsky A (1992): The Ca^{2+} ion and membrane binding structure of the Gla domain of Ca-prothrombin fragment 1. *Biochemistry* 31:2554–2566

Sundell S, Hansen S, Hough E (1994): A proposal for the catalytic mechanism in phospholipase C based on interaction energy and distance geometry calculations. *Prot Engin* 7:571–577

Sunnerhagen M, Forsén S, Hoffren A-M, Drakenberg T, Teleman O, Stenflo J (1995): Structure of the Ca^{2+}-free Gla domain sheds light on membrane binding of blood coagulation proteins. *Nature Struct Biol* 2:968–974

Swairjo MA, Seaton BA (1995): unpublished data

Swairjo MA, Seaton BA (1994): Annexin structure and membrane interactions: a molecular perspective. *Ann Rev Biophys Biomol Struct* 23:193–213

Swairjo MA, Concha NO, Kaetzel MA, Dedman JR, Seaton BA (1995): Ca^{2+}-bridging mechanism and phospholipid head group recognition in the membrane-binding protein annexin V. *Nature Struct Biol* 2:968–974

Swairjo MA, Roberts MF, Campos M-B, Dedman JR, Seaton BA (1994): Annexin V binding to the outer leaflet of small unilamellar vesicles leads to altered inner-leaflet properties: ^{31}P- and ^{1}H-NMR studies. *Biochemistry* 33:10944–10950

Taguchi R, Ikezawa H (1978): PI-specific PLC from colestium. *Arch Biochem Biophys* 186:196–201

Tait JF, Gibson D, Fujikawa K (1989): Phospholipid binding properties of human placental anticoagulant protein-I, a member of the lipocortin family. *J Biol Chem* 264:7944–7949

Tanaka T, Ames JB, Harvey TS, Stryer L, Ikura M (1995): Sequestration of the membrane-targeting myristoyl group of recoverin in the calcium free state. *Nature* 376:444–447

Thunnissen MMGM, Kalk KH, Drenth J, Dijkstra BW (1990): Structure of an engineered porcine phospholipase A_2 with enhanced activity at 2.1 Å resolution. *J Mol Biol* 216:425–439

Uppenberg J, Hansen MT, Patkar S, Jones TA (1994): The sequence, crystal structure determination and refinement of two crystal forms of lipase B from *Candida antarctica*. *Structure* 2:293–308

Van Blitterswijk, Hilkmann H (1993): Rapid attenuation of receptor-induced diacylglycerol and phosphatidic acid by phospholipase D-mediated transphosphorylation: Formation of bisphosphatidic acid. *EMBO J* 12:2655–2662

van Tilbeurgh H, Egloff M-P, Martinez C, Rugani N, Verger R, Cambillau C (1993): Interfacial activation of the lipase-procolipase complex by mixed micelles revealed by X-ray crystallography. *Nature* 362:814–820

van Tilbeurgh H, Roussel A, Lalouel J-M, Cambillau C (1994): Lipoprotein lipase. Molecular model based on the pancreatic lipase X-ray structure: Consequences for heparin binding and catalysis. *J Biol Chem* 269:4626–4633

van Tilbeurgh H, Sarda L, Verger R, Cambillau C (1992): Structure of the pancreatic lipase-procolipase complex. *Nature* 359:159–162

Vega QC, Cochet C, Filhol O, Chang CP, Rhee SG, Gill GN (1992): A site of tyrosine phosphorylation in the C terminus of the epidermal growth factor receptor is required to activate phospholipase C. *Mol Cell Biol* 12:128–135

Verheij HM, Slotboom AJ, De Haas GH (1981): Phospholipase A$_2$: a model for membrane-bound enzymes. *Rev Physiol Pharmacol* 91:91–203

Voges D, Berendes R, Burger A, Demange P, Baumeister W, Huber H (1994): Three-dimensional structure of membrane-bound annexin V, a correlative electron microscopy X-ray crystallography study. *J Mol Biol* 238:199–213

Volwerk JJ, Shashidhar MS, Kuppe A (1990): PI-specific PLC from *Bacillus cereus* combines intrinsic phosphotransferase and cyclic phosphodiesterase activities: A ^{31}-NMR study. *Biochemistry* 29:8056–8062

Wery J-P, Schevitz RW, Clawson DK, Bobbitt JL, Dow ER, Gamboa G, Goodson T, Hermann RB, Kramer RM, McClure DB, Mihelich ED, Putnam JE, Sharp JD, Stark DH, Teater C, Warrick MW, Jones ND (1991): Structure of recombinant human rheumatoid arthritic synovial fluid phospholipase A$_2$ at 2.2 Å resolution. *Nature* 352:79–82

Wells MA (1974): The mechanism of interfacial activation of phospholipase A$_2$. *Biochemistry* 13:2248–2257

Weng X, Luecke H, Song IS, Kang DS, Kim S-H, Huber R (1993): Crystal structure of human annexin I at 2.5 Å resolution. *Protein Science* 2:448–458

White SP, Scott DL, Otwinowski Z, Gelb MH, Sigler PB (1990): Crystal structure of cobra venom phospholipase A$_2$ in a complex with a transition state analogue. *Science* 250:1560–1563

Winkler FK, D-Arcy A, Hunziker W (1990): Structure of human pancreatic lipase. *Nature* 343:771–774

13

Thermodynamics of the Interaction of Proteins with Lipid Membranes

THOMAS HEIMBURG AND DEREK MARSH

Introduction

Peripheral proteins associated at the lipid surface are one of the major components of biological membranes. They may function in situ as electron carriers (e.g., cytochrome c), as enzymes (e.g., protein kinase C), as signal transduction proteins (e.g., G-proteins), or primarily as structural elements (e.g., spectrin and myelin basic protein). The protein density at the membrane surface can be relatively high, and the peripheral proteins may also interact with the exposed portions of integral proteins embedded within the membrane (e.g., with redox enzymes of the respiratory chain, or with receptors such as those to which G-proteins are coupled). The association with the membrane is most frequently of electrostatic origin but may also include surface adsorption and hydrophobic components. The interactions of the isolated peripheral proteins with lipid bilayer membranes, therefore, are of direct relevance to the structure and function of biological membranes, and the determination of binding isotherms has proved to be particularly useful in the study of such interactions. Analysis of the latter constitutes the first and an important part of this chapter that is directly relevant to the thermodynamics of binding.

From a biophysical point of view, the binding of peripheral proteins to lipid membranes has a profound influence on parameters such as the electrostatic surface potential or the enthalpy and the heat capacity of the lipid-protein system. These parameters determine the phase transition behavior of lipid membranes as well as the energetics of the binding reaction itself. At temperatures close to a phase transition, the thermodynamic properties of protein binding cannot be considered separately from, for example, the chain-melting reactions. It is evident from calorimetric heat capacity profiles that the chain-melting of

Biological Membranes
K. Merz, Jr. and B. Roux, Editors
© Birkhäuser Boston 1996

lipids is influenced by the binding of proteins. As a direct consequence, the binding of proteins must be influenced by changes in the lipid state. Therefore, the change in lipid state shows up in the behavior of binding isotherms of proteins to membranes. Correspondingly, the binding reaction results in shifts, broadening, or enthalpy changes of the calorimetric heat capacity profiles of lipid chain-melting. Also, structural changes in the lipid matrix might occur as a consequence of protein binding due to the change in the overall physical properties of the lipid-protein complex. Consideration of the lipid chain-melting transition, and its coupling to the lipid binding and lipid-protein interactions, from a thermodynamic point of view constitutes the second part of this chapter.

Although most principles described below are generally applicable, in the following we shall restrict the thermodynamic treatment of protein binding at the membrane to extended planar lipid surfaces. A goal of this chapter is to link the thermodynamics of binding with that of chain-melting and to discuss the consequences for protein function. For this, we first outline a general description of the electrostatics of binding as a function of ionic strength and lipid composition with emphasis particularly on binding isotherms for continuous membrane surfaces. We then apply some of the results to the description of the chain-melting reaction in the presence of proteins. It will be shown that changes of the state of the membrane can result in a change of the protein distribution on the surface, as well as in changes in the forces acting on the bound proteins. These changes correlate with functional changes observed in some proteins. We shall discuss several examples for integral and peripheral proteins in which lipid chain-melting and function of the protein are in a clear relation to each other.

Binding of Peripheral Proteins to Lipid Surfaces

The binding of peripheral proteins to membranes is controlled by a variety of different interactions. Most important are the electrostatic interactions between charges on the bound protein and the individual charges on the lipids. Also of relevance are nonelectrostatic or hydrophobic contributions to the binding introduced by interactions of the protein with the hydrocarbon chains of the membrane lipids. A third important factor is the interaction between proteins bound to the surface which can lead to dimerization or aggregation of individual protein molecules. Furthermore, the degree of binding is affected by the finite size of the proteins, and, in lipid mixtures, the lateral distribution of lipids is very likely to be affected by the specificity of interaction of the various lipids with the proteins. These interactions are, in general, dependent on the degree of binding, particularly because of the special nature of ligand binding to continuous surfaces.

Binding Isotherms

The simplest and most frequently used approach to describe ligand binding to surfaces is the Langmuir isotherm. It describes how much of a ligand (protein) is bound to a surface of given size as a function of the free ligand concentration. In this model, the concentrations (or activities) of free ligand $[L]$, occupied binding sites $[S_b]$, and free binding sites $[S_f]$ are related to the binding constant, K, by the mass action law:

$$K = \frac{[S_b]}{[S_f] \cdot [L]} \tag{1}$$

which gives rise to the following binding isotherm:

$$[L] = \frac{1}{K} \cdot \frac{\theta}{(1 - \theta)} \tag{2}$$

where $\theta = [S_b]/([S_b] + [S_f])$ is the fraction of occupied sites. A model that is compatible with the law of mass action implies the existence of localized, independent binding sites on the membrane surface (see Figure 1, top). In general, this assumption is not appropriate for a lipid membrane because a lipid bilayer represents a continuous surface. Proteins are free to arrange in such a way that the free surface area between two adjacent bound proteins is greater than zero but smaller than a protein cross section. This means that, in the continuous case, the surface accessible for binding is smaller than the total free surface (see Figure 1, bottom). A further aspect is that there might also exist specific attractive or repulsive interactions between ligands bound to the surface which could lead to aggregation, as indicated schematically by the arrows in Figure 1 (bottom).

These features that are specific to the association with continuous surfaces become most important at higher degrees of binding and are often neglected in considering the binding of peripheral proteins to membranes. Because the surface concentration of proteins in natural membranes can be quite high, it is worthwhile to consider these particular aspects specifically here before going on to a more complete and general treatment. It has been found, for instance, that such effects are important in interpreting the displacement of peripheral proteins from the membrane surface at high ionic strength (Heimburg and Marsh, 1995).

The lateral interactions between proteins that control their distribution on the membrane surface give rise to a two-dimensional pressure, $\Pi(i)$, where i is the number of proteins bound. It is therefore convenient, for illustrative purposes, to analyse these in terms of the Gibbs absorption isotherm (Aveyard and Haydon, 1973):

$$\frac{d\Pi(i)}{d \ln[L]} = \frac{ikT}{n\Delta A} \tag{3}$$

where n is the maximum number of proteins that ideally can be accommodated on the surface, ΔA is the excluded area per protein, and $[L]$ is the free pro-

LANGMUIR-TYPE ISOTHERM :

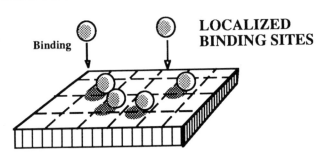

CONTINUOUS SURFACE:

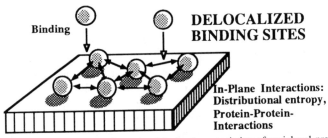

Figure 1. Schematic indication of different models for the association of peripheral proteins with membranes. Top: Binding to a surface lattice with fixed independent binding sites, as assumed for the Langmuir isotherm. Bottom: Binding to a continuous surface allowing for in-plane protein rearrangement and protein-protein interaction, indicated by the arrows.

tein concentration (or activity). The simplest expression for the lateral pressure between proteins bound to a continuous surface (i.e., for nonlocalized binding sites) is the Volmer equation of state:

$$\Pi(i) = \frac{ikT}{(n-i)\Delta A}. \tag{4}$$

This equation takes explicit account of the finite size of the protein ligands, but does not include specific interactions between the bound proteins. It is straightforward to show by differentiating Equation 4 with respect to i, substituting from Equation 3, and then integrating, that the corresponding isotherm is given by (Haydon and Taylor, 1960):

$$[L] = \frac{1}{K} \cdot \frac{\theta}{1-\theta} \exp\left(\frac{\theta}{1-\theta}\right) \tag{5}$$

where $\theta = i/n$ is the degree of surface coverage and K, which is derived from the integration constant, is the effective binding constant. This equation differs by the exponential factor from the standard Langmuir isotherm (Equation 2), which is appropriate for binding to fixed localized sites. The difference arises from the

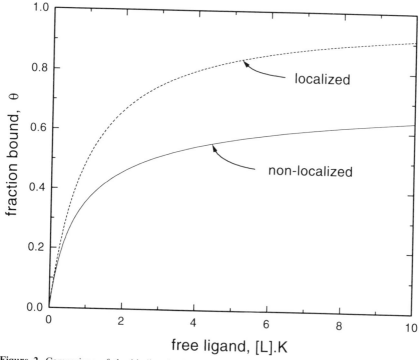

Figure 2. Comparison of the binding isotherm for a continuous surface calculated according to Equation 5 (full line) with the Langmuir isotherm (Equation 2) for binding to fixed sites (dashed line). The degree of ligand binding, θ, is plotted as a function of the free ligand concentration, $[L]$, multiplied by the binding constant, K.

modification of the distributional entropy for nonlocalized binding to continuous surfaces compared with that to fixed binding sites (Figure 1). The result of this is that, at high concentrations of free ligand, less protein is bound to continuous surfaces (Figure 2). This difference also becomes particularly pronounced for high binding constants. Only at very low degrees of binding ($\theta \ll 1$), for which steric interactions are unimportant, is the binding to continuous surfaces similar to that described by the Langmuir isotherm.

In mixed lipid membranes, the protein additionally can display a preferential affinity for different lipids, e.g., the protein may prefer charged lipids over uncharged lipids. A consequence of this may be a local rearrangement of the lipids to a degree that depends also on the mixing properties of the lipids. Figure 3 shows the three different arrangements that arise if the lipids do not mix at all (bottom), if they distribute according to the relative affinities for the protein (center), or if they mix in a fashion that results in a homogeneous distribution (top). The first case occurs either if unlike lipid species have energetically very unfavorable interactions with unlike nearest neighbor lipids or if the affinity of the protein for one lipid type is extremely high relative to the other lipid

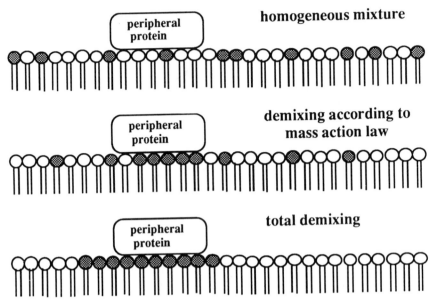

Figure 3. Schematic representation of the lipid rearrangement in a mixed lipid membrane upon the binding of peripheral proteins. Top: homogeneous lipid mixture in which no lipid rearrangements take place. Middle: lipids with no preferential interaction which redistribute according to the different affinities for the protein. Bottom: totally unmixed lipids forming in-plane membrane domains.

type. The lipids then either form preexisting domains of different composition to which the protein binds with different affinity or form domains that are induced by the protein binding (Figure 3, bottom). In the case of homogeneous mixtures, the interactions between unlike lipids are favored so strongly that the formation of domains is inhibited even in the presence of proteins (Figure 3, top). In this case, the binding is generally weaker than when domains are formed. In the intermediate case, there are no preferential interactions between lipids and the lipids alone mix statistically. However, for binding of a protein that associates preferentially with one of the lipid species, a local rearrangement of the lipids occurs, which increases the local binding strength (Figure 3, center). In general, different lipid mixing behavior reflects itself in the form of the binding isotherms.

General Statistical Mechanical Model

A general way to express the binding is given by the following statistical mechanical expression for the mean number, $\langle i \rangle$, of ligands bound to the membrane (Cantor and Schimmel, 1980):

$$\langle i \rangle = \frac{\displaystyle\sum_{i=1}^{n} i \cdot \frac{1}{i!} K^i [L]^i \exp\left(-\frac{\Delta F(i)}{kT}\right)}{\displaystyle\sum_{i=0}^{n} \frac{1}{i!} K^i [L]^i \exp\left(-\frac{\Delta F(i)}{kT}\right)} \tag{6}$$

where $\Delta F(i)$ is the overall free energy change upon binding i ligands. It is possible to show that this equation can be expressed in the following condensed form (Heimburg and Marsh, 1995):

$$\langle i \rangle = K[L] \exp\left(-\frac{d}{di}\left[\frac{\Delta F(i)}{kT}\right]\right) \tag{7}$$

where $d\Delta(F(i)/di$ is the free energy of binding a single ligand when i ligands are already bound. If the dependence of the free energy change, $\Delta F(i)$, on the number of bound ligands is known, one can then derive the binding isotherms for any specific situation. The free energy change, $\Delta F(i)$, in the case of the Langmuir isotherm is given by the distributional entropy contribution, which is Boltzmann's constant times the logarithm of the number of ways of arranging i ligands on a surface with n sites. This results in $\Delta F(i) = -kT ln[n!/(n-i)!]$ for a model with fixed sites. By using Stirling's formula and Equation 7, one can readily verify that this yields the Langmuir isotherm (i.e., Equation 2).

   In a protein-membrane complex, the expression for the free energy is more complicated. As well as distributional terms, it also contains contributions from ligand-ligand interactions, from electrostatics, and from redistribution of the different lipids in the membrane plane upon ligand binding. These contributions are designated by $\Delta F_{dist}(i)$, $\Delta F_{LL}(i)$, $\Delta F_{el}(i)$, and $\Delta F_{LD}(i)$, respectively. The total free energy change with i ligands bound is then:

$$\Delta F(i) = \Delta F_{dist}(i) + \Delta F_{LL}(i) + \Delta F_{el}(i) + \Delta F_{LD}(i). \tag{8}$$

The different terms in this equation are treated separately in the sections following.

Distributional Free Energy

The distributional free energy term, ΔF_{dist}, has already been considered above. It may be generalized to include the lateral interactions between bound proteins, ΔF_{LL}, by using an equation of state for the two-dimensional surface pressure, $\Pi(i)$, that explicitly accounts for the ligand-ligand interactions. This combined free energy change is then given by the work done against the surface pressure on bringing the proteins to their equilibrium surface density:

$$\Delta F_{dist}(i) + \Delta F_{LL}(i) = -\int \Pi(i)dA \tag{9}$$

where the integration extends from a reference area that corresponds to infinite ligand separation to the area of the membrane. As a first approximation for the lateral interactions, one can take the Van der Waals equation of state for a two-dimensional gas, which includes also the excluded area term:

$$\Pi(i) = \frac{ikT}{(n-i)\Delta A} - a\left(\frac{kT}{\Delta A}\right)\left(\frac{i}{n}\right)^2. \tag{10}$$

Here a is an empirical parameter characterizing the attractive ($a > 0$) or repulsive interactions ($a < 0$) between bound proteins. In the absence of membrane surface electrostatics, this results (from Equations 7 and 8) in the following binding isotherm for a lipid surface (Heimburg and Marsh, 1995):

$$[L] = \frac{1}{K}\frac{\theta}{1-\theta}\exp\left(\frac{\theta}{1-\theta} - 2a\theta\right) \tag{11}$$

where the integration reference state from Equation 9 is absorbed into the binding constant, K. This isotherm is identical to the isotherm obtained previously by Hill (1946) by using other means, and differs from that derived from the Volmer equation of state (Equation 5) by inclusion of the interaction term.

Electrostatic Free Energy

Now we consider the binding of a charged ligand to a lipid membrane in the presence of surface electrostatics. The electrostatic interaction of ligands with surfaces depends on the number of charges involved in the binding reaction. It is assumed that each surface binding region for a protein corresponds to α lipid molecules of which a fraction f are charged, and that each protein ligand bears Z net charges of opposite sign. The change in electrostatic free energy, $\Delta F_{el}(i)$, is given by the difference between the electrostatic free energy of the system before and after binding:

$$\Delta F_{el}(i) = F_{el}^S(i) - F_{el}^S(0) - i \cdot F_{el}^L \tag{12}$$

where the first two terms on the right-hand side represent the surface electrostatic free energy and the last term that of the free ligand in the bulk medium. By using Gouy-Chapman double-layer theory to obtain the electrostatic free energy of the surface (Jähnig, 1976) and Debye-Hückel theory for the electrostatic self energy of the free ligand (Tanford, 1955), one then can obtain an analytical expression for the change in total electrostatic free energy on binding of the charged ligand.

In the high potential limit, the electrostatic free energy of a surface, F_{el}^S, derived from standard electrostatic double-layer theory, is given by (Jähnig, 1976):

$$F_{el}^S = -2q \cdot \left(\frac{kT}{e}\right) \cdot \ln\left(-\frac{\Lambda_0 \sigma}{\sqrt{c}}\right) \tag{13}$$

where $\Lambda_0 = (1000\pi/2\varepsilon N_A kT)^{1/2}$ is a constant, σ is the surface charge density, q is the total surface charge, and c is the ionic strength of the (assumed) monovalent salt, e is the elementary charge, and N_A is Avogadro's number. For a surface of total area equal to n protein cross sections, with i proteins bound, the total charge, q, and the mean charge density, σ, are given by:

$$q = -(n\alpha f - iZ) \cdot e \tag{14}$$

$$\sigma = -\left(1 - \frac{iZ}{n\alpha f}\right) \cdot \frac{e}{a_0} f \tag{15}$$

where a_0 is the area per lipid molecule. It is assumed here that the charged lipids bear a single negative charge, that the proteins are positively charged, and also that charged and uncharged lipids are mixed homogeneously. The implications and the limitations of the latter assumption are discussed later.

A detailed calculation results finally in the following form for the absorption isotherms of charged lipid membranes (Heimburg and Marsh, 1995):

$$[L] = \frac{1}{K(0, f)} \left(1 - \theta \frac{Z}{f \cdot \alpha}\right)^{-2Z} \frac{\theta}{1 - \theta} \exp\left(\frac{\theta}{1 - \theta} - 2a\theta\right) \tag{16}$$

where $K(0, f)$ is the binding constant for initial ligand binding, i.e., with no appreciable electrostatic neutralization. This initial binding constant has the following dependence on the ionic strength, c, provided that the lipid distribution is homogeneous (Heimburg and Marsh, 1995):

$$K(0, f) = K' \cdot c^{-(Z+\Lambda Z^2)} \cdot f^{2 \cdot Z} \tag{17}$$

where $\Lambda = e^2/16\varepsilon r_0 kT$ arises from the electrostatic self energy of the protein (r_0 is the protein radius) and is relatively small. Equation 17 results in a linear dependence of the initial binding on ionic strength in a double logarithmic plot, from which the effective charge, Z, on the protein and, K', which contains the intrinsic binding constant, K, may be determined. Having obtained these constant parameters, the entire binding curves, at different ionic strengths, may then be analyzed by combining Equations 16 and 17.

Free Energy of Lipid Redistribution in Mixed Lipid Systems

As discussed earlier, the strength of protein binding to mixed lipid membranes depends on the mixing properties of the component lipids and, in general, on the redistribution of the lipids in response to the protein binding. If the charged and uncharged lipids are mixed homogeneously and remain so on protein binding (Figure 3, top), then the binding isotherm can be described by Equations 16 and 17 above, in which f is the fraction of charged lipid. If, at the other extreme, the charged and uncharged lipids separate completely forming macroscopic domains (Figure 3, bottom), then the protein binds only to the domains consisting of charged lipids. The effective membrane area for binding is correspondingly reduced, but the strength of binding is that characteristic of a bilayer consisting wholly of charged lipids. In this case, the binding isotherm can be described again by Equations 16 and 17, but with the replacements $f \rightarrow 1$ and $\theta \rightarrow \theta/f$ at all occurrences of these quantities.

Protein binding to most real mixed lipid membranes generally shows a behavior intermediate between these two limiting cases described above. If there are no preferential interactions between any of the lipids, then the binding of charged proteins tends to recruit the lipids of opposite charge into the vicinity of the protein binding site (Figure 3, middle). The energetic effects of this lipid redistribution are represented by the contribution, $\Delta F_{\mathrm{LD}}(i)$, to the total free energy in Equation 8. To determine this term, the binding can be split into two consecutive steps: an absorption to a homogeneously charged surface as described above and an additional lipid redistribution step, arising from the local binding properties. As an approximation, the latter can be described by an in-plane multiple binding reaction according to the mass action law (Cutsforth et al, 1989). Adopting this approach, the resultant protein binding constant is $K_{eff} = K(0, f)(1 + K_s f_i)^{\alpha}$, and the free energy of binding the charged lipids in the redistribution process is given by (Cutsforth et al, 1989):

$$\Delta F_{\mathrm{LD}}(i) = -\alpha k T \ln(1 + K_s f_i) \tag{18}$$

where K_s is the in-plane surface binding constant and f_i is the surface concentration of charged lipid that is unbound when i proteins are associated with the lipid surface. This contribution to the binding isotherm is then given from Equation 7 by the dependence on i of the fraction, f_i, of free lipids that are charged.

Experimental Binding Isotherms

As shown in the previous paragraphs, the study of binding isotherms can give information on the energetics of binding, the aggregation of ligands due to ligand-ligand interactions, and the mixing of lipids in systems containing more than one lipid species. These aspects are illustrated in the following subsections by experimental results on the binding of the peripheral protein cytochrome c to anionic lipids and their mixtures with zwitterionic lipids.

CYTOCHROME c BINDING TO NEGATIVELY CHARGED LIPID MEMBRANES

Binding of cytochrome c to charged lipid surfaces leads to a reduction in the protein denaturation temperature. Therefore, one can study the surface binding of both native and denatured forms with the protein still native in solution. This is done by choosing temperatures below and above that for surface denaturation, respectively, with both below that for denaturation of the free protein. Denatured cytochrome c tends to aggregate on surfaces. By comparing the native and denatured forms, one can determine the effects of surface ligand-ligand interactions on the binding isotherms.

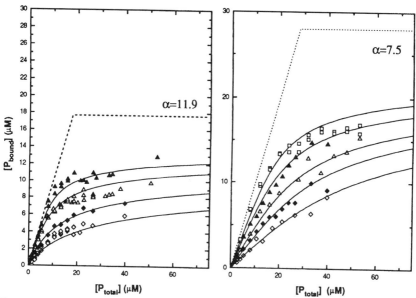

Figure 4. Binding isotherms for cytochrome c with dioleoyl phosphatidylglycerol membranes at neutral pH and different ionic strengths. Left: Native cytochrome c ($T = 20°C$) at ionic strengths of 41.9, 54.4, 79.4, and 104.4 mM (from top to bottom) . Solid lines represent a global fit of the isotherms at different ionic strengths to Equation 16 without protein-protein interactions. Right: Surface-denatured cytochrome c ($T = 65°C$) at ionic strengths of 104.4, 125, 154.4, 175, and 204.4 mM (from top to bottom). Solid lines represent a global fit of the isotherms at the different ionic strengths to Equation 16 using attractive protein-protein interactions. Data from Heimburg and Marsh (1995).

Examples of experimental isotherms for binding of cytochrome c to anionic dioleoyl phosphatidylglycerol membranes are given in Figure 4. The intrinsic binding constant $K = 1.9 \times 10^{-5}$ l/mole lipid and the effective charge $Z = +3.8$ for native cytochrome c are determined from Equation 17 by measuring the binding at very low degrees of surface coverage. Using these parameters, consistent fits of the isotherms in Figure 4 (left panel) at different ionic strengths to Equation 16 yield a lipid stoichiometry of $\alpha = 11.9$ per protein with the parameter, a, for the protein-protein interactions being close to zero. This means that the cross-section of the protein is equivalent to that of 12 lipids in a fluid bilayer which correlates reasonably well with crystallographic data for cytochrome c (Dickerson et al, 1971). Also, the binding isotherms indicate that there are few energetically significant interactions between the bound native proteins on the surface, other than steric exclusion (i.e., $a = 0$). The effective charge on the protein is considerably lower than the physical net charge of the protein of $+7$, including the heme charge, as deduced from the amino acid sequence (Cheddar and Tollin, 1994). Possible reasons for this low effective charge, which is found relatively frequently in the binding of extended multiply charged ligands, have been discussed previously (Heimburg and Marsh, 1995).

An interesting feature of the isotherms in Figure 4 (left panel) is the tendency to flatten off at values well below that corresponding to maximum surface coverage. This tendency becomes more pronounced with increasing ionic strength and is the basis for the removal of peripheral proteins from membranes by washing in high salt. The origin of this effect lies in the nature of binding to continuous surfaces. As the electrostatic component of the binding becomes weaker at high ionic strength, the effective lateral surface pressure between the surface-associated proteins still remains equally effective in preventing further binding (Equations 2 and 5 and Figure 2).

The situation is somewhat different for the binding of surface-denatured cytochrome c (Figure 4, right panel). The effective charge on the protein $Z = +3.3$, obtained from the initial parts of the binding isotherms, is similar to that for the native protein, but the intrinsic binding constant $K = 1.6 \times 10^{-3}$ l/mole lipid is considerably larger, corresponding to an increase in binding free energy by approximately -2.5 kcal/mol. For surface-denatured cytochrome c, the entire binding isotherms at different ionic strengths only can be fit with a nonzero value for the protein-protein interaction parameter. Global fits to the isotherms yield a reduced stoichiometry of $\alpha = 7.5$ lipids per protein, and a positive interaction parameter of $a = +2.3$. This indicates both that the protein decreases its cross-section on surface denaturation and that at the same time attractive interactions occur between the bound proteins, reflecting a tendency for the denatured proteins to aggregate on the lipid surface. There is independent evidence suggesting aggregation of surface-denatured cytochrome c from infrared spectroscopic studies (Heimburg and Marsh, 1993; Muga et al, 1991).

Qualitatively, the effects of the surface aggregation of the denatured cytochrome c are seen from comparison of the binding isotherms with those of the native protein in Figure 4. The isotherms for the surface-denatured protein are mostly at higher ionic strength than those given for the native protein. However, for comparable initial binding strengths, the isotherms for the denatured protein display less tendency to flatten off than do those for the native protein. This is because the attractive interactions between the surface-denatured proteins decrease the effective lateral pressure exerted between them, hence allowing more protein to bind.

CYTOCHROME c BINDING TO MIXTURES OF CHARGED
AND ZWITTERIONIC LIPIDS

For electrostatic reasons, the admixture with zwitterionic lipids decreases the strength of binding of peripheral proteins to negatively charged lipids. As discussed above, the extent of this effect can differ depending on the mixing properties of the lipids. This is illustrated in Figure 5, which gives the binding isotherm for native cytochrome c to a mixed membrane composed of dioleoyl phosphatidylglycerol (60 mol %) and dioleoyl phosphatidylcholine (40 mol %). The experimental binding isotherm is compared with theoretical pre-

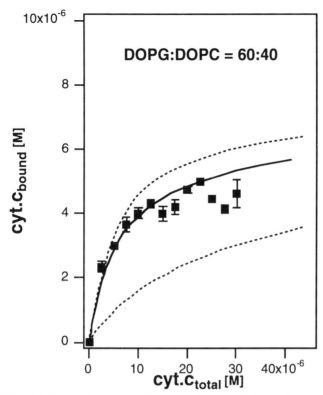

Figure 5. Binding isotherm of native cytochrome c with mixed lipid membranes composed of dioleoyl phosphatidylglycerol (60 mol %) and dioleoyl phosphatidylcholine (40 mol %) at an ionic strength of 45 mM. The dashed lines correspond to isotherms predicted from Equation 16 for a homogeneous lipid mixture (lower) and for complete separation of the two lipids (upper), using parameters obtained from fitting the data in Figure 4 (left panel). The full line represents a fit to the experimental data that takes into account the redistribution of the lipids on protein binding.

dictions (dashed lines) of the isotherms expected for binding to a homogeneous distribution of the two lipids (Figure 3, top) and to completely separated domains of the negatively charged component (Figure 3, bottom). The predicted isotherms are calculated from Equation 16, using the methods described above. The parameters for binding to dioleoyl phosphatidylglycerol are taken from Figure 4 for native cytochrome c, and it is assumed that the binding to zwitterionic phosphatidylcholine is negligible (for which there is experimental evidence). The experimental isotherm lies between those predicted for the two extreme cases, indicating that an in-plane redistribution of the negatively charged lipids in response to protein binding (Figure 3, middle) does in fact take place. The binding is weaker than to separated domains of phosphatidylglycerol but is considerably stronger than to a homogeneous lipid mixture. The experimental binding isotherm can be described adequately by taking into account the free energy of lipid redistribution with a simple mass action formalism as discussed above (Equation 18). This is indicated by the full line in Figure 5, in which the effective surface lipid binding constant, K_S, is related directly to the binding constant, $K(0, 1)$, of native cytochrome c to phosphatidylglycerol obtained above (Figure 4), as expected on energetic grounds.

This example illustrates the way in which analysis of the binding isotherms may be used to determine the rearrangements of the lipid distribution on protein binding, to study the selectivity of interaction of lipids with peripheral proteins (Sankaram and Marsh, 1993), and to obtain information on the mixing properties of lipids in bilayers (Figure 3). The latter aspect bears certain similarities to the determination of the activity of negatively charged phosphatidylserine in lipid mixtures by means of the Ca^{2+}-binding properties (Huang et al, 1993).

Difference in Protein Binding to Gel and Fluid Membrane States

In general, it is to be expected that the strength of binding of peripheral proteins will be dependent not only on the lipid species but also on the state of the lipid, particularly on whether it is in a gel or a fluid phase. Because the area per lipid molecule, a_0, increases appreciably at the gel-to-fluid chain-melting transition of lipid bilayers (Cevc and Marsh, 1987), the surface electrostatic free energy will be lower for fluid bilayers than for gel-phase bilayers, and the strength of binding will decrease correspondingly. According to Equations 13–15, in the high potential limit, the electrostatic free energy of the lipid surface in the presence of protein changes on chain-melting by:

$$\delta[\Delta F_{el}^S(i)]^{\text{melting}} = -2q\left(\frac{kT}{e}\right)\ln\left(\frac{\sigma_{\text{fluid}}}{\sigma_{\text{gel}}}\right) = +2(n\alpha f - iZ)kT\ln\left(\frac{a_0^{\text{gel}}}{a_0^{\text{fluid}}}\right)$$

$$(19)$$

where $\sigma_{\text{gel}}, \sigma_{\text{fluid}}$ are the surface charge densities, and $a_0^{\text{gel}}, a_0^{\text{fluid}}$ are the areas per lipid molecule in gel and fluid phase bilayers, respectively. The difference in electrostatic free energy of binding a protein, $d\Delta F_{el}^S(i)/di$, to the fluid and gel lipid states is then given by (Heimburg and Marsh, 1995):

$$\delta[d\Delta F_{el}^S(i)/di]^{\text{melting}} = -2kTZ\ln\left(\frac{a_0^{\text{gel}}}{a_0^{\text{fluid}}}\right).$$

$$(20)$$

In the high-potential regime of electrostatic double-layer theory that is used here, this free energy is almost totally entropic (Jähnig, 1976). Assuming an increase in surface area on lipid chain-melting of approximately 30% (Marsh, 1990), the change in electrostatic binding energy amounts to $\approx +1.2$ kcal/mol protein. This would correspond to a decrease in binding constant of the protein by a factor of approximately eight on chain-melting. Expressing the electrostatic surface free energy difference in terms of the number of lipids yields a value of $\approx +100$ cal/mol per lipid, assuming a lipid/protein stoichiometry of $\alpha = 12$ that is appropriate to native cytochrome *c*. Energetically speaking, this is a relatively small change, but because of the highly cooperative nature of lipid chain-melting it is capable of changing the gel-fluid equilibrium. The partial neutralization of the membrane electrostatic charge by binding the protein results in a shift in the lipid transition to higher temperatures by an amount that, from perturbation theory for first-order transitions, is given by (Cevc and Marsh, 1987):

$$\Delta T_t = \frac{\delta[d\Delta F^S_{el}(i)/di]^{melting}}{\alpha \Delta S_t} \qquad (21)$$

where ΔS_t is the chain-melting entropy. Taking a value for the latter of ≈ 23 cal/mol/K appropriate to dimyristoyl phosphatidylgylcerol (Marsh, 1990) results in an upward shift in transition temperature of $\approx +4\text{--}5°C$. This illustrates the coupling of the lipid chain-melting equilibrium with the protein binding equilibrium via the reciprocal influence of protein binding and membrane surface electrostatics on one another.

Irrespective of mechanism, the different strength of binding of peripheral proteins to different states of the lipid results quite generally in a direct coupling between the lipid-protein interaction and the lipid phase transition. The overall combined association and phase equilibria at fixed temperature can be depicted by the following scheme (Heimburg and Biltonen, 1996):

$$
\begin{array}{ccc}
 & K^L & \\
P+G & \rightleftharpoons & P+F \\
K_G \;\; \Updownarrow & & \Updownarrow \;\; K_F \\
P\cdot G & \rightleftharpoons & P\cdot F \\
 & K^{LP} &
\end{array}
$$

where K_G and K_F are the association constants of the protein (P) with fluid (F) and gel (G) phase lipids. The apparent equilibrium constants for the gel-fluid equilibria of the free and protein-bound lipid are $K_L = \theta^L/(1-\theta^L)$ and $K^{LP} = \theta^{LP}/(1-\theta^{LP})$, respectively. From the cyclic nature of the equilibria (i.e., $K_G \cdot K^{LP} = K^L \cdot K_F$), the degree of conversion to the fluid phase for the protein-bound lipid (θ^{LP}) is related to that for the free lipid (θ^L) by:

$$\theta^{LP} = r\theta^L/[1 + (r-1)\theta^L] \qquad (22)$$

where $r = K_F/K_G$ is the ratio of the association constants with the fluid- and gel-state lipids. This general relation demonstrates that both the mid-point of the transition will be shifted and the shape of the transition curve will be skewed, in the presence of bound protein. In particular, the transition mid-point ($\theta^{LP} = 0.5$) occurs at a temperature for which the degree of transition of the free lipid is: $\theta^L = 1/(r+1)$, which results in downward shifts for $r < 1$ and upward shifts for $r > 1$.

To illustrate simply the effects of protein binding on the gel-fluid lipid coexistence, the cooperativity of the phase transition can be depicted by a one-dimensional two-state Ising model (Marsh et al, 1976). The degree of transition of the free lipid is then approximated by (Zimm and Bragg, 1959):

$$\theta^L = \frac{1}{2}\left[1 + \frac{s-1}{\sqrt{(s-1)^2 + 4\sigma_L s}}\right] \tag{23}$$

where s is a scaled temperature parameter, and the cooperativity parameter, σ_L, is determined by the boundary free energy between gel and fluid lipid domains. The transition curves for the protein-bound lipid deduced from Equations (22) and (23) are given for various values of the ratio of fluid and gel association constants in Figure 6. The transition curve for $r = 1$ is identical with that for the free lipid. An increasing preference of the protein for binding to one of the states of the lipid results in a progressive broadening and shift of the transition towards the side of transition that favors the other phase. Under these conditions, the range of phase coexistence is broadened, and the domains of the lipid phase

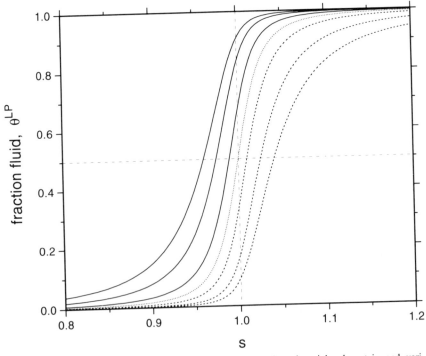

Figure 6. Chain-melting transition curves for lipids with bound peripheral protein and various ratios of the protein association constants with gel and fluid lipid. The degree of transition, θ^{LP}, is obtained from Equations 20 and 21 with $\sigma_L = 2.10^{-4}$ and from left to right: $r = 0.1, 0.2, 0.5, 1.0, 2.0, 5.0, 10.0$. The x-axis is the dimensionless temperature parameter, $s \approx 1 + (\Delta H_t/RT_t^2)(T - T_t)$, where T_t is the transition temperature of the free lipid (i.e., $s = 1$) and ΔH_t the corresponding transition enthalpy (Marsh, 1995b).

favoring protein binding are preferentially stabilized. This inevitably gives rise to a nonhomogeneous distribution of protein molecules on the membrane surface, as will be seen in detail later.

Lipid Chain-Melting and Lipid-Protein Interaction

In the following sections, the lipid chain-melting phase transition and its reciprocal coupling with protein binding and lipid-protein interactions are considered in some detail. The emphasis here is on the formation of spatially separated domains of lipids in the gel and fluid states, and on the ensuing spatial heterogeneity of the in-plane distribution of the membrane-bound proteins. For this purpose, a two-state model (i.e., gel and fluid) is adopted for the lipid phase transition in which only interactions between nearest-neighbor lipids are considered, i.e., a two-dimensional Ising model. In such a model, cooperativity arises in the phase transition because of energetically unfavorable interactions between lipids in different states, which are located principally at the boundaries of the domains composed of lipids in a single state. Monte Carlo calculations based on a fixed two-dimensional lipid lattice are used in order to allow for the entropy of the domain boundaries in a proper manner and at the same time to give direct information on the spatial extent of the domains and their fluctuations. The distribution of proteins bound to the membrane surface is also obtained by the same method.

Lipid Transitions and Lipid Domain Formation

First we consider the chain-melting transition of lipid membranes in the absence of proteins. This is a highly cooperative process with a transition half-width that can be as small as 0.07 degrees (Albon and Sturtevant, 1978). Cooperativity implies that the lipid chains do not melt independently; instead the state of a lipid is dependent on that of the surrounding lipids. In the following it is assumed that interactions take place only between nearest-neighbor lipids that are in one of two states (gel or fluid), i.e., a two-dimensional Ising model.

The nearest-neighbor lipid-lipid interaction free energies are designated by ε_{gg}, ε_{ll}, and ε_{gl}, corresponding to the interaction between two gel lipids, two fluid lipids, and a gel and a fluid lipid, respectively. At temperature T, the chain-melting free energy of the bilayer in a particular configuration, for which the number of lipids in the fluid state is n_l and the number of unlike lipid contacts is n_{gl}, is then given by:

$$\Delta G(T) = n_l(\Delta H_t - T\Delta S_t) + \frac{1}{2}n_{gl}\omega_{gl} \qquad (24)$$

where ΔH_t and ΔS_t are the total transition enthalpy and entropy, respectively, and $\omega_{gl} = \varepsilon_{gl} - (\varepsilon_{gg} + \varepsilon_{ll})/2$ is the excess interaction free energy of a pair of unlike lipids. The mid-point of the transition (strictly speaking, the isoenergetic point) is given by $\Delta G(T) = 0$, and occurs at a temperature $T_m = \Delta H_t / \Delta S_t$, ignoring the small term of order $(n_{gl}/n_l)\omega_{gl}$.

The mean value for any observable, $\langle X \rangle$, can be obtained by averaging over all bilayer configurations at a particular temperature, using the statistical thermodynamic expression:

$$\langle X(T) \rangle = \sum_{n_l} \sum_{n_{gl}} X(n_l, n_{gl}) \cdot P(T, n_l, n_{gl})$$

$$= \frac{\displaystyle\sum_{n_l} \sum_{n_{gl}} X(n_l, n_{gl}) \cdot \Omega(n_l, n_{gl}) \exp\left(-\frac{n_l \cdot (\Delta H_t - T\Delta S_t) + \frac{1}{2}n_{gl}\omega_{gl}}{RT}\right)}{\displaystyle\sum_{n_l} \sum_{n_{gl}} \Omega(n_l, n_{gl}) \exp\left(-\frac{n_l \cdot (\Delta H_t - T\Delta S_t) + \frac{1}{2}n_{gl}\omega_{gl}}{RT}\right)} \qquad (25)$$

where $P(T, n_l, n_{gl})$ is the probability at temperature T of finding a bilayer configuration with a given fraction of lipids in the fluid state and of gel-fluid contacts, and $\Omega(n_l, n_{gl})$ is the number of independent ways of generating such a configuration (i.e., the degeneracy). Possible observables are, for example, the enthalpy, $\langle H \rangle$ or $\langle H^2 \rangle$, or the mean surface area of the bilayer, $\langle A \rangle$. Of particular interest is the mean fraction of fluid lipids, $\langle f \rangle$, where $f = n_l/n$ for a bilayer composed of n lipids.

The main problem in calculating the configurational partition function and the mean value of an observable lies in the determination of the distribution function $\Omega(n_l, n_{gl})$. Because an analytical method is not available, Monte Carlo simulations are used. The simulations are performed numerically by defining a lipid matrix and switching the state of each individual lipid according to the statistical mechanical probability for such a change. If this probability is greater than a randomly generated probability, it is accepted, and if below this value, it is rejected. The relevant probability for an individual lipid to change from the gel state to the fluid state is given by:

$$P = \frac{K(T)}{1 + K(T)} \qquad (26)$$

where the statistical weight is given by:

$$K(T) = \exp\left(-\frac{\Delta H_t - T\Delta S_t + \Delta n_{gl} \cdot \omega_{gl}}{RT}\right). \qquad (27)$$

Here Δn_{gl} is the increase in the number of unlike nearest neighbors upon the change from the gel to the fluid state, where for lipids on a triangular lattice: $-6 \leq \Delta n_{gl} \leq 6$ (see Figure 7). For further details see Mouritsen and Biltonen (1992), Sugar et al (1994) and Heimburg and Biltonen (1996). The Monte Carlo steps are repeated many times so that the distribution function $\Omega(n_l, n_{gl})$ can be determined by counting the relative numbers of the respective configurations generated. The Metropolis procedure used for obtaining $\Omega(n_l, n_{gl})$ can be described as a representative random walk through the phase space and is

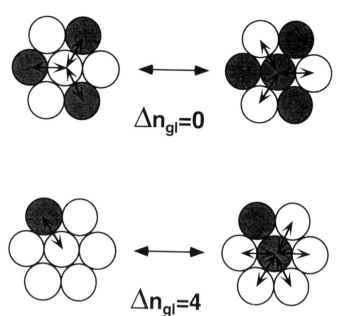

Figure 7. Illustration of the change in state of a lipid during a Monte Carlo step. The lipids are arranged on a triangular (i.e., centered hexagonal) lattice, and gel-state and fluid-state lipids are depicted by open and shaded circles, respectively. Two cases are given in which the number of unlike nearest-neighbors of the central test lipid changes by $\Delta n_{gl} = 0$ and $\Delta n_{gl} = 4$. In general: $\Delta n_{gl} = z - 2n^{\circ}_{gl}$, where z is the total number of nearest-neighbors (i.e., the coordination number) and n°_{gl} is the original number of unlike nearest-neighbors before the change of state (Heimburg and Biltonen, 1996).

required to be performed only once. When the distribution $\Omega(n_l, n_{gl})$ is known, the mean value of an observable $\langle X \rangle$ can be derived for any given set of values for the parameters ω_{gl}, ΔH_t and ΔS_t, and at a particular temperature T, by using Equation 25 (Ferrenberg and Swendson, 1988).

By differentiating Equation 25 for $\langle H \rangle$ with respect to temperature, the molar heat capacity at constant pressure, C_p, can be determined, yielding the well-known fluctuation-dissipation theorem from statistical mechanics (Hill, 1960):

$$C_p = \left(\frac{\partial \langle H \rangle}{\partial T} \right)_p = \frac{\langle H^2 \rangle - \langle H \rangle^2}{RT^2}. \tag{28}$$

The required averages, $\langle H^2 \rangle$ and $\langle H \rangle$, are obtained from Equation 25. In the two-state model, each is related to the corresponding mean values for the fraction of fluid lipids, f, i.e.:

$$C_p = [\langle f^2 \rangle - \langle f \rangle^2] \frac{(\Delta H_t)^2}{RT^2} \tag{29}$$

because the molar chain-melting enthalpy at fluid fraction f is given simply by: $\Delta H(f) = f \Delta H_t$. This latter holds as long as the enthalpic contribution

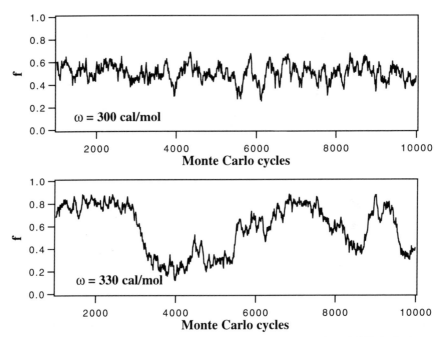

Figure 8. Fluctuations in the percentage of fluid lipids about a mean value of 50% obtained over a range of 9000 Monte Carlo cycles with a 41 × 41 lipid matrix. The initial 1000 simulation steps were discarded in order to ensure that a representative configuration had been achieved. The parameters used for the simulation were $\Delta H_t = 8.7$ kcal/mol, $\Delta S_t = 28$ cal/mol/K, and an enthalpic interaction energy of $\omega_{gl} = 300$ cal/mol (upper) or 330 cal/mol (lower) with a triangular lattice and a temperature corresponding to the mid-point of the gel-to-fluid transition of $T_m = 310.3$ K.

from the interfacial free energy, ω_{gl}, which is relatively small, can be neglected. From the properties of mean values, it can readily be shown that $\langle f^2 \rangle - \langle f \rangle^2 = \langle (f - \langle f \rangle)^2 \rangle$, i.e., that the quantities of interest are the mean-square fluctuations about the mean value.

The fluctuations in the fraction of fluid lipids obtained from a Monte Carlo simulation for lipids on a triangular lattice are illustrated in Figure 8. The temperature chosen for the simulation corresponds to the mid-point of the gel-to-fluid bilayer transition, and, therefore, the fluctuations about the mean value of 50% are especially large. Simulations are given for two values of ω_{gl}; for the moment we concentrate on the upper one corresponding to the lower degree of cooperativity. This simulation allows the calculation of the heat capacity maximum C_p^{max} for the transition, by using Equation 29. Having established the relevant parts of the distribution function, $\Omega(n_l, n_{gl})$, in this way by a single simulation, it is then possible to calculate the heat capacity profile as a function of temperature from Equation 25. The results for various values of the interfacial free energy parameter, ω_{gl}, are given in Figure 9. It can be seen that increasing

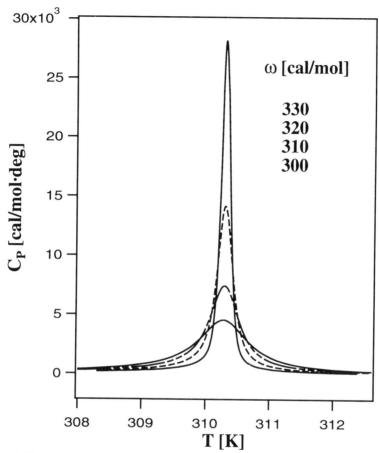

Figure 9. Heat capacity profiles for the chain-melting of a lipid bilayer with $\Delta H_t = 8.7$ kcal/mol, $T_m = 310.3$ K, calculated with a two-dimensional Ising model and unlike nearest-neighbor interaction free energies of $\omega_{gl} = 300, 310, 320, 330$ cal/mol, which are assumed to be purely enthalpic. Calculations were performed for a triangular lipid matrix with 31×31 sites by using a single Monte Carlo simulation (for $\omega_{gl} = 320$ cal/mol and $T = T_m$) and the Ferrenberg-Swendsen (1988) method with Equation 25.

the interfacial free energy parameter increases the cooperativity of the transition, resulting in both a decreasing transition half-width and an increasing heat capacity maximum.

Figure 10 shows representative lipid configurations obtained from Monte Carlo simulations for temperatures below, at, and above the heat capacity maximum of the chain-melting transition. As can be seen, the lipids form spatially separated domains of various sizes which are composed essentially of lipids in a single state (either gel or fluid). As the temperature changes towards the mid-point of the transition, the domains composed of lipids in the

melting of single lipid membrane

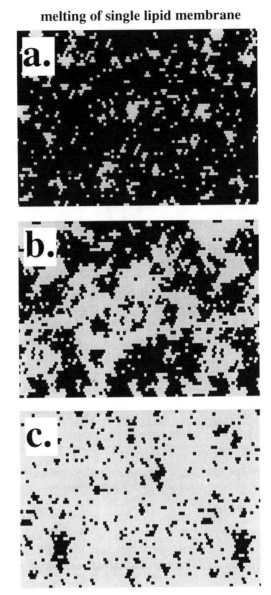

Figure 10. Snapshots of the lipid configuration at temperatures: (a) below ($T = 308.3$ K), (b) at ($T = 310.3$ K), and (c) above ($T = 312.3$ K) the heat capacity maximum for a lipid matrix with 61×61 sites, $\Delta H_t = 8.7$ kcal/mol, $\omega_{gl} = 310.3$ cal/mol and $\Delta S_t = 28$ cal/mol/K. For all these Monte Carlo simulations a triangular lattice with periodic boundary conditions was used. Gel-state lipids are black and fluid-state lipids are grey (Heimburg and Biltonen, 1996).

minority state grow in size at the expense of the majority state, achieving their maximum size at the transition mid-point. The tendency to domain formation results from the unfavorable interfacial energy between the gel and fluid lipid states and becomes increasingly more pronounced as the interfacial energy increases and the transition becomes more cooperative. On the other hand, as can be seen from Figure 10, the domain surface is complex and approximately fractal. This is a result of the energetically favorable increased entropy of a more extended interface. The unfavorable interfacial free energy (assumed here to be totally enthalpic) and the favorable entropy of an extended interface balance each other in a temperature-dependent manner. However, the higher the value of the enthalpic parameter ω_{gl}, the larger the average domain size and the shorter the overall domain interface become. This is of importance in considering the distribution of bound proteins as described in the next section.

With increasing cooperativity one expects eventually to get into a regime where a continuous change in the lipid state, such as considered above, converts to first-order behavior. First-order transitions show a coexistence of two macroscopic states in the same canonical ensemble at the transition point which means that at the chain-melting point two membrane configurations with different fractions of fluid lipids are present with equal probability. This is demonstrated in Figure 11, in which the probability $P(f, \bar{n}_{gl})$ of finding a membrane with a fluid lipid fraction f and a mean number of unlike contacts per lipid, $\bar{n}_{gl}$, at the transition point is plotted as a function of these two variables. The probabilities are obtained from the individual terms in the summation in Equation 25 with $X = 1$, $f = n_l/n$, and $\bar{n}_{gl} = n_{gl}/n$, where n is the total number of lipids. As the cooperativity parameter increases, the distribution of states progressively develops two maxima (Figure 11d), which indicates first-order behavior (Lee and Kosterlitz, 1991). The fluctuations in such a highly cooperative system are shown in the lower panel of Figure 8. Rapid fluctuations take place about one of the maxima of Figure 11d, and then, less frequently, large-scale fluctuations take place between the two maxima. As the size of the system increases, the distributions in Figure 11d become sharper, and the large-amplitude fluctuations become more abrupt but less frequent. In the case of a low cooperativity (Figure 11a), only one maximum exists in the probability distribution. This corresponds to a second- or higher-order transition, because the system does not switch from one state to the other at the transition but undergoes a continuously varying change of state. The fluctuations in this type of low-cooperativity system correspond to those given in the upper panel of Figure 8.

Finally, it will be noted that the lateral compressibility of the lipid matrix also can be obtained from the Monte Carlo simulations for the two-state model. This is done in a manner similar to that described for the heat capacity (Hill, 1960), because the fluctuations in lipid surface area also are proportional to those in the fraction of fluid lipids. These elastic properties of the lipid membrane are considered in detail in a later part of the chapter.

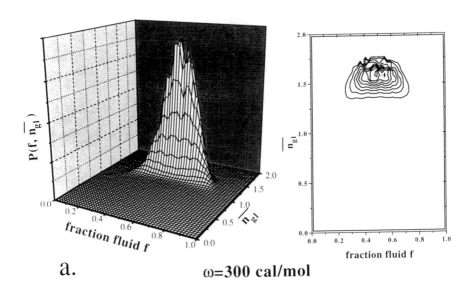

a.   ω=300 cal/mol

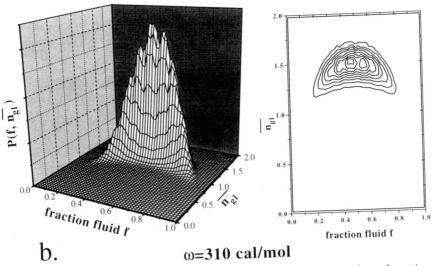

b.   ω=310 cal/mol

Figure 11. Distribution $P(f, \bar{n}_{gl})$ of lipid configurations at the heat capacity maximum for a triangular lipid matrix with 31×31 sites with increasing values (a to d) of the interaction energy, ω_{gl}, between unlike lipids. Parameters used in the Monte Carlo simulations are those given in the legend to Figure 9. The probabilities $P(f, \bar{n}_{gl})$ are obtained from the individual terms in the summation of Equation 25, with $X = 1$, $f = n_l/n$ and $\bar{n}_{gl} = n_{gl}/n$, where n ($= 31 \times 31$) is the total number of lipids in the matrix. At higher values of ω_{gl} the distribution curve displays two maxima (panel d), indicative of a first-order transition, whereas a single maximum (panel a) is typical for second- or higher-order transitions.

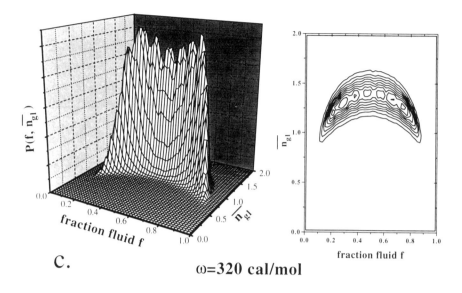

c. ω=320 cal/mol

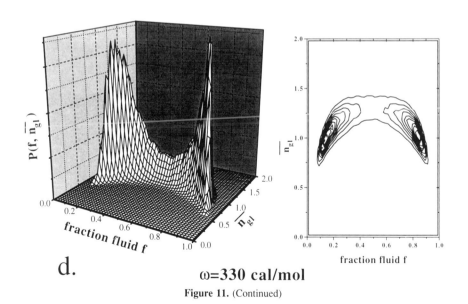

d. ω=330 cal/mol

Figure 11. (Continued)

Binding of Peripheral Proteins

This section deals with the effects of binding peripheral proteins on the phase transition behavior of lipid bilayers and the coupling of protein binding and lipid phase state. The results can be analyzed using the two-state Ising model that was given for the lipid chain-melting transition in the previous section, together with the thermodynamics of protein binding discussed earlier. In addition, the Monte Carlo methods of the previous section may also be applied to investigate the protein distribution on the lipid surface.

CALORIMETRIC DATA

First we begin with some experimental results. In Figure 12, the heat capacity curves in the temperature range of the lipid chain-melting transition are given for bilayers of two different anionic lipids in the presence and absence of basic peripheral proteins. These thermograms have been measured by using differential scanning calorimetry. Concentrating on the left side of Figure 12 for dimyristoyl phosphatidylglycerol bilayers in the presence of cytochrome *c*, it can be seen that

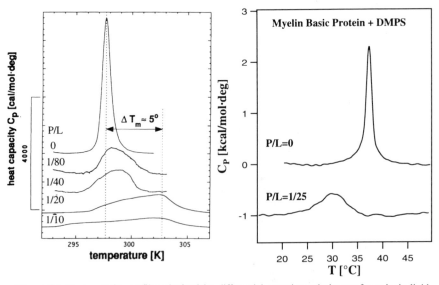

Figure 12. Heat capacity profiles obtained by differential scanning calorimetry for anionic lipid bilayers and lipid/protein complexes with peripheral proteins. Left: Dimyristoyl phosphatidylglycerol bilayers with cytochrome *c* bound at different degrees of surface saturation (from upper to lower: 0, 25, 50, 75, and 100%, assuming a lipid/protein stoichiometry of 10 mol/mol for saturation). The change in total integrated heat of transition upon saturation binding is −4.3 kcal/mol. (Adapted from Heimburg and Biltonen, 1996). Right: Dimyristoyl phosphatidylserine bilayers in the absence (upper) and in the presence (lower) of myelin basic protein (lipid/protein = 25 mol/mol). The change in total integrated heat of transition upon protein binding is −1.6 kcal/mol. (Adapted from Ramsay et al, 1986 with permission).

binding of increasing amounts of protein results in an asymmetric broadening of the transition and a shift in the heat capacity maximum to higher temperatures by approximately $\Delta T_m \approx +5°C$. These results are in qualitative agreement with the general considerations of the coupled protein binding and lipid phase equilibria given earlier (Figure 6). In addition, it has been found that the protein binding has a large effect on the transition enthalpy, which decreases from $\Delta H_t = 6.0$ kcal/mol for the lipid bilayers alone to $\Delta H_t = 1.7$ kcal/mol lipid in the presence of a saturating amount of protein bound. This represents a very significant energetic effect of protein binding.

The enthalpies of protein binding at saturation obtained in the cytochrome c/dimyristoyl phosphatidylglycerol system by titration calorimetry are given for different temperatures in Figure 13. It can be seen that the binding reaction is strongly endothermic, indicating that it must be entropy driven, at least in part by surface electrostatics. The binding enthalpy decreases strongly from $\Delta H_p = +6.8$ kcal/mol lipid in the gel phase to $\Delta H_p = +2.8$ kcal/mol lipid in the fluid phase, correlating with the large decrease in chain-melting enthalpy on

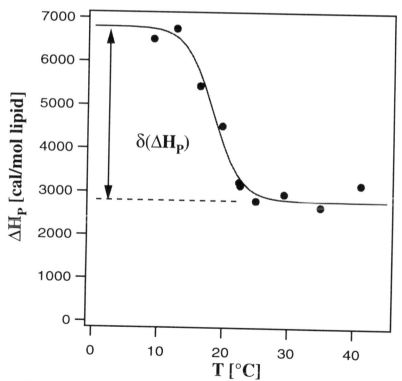

Figure 13. Total heat of binding a saturating amount of cytochrome c to dimyristoyl phosphatidylglycerol bilayers as a function of temperature, determined in a titration calorimeter. Note the strong change in enthalpy of the endothermic binding reaction in a region close to the lipid phase transition. (Adapted from Heimburg and Biltonen, 1994).

protein binding in Figure 12. This change is unrelated to the surface free energy of the electrostatic double layer, which in the high potential limit is entropic in origin and most probably arises from structural and hydrational changes in the lipid bilayer on protein binding. In addition, the temperature dependence of the enthalpy of protein binding (Figure 13) does not correspond in a one-to-one fashion with the degree of transition in the lipid state as deduced from Figure 12. This suggests a strong contribution from the heat capacity of binding to one or both of the lipid states (probably to the gel state because the large changes occur at lower temperature than the lipid transition mid-point and the binding enthalpy is essentially independent of temperature in the fluid state). The changes in heat of transition on protein binding and the changes in heat of binding on lipid transition may be related by the following thermo-dynamic scheme:

$$
\begin{array}{ccc}
L(9°C) + P & \xrightarrow{\Delta H_t^L} & L(41°C) + P \\
\Delta H_P^G \downarrow & & \downarrow \Delta H_P^F \\
LP(9°C) & \xrightarrow[\Delta H_t^{LP}]{} & LP(41°C)
\end{array}
$$

where ΔH_t^L and ΔH_t^{LP} are the lipid transition enthalpies for the lipid bilayers alone (L) and for the lipid-protein complex (LP), respectively, and ΔH_P^G and ΔH_P^F are the enthalpies of binding the protein (P) to gel-phase lipid (9°C) and to fluid-phase lipid (41°C). Because enthalpy is a thermodynamic state function, independent of pathway: $\Delta H_t^L + \Delta H_P^F = \Delta H_P^G + \Delta H_t^{LP}$, i.e., the decrease in protein binding enthalpy $(\Delta H_P^F - \Delta H_P^G = -4.0$ kcal/mol) should compensate that of the lipid transition $(\Delta H_t^{LP} - \Delta H_t^L = -4.3$ kcal/mol). This consistency suggests that the strongly endothermic enthalpy of protein binding is associated with structural and other changes in the lipid which are reflected in the strongly reduced chain-melting enthalpy (Heimburg and Biltonen, 1994).

Finally, we consider the shift in temperature of the lipid transition. For the lipid bilayers alone, the mid-point of the transition occurs at a temperature given by (Cevc and Marsh, 1987):

$$
T_m^L = \frac{\Delta H_t}{\Delta S_t} \tag{30}
$$

where $\Delta H_t = 6.0$ kcal/mol and $\Delta S_t = 20$ cal/mol/K for dimyristoyl phosphatidylglycerol. In the presence of a saturating amount of proteins, almost all lipids are in contact with a bound protein. It can be seen from the left-hand panel of Figure 12 that, for saturation binding of cytochrome c to dimyristoyl phosphatidylglycerol, the heat capacity profile of the lipid transition is not symmetric. Nevertheless, the mid-point of the transition in the presence of a saturating amount of protein might be expected to occur at a temperature given approximately by:

$$
T_m^{LP} = \frac{\Delta H_t + \delta \Delta H_p}{\Delta S_t + \delta \Delta S_p} \tag{31}
$$

where $\delta\Delta H_p$ and $\delta\Delta S_p$ are the differences in enthalpy and entropy per lipid, respectively, for protein binding to the fluid and gel states (Heimburg and Biltonen, 1996). For cytochrome c binding to dimyristoyl phosphatidylglycerol: $\Delta T_m \approx +5°$ (Figure 12) and $\delta\Delta H_p \approx -4.3$ cal/mol (Figure 13), and, therefore, from Equation 31 one obtains a change in the entropy of protein binding on lipid chain-melting of $\delta\Delta S_p \approx -14.6$ cal/mol/K. The difference in free energy of protein binding to the fluid and gel lipid states, $\delta\Delta G_p = \delta\Delta H_p - T \cdot \delta\Delta S_p$, therefore exhibits a strong degree of entropy-enthalpy compensation. It has a net entropic value of $\delta\Delta G_p = +100$ cal/mol at 302 K (the approximate mid-point for the transition of the lipid-protein complex) and decreases to give complete compensation ($\delta\Delta G_p = 0$) at 295 K, i.e., 2 degrees below the transition of the lipid in the absence of protein. This sensitive temperature dependence of the free energy leads to a considerably asymmetric broadening of the lipid chain-melting transition (see Figure 12, left side). Indeed, experimentally observed transitions often are broadened to such an extent that it is difficult to resolve the transition maximum, suggesting that the binding of proteins involves considerable entropic contributions.

Applying the same principles of analysis to the data for the binding of myelin basic protein to dimyristoyl phosphatidylserine that are given in the right-hand panel of Figure 12, it can be inferred that the difference in enthalpy of protein binding is $\delta\Delta H_p \approx -1.6$ kcal/mol lipid, and, from the transition shift, that the difference in entropy of binding is $\delta\Delta S_p \approx -4.8$ cal/mol/K. Here also, there is a considerable degree of entropy-enthalpy compensation but the difference in net free energy of protein binding is enthalpic, $\delta\Delta G_p \approx -135$ cal/mol lipid, at the transition temperature of the lipid-protein complex. In this case, the free energy difference is negative, corresponding to the downward shift in temperature of the transition on protein binding.

These two experimental examples demonstrate that considerable insight can be gained into the nature of the lipid-protein interaction by a detailed study of the thermodynamic behavior. It is clear that there can be very sizeable enthalpic contributions to the binding free energy that do not arise from changes in the electrostatic double layer free energy, which in the high potential limit is almost totally entropic. Also, there must be considerable entropic contributions that do not have their origin in the electrostatic double layer because the difference in entropy of binding is found to be much greater than that predicted by the entropic contribution in Equation 20 from double-layer theory. It is noted that these differences are unlikely to arise from the short-comings of Gouy-Chapman electrostatic double-layer theory because, if anything, this theory tends to overestimate the energetics of lipid-chain melting (Cevc et al, 1981; 1980).

Generally, the nonelectrostatic contributions to the binding energy express themselves in the intrinsic binding constant, K, that was considered earlier. Possible sources for these contributions are structural and conformational changes in both protein and lipid, changes in hydration and hydrogen bonding, and possibly even hydrophobic interactions. Tertiary structural and conformational changes in cytochrome c have been found on binding to anionic lipids, and there can be morphological changes in the bilayer as well (Heimburg and Marsh, 1993; Heimburg et al, 1991; Muga et al, 1991). Considerable changes occur in the secondary structure of myelin basic protein on binding to negatively charged lipids (Surewicz et al, 1987). Evidence for dehydration of the lipid surface on binding peripheral proteins has come from other spectral studies (Jain and Vaz, 1987; Sankaram et al, 1990). Binding of cytochrome c (Görrisen et al, 1986), myelin basic protein (Sankaram et al, 1989a), as well as other proteins (Sankaram et al, 1989b), has a marked effect on lipid mobility and hence potentially on the energetics of chain packing and possibly also of the chain configuration. Additionally, there is evidence for penetration of myelin basic protein into the hydrophobic region of the lipid bilayer (Boggs et al, 1988; London et al, 1973; Sankaram et al, 1989a). Thermodynamic studies, as described above, can give some indication of the relative energetic signif- icance of these different structural and hydrational changes. However, con- siderable entropy-enthalpy compensation may also take place in the individ- ual binding reactions, as is found for the differences in binding to the gel and fluid states.

SIMULATIONS

Now we return to the two-state Ising model by considering the effects of protein binding on the lipid state. A strictly local model is taken for the lipid-protein interaction (see Figure 14, upper); the peripheral proteins are assumed to affect only those lipids situated directly under them. (For ionic strengths of 10 mM and above, at which the Debye length of the electrolyte is of the order or less than a typical protein diameter, such an assumption is not strongly at variance with the nonlocalized double-layer approach used for the surface electrostatic interactions earlier.) For a lipid in contact with a peripheral protein, the statistical weight $K(T)$ that determines the probability to switch from the gel to the fluid state in a Monte Carlo step becomes:

$$K(T) = \exp\left(-\frac{\Delta H_t - T\Delta S_t + \Delta n_{gl} \cdot \omega_{gl} + \Delta E_P}{RT}\right) \tag{32}$$

whereas it is given by Equation 27 in the absence of proteins. Here, the dif- ferential free energy of protein binding, $\delta\Delta G_P$, is approximated by a simple enthalpic term, ΔE_P. In addition to the switching of the lipid states, the pro- teins are also allowed to rearrange laterally on the lipid surface. Translational steps for a protein are favored in a direction for which the strength of binding increases. This, in turn, depends on the distribution of gel and fluid lipids. For a given lipid arrangement the probability for translation in one of the six possible directions on a triangular lattice is given by:

$$P_m(i) = \frac{K_m(i)}{1 + K_m(i)} \tag{33}$$

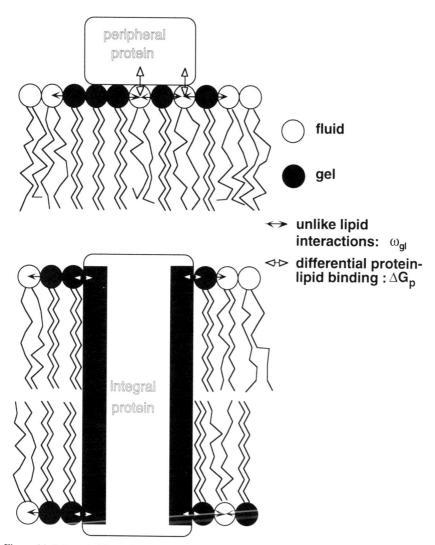

Figure 14. Schematic illustration of the local interactions between proteins and lipids, and the coupling with the interactions between unlike lipids. Upper: peripheral protein; lower: integral protein. Energetically less favorable interactions, characterized by the excess free energies ω_{gl} and $\delta\Delta G_P$, are indicated by arrows (Heimburg and Biltonen, 1996).

where the index i represents the six different directions. The statistical weights are given by:

$$K_m(i) = \exp\left(-\frac{\Delta n_l(i) \cdot \Delta E_P}{RT}\right) \tag{34}$$

where $\Delta n_l(i)$ is the increase in the number of fluid lipids with which the protein is associated. Using a random number generator the proteins are moved from

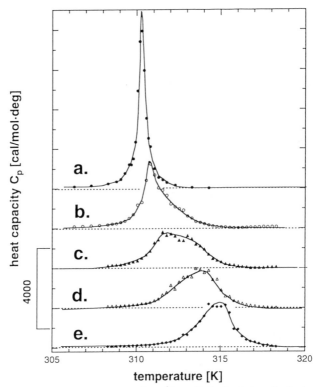

Figure 15. Heat capacity profiles calculated from Monte Carlo simulations for the two-state chain-melting of a lipid membrane with varying amounts of peripheral protein bound. Calculations were performed for a lipid matrix of 61×61 sites with $\Delta H_t = 8.7$ kcal/mol, $T_m = 310.3$ K, $\omega_{gl} = 310.3$ cal/mol. Each protein covers 19 lipid sites and has a differential binding energy with the two lipid states of $\Delta E_P = 200$ cal/mol. The five different profiles correspond to degrees of surface saturation (from upper to lower) of 0, 16, 32, 48, and 64%, or protein/lipid ratios of 0, 1/125, 1/62, 1/41, and 1/31 mol/mol, respectively. (Adapted from Heimburg and Biltonen, 1996).

an initial statistical distribution in one of the six possible directions with the probability $P_m(i)$ (Heimburg and Biltonen, 1996).

Simulations are performed by alternating readjustment of single lipid states and recalculation of the protein positions. Figure 15 shows the heat capacity profiles calculated from Monte Carlo simulations for a peripheral protein with a cross section equal to that of 19 lipid molecules (approximately equivalent to phospholipase A_2) and a differential gel-fluid protein binding energy of $\Delta E_P = +200$ cal/mol (enthalpic). The parameters for the lipid matrix are: $\Delta H_t = 8.7$ kcal/mol, $T_m = 310.3$ K and $\omega_{gl} = 310.3$ cal/mol, which are similar to those found for dipalmitoyl phosphatidylcholine small unilamellar vesicles (Mouritsen and Biltonen, 1992; Sugar et al, 1994). It can be seen that the heat capacity maxima are shifted progressively to higher temperatures, and the profiles are broadened asymmetrically as increasing numbers of proteins are bound. Although the simulation parameters are rather different from those found for the binding of cytochrome c to dimyristoyl phosphatidylgylcerol, there is a remarkable qualitative similarity between the calculated heat capacity profiles in Figure 15 and the experimental data in Figure 12 (left panel). The differential binding energy was chosen in the simulations to give a shift of the transition mid-point at saturation binding of $+7°$ and therefore produces a shift in the transition comparable to that in Figure 12, at less than saturation binding. To produce a downward shift in the transition on protein binding, such as that observed in the right-hand panel of Figure 12, would require ΔE_P to have a negative sign. Inclusion of entropic contributions to the differential free energy of binding, such as is suggested by the experimental data, would produce a more pronounced asymmetric broadening of the heat capacity profiles.

The distribution of the proteins on the lipid surface is shown in Figure 16, which gives representative configurations of the protein and lipid at temperatures below, within, and above the chain-melting transition. It can be seen that the proteins preferentially accumulate and cluster on the lipid gel domains that are present in the transition region. This is because the free energy of binding to the gel domains is lower than that to the fluid domains, for a positive value of ΔE_P. Correspondingly, gel domains form preferentially at the locations of the protein molecules. The degree of protein clustering can be characterized by the deviation of the mean separation between adjacent proteins obtained in a simulation from that expected for a random distribution:

$$C(T) = \frac{d_{\text{random}} - d(T)}{d_{\text{random}} - d_{\text{minimum}}} \tag{35}$$

where d_{random}, d_{minimum} and $d(T)$ are the mean distances between neighboring proteins for a random distribution, for closest packing, and from the simulation results at temperature T, respectively. As shown in Figure 17, the protein cluster parameters $C(T)$, derived from distributions such as those in Figure 16, show a distinct maximum at temperatures close to, or on the gel-phase side of, those of the heat capacity maxima. This aggregation is a necessary accompaniment to the shift in the heat capacity curve.

Integral Proteins

In this section, we briefly consider lipid interactions with integral membrane proteins. The intention is not to give a comprehensive discussion (Marsh, 1995a, 1985) but rather to concentrate on the different interaction of the integral protein with lipids in the different phases, within the context of the two-state model (see Figure 14, lower panel). In general, it is to be expected that integral proteins will interact and mix differently with lipids in the fluid and gel states. There is considerable evidence from freeze-fracture electron microscopy that integral proteins frequently tend to aggregate in gel-phase lipid membranes, whereas they are more randomly distributed within the fluid lipid phase. Clear examples are, for instance, the Ca^{2+}-ATPase (Kleeman and McConnell, 1976),

melting in the presence of peripheral proteins

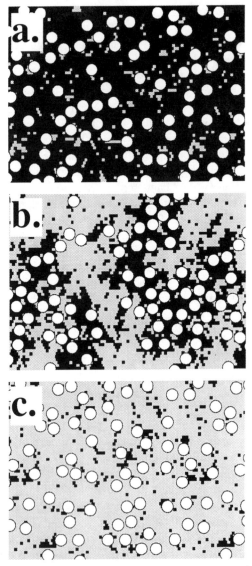

Figure 16. Snapshots of the configuration of a 61×61 lipid matrix with 60 surface-bound peripheral proteins (each covering 19 lipid cross sections) at temperatures (a) below ($T = 306.3$ K), (b) at ($T = 312.3$ K), and (c) above ($T = 318.3$ K) the heat capacity maximum. Monte Carlo simulations were performed as for Figure 15 with $\Delta H_t = 8.7$ kcal/mol, $\omega_{gl} = 310.3$ cal/mol, $T_m = 310.3$ K, and $\Delta E_P = 200$ cal/mol. Gel-state lipids are black, fluid-state lipids are grey, and proteins are represented by white circles (Heimburg and Biltonen, 1996).

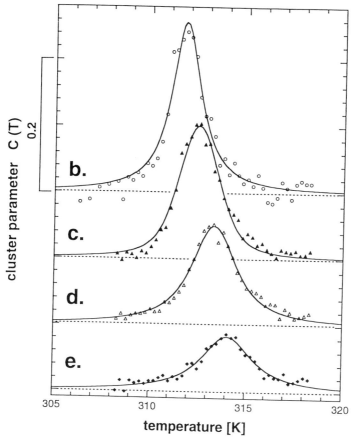

Figure 17. Aggregation profiles for peripheral proteins, corresponding to the calculated heat capacity profiles of Figure 15. The protein cluster parameters calculated from Monte Carlo simulations (Figure 16) are for protein/lipid ratios of (from upper to lower): 1/125, 1/62, 1/41, and 1/31 mol/mol. (Adapted from Heimburg and Biltonen, 1996).

cytochrome oxidase (Fajer et al, 1989), myelin proteolipid protein (Boggs et al, 1980), and glycophorin (Grant and McConnell, 1974), as well as membranes from fatty acid-supplemented *Escherichia coli* auxotrophs (Kleeman and Mc-Connell, 1974). Matching of the hydrophobic span of the integral protein with that of the lipid chains in the different states is likely to play an important role (Marsh, 1995a,b). Hydrophobic matching may also play a role in the mixing properties with lipids of different chainlength in a single lipid phase, as is found for rhodopsin (Chen and Hubbell, 1973; Ryba and Marsh, 1992). In addition, any preferential interaction between proteins, relative to the lipid-protein interactions, will tend to induce protein aggregation, which may again be dependent on the lipid state. An example of this is the well-known propensity of the band

3 anion transport protein to aggregate in fluid-phase lipids (Mühlebach and Cherry, 1985).

If, in the model of the previous section, the differential interaction energy, ΔE_P between the peripheral protein and the gel and fluid states of the lipid membrane is allowed to become extremely large, all lipids beneath the protein remain in the gel state, even at high temperatures. Conversely, with a very strongly negative value for the differential binding energy, domains consisting entirely of fluid lipid form beneath the peripheral protein. This can be used to mimic the thermodynamic behavior of integral proteins, if it is assumed that these all-gel or all-fluid domains beneath the protein represent the membrane-spanning core of the protein. In determining the thermodynamic observables (heat capacity, compressibility, etc.) the lipids in these cores must be excluded because they are considered to be part of the integral protein. This integral core interacts with the surrounding lipid in the same manner as do gel-phase lipids (in the case of a strongly positive value of ΔE_P), or as do fluid-phase lipids (in the case of a strongly negative ΔE_P). Clearly this is a considerable oversimplification of the actual situation because it assumes that the interaction of integral proteins with the two lipid states is identical with that of lipids in the complementary state, i.e., gel with fluid or fluid with gel, as described by the excess free energy ω_{gl}. Nevertheless, the model contains all the qualitative features of a different protein interaction with the two lipid states (Heimburg and Biltonen, 1996). In the case of peripheral proteins with moderate values of the differential interaction energy ΔE_P, the aggregation of such proteins at the phase transition (see Figure 16) is caused by different interactions with the lipid domains, which are created primarily by the different lipid chain interactions. This type of interaction disappears outside the transition range, and the distribution of such proteins then becomes random, in the absence of specific protein-protein interactions. In contrast to this, the lipid interface with integral proteins (or with the lipids beneath a peripheral protein to which they bind extremely strongly) exists at all temperatures. As will be seen in the following, this affects the aggregation profile of the protein, which in turn shows up in the heat capacity profiles for lipid chain-melting.

We consider here the simulated heat capacity profiles for a hypothetical integral protein corresponding to a highly positive value of ΔE_P, i.e., with a preference for the gel phase. The cross section of the protein is assumed to correspond to 19 lipid cross sections and, for a compact structure, will have a comparable number of nearest-neighbor true lipids in its boundary shell. The calculation is performed by using the same approach as that used for peripheral proteins, as outlined in the previous section. The integral cores of the hypothetical protein consist of gel-phase lipids that do not undergo a change of state and are not counted in calculating the heat capacity. As can be seen from Figure 18, the heat capacity profiles calculated from the Monte Carlo simulations for this model shift to higher temperatures with increasing effective protein/lipid ratio. This aspect of the behavior is similar to that for peripheral proteins with smaller positive values of ΔE_P (see Figure 15). However, the shape of the heat capac-

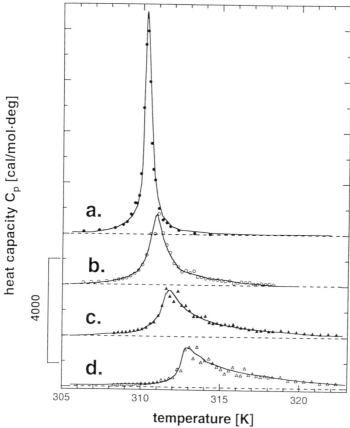

Figure 18. Heat capacity profiles calculated for a lipid matrix containing different amounts of a hypothetical integral protein that is modeled as a cluster of 19 permanently gel-state lipids. Monte Carlo calculations were performed for a lipid matrix of 61 × 61 sites with ΔH_t = 8.7 kcal/mol, T_m = 310.3 K, and ω_{gl} = 310.3 cal/mol (equal also to the excess interaction energy of the protein with a fluid-state lipid). The four different profiles correspond to protein/lipid ratios of (from upper to lower) 0, 1/105, 1/43, and 1/22 mol/mol. (Adapted from Heimburg and Biltonen, 1996).

ity profiles is different from the latter case. In the case of Figure 15, the heat capacity profiles display a maximum at the lower end of the transition for low degrees of surface coverage by the protein, whereas the maximum shifts to the upper end of the transition profile for high degrees of protein occupancy. This change arises from the change in number of lipids in contact with the protein for increasing numbers of proteins bound. The heat capacity profiles in Figure 18, however, always display a maximum at the low temperature end of the transition (in the case of a highly positive value of ΔE_P). In the case of a preference of the protein for the fluid-phase lipids (highly negative values of ΔE_P), the heat capacity profiles would shift downwards and display the heat capacity maximum at the upper end of the transition profile.

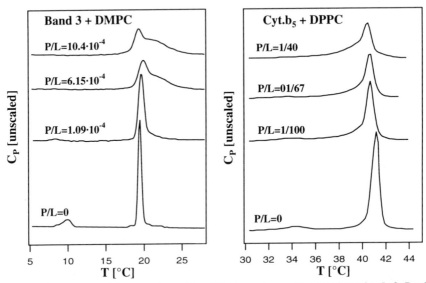

Figure 19. Experimental heat capacity profiles of lipid complexes with integral proteins. Left: Band 3 anion transport protein from erythrocytes reconstituted with dimyristoyl phosphatidylcholine at protein/lipid ratios of (from lower to upper): 0, 1/9200, 1/1630, and 1/960 mol/mol. (Adapted from Morrow et al, 1986). Right: Cytochrome b_5 reconstituted with dipalmitoyl phosphatidylcholine at protein/lipid ratios of (from lower to upper): 0, 1/100, 1/67, and 1/40 mol/mol. (Adapted from Freire et al, 1983 with permission).

The results of the model calculations that are given in Figure 18 can be compared with experimental heat capacity profiles for the chain-melting of lipid membranes reconstituted with purified integral membrane proteins. The left panel of Figure 19 shows heat capacity profiles for bilayers of dimyristoyl phosphatidylcholine that contain varying amounts of the band 3 protein of erythrocytes. These display a peak at a temperature close to that for the lipid alone and a high-temperature shoulder, which becomes progressively more pronounced as the protein concentration in the membrane increases. The right panel of Figure 19 gives heat capacity profiles for bilayers of dipalmitoyl phosphatidylcholine containing increasing amounts of cytochrome b_5 which display a similar but inverse behavior to that in the left panel. The heat capacity maximum here again does not shift with increasing protein content, but a shoulder develops now to the low-temperature side of the transition curve. These results resemble qualitatively the behavior of the model calculations in Figure 18 (and corresponding ones for highly negative ΔE_P). As can be seen in the experimental examples, however, the shift of the heat capacity maximum can be small. Whereas the upward shift in heat capacity maximum in Figure 18 is consistent with the preferential affinity of the hypothetical integral protein for the gel-phase lipid, the near constancy of the lower boundary of the experimental heat capacity curves in the left panel of Figure 19 could be consistent with immiscibility of the band

melting in the presence of integral proteins

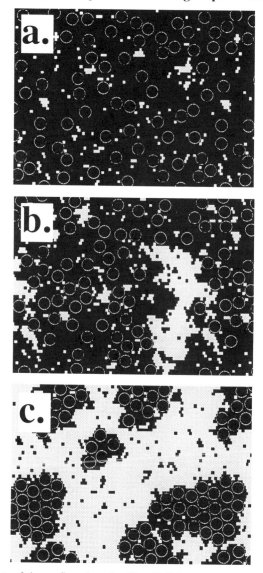

Figure 20. Snapshots of the configuration of a 61×61 lipid matrix containing 60 hypothetical integral proteins (modeled as a cluster of 19 gel-state lipids) at temperatures (a) below ($T = 306.3$ K), (b) at ($T = 311.3$ K), and (c) above ($T = 318.3$ K) the heat capacity maximum. Monte Carlo simulations were performed as for Figure 18 with $\Delta H_t = 8.7$ kcal/mol, $T_m = 310.3$ K, and $\omega_{gl} = 310.3$ cal/mol (equal also to the protein excess interaction energy with a fluid-state lipid). Gel-state lipids are black (stippled), fluid-state lipids are grey, and integral proteins are represented by black circles (Heimburg and Biltonen, 1996).

3 protein in the gel phase. Clearly, effects such as the latter would require the additional specification of lipid-protein and protein-protein interaction energies that are distinct from those which are represented in the current model solely in terms of lipid-lipid interactions.

In Figure 20, representative views of the lipid configuration and the protein distribution obtained from the model of Figure 18 are given for temperatures below, within, and above the chain-melting transition region. The protein has a near-random distribution at lower temperatures for which the lipid is predominantly in the gel state. As the temperature is increased, the proteins then cluster progressively in the domains of gel-state lipid that remain above the chain-melting point. This is reflected in the temperature dependence of the protein cluster parameter (Equation 35) that is given in Figure 21 for results calcu-

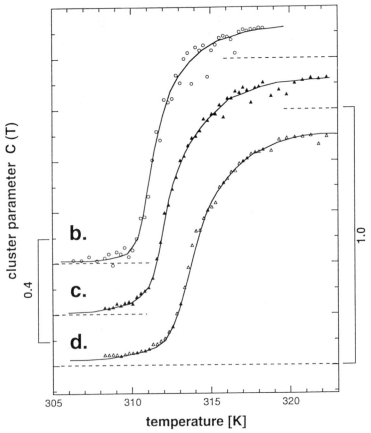

Figure 21. Calculated aggregation profiles for hypothetical integral proteins modeled as a cluster of 19 permanently gel-state lipids, using the simulation parameters from Figure 18. The protein cluster parameters are calculated from Monte Carlo simulations (Figure 20) with effective protein/lipid ratios of (from upper to lower): 1/105, 1/43, and 1/21 mol/mol. (Adapted from Heimburg and Biltonen, 1996).

lated with the model of Figure 18 at different protein contents. As can be seen, the proteins rapidly become progressively more aggregated (or clustered) as the temperature increases and the bulk of the lipid converts to the fluid state. The clustering profiles are shifted to higher temperatures as the protein concentration is increased. This can be rationalized as follows: when the chains of lipids in the presence of proteins that prefer gel-state lipids progressively melt as the temperature is increased, the proteins become more and more restricted to the few remaining gel domains. Simultaneously, protein-protein contacts become more favorable in order to avoid contacts of the protein with fluid lipids. This is a consequence of the hydrophobic core of the integral proteins being modeled as gel-state lipids and having a permanent interface with the free lipids. Quite generally, irrespective of model or mechanism, one will expect the aggregation profiles for integral proteins to reflect the preference for partitioning between the different lipid phases, and this will show up in the behavior of the heat capacity profiles.

Within the context of the model proposed in this section, the lipid interactions with integral membrane proteins are represented as an extreme case of the interactions with peripheral proteins. Nevertheless, comparison of Figure 18 with Figure 15, together with the energetic considerations discussed earlier, does indicate that there can be qualitative differences in the heat capacity profiles for lipid chain-melting between membranes containing integral proteins and those with surface-bound peripheral proteins that interact purely by their effects on the electrostatic double layer. As already discussed (Figures 17 and 21), this reflects a very different profile of protein clustering for peripheral proteins that do not interact extremely strongly with the lipid surface. In the case of complex lipid mixtures, of interactions with peripheral proteins that have a strong degree of enthalpy-entropy compensation, or particularly of peripheral proteins that bind extremely strongly, these distinctions might not be quite so clear.

Discussion

A major goal of this chapter has been to demonstrate the thermodynamic linkage between protein binding, lipid mixing, lipid chain-melting, and domain formation, as well as protein aggregation, in membranes. In this final main section, we first review some of the consequences that lipid chain-melting and different interactions of proteins with the different lipid states may have on the function of membrane-associated proteins. This is done by taking examples of membrane proteins reconstituted with defined phospholipids. We then go on to consider the elastic properties of membranes, particularly in situations in which there is a coexistence of gel and fluid lipid states and the lateral membrane compressibility differs radically from that for membranes in a single lipid state. The lateral membrane compressibility most likely controls the readiness with which conformational changes may take place in integral proteins, those for which a conformational change is accompanied by a change in cross-sectional area.

Additionally, compressional waves within the membrane constitute a possible channel of communication between both surface-bound and integrally incorporated proteins.

Coupling of Protein Function to Lipid State

The study of proteins in association with membranes of a single lipid species provides insight into the nature of the thermodynamic interactions that may control protein function. Examples are given here of various studies with such well-defined systems that contain a single type of peripheral or integral protein. The highly cooperative nature of the lipid chain-melting transition, and the clearly characterized structural and thermodynamic changes that accompany it, afford an unambiguous illustration of the dependence of protein function on the state of the lipid.

FUNCTIONAL ASPECTS: PERIPHERAL PROTEINS

As examples of the conformational response of peripheral proteins to the state of lipid membranes, mitochondrial cytochrome c, and pancreatic or snake venom phospholipase A_2 will be discussed.

Cytochrome c is a peripheral membrane protein that functions as an electron carrier in the mitochondrial respiratory chain. This protein is associated at the outer face of the inner membrane, which contains a relatively high concentration of charged lipids, most notably of the unique mitochondrial lipid cardiolipin (Daum, 1985). The affinity of cytochrome c for electrons is determined by its redox potential. It has been found that the protein can assume at least two different conformational states, which have very different redox potentials (Hildebrandt and Stockburger, 1989a,b). The equilibrium between these two states is influenced by the potential applied to metal electrodes to which the protein is absorbed. Furthermore it has been found that a similar change in conformational state can be induced by the binding of cytochrome c to charged lipid membranes (Hildebrandt et al, 1990). Figure 22a shows the temperature dependence of the ratio of the two conformational states of cytochrome c bound to membranes of dimyristoyl phosphatidylglycerol, as recorded by the resonance Raman spectra of the haem group. As can be seen from the figure, the relative population of the conformational state II, which is characterized by a more negative redox potential, increases considerably in the temperature interval around 25°C, as a result of the chain-melting in the lipid/protein complex (Figure 12, left). Cytochrome c bound to dioleoyl phosphatidylglycerol, which is entirely in the fluid state throughout this range, displays no temperature dependence of this conformational equilibrium (Heimburg et al, 1991). In addition, it has been found that the states of cytochrome c bound to dimyristoyl phosphatidylglycerol above and below the chain-melting transition differ in their tertiary structure because they are characterized by significantly different hydrogen exchange kinetics of the amide backbone (Heimburg and Marsh, 1993). In Figure 22b, the

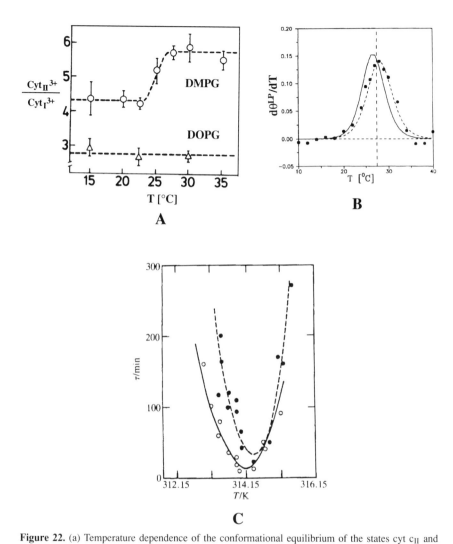

Figure 22. (a) Temperature dependence of the conformational equilibrium of the states cyt c_{II} and cyt c_I of cytochrome c bound to bilayers of dimyristoyl (O) and dioleoyl (△) phosphatidylglycerol, as determined by resonance Raman spectroscopy. (Adapted from Heimburg et al, 1991). (b) Temperature dependence of derivative curves for the degree of transition θ^{LP} between the tertiary structural states of cytochrome c bound to dimyristoyl phosphatidylglycerol that have different amide proton exchange kinetics (solid line), and for the chain-melting transition of the lipid (dashed line). Exchange kinetics were measured by monitoring the protein amide I band and chain-melting by monitoring the lipid carbonyl band in the Fourier-transform infrared spectrum. (Adapted from Heimburg and Marsh, 1993). (c) Temperature dependence of the lag time, τ, for achieving the maximum hydrolysis rate of 1.2 mM (●) or 0.12 mM (O) dipalmitoyl phosphatidylcholine large unilamellar vesicles by porcine pancreatic phospholipase A_2 (25 μg/ml). The downward shift in the minimum lag time with increasing enzyme/lipid ratio parallels that of the lipid phase transition temperature. (Adapted from Lichtenberg et al, 1986 with permission).

profile of lipid chain-melting is superimposed on that for the transition between the states with different amide proton exchange kinetics, both determined by Fourier transform infrared spectroscopy. The two transitions are essentially co-incident. The transition in tertiary structural state is found to depend on the lipid to which cytochrome c is bound; for other lipids it takes place wholly in the fluid state (Heimburg and Marsh, 1993). In the case of dimyristoyl phos-phatidylglycerol, the transition occurs immediately on conversion to the fluid state and is controlled by the chain-melting of this lipid (Figure 22b). Because the two conformational states in Figure 22a possess different redox potentials, the shift in conformational equilibrium with change in state of the lipid, and also with change in the lipid species (Heimburg et al, 1991), will affect the electron transfer efficiency of the lipid-bound protein. As discussed earlier, the shift in the heat capacity profile suggests a stronger binding of cytochrome c to the gel state of dimyristoyl phosphatidylglycerol than to the fluid phase. The difference of the binding affinity, ΔG_P, may contribute to the free energy required for the conformational change, and correspondingly ΔG_P may contain contributions from the structural change in the protein.

Phospholipase A_2 is a peripherally bound lipolytic enzyme that catalyzes the hydrolysis of phospholipids by cleavage of the ester bond of the sn-2 acyl chain, producing as hydrolysis products, fatty acids and lysolipids. The enzy-matic activity is controlled by a number of factors, including the presence of calcium and hydrolysis products, and, depending on the lipid system, also can depend rather strongly on temperature. With membranes composed of zwitteri-onic phospholipids such as phosphatidylcholine, the protein is hardly active until a certain lag time has passed. After this time, a burst in activity is observed. A possible explanation for the lag time is that the aggregation kinetics of the membrane-associated enzyme are slow and only the protein aggregates are ac-tive (Biltonen, 1990). It has been found that this lag time is highly temperature dependent. In the case of pancreatic phospholipase A_2 binding to dipalmitoyl phosphatidylcholine bilayers, a minimum in the lag time (and a maximum in the activity) is reached close to the chain-melting phase transition temperature (see Figure 22c). At higher enzyme/lipid ratios, the minimum in lag time is shifted to lower temperatures. This parallels the aggregation behavior predicted for a peripheral protein that prefers the fluid phase over the gel phase (Figure 17 for the reverse situation). However, epifluorescence measurements on dipalmitoyl phosphatidylcholine monolayers have localized the site of action of snake venom phospholipase A_2 to the gel-fluid interfaces, rather than to a particular domain (Grainger et al, 1990). In this case, principles of preferential affinity similar to those for domains of lipids in a particular state could apply, but for interfacial sites considerations of enhanced lateral compressibility also may be important (see below). Unlike that of the snake venom enzyme, the binding of pancreatic phospholipase A_2 to zwitterionic lipids is, however, rather weak at all temper-atures (Apitz-Castro et al, 1982; Jain et al, 1982). Alternative explanations for the lag time, therefore, include the generation of negatively charged fatty acids and other hydrolysis products, which increase the binding of the enzyme to the membrane (Jain et al, 1982), and specifically the lateral phase separations induced by the hydrolysis products (Burrack et al, 1993; Jain et al, 1989). In these cases, the minimum in the lag time could be related to the ease of insertion of the protein, or conversely to the increased accessibility of the lipid substrate which arises from the enhanced lateral compressibility of the membrane in the transition region, particularly at the gel-fluid interface, as is discussed later.

A different illustration of the way in which the state of the lipids may affect the activity of peripheral proteins is afforded by the dependence of the activation of lipid-requiring enzymes on the physical form in which the activating lipid is presented. The activation of the mitochondrial enzyme D-β-hydroxybutyrate dehydrogenase by phosphatidylcholine is noncooperative with soluble monomeric lipid, but is cooperative with lipid vesicles (Cortese et al, 1987). The specific activation of the cellular regulatory enzyme protein kinase C by phosphatidylserine in the presence of diacylglycerol is cooperative for both micellar and bilayer lipid systems, but is more highly cooperative when the phosphatidylserine is presented in micelles (Newton, 1993). In general, the preferential interaction with a particular lipid species can be rather similar to that with a particular state of a single lipid, and can be analyzed in a similar manner. The specificity of protein kinase C for phosphatidylserine is an interesting case. This enzyme binds in an electrostatic manner to negatively charged lipids in general but, on binding the second messenger lipid diacylglycerol, the affinity specifically for phosphatidylserine increases greatly and to an extent that is no longer dependent on ionic strength (Orr and Newton, 1992a,b).

FUNCTIONAL ASPECTS: INTEGRAL PROTEINS

There are many well-known examples in which the activity of integral membrane enzymes or transport systems are dependent on the state of the lipid. Among these are the Ca^{2+}-ATPase (Hesketh et al, 1976; Hidalgo et al, 1978; Starling et al, 1995), Na,K-ATPase (Kimelberg and Papahadjopoulos, 1974), cytochrome oxidase (Fajer et al, 1989), the erythrocyte hexose transporter (Carruthers and Melchior, 1984), and sugar transport systems in *E. coli* (Overath and Thilo, 1978). Most often, it has been found that the activity is lower in gel-phase lipids than in the corresponding fluid-phase lipids, and nonlinear Arrhenius kinetics are obtained for the temperature dependence of the activity. The form of the temperature dependence is determined by several factors, including the difference in activation enthalpies and heat capacities in the two lipid phases, and most importantly on the preferential partitioning of the protein between the coexisting lipid states (Silvius and McElhaney, 1981). In the case of sugar transport in fatty-acid supplemented auxotrophs of *E. coli*, the temperature dependence of the transport rates has been interpreted specifically in terms of different preferential phase-partitioning of the β-glucoside and β-galactoside transport systems in membranes of various lipid compositions, by comparison with the degree of lipid phase separation determined by independent means (Thilo et al, 1977). The estimated protein distribution coefficients are in the range 2–20 in favor of the

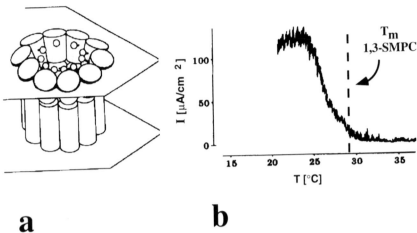

a   b

Figure 23. (a) Schematic structure of the alamethicin pore consisting of a complex of several single peptides. (Adapted from Fox Jr. and Richards, 1982 with permission). (b) Temperature dependence of the transmembrane current density in bilayers of 1-stearoyl-3-myristoyl phosphatidylcholine (1,3-SMPC) containing alamethicin. The vertical line indicates the gel-to-fluid phase transition temperature of 1,3-SMPC bilayers. (Adapted from Boheim et al, 1980).

fluid lipid phase. It also has been found that activity can depend on the lipid species (Bennett et al, 1978; Teft et al, 1986), which in the case of spatially separated domains of different lipid composition would again be controlled by the protein distribution.

A striking example of the effect of the lipid chain-melting transition on the activity of an ion channel is afforded by the pore-forming peptide alamethicin in planar bilayer membranes of 1-stearoyl-3-myristoyl phosphatidylcholine (see Figure 23). This unusual isomer of an asymmetric-chain phospholipid is capable of forming stable unsupported bilayers in the gel phase. The formation of ion-conducting pores by oligomerization of alamethicin monomers (Figure 23a) is very strongly concentration dependent: the conductance depends on the 9th to 10th power of the alamethicin concentration (Boheim et al, 1980). Therefore, it is to be expected that the membrane conductance would change strongly in the region of the bilayer chain-melting transition, if the degree of aggregation and hence the local concentration of alamethicin was dependent on the state of the lipid. This is exactly what is observed in Figure 23b. The current density increases from a level corresponding to only 1 pore/cm^2 in the fluid phase at 34°C to that corresponding to approximately 10^6 pores/cm^2 in the gel phase at 24°C. Qualitatively, this parallels the aggregation behavior expected for an integral protein that exhibits a preference for the fluid phase and therefore would have a cluster profile that is the inverse of those shown in Figure 21.

Compressibility, Protein Insertion, and Membrane Permeability

The lateral compressibility of membranes is an important feature that may control conformational changes of embedded proteins, the insertion of proteins, peptides, presequences and other amphiphiles, as well as determining the elastic properties of the membrane (Marsh, 1996). The functional implications and possible advantages arising from the enhanced lateral compressibility of membranes in a state of lateral phase separation have long been recognized (Linden et al, 1973). Specifically in this regard, as already mentioned, the methods introduced earlier may be used also to obtain the isothermal area compressibility of membranes with the two-state model. By introducing a term dependent on the lateral pressure (or membrane tension) into the free energy, an expression for the isothermal compressibility, κ_T, can be obtained in terms of the fluctuations in the membrane area, in much the same way as was done for the heat capacity earlier (Hill, 1966):

$$\kappa_T = -\frac{1}{\langle A \rangle} \left(\frac{\partial \langle A \rangle}{\partial \pi} \right)_T = \frac{1}{\langle A \rangle} \frac{\langle A^2 \rangle - \langle A \rangle^2}{RT} \tag{36}$$

where π is the lateral pressure (or strictly speaking surface tension) in the membrane.

Within the framework of the two-state model, the membrane area is related directly to the fraction of fluid lipids, f, by: $A = A_g + f\Delta A_t$, where A_g and A_l are the molar lipid areas in the gel and fluid phases, respectively, and $\Delta A_t = (A_l - A_g)$. The isothermal compressibility then becomes:

$$\kappa_T = \frac{\langle f^2 \rangle - \langle f \rangle^2}{RT} \frac{(\Delta A_t)^2}{A_g + \langle f \rangle \Delta A_t}. \tag{37}$$

This latter equation has strong similarities to that for the heat capacity in the two-state model that was given in Equation 29. It is, therefore, also possible to relate experimentally measured heat capacity profiles to those for the compressibility because, from Equations 29 and 37:

$$\kappa_T = \frac{\gamma^2 C_P \cdot T}{A_g + \gamma \cdot \langle \Delta H(T) \rangle} \tag{38}$$

where $\gamma = \Delta A_t / \Delta H_t$ is the proportionality constant between the transition enthalpy and the change in molar lipid area at the transition, and it will be remembered that $\Delta H(T) \approx f\Delta H_t$. For $\Delta A_t = N_A \times 18\text{Å}^2/\text{mol}$ ($\approx 37\%$ area change), and $\Delta H_t = 8.7$ kcal/mol, the value of the proportionality constant is $\gamma = 12.5$ m^2/cal. The compressibility also has units of m^2/cal (or m/N, the units of an inverse lateral pressure).

The heat capacity profile and the corresponding temperature profile of the lateral compressibility, derived from a single series of Monte Carlo simulations for the two-state model by using Equations 29 and 37, are given in Figure 24. The parameters used are those appropriate to small unilamellar vesicles of di-

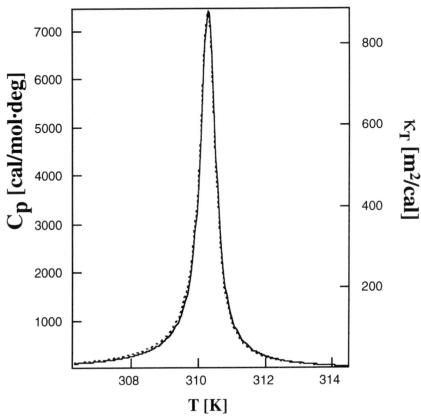

Figure 24. Temperature dependence of the isothermal lateral compressibility, κ_T (dotted line) and of the heat capacity, C_P (solid line) calculated from Monte Carlo simulations by using Equations 39 and 27, respectively. A 61×61 lipid lattice was used with $\Delta H_t = 8.7$ kcal/mol, $T_m = 310.3$ K and $\omega_{gl} = 310$ cal/mol. Surface areas were assumed to be 60 Å²/lipid and 78 Å²/lipid in the gel and fluid states, respectively.

palmitoyl phosphatidylcholine. The temperature dependence of the compressibility is nearly superimposable on that of the heat capacity, differing from the latter only in the temperature dependence of the mean area (divided by T) in the denominator of Equation 36. (The latter is introduced conventionally in defining a compressibility modulus rather than a susceptibility or response function.) The calculated heat capacity maximum in Figure 24 is $C_p^{\max} \approx 7400$ cal/mol/K, and the corresponding maximum value of the compressibility is approximately $\kappa_T^{\max} = 880$ m²/cal, or 210 m/N (Equation 38). The temperature dependence of the compressibility modulus of a dimyristoyl phosphatidylcholine bilayer has been measured in the chain-melting region by pipette aspiration of giant unilamellar vesicles (Evans and Kwok, 1982). A pronounced maximum is found in the compressibility modulus at the gel-to-fluid phase transition temperature of

24°C and the compressibility decreases steeply away from the transition region on going to either the gel or fluid phase, as is predicted by the two-state model in Figure 24. The compressibility modulus has a value of $\kappa_T^{max} = 30$–50 m/N at the phase transition temperature, and values of $\kappa_T \approx 1$ m/N at 8°C and ≈ 7 m/N at 30°C in the gel and fluid states, respectively. The higher value at 30°C than at 8°C corresponds to the intrinsic lateral compressibility of the fluid state that is not included in the two-state model.

The enhanced lateral compressibility, and accompanying fluctuations in the lipid area (Equation 36) that occur in the gel-fluid coexistence region promote the formation of defects and transient pores in the lipid matrix. The nucleation site for such defects is the interfacial region between the gel and fluid domains. This is the site of locally higher free energy and the point at which the conversion between gel and fluid states, which is the origin of the high compressibility, primarily takes place (Figure 10). One of the consequences of this can be, depending on solute size (Van Hoogevest et al, 1984), an increased permeability of the membrane to polar solutes, which otherwise have an intrinsically low permeability in the isolated lipid states. The permeability to alkali ions and the ionic conductivity of lipid bilayers, for instance, have been found to have a maximum in the region of the gel-fluid phase transition (Blok et al, 1975; Papahadjopoulos et al, 1973; Wu and McConnell, 1973).

The temperature dependence of the permeability of monodisperse, small unilamellar vesicles of dimyristoyl phosphatidylcholine to a small spin-labeled cation (Tempocholine) is given in Figure 25. Data are presented for both uptake and release, in which the latter corresponds to the somewhat lower of the biphasic transport rates that is obtained at longer times (Marsh et al, 1976). The rates of both permeability processes display a maximum at the phase transition region, which in this case is rather broad because of the low cooperativity of the phase transition in small unilamellar lipid vesicles (Marsh et al, 1977). The permeability profiles are compared with calculations of the lateral compressibility and the fraction of interfacial lipids, using for analytical simplicity a one-dimensional two-state Ising model rather than the rigorous Monte Carlo simulations for a two-dimensional system that were given above. The compressibility is calculated from Equations 23 and 37. The degree of cooperativity of the transition ($\sigma_L = 4.10^{-3}$) is determined from independent measurements of the degree of transition. As expected, both the compressibility and the fraction of interfacial lipids display a maximum at the center of the phase transition, but the temperature profile for the compressibility is narrower than that for the fraction of interfacial lipids. The permeabilities on the low temperature side of the transition correspond more closely to the compressibility and those on the high temperature side to the fraction of interfacial lipids. This asymmetry possibly corresponds in part to a higher intrinsic permeability for the defects at the interfacial regions in a predominantly fluid environment. The compressibility calculated at the center of the transition is $\kappa_T^{max} \approx 20$ m/N, which is comparable to that measured experimentally for dimyristoyl phosphatidylcholine bilayers, although the cooperativity is lower for small vesicles.

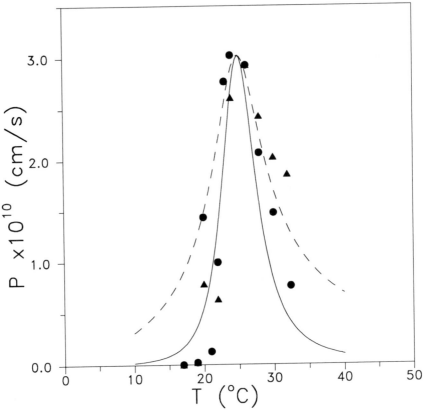

Figure 25. Temperature dependence of the permeability, P, of small unilamellar vesicles of dimyris-toyl phosphatidylcholine to the cation, Tempocholine[+]. Permeabilities were determined from the initial rate of uptake (●) and from release over a longer timescale (▲). For the latter, the ordinate is to be multiplied by a factor of 0.23. (Initial rates of release gave similar permeabilities to those from uptake.) The lateral compressibility (solid line) and fraction of interfacial lipids (dashed line) were calculated for a one-dimensional Ising model as described in the text. The maximum compressibility is 19 m/N and the maximum fraction of interfacial lipids is 3%. (Adapted from Marsh et al, 1976 with permission).

A range of other experiments have also indicated different aspects of lipid membrane function for which a maximum in activity is exhibited at the chain-melting transition or in regions of gel-fluid lateral phase separation. A maximum in the partitioning of amphiphiles, such as the fluorescence probe anilinonaptha-lene sulphonic acid (Jacobsen and Papahadjopoulos, 1976; Tsong, 1975) and lysolipids (Harlos et al, 1977), into lipid vesicles has been demonstrated at the gel-to-fluid phase transition. The incorporation of the bacteriophage M13 coat protein into synthetic lipid membranes is maximal (and complete) at the chain-melting transition, and only in the transition region is the incorporated coat protein able to adopt the correct transmembrane orientation (Wickner, 1976).

A discontinuous increase in sugar transport activity with decreasing temperature has been observed in fatty-acid supplemented auxotrophs of *E. coli* on entering the lateral phase separation region of the membrane lipids from the fluid phase (Linden et al, 1973). As mentioned previously, the lipolytic activity of pancreatic phospholipase A_2 is maximum at the bilayer phase transition of the lipid substrate, and this extends also to the gel-fluid coexistence region of lipid mixtures (Op den Kamp et al, 1975). Experiments with monolayers have located the site of action of snake venom phospholipase A_2 at the interfacial regions between coexisting gel and fluid domains, i.e., exactly in those regions where the enhanced lateral compressibility is expressed (Grainger et al, 1990). Measurements of trapped volume have shown that the extent of fusion of freshly sonicated dipalmitoyl phosphatidylcholine vesicles, essentially an annealing process, is maximal at the phase transition (Marsh and Watts, 1981). All these results suggest that the high lateral compressibility of lipid bilayers in their regions of gel-fluid coexistence, and the accompanying fluctuations and creation of transient membrane defects, facilitates the insertion of membrane molecules and other amphiphiles and enhances the function of surface-active enzymes.

Propagation of Compressional Waves

A further feature of the elastic properties of surfaces is the possibility for propagation of mechanical distortions through the planar membrane which could be considered as a two-dimensional compression wave. This might serve as a possible means of communication between membrane proteins, the conformational changes of which are sensitive to, and could generate, propagating changes in the membrane lateral pressure. A local change in lipid surface density will lead to a local increase in temperature because the work of compression will convert to an equivalent thermal energy. If the distortion is so short that this thermal energy cannot dissipate by diffusion, the local compression is described by the adiabatic compressibility, κ_S, rather than by the isothermal compressibility, κ_T. The adiabatic compressibility depends on the heat capacity and the isothermal compressibility, and for simplicity we assume that the adiabatic and the isothermal compressibility have approximately the same temperature dependence and comparable quantitative values. The bulk compressibility of dipalmitoyl phosphatidylcholine bilayers deduced from ultrasonic velocity measurements (Mitaku et al, 1978) is found to be a factor of approximately 2.5 times smaller than that obtained from static isothermal measurements (Liu and Kay, 1977; Marsh, 1990). The velocity, c_0, of the compression wave, at frequencies well below the rates of processes contributing to the velocity dispersion, is related to the adiabatic area compressibility by an inverse square root dependence:

$$\frac{1}{c_0^2} = \left(\frac{\partial \rho_A}{\partial \pi}\right)_S = -\frac{\rho_A}{A}\left(\frac{\partial A}{\partial \pi}\right)_S = \rho_A \cdot \kappa_S \tag{39}$$

where ρ_A is the surface density. From the results of the previous section, it is therefore expected that the speed of sound in a membrane is slowest at the transition mid-point and increases strongly outside the transition region. Such slowing-down effects have been found experimentally for ultrasonic velocity measurements in bulk membrane dispersions. An abrupt decrease in ultrasonic velocity in the bilayers by approximately 270 m/s has been found at the phase transition of dipalmitoyl phosphatidylcholine (Mitaku et al, 1978). In this latter situation, however, it is the three-dimensional propagation and the volume compressibility that are involved.

The density ρ_A used in Equation 39 is the mass per unit area, which is approximately $4.4 \cdot 10^{-3}$ g/m^2 (assuming an area of 60 Å^2 per lipid and a molar mass of 800 g/mol). Using the isothermal compressibility, instead of the adiabatic compressibility, yields a velocity for the compression wave at the lipid transition point of approximately 33 m/s. (Correction to the adiabatic compressibility, by scaling according to the ratio of the bulk values given above, yields approximately 52 m/s.) The velocity is slower for more cooperative transitions and faster for less cooperative transitions, and increases steeply outside the transition range. In large membranes, this propagation could, in principle, play a role in signal transduction within the membrane plane. In charged membranes, these signals might propagate not only changes in temperature and density, but also changes in electrostatic surface potential. As discussed earlier, changes of surface potential play a role in binding affinities and in structural equilibria of the bound proteins. In general, therefore, it is possible that structural changes in the proteins could reciprocally trigger the propagation of electrostatic and density waves, resulting in a communication between membrane proteins.

Conclusions

This chapter has been concerned with the thermodynamics of interactions of lipid membranes with, principally, peripheral proteins. First, the binding has been analyzed, which gives information on the free energy of interaction, in which electrostatics plays a very major role. Additionally, it has been shown that binding isotherms can give information on the interactions between the surface-bound proteins, on the interactions between the lipids to which they are bound, and on their redistribution in response to the binding of proteins. The second aspect considered has been the preferential interaction of membrane proteins with lipids in different physical states at the chain-melting transition, and by implication more generally, in regions of lateral lipid phase separation. The thermodynamics of lipid chain-melting has been shown to reflect also that of protein binding and constitutes a viable alternative method for studying the difference in energetics of protein binding to lipid membranes in different states. By combining experimental thermodynamic measurements with theoretical Monte Carlo simulations, information has been obtained on the distribution of proteins at the surface of the membrane, or within the membrane (in the case of integral proteins). These protein distributions are important not only in cases in which function specifically requires aggregation or clustering of proteins, but also quite generally in determining possible structural interactions, the likelihood of encounters, and the spatial relations between the different membrane components. Finally, the coupling between thermodynamic fluctuations and structural fluctuations linked to changes in membrane area has been addressed, leading to consideration of the possible functional consequences of enhanced membrane compressibility.

It should be noted that, although a considerable part of this chapter has concentrated on gel-to-fluid lipid phase transitions, the principal results are generally applicable to membranes with any form of coexistence of lipid states, in particular those arising from a heterogeneity in lipid composition or from specific binding of divalent ions. Relevant thermodynamic parameters in all cases are the heat capacities, enthalpies, interfacial energies between lipids of different type, and the differential binding free energies. These general results, therefore, may form some physical rationale for the complex chemical composition of the lipids in biological membranes.

REFERENCES

Albon N, Sturtevant JM (1978): Nature of the gel to liquid crystal transition of synthetic phosphatidylcholines. *Proc Natl Acad Sci USA* 75:2258–2260

Apitz-Castro R, Jain MK, de Haas GH (1982): Origin of the latency phase during the action of phospholipase A_2 on unmodified phosphatidylcholine vesicles. *Biochim Biophys Acta* 688:349–356

Aveyard R, Haydon DA (1973): *An Introduction to the Principles of Surface Chemistry.* Cambridge: Cambridge University Press

Bennett JP, Smith GA, Houslay MD, Hesketh TR, Metcalfe JC, Warren GB (1978): The phospholipid headgroup specificity of an ATP-dependent calcium pump. *Biochim Biophys Acta* 513:310–320

Biltonen RL (1990): A statistical-thermodynamic view of cooperative structural changes in phospholipid bilayer membranes: Their potential role in biological function. *J Chem Thermody* 22:1–19

Blok MC, van der Neut-Kok ECM, van Deenen LLM, de Gier J (1975): The effect of chainlength and lipid phase transitions on the selective permeability properties of liposomes. *Biochim Biophys Acta* 406:187–196

Boggs JM, Clement IR, Moscarello MA (1980): Similar effect of proteolipid apoproteins from human myelin (lipophilin) and bovine white matter on the lipid phase transition. *Biochim Biophys Acta* 601:134–151

Boggs JM, Rangaraj G, Kostry KM (1988): Photolabeling of myelin basic protein in lipid vesicles with the hydrophobic reagent 3-(trifluoromethyl)-3-(m-[^{125}I]iodophenyl) diazirine. *Biochim Biophys Acta* 937:1–9

Boheim G, Hanke W, Eibl H (1980): Lipid phase transition in a planar bilayer and its effect on carrier- and pore-mediated transport. *Proc Natl Acad Sci USA* 77:3403–3407

Burack WR, Yuan Q, Biltonen RL (1993): Role of lateral phase separation in the modulation of phospholipase A_2 activity. *Biochemistry* 31:583–589

Cantor CR, Schimmel PR (1980): *Biophysical Chemistry, Vol. 3*.New York: Freeman

Carruthers A, Melchior DL (1984): Human erythrocyte hexose transporter activity is governed by bilayer lipid composition in reconstituted vesicles. *Biochemistry* 23:6901–6911

Cevc G, Marsh D (1987): *Phospholipid Bilayers. Physical Principles and Models.* New York: Wiley-Interscience

Cevc G, Watts A, Marsh D (1981): Titration of the phase transition of phosphatidylserine bilayer membranes. Effects of pH, surface electrostatics, ion binding and head-group hydration. *Biochemistry* 20:4955–4965

Cevc G, Watts A, Marsh D (1980): Non-electrostatic contribution to the titration of the ordered-fluid phase transition of phosphatidylglycerol bilayers. *FEBS Lett* 120:267–270

Cheddar G, Tollin G (1994): Comparison of electron transfer kinetics between redox proteins free in solution and electrostatically complexed to a lipid bilayer membrane. *Arch Biochem Biophys* 310:392–396

Chen YS, Hubbell WL (1973): Temperature- and light-dependent structural changes in rhodopsin-lipid membranes. *Exp Eye Res* 17:517–532

Cortese JD, Fleischer S (1987): Noncooperative vs. cooperative reactivation of D-β-hydroxybutyrate dehydrogenase: Multiple equilibria for lecithin binding are determined by the physical state (soluble vs. bilayer) and composition of the phospholipids. *Biochemistry* 26:5283–5293

Cutsforth GA, Whitaker RN, Hermans J, Lentz BR (1989): A new model to describe extrinsic protein binding to phospholipid membranes of varying composition: Application to human coagulation proteins. *Biochemistry* 28:7453–7461

Daum G (1985): Lipids of mitochondria. *Biochim Biophys Acta* 822:1–42

Dickerson RE, Takano T, Eisenberg D, Kallai OB, Samson L, Cooper A, Margoliash EJ (1971): Ferricytochrome *c* 1. General features of the horse and bonito proteins at 2.8 Å resolution. *J Biol Chem* 246:1511–1535

Evans EA, Kwok R (1982): Mechanical calorimetry of large dimyristoyl phosphatidylcholine vesicles in the phase transition region. *Biochemistry* 21:4874–4879

Fajer P, Knowles PFK, Marsh D (1989): Rotational motion of yeast cytochrome oxidase in phosphatidylcholine complexes studied by saturation-transfer electron spin resonance. *Biochemistry* 28:5634–5643

Ferrenberg AM, Swendsen RH (1988): New Monte Carlo technique for studying phase transitions. *Phys Rev Lett* 61:2635–2638

Fox RO, Jr., Richards FM (1982): A voltage-gated ion channel model inferred from the crystal structure of alamethicin at 1.5-A resolution. *Nature* 30:325–330

Freire E, Markello T, Rigell C, Holloway PW (1983): Calorimetric and fluorescence characterization of interactions between cytochrome b_5 and phosphatidylcholine bilayers. *Biochemistry* 22:1675–1680

Görrisen H, Marsh D, Rietveld A, de Kruijff B (1986): Apocytochrome *c* binding to negatively charged lipid dispersions studied by spin-label electron spin resonance. *Biochemistry* 25:2904–2910

Grainger DW, Reichert A, Ringsdorf H, Salesse C (1990): Hydrolytic action of phospholipase A_2 in monolayers in the phase transition region: Direct observation of enzyme domain formation using fluorescence microscopy. *Biochim Biophys Acta* 1023:365–379

Grant CWM, McConnell HM (1974): Glycophorin in lipid bilayers. *Proc Natl Acad Sci USA* 71:4653–4657

Harlos K, Vaz WLC, Kovatchev S (1977): Effect of desoxylysolecithins on dimyristoyl-lecithin vesicles. Influence of the lipid phase transition. *FEBS Lett* 77:7–10

Haydon DA, Taylor FH (1960): On adsorption at the oil/water interface and the calcula-tion of electrical potentials in the aqueous surface phase. I. Neutral molecules and a simplified treatment for ions. *Phil Trans Roy Soc* A 252:225–248

Heimburg T, Biltonen RL (1996): A Monte-Carlo simulation study of protein-induced heat capacity changes and lipid-induced protein clustering. *Biophys J* 70:84–96

Heimburg T, Biltonen RL (1994): Thermotropic behavior of dimyristoylphosphatidyl-glycerol and its interaction with cytochrome *c*. *Biochemistry* 33:9477–9488

Heimburg T, Marsh D (1995): Protein surface-distribution and protein-protein interactions in the binding of peripheral proteins to charged lipid membranes. *Biophys J* 68:536–546

Heimburg T, Marsh D (1993): Investigation of secondary and tertiary structural changes of cytochrome *c* in complexes with anionic lipids using amide hydrogen exchange measurements: An FTIR study. *Biophys J* 65:2408–2417

Heimburg T, Hildebrandt P, Marsh D (1991): Cytochrome *c*-lipid interactions studied by resonance Raman and ^{31}P-NMR spectroscopy. Correlation between the conformational changes of the protein and the lipid bilayer. *Biochemistry* 30:9084–9089

Hesketh TR, Smith GA, Houslay MD, McGill KA, Birdsall, NJM, Metcalfe JC, Warren GB (1976): Annular lipids determine the ATPase activity of a calcium transport protein complexed with dipalmitoyllecithin. *Biochemistry* 15:4145–4151

Hidalgo C, Thomas DD, Ikemoto N (1978): Effect of the lipid environment on protein motion and enzymatic activity of the sarcoplasmic reticulum calcium ATPase. *J Biol Chem* 253:6879–6887

Hildebrandt P, Stockburger M (1989a): Cytochrome *c* at charged interfaces. 1. Confor-mational and redox equilibria at the electrode/electrolyte interface probed by surface-enhanced Raman spectroscopy. *Biochemistry* 28:6710–6721

Hildebrandt P, Stockburger M (1989b): Cytochrome *c* at charged interfaces. 2. Com-plexes with negatively charged macromolecular systems studied by resonance Raman spectroscopy. *Biochemistry* 28:6722–6728

Hildebrandt P, Heimburg T, Marsh D (1990): Quantitative conformational analysis of cy-tochrome *c* bound to phospholipid vesicles studied by resonance Raman spectroscopy. *Eur Biophys J* 18:193–201

Hill TL (1960): *An Introduction to Statistical Thermodynamics.* New York: Dover

Hill TL (1946): Statistical mechanics of multimolecular absorption. II. Localized and mobile adsorption and absorption. *J Chem Phys* 14:441–453

Huang J, Swanson JE, Dibble ARG, Hinderliter AK, Feigenson GW (1993): Nonideal mixing of phosphatidylserine and phosphatidylcholine in the fluid lamellar phase. *Bio-phys J* 64:413–425

Jacobson K, Papahadjopoulos D (1976): Effect of phase transition on the binding of 1-anilino-8-naphthalenesulfonate to phospholipid membranes. *Biophys J* 16:549–560

Jähnig F (1976): Electrostatic free energy and shift of the phase transition for charged lipid membranes. *Biophys Chem* 4:309–318

Jain MK, Vaz WLC (1987): Dehydration of the lipid-protein microinterface on binding phospholipase A_2 to lipid bilayers. *Biochim Biophys Acta* 905:1–8

Jain MK, Egmond MR, Verheij H, Apitz-Castro R, Dijkman R, de Haas GH (1982): Inter-action of phospholipase A_2 and phospholipid bilayers. *Biochim Biophys Acta* 688:341–348

Jain MK, Yu B-Z, Kozubek A (1989): Binding of phospholipase A_2 to zwitterionic bilayers is promoted by lateral segregation of anionic amphiphiles. *Biochim Biophys Acta* 980:23–32

Kimelberg HK, Papahadjopoulos D (1974): Effects of phospholipid acyl chain fluidity, phase transitions and cholesterol on $(Na^+ + K^+)$-stimulated adenosine triphosphatase. *J Biol Chem* 249:1071–1080

Kleeman W, McConnell HM (1976): Interactions of proteins and cholesterol with lipids in bilayer membranes. *Biochim Biophys Acta* 419:206–222

Kleeman W, McConnell HM (1974): Lateral phase separations in Escherichia coli membranes. *Biochim Biophys Acta* 345:220–230

Lee J, Kosterlitz JM (1991): Finite-size scaling and Monte Carlo simulations of first-order phase transitions. *Phys Rev* B 43:3265–3277

Lichtenberg D, Romero G, Menashe M, Biltonen RL (1986): Hydrolysis of dipalmitoyl phosphatidylcholine large unilamellar vesicles by porcine pancreatic phospholipase A_2. *J Biol Chem* 261:5334–5340

Linden C, Wright K, McConnell HM, Fox CF (1973): Phase separations and glucoside uptake in *E. coli* fatty acid auxotrophs. *Proc Natl Acad Sci USA* 70:2271–2275

Liu NI, Kay RL (1977): Redetermination of the pressure dependence of the lipid bilayer phase transition. *Biochemistry* 16:3484–3486

London Y, Demel RA, Guerts van Kessel WSM, Vossenberg FGA, van Deenen LLM (1979): The protection of A_1 myelin basic protein against the action of proteolytic enzymes after injection of the protein with lipids at the air-water interface. *Biochim Biophys Acta* 311:520–530

Marsh D (1996): Lateral pressure in membranes. *Biochim Biophys Acta*: in press

Marsh D (1995a): Specificity of lipid-protein interactions. In: *Biomembranes*, Vol. 1, Lee AG, ed. Greenwich, CT: JAI Press

Marsh D (1995b): Lipid-protein interactions and heterogeneous lipid distribution in membranes. *Mol Memb Biol* 12:59–64

Marsh D (1990): *Handbook of Lipid Bilayers*. Boca Raton, FL: CRC Press

Marsh D (1985): ESR spin label studies of lipid-protein interactions. In: *Progress in Protein-Lipid Interactions*, Vol. 1, Watts A, de Pont JJHHM, eds. Amsterdam: Elsevier

Marsh D, Watts A (1981): ESR spin label studies. In: *Liposomes: from Physical Structure to Therapeutic Applications*, Knight CG, ed. Amsterdam: Elsevier/North-Holland Biomedical Press

Marsh D, Watts A, Knowles PF (1977): Cooperativity of the phase transition in single- and multibilayer lipid vesicles. *Biochim Biophys Acta* 465:500–514

Marsh D, Watts A, Knowles PF (1976): Permeability of Tempo-choline into dimyristoyl phosphatidylcholine vesicles at the phase transition. *Biochemistry* 15:3570–3578

Mitaku S, Ikegami A, Sakanishi A (1978): Ultrasonic studies of lipid bilayer. Phase transition in synthethic phosphatidylcholine liposomes. *Biophys Chem* 8:295–304

Morrow MR, Davis JH, Sharom FJ, Lamb MP (1986): Studies of the interaction of human erythrocyte band 3 with membrane lipids using deuterium nuclear magnetic resonance and differential scanning calorimetry. *Biochim Biophys Acta* 858:13–20

Mouritsen OG, Biltonen RL (1992): Protein-lipid interactions and membrane heterogeneity. In: *Protein-Lipid Interactions. New Comprehensive Biochemistry*, Vol. 25, Watts A, ed. Amsterdam: Elsevier

Muga A, Mantsch HH, Surewicz WK (1991): Membrane binding induces destabilization of cytochrome *c* structure. *Biochemistry* 30:7219–7224

Mühlebach T, Cherry RJ (1985): Rotational diffusion and self-association of band 3 in reconstituted lipid vesicles. *Biochemistry* 24:975–983

Newton AC (1993): Interaction of proteins with lipid headgroups: Lessons from protein kinase C. *Annu Rev Biophys Biomol Struct* 22:1–25

Op den Kamp JAF, Kauerz MTh, van Deenen LLM (1975): Action of pancreatic phospholipase A_2 on phosphatidylcholine bilayers in different physical states. *Biochim Biophys Acta* 406:169–177

Orr JW, Newton AC (1992a): Interaction of protein kinase C with phosphatidylserine 1. Cooperativity in lipid binding. *Biochemistry* 31:4661–4667

Orr JW, Newton AC (1992b): Interaction of protein kinase C with phosphatidylserine. 2. Specificity and regulation. *Biochemistry* 31:4667–4673

Overath P, Thilo L (1978): Structural and functional aspects of biological membranes revealed by lipid phase transitions. In: *Biochemistry of Cell Walls and Membranes II*, Metcalfe JC, ed. London: Butterworth and University Park Press

Papahadjopoulos D, Jacobson K, Nir S, Isac T (1973): Phase transitions in phospholipid vesicles. Fluorescence polarization and permeability experiments concerning the effects of temperature and cholesterol. *Biochim Biophys Acta* 311:330–348

Ramsay G, Prabhu R, Freire E (1986): Direct measurement of the energetics of association between myelin basic protein and phosphatidylserine vesicles. *Biochemistry* 25:2265–2270

Ryba NJP, Marsh D (1992): Protein rotational diffusion and lipid/protein interactions in recombinants of bovine rhodopsin with saturated diacyl-phosphatidylcholines of different chainlengths studied by conventional and saturation-transfer electron spin resonance. *Biochemistry* 31:7511–7518

Sankaram MB, Marsh D (1993): Protein-lipid interactions with peripheral membrane proteins. In: *Protein-Lipid Interactions. New Comprehensive Biochemistry*, Vol. 25, Watts A, ed. Amsterdam: Elsevier

Sankaram MB, Brophy PJ, Jordi W, Marsh D (1990): Fatty acid pH titration and the selectivity of interaction with extrinsic proteins in dimyristoylphosphatidylglycerol dispersions. Spin label ESR studies. *Biochim Biophys Acta* 1021:63–69

Sankaram MB, Brophy PJ, Marsh D (1989a): Spin-label ESR studies on the interaction of bovine spinal cord myelin basic protein with dimyristoylphosphatidylglycerol dispersions. *Biochemistry* 28:9685–9691

Sankaram MB, de Kruijff B, Marsh D (1989b): Selectivity of interaction of spin-labelled lipids with peripheral proteins bound to dimyristoyl phosphatidylglycerol bilayers, as determined by ESR spectroscopy. *Biochim Biophys Acta* 986:315–320

Silvius JR, McElhaney RN (1991): Non-linear Arrhenius plots and the analysis of reaction and motional rates in membranes. *J Theoret Biol* 88:135–152

Starling AP, East JM, Lee AG (1995): Effects of gel phase phospholipid on Ca^{2+}-ATPase. *Biochemistry* 34:3084–3091

Sugar IP, Biltonen RL, Mitchard N (1994): Monte Carlo simulations of membranes: Phase transition of small unilamellar dipalmitoylphosphatidylcholine vesicles. *Methods Enzymol* 240:569–593

Surewicz WA, Moscarello MA, Mantsch HH (1987): Fourier transform infrared spectroscopic investigation of the interaction between myelin basic protein and dimyristoylphosphatidylglycerol bilayers. *Biochemistry* 26:3881–3886

Tanford C (1955): The electrostatic free energy of globular protein ions in aqueous solution. *J Phys Chem* 59:788–793

Teft RC Jr, Carruthers A, Melchior DL (1986): Reconstituted human erythrocyte sugar transporter activity is determined by bilayer lipid headgroups. *Biochemistry* 25:3709–3718

Thilo L, Träuble H, Overath P (1977): Mechanistic interpretation of the influence of lipid phase transitions on transport functions. *Biochemistry* 16:1283–1290

Tsong TY (1975): Effect of a phase transition on the kinetics of dye transport in phospholipid bilayer structures. *Biochemistry* 14:5409–5414

Van Hoogevest P, de Gier J, de Kruijff B (1984): Determination of the size of the packing defects in dimyristoyl phosphatidylcholine bilayers, present at the phase transition temperature. *FEBS Lett* 171:160–164

Wickner W (1976): Asymmetric orientation of phage M13 coat protein in *Escherichia coli* cytoplasmic membranes and in synthetic lipid vesicles. *Proc Natl Acad Sci USA* 73:1159–1163

Wu SH, McConnell HM (1973): Lateral phase separations and perpendicular transport in membranes. *Biochem Biophys Res Commun* 55:484–491

Zimm BH, Bragg JK (1959): Theory of the phase transition between helix and random coil in polypeptide chains. *J Chem Phys* 31:526–535

14

Role of Lipid Organization and Dynamics for Membrane Functionality

OLE G. MOURITSEN AND PAAVO K.J. KINNUNEN

Introduction: The Need for a New Membrane Model

For almost 25 years the Singer-Nicolson fluid-mosaic model of biological membranes (Singer and Nicolson, 1972) has been the central paradigm of membrane science. This model, and the powerful and rather simple conceptual framework it provided to membrane scientists, have had a tremendous impact on advances in the broad field of biological membranes over the past more than two decades.

When the fluid-mosaic model was proposed in 1972, it had become clear from a large body of solid experimental evidence that older views of the membrane as a rigid lipid structure with the proteins attached to it in a fairly structureless composite could not be maintained. The Singer-Nicolson model imparted to the membrane a certain degree of fluidity, a concept which was not uniquely defined but which was meant to characterize the lipid-bilayer component of the membrane as a kind of pseudo-two-dimensional liquid in which the lipids display a considerable degree of lateral mobility. Similarly, the peripheral or integral membrane proteins associated with the membrane were assumed to diffuse around at or in the fluid lipid sea. Fluidity was proposed as an essential requirement for biological viability and for supporting membrane-associated functions (Gennis, 1989). The overall random appearance of the lipid-protein fluid composite made the membrane look like a mosaic structure (Singer and Nicolson, 1972). Except for unsaturated lipid acyl chains imparting enhanced fluidity to membranes, the impressive diversity of the chemical structures of lipids, actively maintained by cells, had little significance in the fluid mosaic model. Yet, the biological significance of the complex arrays of lipids and their cell type and organelle specific distributions became even more puzzling when the variation in their physical properties began to be understood.

Biological Membranes
K. Merz, Jr. and B. Roux, Editors
© Birkhäuser Boston 1996

The fluid nature imparted to the membrane within the fluid-mosaic model has over the years lead many workers to tacitly assume that fluidity also implied randomness. This viewpoint obviously neglects the fact that fluids may be structured on length scales that lie between the molecular scale and those length scales that are accessible by conventional microscopic or spectroscopic techniques commonly applied to study the lateral organization of membranes. Similarly, and somewhat paradoxically, the membrane structure in the time regime, i.e., the dynamic modes, in particular the correlated dynamical phenomena, often were not considered as existing or as being important for membrane function, despite the fact that the lively dynamics of membranes (Robertson, 1983), as we now recognize it, is perhaps the most obvious physical feature of a fluid membrane.

The picture of the biological membrane that we want to advocate in this chapter takes its starting point in the fluid-bilayer component of the biological membrane as a physical entity with a considerable dynamics and a highly non-trivial lateral organization of the membrane components on many different length and time scales. This picture anticipates the physical fact that the membrane assembly is a many-particle system which, by basic laws of Nature, displays correlated and cooperative modes involving many molecules. These modes give rise to a rich dynamic structure. It is our hypothesis that this dynamic structure and the corresponding lipid organization are important features for membrane function and that it is essential to characterize this dynamical structure in a quantitative manner in order to fully understand membrane functionality as well as how to modulate it (Kinnunen and Mouritsen, 1994).

We contend that the development of a new membrane model, somewhat paradoxically, has been hampered by our preoccupation with molecular structure as a predominantly static entity in the way traditional biochemistry as well as molecular and structural biology has so successfully used it to investigate the structure-function relationships of proteins and polynucleotides. As it appears now, thinking about membrane structure and how it controls function in a static way is not going to provide the right questions. If the question asked is if and how the static lipid membrane structure couples to functionality, the answer is likely to be that there is little if any coupling and that the lipids are probably just a convenient and not too specific solvent for the membrane proteins. If, on the other hand, one questions how the dynamic organization of a lipid bilayer couples to functionality, the answer is likely to be different. In fact, there is now ample evidence for a number of specific systems in which this question has been addressed in a quantitative fashion that lipids indeed play a significant role in controlling membrane functionality.

Perhaps the most intriguing example available at present of the importance of membrane lipid dynamics to cellular functions concerns the biological significance of those lipids, which, in isolation, form nonlamellar phases, such as the inverted hexagonal phase H_{II}. It has been demonstrated in several laboratories that the optimal growth of microorganisms is achieved when their membranes have lipid compositions such that, under the prevailing conditions (e.g., temperature, presence of ions), a certain tendency to form nonbilayer phases is evident (Goldfine, 1984; Rietveld et al, 1994; Wieslander et al, 1986). Notably, this tendency to form (but not the actual formation of) nonbilayer phases is also needed for the activation of enzymes such as protein kinase C (Epand, 1991; Slater et al, 1994) as well as CTP:phosphocholine cytidyltransferase (Jamil et al, 1993). The former is centrally involved in cellular signal transmission cascades while the latter is the rate-limiting enzyme in the synthesis of the major membrane lipid, phosphatidylcholine. We will revert to this subject in more detail below.

A wealth of experimental information on all conceivable properties of natural membranes as well as model membranes forms the empirical foundation for a new and more refined model of the biological membrane which we contend is about to emerge and on which a new level of understanding can be built. In a review of the present type it is impossible to give full credit to all the important discoveries that reveal the need for a new membrane model. We shall make a modest attempt to review some of the main findings as we see them and give references to key papers where details can be found. We wish to stress that in the present chapter we have not attempted to present a conventional review but rather to give a topical perspective, possibly flavored by our own personal preoccupation, on a vast, complicated, and rapidly developing field of considerable current scientific excitement.

Levels of Lipid Organization and the Material Properties of Lipid Bilayers

Lipid Aggregates and Three-Dimensional Phases

In aqueous solution, due to their amphiphilic nature, lipids undergo self-assembly processes and organize themselves spontaneously on many different levels. Each level corresponds to a different energy scale. The central level of organization are lipid aggregates, which correspond to an energy only slightly larger than $k_B T$, and may be seen as supramolecular structures bound by weak physical forces. Different types of aggregates with different symmetry exist (Cevc and Marsh, 1987), such as micelles, bilayers (e.g., uni- and multilamellar liposomes), and hexagonal and inverted hexagonal rods.

In the aqueous solution these lipid aggregates organize themselves in three-dimensional space in order to form a higher level of organization and eventually a thermodynamic phase, which represents the highest level of lipid organization. Some of these phases display long-range order, such as lamellar bilayers, cubic and hexagonal phases, or correspond to structures with a considerably degree of disorder, such as sponge phases. The self-assembly and organization principles at this level are highly complex and depend on many different inter- as well as intra-aggregate forces of both enthalpic as well as entropic nature (Israelachvili,

1992). Transformations between the different phases may arise in response to changes in composition or thermodynamic conditions leading to a very rich polymorphism.

On the subaggregate level, the lipids organize themselves in surfaces with the polar-head groups facing the water regions, e.g., as bilayer structures in lamellar phases that may be planar or display local curvature. The tendency for formation of one or the other type of aggregates is to a large extent controlled by the shape of the lipid molecules (Israelachvili, 1992). On this level, lipid organization also involves a possible asymmetry between the distribution of different lipids in the two leaflets of a bilayer as well as a specific static lateral organization, e.g., thermodynamic phase separation, of differerent lipid species within the lipid layer. We shall return to these type of phase equilibria below.

In this chapter we shall mainly be concerned with lamellar lipid bilayer aggregates. We shall be predominantly interested in nonlamellar phases as virtual phases, for which certain types of lipids, under conditions to be discussed below, show specific propensity in the sense that they produce local modifications of the bilayer due to their ability to promote bilayer instabilities without corrupting the overall lamellar symmetry. Nonbilayer forming lipids, e.g., lipids with propensity for H_{II} hexagonal phases, are possibly among the more peculiar molecular membrane constituents whose distinct effects on membrane dynamics and organization have revealed the need for a new membrane model.

Phase Equilibria in Lamellar Membranes

Lipid bilayers in the lamellar phase undergo a number of differerent lyotropic and thermotropic phase transitions that involve major in-plane structural reorganizations (Kinnunen and Laggner, 1991). These phase transitions are possibly the most conspicuous consequences of the basic fact that membranes are many-particle systems that can display highly nontrivial collective phenomena. These phenomena are nontrivial in the sense that they can not be understood, described, or even predicted on the basis of the properties of individual molecules. We believe, that although this fact is well-established in the physical sciences, it is often overlooked in the traditional biochemistry of membranes in which a substantial effort is made to characterize lipid compositional profiles and the properties and function of individual macromolecules, e.g., membrane proteins, without fully anticipating the fact that the solvent for these macromolecules, the lipid bilayer, is an active player in the game because of its correlated and collective structural organization, which in some cases is highly sensitive to the environment. Even though phase transitions as such may not be directly relevant to membrane function (Jørgensen and Mouritsen, 1994; Mouritsen, 1991) the fact remains that a many-component membrane has an underlying phase structure and that the associated cooperative modes may establish themselves as differentiated lipid domains, which are dynamically maintained on scales of the order of 10–1000 Å (Mouritsen and Jørgensen, 1994; 1995). In relation to the viewpoint of the present chapter advancing a new membrane model, our description of lipid phase transitions and phase equilibria mainly serves as to lay the foundation for understanding the sources for lipid dynamics and formation of dynamic lipid domains.

Here we shall focus mainly on the so-called main phase transition of lipid bilayers (Mouritsen, 1991). This transition can be driven by temperature, as well as by changes in environmental conditions, such as pH, ionic strength, electric fields, etc., depending on the nature of the polar-head group (Kinnunen and Laggner, 1991). The main transition is predominantly a chain-melting transition which at a temperature, T_m, takes the lipid bilayer from a low-temperature solid phase (gel phase) with highly ordered lipid-acyl-chains to a high-temperature fluid phase characterized by a considerable degree of acyl-chain disorder. Since the order of the bilayer changes in two fundamentally different ways during the transition, one way being chain disordering and the other being loss of translational order, it was realized some years ago (Ipsen et al, 1987) that a full description of the lamellar lipid phases requires at least two labels refering to the translational as well as conformational degrees of freedom of the lipid molecules. Hence, the transition may be said to proceed from a solid-ordered phase to a liquid-disordered phase. This finer distinction between the phases becomes of seminal importance when the modification of the lipid phase behavior due to cholesterol is at issue.

When we proceed from one-component lipid bilayers to mixtures, e.g., binaries, a rich phase behavior emerges, involving thermodynamic states of phase separation. Equilibrium phase separation is a macroscopic phenomenon which implies that the bilayer develops macroscopically large regions of coexisting phases. The thermodynamically controlled phase coexistence is a source of static membrane heterogeneity, which in turn is a consequence of both the entropy of mixing as well as the nonideal mixing properties governed by the molecular interactions. The more nonideal the interactions are, i.e., the more incompatible the mixed lipid species, the more dramatic and rich are the phase equilibria (Lee, 1977). In case of imperfections, inhomogeneities, and couplings between the fluid lipid bilayer and the cytoskeleton, the phase separation may not be complete, and segregated regions or static lipid domains on submacroscopic scales can result. In Figure 1 we show some typical phase diagrams for a binary lipid mixture of DMPC-DSPC and for a mixture of DPPC with cholesterol (DMPC, dimyristoyl phosphatidylcholine; DPPC, dipalmitoyl phosphatidylcholine; DSPC, distearoyl phosphatidylcholine). The phase diagram for the DMPC-DSPC mixture is highly nonideal displaying a broad region of fluidgel coexistence. The DPPC-cholesterol phase diagram, whose structure is likely to be generic for a large class of phospholipid-cholesterol mixtures (Bloom et al, 1991), contains a liquid-ordered phase, which is a liquid with a high degree of acyl-chain order (Ipsen et al, 1987).

Many lipids display, in addition to the main phase transition, both a subtransition as well as a pretransition. Despite the fact that these transitions take place within the bilayer, i.e., predominantly in two dimensions, they couple to the interlamellar interactions and the three-dimensional organization of multilamellar

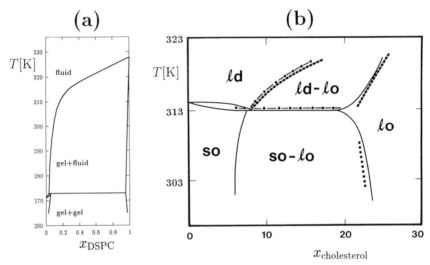

Figure 1. (a): Phase diagram of a binary DMPC-DSPC lipid mixture (Jørgensen et al, 1993b). (b): Phase diagram of DPPC-cholesterol mixtures (Ipsen et al, 1987). The phase labels refer to the solid-ordered (**so**), the liquid-disordered (**ld**), and the liquid-ordered (**lo**) phase.

arrays. Therefore, it is not possible to completely separate the two-dimensional phase equilibria from the fully three-dimensional behavior. This is clearly reflected in a number of observations, e.g., the calorimetric appearance of the main phase transition in unilamellar vesicles is rather different from that in multilamellar vesicles (Biltonen, 1990). Similarly, the characteristic interlamellar spacing (the swelling) in multilamellar arrays is strongly influenced by changes in the in-plane bilayer structure, and visa versa. We shall return to this phenomenon.

Lipid Bilayer Instability to Nonbilayer-Forming Lipids

The presence in cellular membranes of lipids (e.g., phosphatidyl ethanolamine and diacylglycerol in eucaryotes and in some bacteria as well as monoglucosyl diacylglycerol (MGDG) in microorganisms such as *Acholeplasma*) which in isolation form nonbilayer phases, such as the cubic phase and the inverted hexagonal phase, H_{II}, represents a perplexing enigma, see Figure 2. While the contents of these types of lipids in different cells as well as cell organelle membranes are rather strictly controlled, no evidence for the actual formation of nonbilayer phases per se in living cells has been obtained.

The physical basis for the formation of different three-dimensional assemblies was originally related to the molecular geometries of lipids, more specifically to the relative sizes of the polar headgroups and the hydrophobic parts (Israelachvili et al, 1980). This view has been refined and made more quantitative by emphasizing the spontaneous curvature of the structures formed by different lipids in aqueous environments (Hui and Sen, 1989; Tate et al, 1991).

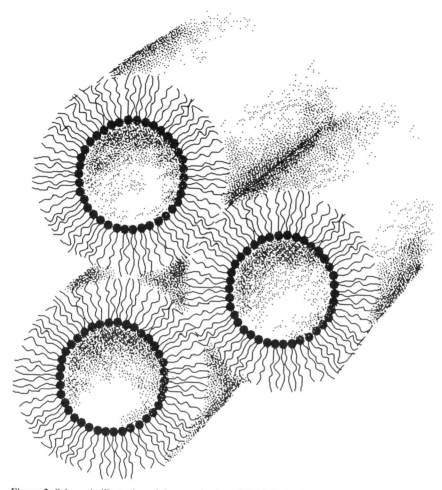

Figure 2. Schematic illustration of the organization of lipids in the inverted hexagonal phase H_{II}. The dimensions are arbitrary.

It is at present clear that a lamellar phase formed by proper lipid(s) can be induced to undergo a transition, for example, into the H_{II} phase by increase in temperature, reduction in pressure, dehydration, or introduction of hydrophobic solutes into the hydrocarbon region of the membrane. In brief, anything diminishing the effective size of the polar headgroup or increasing the effective size of the hydrocarbon chain region will increase the spontaneous curvature of the membrane which, ultimately, upon exceeding a critical energy barrier, results in the transformation of the lamellar phase into a nonlamellar assembly. Molecular mechanisms of lamellar → nonlamellar phase transitions have been discussed elsewhere (e.g., Kinnunen, 1995; Kinnunen, 1992; Laggner and Kriechbaum, 1991; Tate et al, 1991). Obviously, H_{II} propensity can be achieved for instance by approaching the temperature of the lamellar $\rightarrow H_{II}$ transition, or introduction of H_{II} phase forming lipids (e.g., diacylglycerol or cis-unsaturated PEs) into lamellar phases.

Physical Properties Characterizing Lipid Bilayers:
From Molecules to Material Properties

In principle the microscopic molecular interactions determine uniquely the physical properties of lipid bilayers on all time and length scales (Bloom et al, 1991). However, the general relations between molecular structure and dynamics on the one side and macroscopic physical properties on the other side usually are not very applicable since they are difficult to relate to actual measurements. Any type of physical measurement refers to a time and a length scale, and the specific quantity measured is often difficult to relate to physical properties that refer to measurements corresponding to other length and time scales. For lipid bilayers, that are strongly anisotropic systems, these problems are particularly severe since the bilayer displays structure on many different time and length scales. Being a non-Newtonian liquid, the lipid bilayer may behave as a liquid under some circumstances and as a solid under other circumstances. More specifically the problem arises as one tries to relate the structure and dynamic behavior of individual or a few lipid molecules to large-scale continuum elastic and viscous behavior of the bilayer sheet or even to collective molecular behavior such as microviscosity or fluidity. In fact the very term fluidity is not well defined in the context of membranes, and its use and misuse over the years has to some extent hampered progress in the field. Too often it has not been noted that acyl chain order and molecular mobility (microviscosity and lateral diffusion) is not related in any simple way.

Some experimental techniques applied to study the physical and material properties of lipid bilayers, such as fluorescence spectroscopy or ESR and NMR, are considered as providing molecular information on lipids. Still, each of these techniques refers to its own intrinsic time scale, which together with the time scales of the various molecular motions of the lipid molecules, e.g., lateral diffusion, implies that the properties measured are averages over finite ranges of time and space. If the averaging is over a sufficiently long time span, the property measured may be used to model continuum material properties, such as bilayer surface pressure and tension, surface area, thickness, or curvature at least over some mesoscopic scale (Evans and Skalak, 1980). Over this scale, various constitutive relations can be used to relate microscopic fluctuations to macroscopic mechanical properties. An example is the lateral area compressibility modulus which is related to the fluctuations in the bilayer surface area. When taken to still larger scales, complications arise because of the softness and flexibility of the bilayer, which supports thermally excited undulatory motions. These undulations tend to renormalize the mechanical properties on scales larger that the coherence length of the undulations. Close to phase transitions and within coexistence regions, where strong lateral density and compositional fluctuations prevail, this is particularly troublesome as we shall discuss below.

With the currently available arsenal of physical experimental techniques with each their limitations in time and length scales, it appears that there is a gap in our information on structure of lipid bilayers and membranes on the nanosale, i.e., on scales in the range of 10–1000 Å. Whereas spectroscopic techniques cover smaller scales, and diffraction, scattering, and mechanical measurements cover larger scales, access to the nanoscale regime awaits developments in the application to soft surfaces of atomic-force microscopy and intense neutron and synchrotron X-ray sources.

Dynamic Organization of Lipid Bilayers as Structured Two-Dimensional Fluids: Lipid Domains and Superstructures

Due to the fluid nature of most if not all biological membranes, it is obvious that in order to develop an understanding of their operational principles, the molecular mechanisms causing structure and organization in fluid lipid phases must be established. Accordingly, instead of the well-known, rather static coexistence of solid and fluid membrane domains, this review also focuses on what can be considered as dynamic heterogeneity in fluid membranes.

Dynamic Formation of Lipid Domains and Heterogeneity

It is now well established that the main phase transition in one-component phospholipid bilayers, although a first-order transition, is subject to strong in-plane density fluctuations (Mouritsen, 1991). The density fluctuations manifest themselves microscopically as formation of dynamic lipid domains (or clusters) within the thermodynamic background phase. Experimentally, direct detection of these domains is hampered by the fact that the length scale of the domains is in the nanometer range which is within the gap mentioned in the Introduction as currently inaccessible. However, as we shall see below, the macroscopic consequences of the persistence of these dynamic domains are well known.

Major evidence for the presence of dynamic lipid domains comes from computer-simulation calculations on molecular models of planar lipid bilayers (Mouritsen, 1990; Mouritsen and Jørgensen, 1994; 1995). In Figure 3 are shown snapshots of configurations of phospholipid bilayers in the transition region as derived from computer simulation. Figure 3a shows the lateral organization of a DPPC lipid bilayer for temperatures below and above the phase transition temperature, T_m. It can be seen that in the gel phase below T_m domains of lipid molecules in the fluid state are formed within the gel phase, and conversely above T_m, gel domains are formed within the fluid bilayer phase. Each domain is a dynamic entity that in the course of time nucleates, grows, and disappears again as illustrated in the time series in Figure 3b. This phenomenon has been referred to as dynamic heterogeneity (Mouritsen and Jørgensen, 1992) and may be considered as a kind of dynamic microphase separation. The finite lifetime of a domain is controlled by the line tension of its interface to the bulk phase.

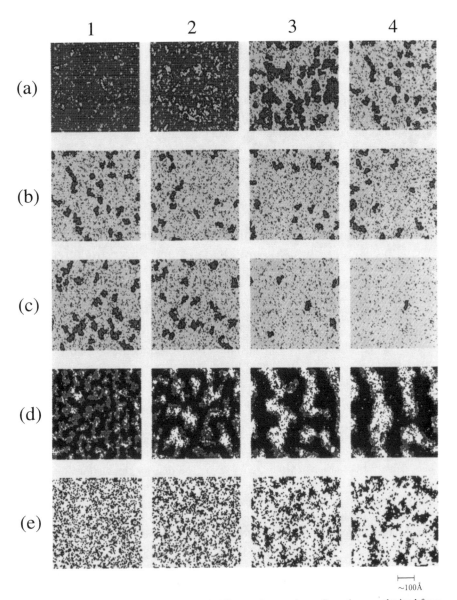

$\longmapsto$ ~100Å

Figure 3. Top-view snapshots of typical lipid bilayer microscopic configurations as obtained from computer-simulation calculations on two-dimensional statistical mechanical lattice models. The configurations show the bilayer heterogeneity in terms of lateral density and compositional fluctuations for pure lipid bilayers as well as certain binary mixtures consisting of DC_nPC lipids with different numbers, n, of carbons in the saturated acyl chains. (a): $DC_{16}PC$ bilayers near the phase transition at temperatures $T = 310K$ (1), 313K (2), 315K (3), and 319K (4). The transition temperature is $T_m = 314K$. Fluid domains are seen to be formed in the gel phase below T_m and gel domains in the fluid phase above T_m. (b): Time evolution of the microstructure of $DC_{16}PC$ at a temperature $T = 319K$ in the fluid phase. (c): Microstructures at the same relative temperature, $T/T_m = 1.016$, in the fluid phase for different lipid chain lengths, $DC_{12}PC$ (1), $DC_{14}PC$ (2), $DC_{18}PC$ (3), and $DC_{20}PC$ (4). (d): Time evolution of the microstructure for a binary mixture of $DC_{12}PC$-$DC_{18}PC$ quenched from a high temperature in the fluid phase to a temperature within the gel-fluid coexistence region, see Figure 1a. (e): Compositional heterogeneity in fluid-phase binary mixtures of $DC_{16}PC$-$DC_{20}PC$ at $T = 338K$ (1), $DC_{14}PC$-$DC_{20}PC$ at $T = 338K$ (2), $DC_{12}PC$-$DC_{20}PC$ at $T = 338K$ (3), and $DC_{12}PC$-$DC_{22}PC$ at $T = 352K$ (4) at a molar fraction, $x = 0.33$, of the longer lipid. (Courtesy of Dr. Kent Jørgensen.).

At any given time there exists an ensemble of dynamic domains characterized by an equilibrium size-distribution function and hence an average domain size. Obviously the average domain size attains a maximum at the transition temperature. Under isothermal conditions the domain formation can be controlled by conditions like degree of hydration and for charged lipids by pH and ionic strength.

The actual degree of dynamic heterogeneity for a given temperature depends on the lipid material. For instance, in the series of di-acyl phosphatidylcholines with saturated acyl chains, the lipid species with the shorter acyl chains leads to stronger dynamic heterogeneity (Ipsen et al, 1990) as illustrated in Figure 3c. The explanation of this is that the main phase transition of the bilayers with the shorter lipid chains is closer to a critical point and hence is subject to stronger lateral density fluctuations (Ipsen et al, 1990).

Hence, the underlying phase transition in phospholipid bilayers provides a natural mechanism for lateral organization of the lipid bilayer on time and length scales that depend on the thermodynamic conditions. It should be noted that this structural organization persists in the fluid phase. Consequently, the fluid phase, which is a disordered phase on macroscopic length scales, develops local order on the nanoscale due to the proximity of a phase transition.

The possibilities for forming local order out of globally disordered lipid-bilayer phases (Mouritsen and Jørgensen, 1994) increase when we turn to many-component systems. In these systems, lateral organization of the bilayer not only occurs in terms of static macroscopic phase separation as given by the phase diagrams in Figure 1. In the various one-phase regions, local order can develop in terms of lateral density fluctuations, e.g., in the fluid phase. This is illustrated by the computer-simulation results presented in Figure 3e, which shows how local structure persists in nonideal binary mixtures of lipids that are miscible in the fluid phase (Jørgensen et al, 1993b). It is noted that the more nonideal the mixture is the more pronounced are the compositional fluctuations. Like the density fluctuations in a one-component system, the compositional fluctuations in a binary mixture correspond to lipid domains of a local composition that is very different from the global composition of the bulk phase. In physical terms, the fluid mixture pictured in Figure 3d may be termed a two-dimensional structured fluid. Hence, we see again that due to the underlying phase equilibria, local order and organization persists in a disordered thermodynamic phase.

So far we have only considered dynamic domain formation in equilibrium. An even larger richness in type of structural organization on a local scale arises when we also consider nonequilibrium phenomena. Evidence is now accumulating that binary lipid bilayers in their gel-fluid phase coexistence region may not be fully phase separated over scales of hours (Jørgensen and Mouritsen, 1995; Klinger et al, 1994; Sankaram et al, 1992; Thompson et al, 1995). Similarly, incomplete phase separation may arise in binary mixtures that are not fully miscible in the fluid phase, such as certain PC-PS mixtures (Huang et al, 1993). In the case of phase separation processes in the gel-fluid phase of phosphatidylcholine mixtures, an example of local structure formation and its evolution in time is provided by the computer-simulation results for the DLPC-DSPC (DLPC, dilauroyl phosphatidylcholine) mixture shown in Figure 3d (Jørgensen and Mouritsen, 1995). This figure shows how the fluid and gel domains evolve very slowly in time. Moreover, the figure indicates that not only does the mixture organize in a convoluted structure of gel and fluid domains but the structure of the fluid (mainly consisting of the low-melting lipid, DLPC) between the gel domains (mainly consisting of the high-melting lipid DSPC) is quite different from the thermodynamic fluid phase in the sense that the acyl chains of the DLPC molecules are conformationally highly ordered as in their gel phase. This type of local organization is related to the phenomenon of capillary condensation (Jørgensen and Mouritsen, 1995). The lifetime of the various types of lipid domains dicussed in this section is difficult to assess from the theoretical model simulations, and only very little experimental data is currently available to shed light on this question. We estimate that a typical lifetime of a nanoscale domain may be as high as 10^{-4} s in a one-component lipid bilayer and considerably longer in many-component mixtures.

Modulation of Lipid Domains by Foreign Compounds

It is obvious that lipid bilayers that are locally ordered and dynamically organized due to fluctuations and cooperative modes controlled by the underlying phase equilibria will be sensitive to molecular agents that are active at the lipid domain interfaces. In physical terms, the only requirement for an agent to be active in structural reorganization is that it is capable of lowering the interfacial tension of the domain interfaces.

There exists a large number of molecules that interact with membranes and are capable of changing the local domain structure described above. Cholesterol is a prominent example that in small amounts appears to enhance the lateral density fluctuations in one-component lipid bilayers (Zuckermann et al, 1993) but that strongly suppresses the fluctuations and the tendency for domain formation in higher concentrations.

Two other large classes of compounds that are water soluble and partition into lipid bilayers, drugs, such as general and local anaesthetics (Jørgensen et al, 1993a; van Osdol et al, 1992), and insecticides (Jørgensen et al, 1991; Sabra et al, 1995) have a particularly dramatic influence on lipid domain formation. Some of these compounds are attracted to the domain interfaces and change the size and the morphology of the domains very much like an emulsifier. The interest in understanding the capacity of drugs, for instance, to manipulate local lipid-bilayer structure is put in a particular perspective when considering that this capacity may be related to the potency of the drug and therefore its molecular mechanism of action (Trudell, 1977).

A special but very prominent case of an interfacially active agent in the sense used here is that of proteins or peptides that either peripherally or integrally interact with lipid bilayers. It is obvious, that such molecules with a special affinity to a certain type of local bilayer structure, domain composition, or interfacial packing will on the one side have a dramatic influence on the bilayer organization, which on the other hand will provide a mechanism for a specific or nonspecific coupling to the protein. We shall return to such mechanisms.

Fluid–Fluid Immiscibility

In addition to the fluctuation-induced domain formation accompanying the phospholipid main transition discussed above, also other mechanisms generating lateral heterogeneity of lipid distribution in fluid (liquid-crystalline) membranes have been described. Immiscibility of certain binary lipid mixtures, driven by differences in intermolecular forces due to mismatch of the acyl chain lengths, has been experimentally verified (Melchior, 1986; Wu and McConnell, 1975). Interestingly, recent Raman spectroscopy studies argue for microscopic heterogeneity also in bilayers of unsaturated phosphatidylcholines (Litman et al, 1991). Evidence for fluid-fluid immiscibility of mixtures of unsaturated phosphatidylserines and phosphatidylcholines has been presented (Hinderliter et al, 1994; Huang et al, 1993). One more mechanism for lipid-domain formation has been reasoned as being related to the hydrophilic nature of the phospholipid headgroup. Diminishing the effective size of the phosphatidylcholine headgroup (reducing the number of water molecules in the headgroup hydration shell) by osmotically decreasing water activity, enhances lipid lateral packing (Wolfe and Brockman, 1988). In keeping with the above, it could be demonstrated that lateral heterogeneity results in fluid, binary mixtures of fluorescent phospholipid analogs with unsaturated phosphatidylcholines in the presence of the strongly hygroscopic polymer, poly(ethyleneglycol), PEG (Lehtonen and Kinnunen, 1995a). Accordingly, osmotic forces and control of phospholipid hydration provide effective means for isothermal control of the lateral distribution of membrane components. These experiments have also provided evidence for a critical temperature, denoted as T^*, well above the main transition, at which the water affinity of phosphatidylcholine and phosphatidylglycerol appears to increase considerably. Thus, above T^* the osmotically driven fluid-fluid immiscibility is abolished. Interestingly, values of this critical temperature found for different lipids are rather close to those reported in relation to the spontaneous formation of bilayers at gas/water interfaces (Gerschfeld, 1989).

The concept of hydrophobic mismatch discussed below in connection with lipid-protein interactions has recently been applied to the ordering in binary mixtures of unsaturated phosphatidylcholines (Lehtonen et al, 1996a). In brief, variation of the chain lenght of PCs with homologous, cis-unsaturated acyl chains with between $N = 14$ and 24 carbon atoms could be shown to result in the formation of domains enriched in the fluorescent phospholipid analog, 1-palmitoyl-2-[10(pyren-1-yl)]decanoyl-sn-glycero-3-phosphocholine (PPDPC), both in thin ($N < 20$) and in thick ($N > 20$) bilayers. Accordingly, the effective length of the probe appears to match better with that of di-20:1-PC.

Finally, it is important to realize that the formation of lateral heterogeneity in membranes could involve nonequilibrium phenomena. To this end, metastable states have been detected in investigations on the effects of cholesterol on lipid lateral distribution (Silvius, 1992).

Superstructures in Lipid Alloys

Recent efforts have provided compelling evidence for the formation of super-structures in binary lipid membranes. The first indications of this phenomenon can be found in the data for the partitioning of the spin-label probe TEMPO between the aqueous phase and cardiolipin/egg PC liposomes, which exhibited seemingly anomalous and strong composition-dependent maxima and minima (Berclaz and Geoffrey, 1984; Berclaz and McConnell, 1981). The partitioning profiles were also dramatically influenced by Ca^{2+}. Later, studies on liposomes and Langmuir-Blodgett films, exploiting phospholipid derivatives containing the covalently linked aromatic hydrocarbon pyrene (PyrPLs) as a fluorescent moiety (Galla and Hartmann, 1980; Kinnunen et al, 1993), revealed strong indications for a nonhomogenous distribution of the fluorescent lipid probes as superstructures in fluid membranes (Kinnunen et al, 1987; Somerharju et al, 1985). The use of these lipid analogs relies on the photophysics of pyrene. In brief, after excitation a monomeric pyrene molecule may relax back to the ground state by emission of photons with fairly sharp energy maxima. Yet, if the concentration of pyrene is sufficiently high, an excited monomeric pyrene may collide with a ground state pyrene so as to form a short-lived, excited state complex, an excimer. This complex then dissociates back to two ground state pyrenes by emitting a photon with a characteristic, broad maximum centered at approximately 480 nm (Förster, 1969). The ratio of the excimer and monomer emission intensities, I_E/I_M, depends on the rate of collisions between the pyrenes. For membranes incorporating PyrPLs, the collision frequency further depends on the rate of lipid lateral diffusion as well as the local concentration of the probe (Galla and Hartmann, 1980). Notably, the spectroscopic signal as a function of the pyrene lipid concentration, x_{PyrPL}, revealed sharp changes at well-defined critical values of x_{PyrPL}. These critical values of x_{PyrPL} were interpreted to arise from the formation of regular superlattices by the pyrene lipids. The driving force causing this long-range ordering is considered to be a repulsive potential between the probes, further caused by packing strain generated in the immediate vicinity of the bulky pyrene moiety of the probe, which represents a substitutional impurity (Kinnunen et al, 1993a). The observed kinks in $[I_E/I_M](x_{PyrPL})$ could be reproduced with a simple mathematical model based on geometrical considerations of distribution patterns in a hexagonal acyl chain lattice, yielding maximal separation of the pyrenedecanoyl chains at different critical stoichiometries (Kinnunen et al, 1987; Virtanen et al, 1988). Subsequently, these results have been extended and clarified in greater detail (Chong et al, 1994; Sugar et al, 1994; Tang and Chong, 1992). A recent investigation by Chong provides evidence for a regular distribution of the cholesterol analog, ergosterol, in liposomal membranes (Chong, 1994). Similar conclusions have been reached from diphenylhexatriene fluorescence polarization measurements for the distribution of cholesterol in phosphatidylcholine bilayers (Tang et al, 1995). Our own data on the membrane association of tacrine, a drug used in the treatment of the symptoms of Alzheimer's disease, suggest that this molecule forms superstructured alloys with acidic phospholipids (Lehtonen et al, 1996c).

Somewhat related to the above, there is also experimental evidence for the formation of regular superlattices by the peripheral membrane protein, cytochrome c, attached to liposome surfaces (Mustonen et al, 1987). This will be discussed briefly below.

Relationship Between Lipid Organization and the Physical Properties of Membranes

Thermodynamic and Thermomechanic Properties

A number of physical properties are strongly dependent on the lateral organization and the structural fluctuations of the lipid bilayer. This is particularly true of thermodynamic and thermomechanic response functions, such as specific heat, area compressibility, and bending rigidity, that express the system response due to changes in temperature, surface area, and curvature, respectively. At phase transitions, these response functions display anomalies. Observation of anomalies in response functions has often been taken as a sign of the presence of strong fluctuations and structural reorganizations in the membrane conformation.

The specific heat, C_p, which is a measure of the system's thermodynamic response in energy due to temperature changes, can, via the fluctuation-dissipation theorem, be expressed by the fluctuations in internal energy, E, as:

$$C_p = \left(\frac{\partial E}{\partial T}\right)_p = (k_B T)^{-2}(\langle E^2 \rangle - \langle E \rangle^2).$$

(1)

The specific heat is found to display a strong peak at the main phase transition of lipid bilayers with a heat content (transition enthalpy), ΔH, that increases with chain length. Early work by Freire and Biltonen (1978) applied a deconvolution procedure to the measured specific heat, in the spirit of the old Zimm-Bragg theory for helix-coil transitions, to produce a lipid-domain (cluster) distribution function that was consistent with the measured specific heat. Within this approach, the energy fluctuations are assigned to the ensemble of lipid-domain interfaces. The relation between lipid-domain formation, see Figure 3, i.e., density fluctuations, and peaks in the specific heat, has been demonstrated directly by computer-simulation calculations (Mouritsen, 1990; 1991). As discussed earlier the closer the main transition is to a critical point, the stronger the density fluctuations are, and the more pronounced is the dynamic heterogeneity. This is reflected in the specific heat as an enhancement in the intensity of the wings in the specific-heat peak around the phase transition. Similarly, for lipid mixtures the compositional fluctuations near the phase boundaries and in the phase coexistence regions (Jørgensen et al, 1993b; Risbo et al, 1995) manifest themselves as an enhancement of the specific heat.

The isothermal area compressibility modulus, K_T:

$$K_T = -A \left(\frac{\partial A}{\partial \Pi} \right)_T^{-1} = k_B T A (\langle A^2 \rangle - \langle A \rangle^2)^{-1} \qquad (2)$$

is a thermomechanic response function that expresses the bilayer response in surface area, A, to changes in a lateral pressure, Π. K_T, which is related to the lateral compressibility, χ_T, as $K_T A = \chi_T^{-1}$, can be measured by micromechanic studies (Evans and Skalak, 1980) or by osmotic stress experiments (Lis et al, 1982). Both experiments and model calculations show that χ_T displays a peak at the main transition indicating that the bilayer gets very soft and compliant. As indicated by the second equation in Equation (2), which is also a version of the fluctuation-dissipation theorem, the lateral compressibility is a thermomechanic response function that is related to the fluctuations in the bilayer surface area. Hence, peaks in χ_T may be interpreted as indications of strong local area fluctuations.

Finally, we shall relate the isothermal bilayer bending rigidity, κ_T (the curvature elasticity modulus) to membrane fluctuations and lateral organization. κ_T is a thermomechanic measure of the bilayer response in curvature elastic energy to changes in curvature. It recently has been proposed (Hønger et al, 1994; Lemmich et al, 1994) on the basis of a simple theory built on Helfrich's theory for fluid membrane conformations (Helfrich, 1973) that lateral density fluctuations in the two monolayer leaflets of the bilayer lead to a thermal renormalization of the bending rigidity:

$$\kappa_T \rightarrow \kappa_T \left(1 - \frac{r^2}{K_T} \right), \qquad (3)$$

where r is a materials constant that determines the strength of the coupling between the density fluctuations and the local bilayer mean curvature. Hence, in the transition region, where the density fluctuations lead to a decrease of K_T, the bending rigidity is diminished. Subsequently it was shown experimentally by flickering-noise analysis that κ_T for a series of phospholipid bilayers indeed decreases dramatically as the main phase transition is approached (Fernandez-Puente et al, 1994). This indicates a progressive softening of the bilayer in the transition region. A direct interpretation of these results is that the lipid organization with regard to dynamic heterogeneity and domain formation provides a mechanism for dynamic formation of spontaneous curvature, which in turn makes the bilayer less rigid over scales larger than the typical domain size. Similarly, it has been proposed for lipid mixtures (Leibler, 1986; Lemmich et al, 1994; Seifert and Langer, 1993) that a coupling exists between mean bilayer curvature and compositional fluctuations that also leads to a renormalization of the bending rigidity and a bilayer softening. By this mechanism, curvature may induce local phase separation, vis a vis local phase separation or formation of lipid domains with different composition. This induces local bilayer curvature, which, for example, can lead to membrane budding (Lipowsky, 1993; Seifert, 1993).

Passive Bilayer Permeability

In the classic work by Papahadjopoulos et al (1973) it was proposed that the observation of an anomalously high passive permeability of Na^+ ions through lipid vesicles in their transition region might be due to formation of regions with bad packing properties that make the bilayer leaky. Several workers have since used variations of this picture to interpret a variety of anomalous passive permeation phenomena in the phase transition region (for a recent review and a list of references, see Mouritsen et al, 1995). Specifically, it has been hypothezised that the interfaces that bound the lipid domains in the dynamically heterogeneous organization of the bilayer in the transition region, see Figure 3a–c, provide a vehicle for passive nonvectorial transport of ions and molecular species across the bilayer (Cruzeiro-Hansson and Mouritsen, 1988). A test of this hypothesis is provided in Figure 4 which shows, in the case of Na^+ permeability, a comparison between experimental data and results obtained from a simple model that assumes that the thermal dependence of the permeability is predominantly controlled by the thermal variation of the degree of dynamic heterogeneity through the transition region as determined from model calculations (Cruzeiro-Hansson and Mouritsen, 1988).

In consequences of this picture, the stronger the density fluctuations are the more permeable is the bilayer. This is indeed found to be the case for cation permeation through di-acyl phosphatidylcholine bilayer of different acyl chain length, where the short-chain lipid bilayers are the more permeable (Ipsen et al, 1990). Furthermore, it can be expected that the larger the molecule is that has to permeate the bilayer, the stronger the fluctuations are that are required to facilitate the transport. This seems also to be borne out by experiments (Cruzeiro-Hansson and Mouritsen, 1988).

Turning then to lipid mixtures, the same picture holds as found for the thermodynamic and thermomechanic response functions discussed above. Specifically, the lateral density fluctuations and the domain organization prevailing in the gel-fluid coexistence region and near the phase boundaries for binary mixtures lead to an enhancement of the passive permeability, as recently has been shown in the case of permeation of glucose and small zwitterionic molecules through lipid vesicles made of DMPC-DPPC mixtures (Clerc and Thompson, 1995).

Lateral Diffusion

The lateral diffusion of lipids (and integral membrane proteins) is strongly dependent not only on the thermodynamic phase of the lipid bilayer but also on

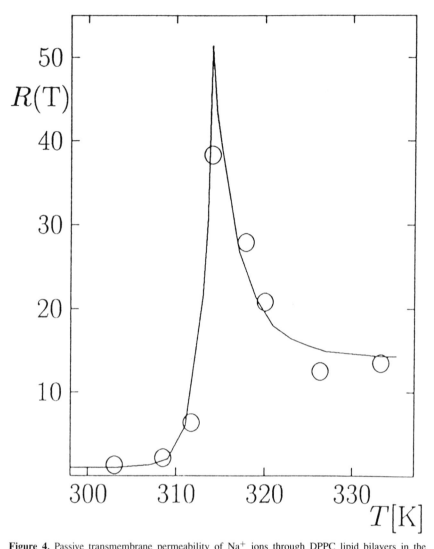

Figure 4. Passive transmembrane permeability of Na^+ ions through DPPC lipid bilayers in the transition region. The solid curve represents results from a theoretical calculation (Cruzeiro-Hansson and Mouritsen, 1988), and the circles represent experimental measurements (Papahadjopoulos et al, 1973).

the bilayer fluctuations and the structural organization of the bilayer. The lateral diffusion constant for lipids increases two to three orders of magnitude upon entry into the fluid phase (Clegg and Vaz, 1985; Lindblom and Orädd, 1994). However, there is some indirect evidence from early NMR relaxation measurements that exactly at the phase transition the diffusion gets anomalously slow (Lindblom and Persson, 1976). This is possibly related to the general phenomenon of critical slowing down and may be caused by the formation of a dynamically heterogenous bilayer. In this heterogeneous structure, the individual diffusing lipid molecule sees the bilayer as an archipelago of obstacles. In general, the obstructed diffusion in a disordered and nonrandom environment leads to anomalous diffusion characteristics, corresponding to time-dependent diffusion constants (Saxton, 1994). Measurements of time-dependent diffusion constants and anomalous diffusion exponents may in some cases be used to gauge the structural properties of the membrane. An interesting corollary to this states that lipid-domain patterns, in particular in relation to percolation phenomena, influence the rate of in-plane membrane reactions that are limited by diffusion, e.g., homo- and heterodimerization processes and enzyme-catalyzed reactions (Thompson et al, 1995).

Membrane Undulations, Swelling, and Intermembrane Forces

A large part of the experimental results for the physical properties of lipid bilayers derive from multilamellar samples. The reason for this is that it is easy to prepare multilamellar vesicles and to obtain the large lipid-to-water ratios necessary for obtaining high intensity in calorimetric, spectroscopic, and diffractometric studies. Some of the more elegant studies of unilamellar liposomes use micromechanic pipette aspiration techniques (Evans and Kwok, 1982). Multilamellar vesicles are usually not as well defined as unilamellar vesicles. Still they are convenient for determining a number of bulk properties of lipid bilayers. Furthermore, they lend themselves readily to investigations of intermembrane interactions and in particular to studies of the relationship between lateral bilayer organization and two-dimensional single-bilayer physical properties on the one side and three-dimensional intermembrane forces and swelling properties on the other side.

In the general case, the interbilayer interactions include a large number of forces of different origin, such as van der Waals forces, electrostatic double-layer forces, hydration forces, surface dipole-dipole forces, and steric forces (Israelachvili, 1992). These forces operate on different length scales, and they are engaged in a subtle and competitive interplay. Whereas the individual bilayer in its tension-free state undergoes large undulations, these undulations are suppressed in a multilamellar array. The undulations provide an entropic mechanism for a repulsive interbilayer force (Helfrich, 1973):

$$F \sim \frac{(k_B T)^2}{\kappa_T d^3},\qquad(4)$$

where κ_T is the bilayer bending rigidity and d is the repeat distance. As discussed above, due to the thermal density fluctuations, κ_T gets renormalized at the main transition, see Equation (3), corresponding to a softening of the bilayer. According to Equation (4), this implies an increased repulsive undulation force. This in turn leads to a larger repeat distance, and consequently an anomalous

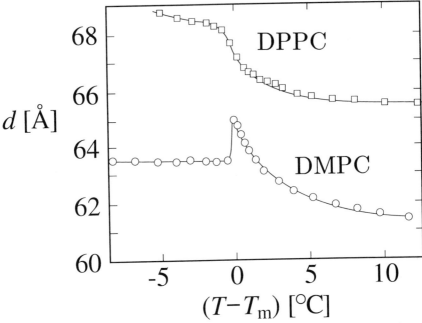

Figure 5. Lamellar repeat distance, d, as a function of temperature in the transition region of multilamellar DMPC and DPPC lipid bilayers in excess water as obtained from small-angle neutron scattering experiments (Lemmich et al, 1995). T_{m} is the appropriate transition temperature.

swelling behavior will develop at the transition (Hønger et al, 1994, Lemmich et al, 1995) as shown in Figure 5 for DMPC and DPPC multilamellar bilayers.

Similarly, for binary multilamellar lipid bilayers the compositional fluctuations and the phase-separation phenomena discussed earlier can lead to a softening of the bilayer rigidity (Leibler, 1986) which also provides a mechanism, via lipid-domain orginazation, for stronger repulsive undulation forces and anomalous swelling behavior (Lemmich et al, 1994).

Membrane Proteins

Two major classes of membrane proteins are generally distinguished. Integral proteins contain one or more hydrophobic segment of sufficient length to span the lipid bilayer. These proteins generally function as receptors, enzymes, and ion channels. Notable examples of the first category are the growth factors receptors, GFRs. Ligand binding is generally thought to result in receptor dimerization, with concomitant activation of their protein tyrosine kinase activity (Heldin, 1995). Subsequently, a complex array of different enzymes, including PI-specific phospholipase C (PLC) and PI-kinase become associated with GFRs via specific sequences in order to elicit a very complicated signal cascade (Fry et al, 1993) which finally culminates in specific changes in the transcriptional activities in nucleus. The molecular details of this chain of events lie at the very heart of understanding the derangement of growth regulation in cancer cells. The second group, peripheral membrane proteins, is equally diverse and important. Prominent examples are, for instance, some of the enzymes of the blood coagulation cascade and proteins such as phospholipases C, D, and A_2, PI-kinase and protein kinase C, all known to be centrally involved in different cellular control mechanisms. Finally, there are also proteins attached to membranes due to covalently linked lipids such as PI and fatty acids (Low, 1989; Schmidt, 1989).

Integral Membrane Proteins

As already stated above, a key feature of the integral membrane proteins is the anchoring to lipid bilayers due to hydrophobic segments traversing through the membrane. These stretches are often α-helical, but also β-sheet structures are known (Sipos and von Heijne, 1993). The number of transmembrane segments (or transmembrane domains, TMDs) in a protein varies considerably. A distinct subclass is composed of the G-protein coupled receptors with seven TMDs (Neubig, 1994). For a long time the TMDs were considered to provide only a hydrophobic membrane anchor. However, this view is readily challenged by the fact that for nearly all integral membrane proteins studied, the evolutionarily most conserved amino acid sequences are found in their TMDs (Kinnunen, 1991). Yet, comparison of different proteins reveals very little sequence homology in their TMDs. Accordingly, since hydrophobic anchorage would not require sequence conservation, it seems reasonable to assume that the structure of the TMDs must be centrally important to the functions of integral proteins. For example, in spite of apparently normal folding of the extramembrane domains of the protein, inversion of the entire TMD of the insulin receptor blocks the insulin stimulated signal transmission (Yamada et al, 1995). Conserved sequences of the TMDs could be important: (1) to the overall conformation of the protein; (2) to intramolecular as well as intermolecular lateral protein-protein interactions; and (3) to lipid-protein interactions. The latter point will be discussed below.

In addition to hydrophobic residues, several TMDs have also basic residues in the interface. Furthermore, the prevalence of Asn and Gln is high (Landolt-Marticorena et al, 1993). The basic residues have been proposed as responsible for attracting acidic phospholipids to the lipid boundary, whereas Asn and Gln could function as sites for liganding protonated acidic phospholipids (Rytömaa and Kinnunen, 1995). Here we shall mainly focus on physical mechanisms influencing the selection of lipids in the immediate vicinity of an integral membrane protein. It should be mentioned that there are at present a number of studies in progress in different laboratories using well-defined synthetic peptides to investigate lipid-protein interactions. We may expect such efforts to contribute to our understanding of the roles of, for example, electrostatic interactions (Horváth et al, 1995), protein insertion and translocation through bilayers

(Moll and Thompson, 1994), as well as the effects of membrane environment on the protein conformation (Li and Deber, 1993; Zhang et al, 1995).

Peripheral Membrane Proteins

Peripheral membrane proteins are attached to the membranes either by binding to integral membrane proteins or directly to lipids. We will limit this discussion to the latter. As was already mentioned above, peripheral proteins are a very heterogenous group of proteins, and it appears that nearly all of the different metabolic pathways or processes in cells seem to involve peripheral proteins. Likewise, their mechanisms of lipid binding are also diverse. On one hand there are proteins such as cytochrome c (Dickerson et al, 1971) and *Escherichia coli* lactate dehydrogenase (Ho et al, 1989), which do not have any clear hydrophobic or amphipathic structures, yet they associate strongly with appropriate phospholipid structures. On the other hand, some of the apolipoproteins of plasma lipoproteins as well as peptide hormones are well characterized in terms of their interactions with phospholipids by amphipathic α-helical segments (Kaiser and Kézdy 1984). Also in this field major new developments utilizing synthetic peptides are underway (Simões et al, 1995; White and Wimley, 1994). For a more detailed discussion of peripheral proteins, the reader is referred to a recent review (Kinnunen et al, 1994).

Relationship Between Lipid Organization and Membrane Functions

Binding of Proteins to Membrane Surfaces and Activation via Lipid Domain Formation

It should be emphasized that the relevance of the understanding of peripheral interactions of proteins with lipid surfaces is not limited to the soluble proteins. The integral membrane protein insulin receptor reconstituted in liposomes was demonstrated to interact with approximately 2000 phospholipid molecules (Sui et al, 1988). This was explained as due to the extramembraneous parts of the receptor contacting the lipid surface. Most intriguing was the finding that upon insulin binding to the receptor these secondary, peripheral contacts between the soluble parts and the lipid surface are drastically reduced. To this end, a mechanism by which peripheral interactions of integral membrane protein receptors could be controlled by their ligands would offer very powerful means of regulating the conformation and thus also the activity of the said receptors. Finally, the strong association of DNA and RNA to membranes containing the natural cationic amphiphile, sphingosine (Kinnunen et al, 1993b), and the high affinity of the basic chromosomal proteins, histones, to acidic phospholipids (Kõiv et al, 1995) raises the possibility that lipids could directly influence chromatin structure and thus also gene expression and mitosis. Understanding of the principles governing the peripheral membrane association of biological macromolecules clearly warrants major research efforts.

In the following we shall briefly summarize lessons learned in the course of studies conducted in the laboratory of the authors of this review, exploiting phospholipase A_2 (PLA_2) and cytochrome c as models for peripheral lipid-protein interactions. Particular emphasis has been put on the emergence of an understanding of how the chemical (lipid) composition of the membrane influences its physical properties and how this is further sensed by these proteins.

PLA_2 activity is ubiquitously found everywhere in cells. Although there are common determinants in their mechanisms of action, it is important to emphasize that this is a heterogenous group of enzymes. Accordingly, results obtained with the most commonly used, commercially available model enzymes, such as snake venom and porcine pancreatic PLA_2, may not be directly applicable to the intracellular enzymes. However, general features of lipid-protein interactions are readily amenable to experimental work. Notably, PLA_2 has been shown to be sensitive to several features characterizing the physico-chemical state of the substrate. Thus, changes in the substrate phospholipid conformation upon exceeding the critical micellar concentration appear to be the mechanism for producing the concomitant pronounced enhancement in the catalytic activity (Thurén et al, 1984). Very large changes in activity occur which are dependent on microstructure and the phase state of the phospholipid (Burack and Biltonen, 1994). An attempt has been made to correlate the dynamic microstructural bilayer heterogeneity derived from computer-simulation calculations to the activity of PLA_2 (Mouritsen and Biltonen, 1993; Mouritsen and Jørgensen, 1992). Figure 6 shows the temperature dependence of the average lipid domain size, $\langle \ell \rangle$, as a measure of the dynamic heterogeneity, see Figure 3. This property exhibits a sharp peak at the transition in a way which is strikingly similar to the strong temperature dependence of the rate of activation, τ^{-1}, as measured experimentally for PLA_2 (Menashe et al, 1986). The ability of certain drugs, like halotane (Mountcastle et al, 1978) and dibucaine (van Osdol et al, 1992), to couple to the bilayer fluctuations and to change the dynamic heterogeneity (Jørgensen et al, 1993a) is likely to be related to their potency for changing the action of the enzyme (Mouritsen and Biltonen, 1993).

One of the physical parameters characterizing lipid bilayers is their equilibrium lateral pressure, Π. Studies utilizing phospholipid monolayers on an air/water interface at specified lateral pressures established a long time ago that the expression of the catalytic activity of PLA_2s exhibits marked sensitivity to the lipid lateral packing and that enzymes from different sources differ in their optimum surface pressure (Demel et al, 1975; Pattus et al, 1979). Similar sensitivity to lipid packing has been described for phospholipase C (Boguslavsky et al, 1994) and protein kinase C (Souvignet et al, 1991). The underlying mechanisms are unclear, and the apparent surface pressure optimum may partly arise from denaturation of the enzyme at lower surface pressures, inability to bind to lipids at high packing densities, or true lipid packing density-dependent changes in protein conformation. Taking into account the pronounced effects of lipid lateral

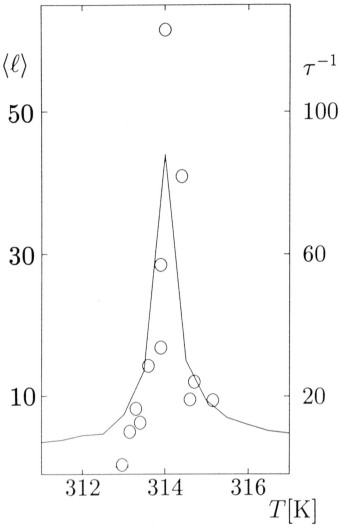

Figure 6. Relationship between degree, $\langle \ell \rangle$, of dynamic heterogeneity of DPPC lipid bilayers and the rate of activation, τ^{-1}, of PLA$_2$ in DPPC vesicles (Menashe et al, 1986). The degree of heterogeneity is determined theoretically by the average lipid domain size, $\langle \ell \rangle$, as obtained from computer-simulation calculations (Mouritsen and Jørgensen, 1992).

pressure on these proteins, it is somewhat unfortunate that values for Π for different cellular membranes are not known. Likewise, mechanisms that could alter Π in cells have not been described. Taking advantage of the sensitivity of PLA$_2$s to Π we have conducted studies using large unilamellar liposomes (LUVs) encapsulating varying concentrations of the hygroscopic polymer, poly(ethylene glycol), PEG, as substrates (Lehtonen and Kinnunen, 1995b). These LUVs then have been subjected to hypotonic conditions in order to produce osmotic water influx and swelling of the liposomes (Lehtonen and Kinnunen, 1994). For *Crodalus adamanteus* PLA_2 with a surface pressure optimum at $12 \ mN \cdot m^{-1}$ a considerable enhancement of enzyme activity can be observed upon osmotic stretching of the bilayer substrate. The magnitude of the osmotic pressure gradient across the membrane producing maximal activation of the enzyme is also dependent on the composition of the phospholipid acyl chains. For saturated lipids higher pressure gradients are required, in keeping with more efficient van der Waals interactions between the chains. These data suggest that the equilibrium lateral pressure of cellular membranes can be altered by osmotic stretching and that membranes can act as mechanosensors. Altered lipid packing also can be expected upon stretching of the plasma membrane due to mechanical forces exerted by the reorganization of the cytoskeleton, for instance. The sensitivity of membrane proteins, e.g., PLA_2s, to lipid packing immediately allows for the generation of signalling molecules such as eicosanoids derived from their fatty acid precursor, arachidonic acid, liberated by PLA_2. Mechanosensitivity of both animal and plant cells is well established and has so far been shown to be mediated by stress-sensitive ion channels and cytoskeleton (Sachs, 1990; Wang et al, 1993).

Regarding the state of most biological membranes, it is important to keep in mind that they are generally subjected to very high electric fields. This is due to the active, energydriven maintenance of concentration gradients of ions across the bilayer. In most cells, membrane potentials of the order of approximately 100 mV are commonly measured. Accordingly, it is interesting to study how electric fields imposed across the substrate influence the action of PLA_2 (Thurén et al, 1987a). This question was assessed using substrate monolayers on an aqueous subphase while residing underneath an alkylated Si-wafer. Thereafter, an electric potential difference was applied between the wafer and the counterelectrode in the subphase, while the action of the enzyme was monitored by measuring the loss of lipid from the monolayer using the so-called zero-order trough. To summarize, it can be shown that electric fields of the order of 100 mV across the cell-membrane-mimicking bilayer can trigger the activity of PLA_2. This simple model clearly demonstrates that signal transmission within cell-to-membrane proteins does not necessarily involve physical complex formation or contact. Accordingly, alterations in membrane potential can have an impact on the functioning of a large array of enzymes simultaneously. Obviously, this is also the case when considering the consequences of osmotically induced changes in the lateral packing of membrane lipids and proteins.

Several commonly used pharmaceutically active compounds bind to lipid surfaces, and their influence on membrane-associated enzymes has been extensively studied. Pancreatic PLA_2 exhibits significantly higher reaction rates with acidic phospholipid substrates (Thurén et al, 1987b). Therefore, it is not unexpected that the expression of its catalytic activity is also sensitive to modulation of the the surface charge by cationic membrane associating compounds, such as polyamines (Thurén et al, 1986), sphingosine (Mustonen et al, 1993), and adriamycin (Mustonen and Kinnunen, 1991). Neutral amphiphiles, platelet activating factor (Thurén et al, 1990) and phorbol esters (Mustonen and Kinnunen, 1992) also have pronounced effects on PLA_2 activity. It has been suggested that in addition to surface charge, surface potential might be an important determinant for the action of this enzyme. Similar conclusions have been made regarding the mechanism of action of protein kinase C (Trudell et al, 1991).

Cytochrome c is generally presented as a paradigm for proteins associating electrostatically with membranes containing acidic phospholipids. Yet, more subtle features of lipid-protein interaction are involved, and evidence for two distinct acidic phospholipid binding sites in cytochrome c recently has come forward (Rytömaa et al, 1992). These sites are nominated as A and C, with the following characteristics. The A-site ligands to deprotonated acidic phospholipids, and this interaction is readily reversed by competing ligands such as, for instance, ATP. In contrast, the C-site interacts with protonated acidic phospholipids, and this interaction is not sensitive to ionic strenght or to nucleotides (Rytömaa and Kinnunen, 1994). Protonation of acidic phospholipids appears to be critically involved also in the binding of the plasma low-density lipoprotein, LDL, to liposomes (Laureaeus et al, 1996).

Cytochrome c bound to liposomes either via the A-site or the C-site is not detached by an excess of subsequently added acidic phospholipid containing liposomes (Rytömaa and Kinnunen, 1995). This apparent irreversibility requires the normal diacylglycerophospholipid structure. In keeping with this, as well as due to experimental evidence from previous studies, a novel mechanism of liganding peripheral proteins to phospholipid surfaces has been proposed. The critical constituent in this mechanism is the intercalation of one of the phospholipid acyl chains into a hydrophobic cavity of the protein (Dickerson et al, 1971), while residing on the lipid surface (Kinnunen et al, 1994). Accordingly, energetics of the peripheral membrane association of the protein now involves a contribution from hydrophobic association, yet without penetration of the protein into the bilayer, Figure 7. This mechanism should be greatly enhanced upon the presence of nonlamellar phase favoring lipids. Comparison of the findings regarding the lipid association of a large number of peripheral proteins suggests that this anchorage via lipids in the extended lipid conformation could be rather general and may apply, for instance, to protein kinase C, CTP:phosphocholine cytidyl transferase, annexin V (Kinnunen et al, 1994), as well as histone H1 (Kõiv et al, 1995).

Membrane binding of cytochrome c appears to require a critical content of acidic phospholipids. Thus, above 10 mol% of phosphatidic acid its association to liposomes is greatly augmented (Mustonen et al, 1987). This property can be used to obtain control of the membrane association of cytochrome c by phase separation. In brief, the content of the acidic lipid with usaturated acyl chains is chosen to be too small in order to constitute protein binding sites. However, if the rest of the membrane is composed of a saturated phosphatidylcholine, lowering the temperature below the chain-melting transition point induces crystallization of this lipid, with concomitant enrichment of the unsaturated acidic phospho-

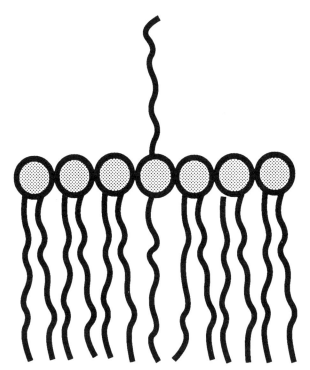

Figure 7. Extreme localization in the aqueous phase of the sn-2 acyl chain of a membrane lipid in the extended conformation. The protruding chain can be accommodated within a hydrophobic cavity of a surface-anchored protein or in an opposing, adhered membrane.

lipid into domains. These domains appear to have high enough negative surface charge density to support the binding of cytochrome c (Mustonen et al, 1987). Interestingly, in an elegant series of experiments Yang and Glaser (1995) recently demonstrated how the lateral co-localization of the membrane-associated substrate with protein kinase C can control the enzyme action.

Lipid-Protein Alloying and Superstructures

The formation of lipid superlattices in fluid membranes was discussed above. Measurements of the quenching of pyrene fluorescence due to resonance energy transfer to membrane associated cytochrome c revealed that $\tau([\text{cyt c}])$, i.e., the lifetime of pyrene emission as a function of [cytochrome c], exhibited stepwise decrements at well-defined cytochrome c/phospholipid stoichiometries (Mustonen et al, 1987). These data were obtained with fluid, liquid-crystalline liposomes and were rationalized as due to the formation of hexagonal superlattices of cytochrome c on liposome surfaces. The driving force for this process is likely to be a repulsive potential between the membrane-associated proteins, presumably due to deformation of the bilayer (Israelachvili, 1977).

Effects on Function Due to Nonbilayer-Forming Lipids

As was already stated earlier, the H_{II} phase is not believed to be of biological relevance in itself, but the tendency to form it, H_{II} propensity, Epand (1991) appears to induce important changes in the properties of such bilayers: (1) H_{II} propensity has been demonstrated to enhance the fusion of bilayers (Ellens et al, 1986); although there are diverging views on this, there is experimental evidence excluding the actual formation of nonlamellar structures during membrane fusion (Allen et al, 1990; Caffrey, 1985). (2) Several enzymes classified as peripheral membrane proteins, such as protein kinase C (Epand, 1991; Newton, 1993; Slater et al, 1994) and CTP:phosphocholine cytidyltransferase (Jamil et al, 1993) have been shown to be activated due to H_{II} propensity. (3) Some integral membrane proteins require H_{II} phase forming lipids for maximal activity; interesting examples are rhodopsin (Brown, 1994), alamethicin (Keller et al, 1993), and insulin receptor tyrosine kinase (McCallum and Epand, 1995). (4) The perhaps most interesting property of H_{II} propensity connects to the growth of microorganisms (Goldfine, 1984; Rietveld et al, 1994; Wieslander et al, 1986) for which optimal growth conditions are provided at temperatures approximately 10–20°C below the lamellar $\rightarrow$ nonlamellar transition of their membranes.

Although no conclusive evidence is available, the possibility has been advanced that of the above listed points, (1), (2), and (4) would all reflect the same property of membranes derived from H_{II} phase propensity (Kinnunen, 1996). In brief, crowding in the hydrocarbon region of H_{II} prone lamellar membranes would, under proper conditions, result in part of the membrane adopting the so-called extended conformation, Figure 7. In this conformation, at the extreme, one of the lipid acyl-chains extends out from the surface, while localization in the interface, parallel to the surface, is also possible. Conditions favoring this conformation would be the adherence of two poorly hydrated, opposing bilayers, as occurs preceeding the actual membrane fusion process (Kinnunen, 1992), as well as the presence of proteins with a hydrophobic, acyl-chain accommodating cavity. These kind of proteins would be firmly anchored to proper lipid surfaces, as has been suggested for cytochrome c and annexin V, for instance (Kinnunen et al, 1994; Rytömaa and Kinnunen, 1995). Finally, it has been proposed, that H_{II} propensity could represent an ancient, common growth signal both in eucaryotes as well as in bacteria (Kinnunen, 1996). The mechanism involved would be membrane association and subsequent activation of sets of proteins required for cellular growth.

Lipid-Bilayer Coupling to Integral Membrane Proteins
via Hydrophobic Matching

Based on the amphiphilic structure of both lipid bilayer membranes and integral membrane proteins, it has been suggested that an essential part of the energetics involved in protein-lipid interactions is related to the degree of mismatch between the hydrophobic bilayer thickness on the one side and the length of the hydrophobic membrane-spanning domains of the integral protein on the other side (Mouritsen and Bloom, 1993; 1984; Mouritsen and Sperotto, 1993). The matching principle points two ways: (1) The embedded protein with its hydrophobic length enforces a boundary condition on the lipids, which in turn have to adopt their conformational states as well as their local composition to this boundary. Consequences of this phenomena may be global thermodynamic phase-separation and micro-phase separation (Sperotto and Mouritsen, 1991a), long-range order parameter profiles (Sperotto and Mouritsen, 1991b), as well as lipid selectivity and specificity near the protein surface (Sperotto and Mouritsen, 1993). The protein simply selects, based on physical forces, the lipid species that adapts best to the hydrophobic interface. The range of perturbation due to these forces can be rather long leading to effective lipid-mediated protein-protein attractive forces that may lead to aggregation or crystallization of the proteins in the plane of the membrane. (2) The tension put on the protein due the the requirements of matching to the lipid bilayer may in some cases lead to conformational changes in the protein and hence provide a link to protein function. Below, specific examples that fit into this general picture are provided.

It is well known that a large number of membrane enzymes and channels display optimum activity at a certain bilayer thickness, which supposedly is the thickness providing the better hydrophobic match. Examples include cytochrome c oxidase (Montecucco et al, 1982), (Na^+-K^+)-ATPase (Johannsson et al, 1981a), and Ca^{2+}-ATPase (Caffrey and Feigenson, 1981; Cornea and Thomas, 1994; Johannsson et al, 1981b). Furthermore, it has been found that the lipid-protein interactions in the case of photosynthetic reaction center and antenna proteins (Pesche et al, 1987; Sperotto and Mouritsen, 1988) as well as bacteriorhodopsin (Piknova et al, 1993) are strongly dependent on the thickness of the lipid bilayer into which the proteins are reconstituted. The presence of the *E. coli* integral membrane protein lactose permease reconstituted in liposomes causes the lateral enrichment of pyrene labeled phospholipids in a manner that readily complies with the hydrophobic matching condition (Lehtonen et al, 1996b).

Possibly the most striking and clear-cut examples of the role played by lipids on function via the hydrophobic matching condition are probably that of gramicidin A (Lundbæk and Andersen, 1994) and rhodopsin (Brown, 1994). In the case of gramicidin A it has been found that the lifetime of the channel can be decreased by increasing the hydrophobic mismatch, and the lifetime can be prolonged by introducing nonbilayer-forming lipids, like lysolipids, that can mediate the mismatch by lowering the cost in elastic energy. Similarly, the transition between the metarhodopsin I and metarhodopsin II conformational states in rhodopsin, that is relevant for triggering the visual process, can be shifted by lipids with H_{II} propensity which presumably mediate the change in mismatch for the meta I and meta II conformations.

The potential importance of the hydrophobic matching condition as a vehicle for modulating protein function becomes obvious when it is considered that

the membrane hydrophobic thickness can be changed by polypeptides, cholesterol, alcohols, hydrocarbons, as well as a great variety of drugs (Mouritsen and
Bloom, 1993). Some evidence has been presented that in the case of the acetylcholine receptor both local and general anaesthetics may couple to the receptor
via alterations in the lipid-protein interface (Arias et al, 1990; Fraser et al, 1990;
Horváth et al, 1990).

Biomembranes Are (1) Cooperative, (2) Dynamic, and (3) Adaptive Supramolecular Assemblies

The properties of lipids and lipid-protein assemblies discussed above have important consequences for the conceptual framework now available for describing
the structure and function of biomembranes. It should be understood that there
are, in addition to immobilization effects due to, for instance, cytoskeleton and
extracellular matrix-membrane interactions, several physical principles generating lateral heterogeneity of lipids and proteins in fluid, liquid-crystalline membrane domains. Biomembrane functions do not depend on random collisions
and interactions of the reactants but are steered in a well-defined manner, while
still allowing for a considerable degree of freedom for the movements of the
individual constituents. In a way this is analogous to the folding of proteins into
an average three-dimensional conformation, fluctuating around the free energy
minimum. An important difference is that the membrane is a vastly complex
many-body system, assembled by noncovalent interactions, hydrophobicity, as
well as van der Waals and Coulombic forces. A key issue in understanding the
features of biomembranes is to keep in mind that they are structured fluid systems, which readily accounts for their cooperative responses and order. What is
summarized above in this review is what we know at present about the molecular mechanisms generating this cooperativity and order in fluid membranes. Due
to the dynamic nature of these systems the ordering is very sensitive to perturbation by both physical (e.g., temperature and pressure) as well as chemical
factors (e.g., drugs and metabolites), Figure 8. We can readily assume biomembranes to be organized in a physiologically meaningful manner. Subsequently,
we can anticipate changes in the organization, brought about by physiological
perturbations, in turn resulting in relevant changes in the functions of the entire biological system, i.e., the living cell (Kinnunen, 1991). The cooperative
nature of the changes in membrane organization would further offer excellent
means for large-scale integration of different metabolic processes (Kinnunen,
1991; Kinnunen et al, 1994). Finally, although seemingly minor changes in
the structure of the constituents, such as a single mutation of a proper amino
acid residue, may result in malignant transformation, for instance, we may also
conceptualize a cell as a self-organizing system, whose physical states and physiological states are strongly correlated. Within this refined model picture of the
biological membrane, transitions of a cell between different metabolic states

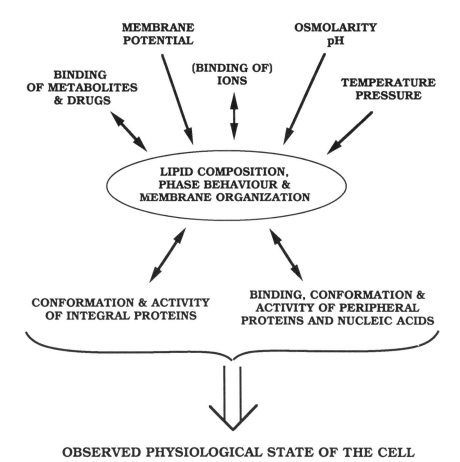

Figure 8. Coupling of environment, organization, and functions in a biomembrane assembly. See text for details.

would involve and in some cases also be determined by the physical state of the membrane (Kinnunen, 1996).

ACKNOWLEDGMENTS

The authors wish to acknowledge Drs. Jukka Lehtonen and Marjatta Rytömaa for numerous discussions on the topics covered in this review. Work in the authors' laboratories is supported by The Danish Natural Science Research Council, The Danish Technical Research Council (O.G.M.), and the Finnish State Medical Research Council, Sigrid Juselius Foundation, as well as Biocenter Helsinki (P.K.J.K.). O.G.M. is a fellow of the Canadian Institute for Advanced Research. Dr. Kent Jørgensen is thanked for providing Figure 3.

REFERENCES

Allen TM, Hong K, Papahadjopoulos D (1990): Membrane contact, fusion, and hexagonal (H_{II}) transitions in phosphatidylethanolamine liposomes. *Biochemistry* 29:2976–2985

Arias HR, Sankaram MB, Marsh D, Barrantes FJ (1990): Effect of local anaesthetics on steroid-nicotinic acetylcholine interactions in native membranes of *Torpedo marmotata* electric organ. *Biochim Biophys Acta* 1027:287–294

Berclaz T, McConnell HM (1981): Phase equilibria in binary mixtures of dimyrisoylphosphatidylcholine and cardiolipin. *Biochemistry* 20:6635–6640

Berclaz T, Geoffrey M (1984): Spin-labeling study of phosphatidylcholine-cardiolipin binary mixtures. *Biochemistry* 23:4033–4037

Biltonen RL (1990): A statistical-thermodynamic view of cooperative structural changes in phospholipid bilayer membranes: Their potential role in biological function. *J Chem Thermodyn* 22:1–19

Bloom M, Evans E, Mouritsen OG (1991): Physical properties of the fluid-bilayer component of cell membranes: A perspective. *Q Rev Biophys* 24:293–397

Boguslavsky V, Rebecchi M, Morris AJ, Jhon D-Y, Rhee SG, McLaughlin S (1994): Effect of monolayer surface pressure on the activities of phosphoinositide-specific phospholipase C-β1, -γ1, and -δ1. *Biochemistry* 33:3032–3037

Brown MF (1994): Modulation of rhodopsin function by properties of the membrane bilayer. *Chem Phys Lipids* 73:159–180

Burack WR, Biltonen RL (1994): Lipid bilayer heterogeneities and modulation of phospholipase A2 activity. *Chem Phys Lipids* 73: 209–222

Caffrey M (1985): Kinetics and mechanism of the lamellar gel/lamellar liquid-crystal and lamellar/inverted hexagonal phase transition in phosphatidylethanol: A real-time X-ray diffraction study using synchrotron radiation. *Biochemistry* 24:4826–4844

Caffrey M, Feigenson GW (1981): Fluorescence quenching in model membranes. 3. Relationship between calcium adenosinetriphosphate enzyme activity and the affinity of the protein for phosphatidylcholines with different acyl chain characteristics. *Biochemistry* 20:1949–1961

Cevc G, March D (1987): *Phospholipid Bilayers. Physical Principles and Models.* New York: Wiley-Interscience

Chong PLG (1994): Evidence for regular distribution of sterols in liquid crystalline phosphatidylcholine bilayers. *Proc Natl Acad Sci USA* 91:10069–10073

Chong PLG, Tang D, Sugar IP (1994): Exploration of physical principles underlying lipid regular distribution: effects of pressure, temperature and radius of curvature on E/M dips in pyrene-labeled PC/DMPC binary mixtures. *Biophys J* 66:2029–2038

Clegg RM, Vaz WLC (1985): Translational diffusion of proteins and lipids in artificial lipid bilayer membranes. A comparison of experiments with theory. In: *Progress in Protein-Lipid Interactions*, Watts A, ed. Amsterdam: Elsevier

Clerc SG, Thompson TE (1995): Permeability of dimyristoyl phosphatidylcholine/dipalmitoyl phosphatidylcholine bilayer membranes with coexisting gel and liquid-crystalline phases. *Biophys J* 68:2333–2341

Cornea RL, Thomas DD (1994): Effects of membrane thickness on the molecular dynamics and enzymatic activity of reconstituted Ca-ATPase. *Biochemistry* 33:2912–2920

Cruzeiro-Hansson L, Mouritsen OG (1988): Passive ion permeability of lipid membranes modelled via lipid-domain interfacial area. *Biochim Biophys Acta* 944:63–72

Demel RA, Geurds von Kessel WSM, Zwaal RFA, Roelofsen B, van Deenen LLM (1975): Relation between various actions on human red cell membranes and the interfacial phospholipid pressure in monolayers. *Biochim Biophys Acta* 406:97–107

Dickerson RE, Takano T, Eisenberg D, Kallai OB, Samson L, Cooper A, Margoliash E (1971): Ferricytochrome c. I. General features of the horse and bonito proteins at 2.8 resolution. *J Biol Chem* 246:1511–1535

Ellens H, Bentz J, Szoka FC (1986): Fusion of phosphatidyl-ethanolamine-containing liposomes and mechanism of the $L_\alpha - H_{II}$ phase transition. *Biochemistry* 25:4141–4147

Epand RM (1991): Relationship of phospholipid hexagonal phases to biological phenomena. *Chem-Biol Interactions* 63:239–247

Evans E, Kwok R (1982): Mechanical calorimetry of large dimyristoylphosphatidylcholine vesicles in the phase transition region. *Biochemistry* 21:4874–4879

Evans E, Skalak R (1980): Mechanics and thermodynamics of membranes. *CRC Crit Rev Bioeng* 3:181–418

Fernandez-Puente L, Bivas I, Mitov MD, Méléard P (1994): Temperature and chain length effects on bending elasticity of phosphatidylcholine bilayers. *Europhys Lett* 28:181–186

Fraser DM, Louro SRW, Horváth LI, Miller KW, Watts A (1990): A study of the effect of general anesthetics on lipid-protein interactions in acetylcholine receptor enriched membranes form *Torpedo nobiliana* using nitroxide spin-labels. *Biochemistry* 29:2664–2669

Freire E, Biltonen RL (1978): Estimation of molecular averages and equilibrium fluctuations in lipid bilayer systems from excess heat capacity function. *Biochim Biophys Acta* 514:54–68

Fry MJ, Panayotou G, Booker GW, Waterfield MD (1993): New insights into protein-tyrosine kinase receptor signaling complexes. *Protein Sci* 2:1785–1797

Förster T (1969): Excimers. *Angew Chem Int Ed* 8:333–343

Galla HJ, Hartmann W (1980): Excimer-forming lipids in membrane research. *Chem Phys Lipids* 27:199–219

Gennis RB (1989): *Biomembranes. Molecular Structure and Function*. London: Springer-Verlag

Gerschfeld NL (1989): The critical unilamellar lipid state: A perspective for membrane bilayer assembly. *Biochim Biophys Acta* 988:335–350

Goldfine H (1984): Bacterial membranes and lipid packing theory. *J Lipid Res* 25:1501–1507

Heldin CH (1995): Dimerization of cell surface receptors in signal transduction. *Cell* 80:213–223

Helfrich W (1973): Elastic properties of lipid bilayers: Theory and possible experiments. *Z Naturforsch* 28c:693–703

Hinderliter AK, Huang J, Feigenson GW (1994): Detection of phase separation in fluid phosphatidylserine/phosphatidylcholine mixtures. *Biophys J* 67:1906–1911

Ho C, Pratt EA, Rule GS (1989): Membrane-bound d-lactate dehydrogenase of *Escherichia coli*: A model for protein interactions in membranes. *Biochim Biophys Acta* 988:173–184

Hønger T, Mortensen K, Ipsen JH, Lemmich J, Bauer R, Mouritsen OG (1994): Anomalous swelling of multilamellar lipid bilayers in the transition region by renormalization of curvature elasticity. *Phys Rev Lett* 24:3911–3914

Horváth LI, Arias HR, Hankovszky HO, Hideg K, Barrantes FJ, Marsh D (1990): Association of spin-labeled local anesthetics at the hydrophobic surface of acetylcholine receptor in native membranes from *Torpedo marmotata*. *Biochemistry* 29:8707–8713

Horváth LI, Heimburg T, Kovachev P, Findlay JBC, Hideg K, Marsh D (1995): Integration of a K^+ channel-associated peptide in a lipid bilayer: Conformation, lipid-protein interactions, and rotational diffusion. *Biochemistry* 34:3893–3898

Huang Y, Swanson JE, Dibble ARG, Hinderliter AK, Feigenson GW (1993): Nonideal mixing of phosphatidylserine and phosphatidylcholine in the fluid lamellar phase. *Biophys J* 64:413–425

Hui SW, Sen A (1989): Effects of lipid packing on polymorphic phase behaviour and membrane properties. *Proc Natl Acad Sci USA* 86:5825–5829

Ipsen JH, Jørgensen K, Mouritsen OG (1990): Density fluctuations in saturated phospholipid bilayer increase as the acyl-chain length decreases. *Biophys J* 58:1099–1107

Ipsen JH, Karlström G, Mouritsen OG, Wennerström H, Zuckermann MJ (1987): Phase equilibria in the phosphatidylcholine-cholesterol system. *Biochim Biophys Acta* 905:162–172

Israelachvili J (1992): *Intermolecular and Surface Forces*. New York: Academic Press

Israelachvili J (1977): Refinement of the fluid-mosaic model of membrane structure. *Biochim Biophys Acta* 469:221–225

Israelachvili JN, Marcelja S Horn RG (1980): Physical principles of membrane organization. *Q Rev Biophys* 13:121–200

Jamil H, Hatch GM, Vance DE (1993): Evidence that binding of CTP:phosphocholine cytidyltransferase to membranes is modulated by the ratio of bilayer- to non-bilayer forming lipids. *Biochem J* 291:419–426

Johannsson A, Keithley CA, Smith GA, Richards CD, Hesketh TR, Metcalfe JC (1981b): The effect of bilayer thickness and *n*-alkanes on the activity of the $(Ca^{2+}\text{-}Mg^{2+})$-dependent ATPase of sarcoplasmic reticulum. *J Biol Chem* 256:1643–1650

Johannsson A, Smith GA, Metcalfe JC (1981a): The effect of bilayer thickness on the activity of $(Na^+\text{-}K^+)$-ATPase. *Biochim Biophys Acta* 641:416–421

Jørgensen K, Mouritsen OG (1995): Phase separation dynamics and lateral organization of two-component lipid membranes. *Biophys J*: 95:942–954

Jørgensen K, Ipsen JH, Mouritsen OG, Zuckermann MJ (1993a): The effects of anaesthetics on the dynamic heterogeneity of lipid membranes. *Chem Phys Lipids* 65:205–216

Jørgensen K, Ipsen JH, Mouritsen OG, Bennett D, Zuckermann MJ (1991): The effects of density fluctuations on the partitioning of foreign molecules into lipid bilayers: Application to anaesthetics and insecticides. *Biochim Biophys Acta* 1067:241–253

Jørgensen K, Sperotto MM, Mouritsen OG, Ipsen JH, Zuckermann MJ (1993b): Phase equilibria and local structure in binary lipid bilayers. *Biochem Biophys Acta* 1152:135–145

Kaiser ET, Kézdy FJ (1984): Amphiphilic secondary structure: design of peptide hormones. *Science* 223:249–255

Karlsson OP, Dahlqvist A, Wieslander Å (1994): Activation of the membrane glucolipid synthesis in *Acholeplasma laidlawii* by phosphatidylglycerol and other anionic lipids. *J Biol Chem* 269:23484–23490

Keller SL, Bezrukov SM, Gruner SM, Tate MW, Vodyanov I, Parsegian VA (1993): Probability of alamethicin conductance states varies with nonlamellar tendency of bilayer phospholipids. *Biophys J* 65:23–27

Kinnunen PKJ (1996): On the mechanisms of the lamellar → hexagonal H_{II} phase transition and the biological significance of the H_{II} propensity. In: *Nonmedical Applications of Liposomes*, Lasic DD, Barenholz Y, eds. Boca Raton, FL: CRC Press

Kinnunen PKJ (1992): Fusion of lipid bilayers: A model involving mechanistic connection to H_{II} phase forming lipids. *Chem Phys Lipids* 63:251–258

Kinnunen PKJ (1991): On the principles of functional ordering in biological membranes. *Chem Phys Lipids* 57:375–399

Kinnunen PKJ, Laggner P, eds. (1991): *Phospholipid Phase Transitions*. Special Issue of *Chem Phys Lipids* 57:109–408

Kinnunen PKJ, Mouritsen OG, eds. (1994): *Functional Dynamics of Lipids in Biomembranes* Special Issue of *Chem Phys Lipids* 73:1–236

Kinnunen PKJ, Kõiv A, Lehtonen JYA, Rytsmaa M, Mustonen P (1994): Lipid dynamics and peripheral interactions of proteins with membrane surfaces. *Chem Phys Lipids* 73:181–207

Kinnunen PKJ, Kõiv A, Mustonen P (1993a): Pyrene-labelled lipids as fluorescent probes in studies on biomembranes and membrane models. In: *Fluorescence Spectroscopy*, Wolfbeis OS, ed. Berlin: Springer-Verlag

Kinnunen PKJ, Rytömaa M, Lehtonen JYA, Kõiv A, Mustonen P, Aro A (1993b): Sphingosine-mediated membrane association of DNA and its reversal by phosphatidic acid. *Chem Phys Lipids* 66:75–85

Kinnunen PKJ, Tulkki A, Lemnetyinen P, Paakkola J, Virtanen J (1987): Characteristics of excimer formation in Langmuir-Blodgett assemblies of 1-palmitoyl-2-pyrenedecanoylphosphatidylcholine and dipalmitoylphosphatidylcholine. *Chem Phys Lett* 136:539–545

Kõiv A, Palvimo J, Kinnunen PKJ (1995): Evidence for ternary complex formation by histone H1, DNA, and liposomes. *Biochemistry* 34:8018–8027

Klinger A, Braiman M, Jørgensen K, Biltonen RL (1994): Slow structural reorganization in binary lipid bilayers in the gel-liquid crystalline coexistence region. *Biophys J* 66:A173

Laggner P, Kriechbaum M (1991): Phospholipid phase transitions: Kinetics and structural mechanisms. *Chem Phys Lipids* 57:121–146

Landolt-Marticorena C, Williams KA, Deber CM, Reithmeier RAF (1993): Non-random distribution of amino acids in the transmembrane segments of human type I single span membrane proteins. *J Mol Biol* 229:602–608

Laureaeus S, Taskinen M-R, Kinnunen PKJ (1996): Aggregation of dimyristoylphosphatidylglycerol liposomes by human plasma low density lipoprotein. *Biophys J*: in press

Lee AG (1977): Lipid phase transitions and phase diagrams. *Biochim Biophys Acta* 472:285–344

Lehtonen JYA, Kinnunen PKJ (1995a): Poly(ethylene glycol)-induced and temperature-dependent phase separation in fluid binary phospholipid membranes. *Biophys J* 68:525–535

Lehtonen JYA, Kinnunen PKJ (1995b): Phospholipase A_2 as a mechanosensor. *Biophys J* 68:1888–1894

Lehtonen JYA, Kinnunen PKJ (1994): Changes in the lipid dynamics if liposomal membranes induced by poly(ethylene glycol). Free volume alterations revealed by inter- and intramolecular excimer forming phospholipid analogs. *Biophys J* 66:1981–1990

Lehtonen JYA, Holopainen JH, Kinnunen PKJ (1996a): Evidence for fluid-fluid immiscibility of phospholipids in large unilamellar vesicles caused by hydrophobic mismatch. *Biophys J*: in press

Lehtonen JYA, Jung K, Kaback RH, Kinnunen PKJ (1996b): Evidence for fluid-fluid immiscibility due to hydrophobic mismatch in liposomes reconstituted with *E. coli* lactose permease. Unpublished

Lehtonen JYA, Rytömaa M, Kinnunen PKJ (1996c): Interaction of tacrine with acidic phospholipids. *Biophys J*: in press

Leibler S (1986): Curvature instability in membranes. *J Phys Paris* 47:507–516

Lemmich J, Ipsen JH, Hønger T, Jørgensen K, Mouritsen OG, Mortensen K, Bauer R (1994): Soft and repulsive: Relationship between lipid membrane in-plane fluctuations, bending regidity, and repulsive undulation forces. *Mod Phys Lett B* 29:1803–1814

Lemmich J, Mortensen K, Ipsen JH, Hønger T, Bauer R, Mouritsen OG (1995): Pseudocritical behavior and unbinding of phospholipid bilayers. *Phys Rev Lett* 75:3958–3961

Li S-C, Deber CM (1993): Peptide environment specifies conformation. *J Biol Chem* 268:22975–22978

Lindblom G, Orädd G (1994): NMR studies of translational diffusion in lyotropic liquid crystals and lipid membranes. *Prog NMR Spectros* 26:483–515

Lindblom G, Persson N-O (1976): Ion binding and water orientation in lipid model membrane systems studied by NMR. *Adv Chem* 152:121–141

Lipowsky R (1993): Domain-induced budding of fluid membranes. *Biophys J* 64:1133–1138

Litman BJ, Lewis EN, Levin IW (1991): Packing characteristics of highly unsaturated bilayers lipids: Raman spectroscopic studies of multilamellar phosphatidylcholine dispersions. *Biochemistry* 30:313–319

Lis LJ, McAlister M, Fuller N, Rand RP, Parsegian VA (1982): Measurement of the lateral compressibility of several phospholipid bilayers. *Biophys J* 37:667–672

Low MG (1989): The glycosyl-phosphatidylinositol anchor of membrane proteins. *Biochim Biophys Acta* 988:427–454

Lundbæk JA, Andersen OS (1994): Lysophospholipids model channel function by altering the mechanical properties of lipid bilayers. *J Gen Physiol* 104:645–673

McCallum CD, Epand RM (1995): Insulin receptor autophosphorylation and signaling is altered by modulation of membrane physical properties. *Biochemistry* 34:1815–1824

Melchior DL (1986): Lipid domains in fluid membranes: A quick-freeze differential scanning calorimetry study. *Science* 234:1577–1580

Menashe M, Romero G, Biltonen RL, Lichtenberg D (1986): Hydrolysis of dipalmitoylphosphatidylcholine small unilamellar vesicles by porcine pancreatic phospholipase A_2. *J Biol Chem.* 261:5328–5333

Moll TS, Thompson TE (1994): Semisynthetic proteins: Model systems for the study of the insertion of hydrophobic peptides into preformed lipid bilayers. *Biochemistry* 33:15469–15482

Montecucco C, Smith GA, Dabbeni-Sala F, Johannsson, A, Galante YM, Bisson R (1982): Bilayer thickness and enzymatic activity in the mitrochondrial cytochrome *c* oxidase and ATPase complex, *FEBS Lett*, 144:145–148

Mountcastle DB, Biltonen RL, Halsey MJ (1978): Effect of anesthetics and pressure on the thermotropic behavior of multilamellar dipalmitoylphosphatidylcholine liposomes. *Proc Natl Acad Sci USA* 75:4906–4910

Mouritsen OG (1990): Computer simulation of cooperative phenomena in lipid membranes. In: *Molecular Description of Biological Membrane Components by Computer Aided Conformational Analysis,* Vol. 1, Brasseur R, ed. Boca Raton, FL: CRC Press

Mouritsen OG (1991): Theoretical models of phospholipid phase transitions. *Chem Phys Lipids* 57:179–194

Mouritsen OG, Biltonen RL (1993): Protein-lipid interactions and membrane heterogeneity. In: *Protein-lipid Interactions,* Watts A, ed. Amsterdam: Elsevier

Mouritsen OG, Sperotto MM (1993): Thermodynamics of lipid-protein interactions in lipid membranes: the hydrophobic matching condition. In: *Thermodynamics of Cell Surface Receptors,* Jackson M, ed. Boca Raton, FL: CRC Press

Mouritsen OG, Bloom M (1993): Model of lipid-protein interactions in membranes. *Annu Rev Biophys Biomol Struct* 22:145–171

Mouritsen OG, Bloom M (1984): Mattress model of lipid-protein interactions in membranes. *Biophys, J* 46:141–153

Mouritsen OG, Jørgensen K (1995): Micro-, Nano-, and Meso-scale heterogeneity of lipid bilayers and its influence on macroscopic membrane properties. *Mol Memb Biol* 12:15–20

Mouritsen OG, Jørgensen K (1994): Dynamical order and disorder in lipid bilayers *Chem Phys Lipids* 73:3–25

Mouritsen OG, Jørgensen K (1992): Dynamic lipid-bilayer heterogeneity: A mesoscopic vehicle for membrane function? *BioEssays* 14:129–136

Mouritsen OG, Jørgensen K, Hønger T (1995): Permeability of lipid bilayers near the phase transition. In: *Permeability and Stability of Lipid Bilayers,* Disalvo EA, Simon SA, eds. Boca Raton, FL: CRC Press

Mustonen P, Kinnunen PKJ (1992): Substrate level modulation of the activity of phospholipase A_2 in vitro by 12-O-tetradecanoyl-phorbol-13-acetate. *Biochem Biophys Res Comm* 185:185–190

Mustonen P, Kinnunen PKJ (1991): Activation of phospholipase A_2 by adriamycin in vitro. Role of drug/lipid interactions. *J Biol Chem.* 266:6302–6307

Mustonen P, Lehtonen JYA, Koiv A, Kinnunen PKJ (1993): Effects of sphingosine on peripheral membrane interactions: comparison of adriamycin, cytochrome c, and phospholipase A_2. *Biochemistry* 32:5373–5380

Mustonen P, Virtanen JA, Somerharju P, Kinnunen PKJ (1987): Binding of cytochrome c to liposomes as revealed by the quenching of fluorescence from pyrene-labeled phospholipids. *Biochemistry* 26:2991–2997

Neubig RR (1994): Membrane organization in G-protein mechanisms. *FASEB J* 18:939–946

Newton AC (1993): Interaction of proteins with lipid headgroups: Lessons from protein kinase C *Annu Rev Biophys Biomol Struct* 22:1–25

van Osdol W, Ye Q, Johnson ML, Biltonen RL (1992): The effects of the anesthetic dibucaine on the kinetics of the gel-liquid crystalline transition of dipalmitoylphosphatidylcholine multilamellar vesicles. *Biophys J* 63:1011–1017

Papahadjopoulos D, Jacobsen K, Nir S, Isac T (1973): Phase transitions in phospholipid vesicles. Fluorescence polarization and permeability measurements concerning the effect of temperature and cholesterol. *Biochim Biophys Acta* 311:330–348

Pattus F, Slotboom AJ, de Haas JH (1979): Regulation of phospholipase A_2 activity by the lipid-water interface: A monolayer approach. *Biochemistry* 18:2691–2697

Peschke J, Riegler J, Möhwald H (1987): Quantitative analysis of membrane distortions introduced by mismatch of protein and lipid hydrophobic thickness. *Eur Biophys J* 14:385–391

Piknova B, Perochon E, Tocanne J-F (1993): Hydrophobic mismatch and long-range protein-lipid interactions in bacteriorhodopsin/phosphatidylcholine vesicles. *Eur J Biochem* 281:385–396

Rietveld AG, Chupin VV, Koorengevel MC, Wienk HLJ, Dowhan W, de Kruijff B (1994): Regulation of lipid polymorphism is eesential for the viability of phosphatidylethanolamine-deficient *Escherichia coli* cells. *J Biol Chem* 269:28670–28675

Risbo J, Sperotto MM, Mouritsen OG (1995): Theory of phase equilibria and critical mixing points in binary lipid bilayers. *J Chem Phys* 103:3643–3656

Robertson RN (1983): *The Lively Membranes.* Cambridge: Cambridge University Press

Rytömaa M, Kinnunen PKJ (1995): Reversibility of the binding of cytochrome c to liposomes. *J Biol Chem* 270:3197–3202

Rytömaa M, Kinnunen PKJ (1994): Evidence for two distinct acidic phospholipid-binding sites in cytochrome c. *J Biol Chem* 269:1770–1774

Rytömaa M, Mustonen P, Kinnunen PKJ (1992): Reversible, non-ionic, and pH-dependent association of cytochrome c with cardiolipin-phosphatidylcholine liposomes. *J Biol Chem* 267:22243–22248

Sabra M, Jørgensen K, Mouritsen OG (1995): Calorimetric and theoretical studies of the effects of lindane on multilamellar DMPC, DPPC, and DSPC bilayers. *Biochim Biophys Acta* 1233:89–104

Sachs F (1990): Ion channels as mechanical transducers. In: *Cell Shape*, Stein WD, Bronner F, eds. San Diego: Academic Press

Sankaram MB, Marsh D, Thompson TE (1992): Determination of fluid and gel domain sizes in two-component, two-phase lipid bilayers. *Biophys J* 63:340–349

Saxton M (1994): Anomalous diffusion due to obstacles: A Monte Carlo study. *Biophys J* 66:394–401

Schmidt MFG (1989): Fatty acylation of proteins. *Biochim Biophys Acta* 988:411–426

Seifert U, Langer SA (1993): Viscous models of fluid bilayer membranes. *Europhys Lett* 23:71–76

Seifert U (1993): Curvature-induced lateral phase separation in two-component vesicles. *Phys Rev Lett* 70:1335–1338

Silvius JR (1992): Cholesterol modulation of lipid intermixing in phospholipid and glycosphingolipid mixtures. Evaluation using fluorescent lipid probes and brominated lipid quenchers. *Biochemistry* 31:3398–3408

Simões AP, Reed J, Schnabel P, Camps M, Gierschik P (1995): Characterization of putative polyphosphoinositide binding motifs from phospholipase $C\beta_2$. *Biochemistry* 34:5113–5119

Singer SJ, Nicolson GL (1972): The fluid mosaic model of the structure of cell membranes. *Science* 173:720–731

Sipos L, von Heijne G (1993): Predicting the topology of eucaryotic membrane proteins. *Eur J Biochem* 213:1333–1340

Slater SJ, Kelly MB, Taddeo FJ, Ho C, Rubin E, Stubbs CD (1994): The modulation of protein kinase C activity by membrane lipid bilayer structure. *J Biol Chem* 269:4866–4871

Somerharju P, Virtanen JA, Eklund KK, Vainio P, Kinnunen PKJ (1985): 1-palmitoyl-2-pyrenedecanoyl glycerophospholipids as membrane probes: Evidence for regular distribution in liquid crystalline phosphatidylcholine bilayers. *Biochemistry* 24:2773–2781

Souvignet C, Pelosin J-M, Daniel S, Chambaz EM, Ransac S, Verger R (1991): Activation of protein kinase C in lipid bilayers. *J Biol Chem* 266:40–44

Sperotto MM, Mouritsen OG (1993): Lipid enrichment and selectivity of integral memnbrane proteins in two-component lipid bilayers. *Eur Biophys J* 22:323–328

Sperotto MM, Mouritsen OG (1991a): Monte Carlo simulation studies of lipid order parameter profiles near integral membrane proteins. *Biophys J* 59:261–270

Sperotto MM, Mouritsen OG (1991b): Mean-field and Monte Carlo simulation studies of the lateral distribution of proteins in membranes. *Eur Biophys J* 19:157–270

Sperotto MM, Mouritsen OG (1988): Dependence of lipid membrane phase transition temperature on the mismatch of protein and lipid hydrophobic thickness. *Eur Biophys J* 16:1–10

Sugar IP, Tang D, Chong PL-G (1994): Monte Carlo simulation of flateral distribution of molecules in a two-component lipid membrane. Effect of long-range repulsive interactions. *J Phys Chem* 98:7201–7210

Sui S, Urumow T, Sackmann E (1988): Interaction of insulin receptors with lipid bilayers and specific and nonspecific binding of insulin to supported membranes. *Biochemistry* 27:7463–7469

Tang D, Chong PL-G (1992): E/M dips. Evidence for lipids regularly distributed into hexagonal super-lattices in pyrene-PC/DMPC binary mixtures at specific concentrations. *Biophys J* 63:903–910

Tang D, van der Meer BW, Chen S-YS (1995): Evidence for a regular distribution of cholesterol in phospholipid bilayers from diphenylhexatriene fluorescence. *Biophys J* 68:1944–1951

Tate MW, Eikenberry EF, Turner DC, Shyamsunder E, Gruner SM (1991): Nonbilayer phases of membrane lipids. *Chem Phys Lipids* 57:147–164

Thompson TE, Sankaram MB, Biltonen RL, Marsh D, Vaz WLC (1995): Effects of domain structure on in-plane reactions and interactions. *Mol Memb Biol* 12:157–162

Thurén T, Tulkki A-P, Virtanen JA, Kinnunen PKJ (1987a): Triggering of the activity of phospholipase A$_2$ by an electric field. *Biochemistry* 26:4907–4910

Thurén T, Vainio P, Virtanen JA, Somerharju P, Blomqvist K, Kinnunen PKJ (1984): Evidence for the control of the action of phospholipases A by the physical state of the substrate. *Biochemistry* 23:5129–5134

Thurén T, Virtanen JA, Kinnunen PKJ (1990): Phospholipid-platelet activating factor interactions probed by monolayers, pyrene fluorescence, and phospholipase A$_2$. *Chem Phys Lipids* 53:129–139

Thurén T, Virtanen JA, Verger R, Kinnunen PKJ (1987b): Hydrolysis of 1-palmitoyl-2-[6-(pyren-1-yl)]hexanoyl-sn-glycero-3-phospholipids by phospholipase A$_2$: Effect of the polar headgroup. *Biochim Biophys Acta* 917:411–417

Thurén T, Virtanen JA, Kinnunen PKJ (1986): Polyamine-phospholipid interaction probed by the accessibility of the phospholipid sn-2 ester bond to the action of phospholipase A$_2$. *J Membrane Biol* 92:1–7

Trudell JR (1977): A unitary theory of anesthesia based on lateral phase separation in nerve membranes. *Anesthesiology* 46:5–10

Trudell JR, Costa AK, Hubbell WL (1991): Inhibition of protein kinase C by local anesthetics. *Ann NY Acad Sci* 625:743–746

Virtanen JA, Somerharju P, Kinnunen PKJ (1988): Prediction of patterns for the regular distribution of soluted guest molecules in liquid crystalline phospholipid. *J Mol Electron* 4:233–236

Wang N, Butler JP, Ingber DE (1993): Mechanotransduction across the cell surface and though the cytoskeleton. *Science* 260:1124–1127

White SH, Wimley WC (1994): Peptides in lipid bilayers: structural and thermodynamic basis for partitioning and folding. *Curr Op Struct Biol* 4:79–86

Wieslander Å, Rilfors L, Lindblom G (1986): Metabolic changes of membrane lipid composition in *Acholeplasma laidlawii* by hydrocarbons, alcohols, and detergents: arguments for effects on lipid packing. *Biochemistry* 25:7511–7517

Wolfe DH, Brockman HL (1988): Regulation of the surface pressure of lipid monolayers and bilayers by the activity of water: derivation and application of an equation of state. *Proc Natl Acad Sci USA* 85:4285–4289

Wu SH, McConnell HM (1975): Phase separations in phospholipid membranes. *Biochemistry* 14:847–854

Yamada K, Goncalves E, Carpentier J-L, Kahn CR, Shoelson SE (1995): Transmembrane domain inversion blocks ER release and insulin receptor signaling. *Biochemistry* 34:946–954

Yang L, Glaser M (1995): Membrane domains containing phosphatidylserine and substrate can be important for the activation of protein kinase C. *Biochemistry* 34:1500–1506

Zhang Y-P, Lewis RNAH, Hodges RS, McElhaney RN (1995): Peptide models of helical hydrophobic transmembrane segments of membrane proteins. 2. Differential scanning calorimetric and FTIR spectroscopic studies of the interaction of Ac-K_2-(LA)$_{12}$-K_2-amide with phosphatidylcholine bilayers. *Biochemistry* 34:2362–2371

Zuckermann MJ, Ipsen JH, Mouritsen OG (1993): Theoretical studies of the phase behavior of lipid bilayers containing cholesterol. In: *Cholesterol and membrane models* Finegold LX, ed. Boca Raton, FL: CRC Press

15

Prediction of the Structure of an Integral Membrane Protein: The Light-Harvesting Complex II of *Rhodospirillum molischianum*

XICHE HU, DONG XU, KENNETH HAMER, KLAUS SCHULTEN, JUERGEN KOEPKE, AND HARTMUT MICHEL

Abstract

We illustrate in this chapter how one proceeds to predict the structure of integral membrane proteins when a highly homologous structure is unknown. We focus here on the prediction of the structure of the light-harvesting complex II (LH–II) of *Rhodospirillum molischianum*, an integral membrane protein of 16 polypeptides aggregating and binding to 24 bacteriochlorophyll a's and 12 lycopenes. Hydropathy analysis was performed to identify the putative transmembrane segments, which were independently verified by multiple sequence alignment propensity analyses and homology modeling. A consensus assignment for secondary structure was derived from a combination of all the prediction methods used. Transmembrane helices were built by comparative modeling. The resulting tertiary structures were then aggregated into a quaternary structure through molecular dynamics simulations and energy minimization under constraints provided by site directed mutagenesis and FT Resonance Raman spectra, as well as conservation of residues. The structure of LH–II, so determined, was an octamer of $\alpha\beta$ heterodimers forming a ring with a diameter of 70 Å. We discuss how the resulting structure may be used to solve the phase problem in X-ray crystallography in a procedure called molecular replacement. We will also discuss the exciton structure which results from the circular arrangement of chlorophyls in LH-II.

Introduction

Membranes are complex structures containing bilayers of amphiphilic phospholipids with proteins either loosely associated on the surface (peripheral membrane proteins, e.g., phospholipase A_2) or embedded (integral membrane pro-

Biological Membranes
K. Merz, Jr. and B. Roux, Editors
© Birkhäuser Boston 1996

teins, e.g., bacteriorhodopsin). This chapter deals with the structure prediction of integral membrane proteins, defined as proteins having peptide chains with substantial tertiary structure within the nonpolar region of the lipid bilayer (Engelman, 1982). With the advent of recombinant DNA technology, determination of primary sequences of proteins is proceeding at a much faster pace than determination of atomic resolution protein structures. While thousands of membrane protein sequences are available, there are few detailed three dimensional membrane protein structures known. Available structures, at present, include the photosynthetic reaction centers from *Rhodopsuedomonas viridis* (Deisenhofer et al, 1985) and *Rhodobacter sphaeroides* (Allen et al, 1987), porins from *Rhodobacter capsulatus* (Weiss et al, 1991) and *Escherichia coli* (Cowan et al, 1992), bacteriorhodopsin from *Holobacterium halobium* (Henderson et al, 1990), and a plant light harvesting complex (Kühlbrandt et al, 1994). Since structural information of membrane proteins is vital to understanding of their cellular functions, a great effort has been made to predict membrane protein structures from their primary sequences (Argos et al, 1982; Arkin et al, 1994; Cramer et al, 1992; Fasman, 1989a; Jähnig, 1989; Popot, 1993; Popot and de Vitry, 1990; Popot and Engelman, 1990; Popot et al, 1994; Tuffery et al, 1994; von Heijne, 1994a; 1994b; 1992; White, 1994).

Most structure prediction algorithms can be categorized into three main classes: statistical (Chou and Fasman, 1978; Garnier et al, 1978; Holley and Karplus, 1989; Kuhn and Leigh, 1985; Levitt, 1978; Lohmann et al, 1994; Persson and Argos, 1994; Rao and Argos, 1986), physico-chemical (Argos et al, 1982; Cornette et al, 1987; Eisenberg, 1984; Engelman et al, 1986; Kyte and Doolittle, 1982; Rees et al, 1989; von Heijne, 1988; White, 1994), and comparative (Blundell et al, 1987; Busetta, 1986; Cohen et al, 1986; Johnson et al, 1994; Presnell et al, 1992; Rooman and Wodak, 1988; Sali and Blundell, 1993; Zvelebil et al, 1987). Presently, prediction of tertiary structure is only of practical use when the structure of a homologous protein is already known. Protein homology modeling typically involves the prediction of side-chain conformations in the modeled protein while assuming a main-chain trace taken from a known tertiary structure of a homologous protein. However, the tertiary structures of proteins have been successfully predicted when experimentally derived constraints are used in conjunction with heuristic methods (Ring and Cohen, 1993). In such a knowledge-based approach information, both from the three-dimensional structures of homologous proteins and from the general analysis of protein structure, is used to derive constraints for modeling a protein of known sequence, but unknown structure.

It is certainly not our intention, and nearly impractical, to give a complete review of this explosively developing field. Interested readers are referred to the excellent review articles assembled in the authoritative books edited by Fasman (1989b) and Creighton (1992) and those listed above. The rest of this chapter will focus on our recent efforts in predicting the structure of the light-harvesting complex II (LH–II) of *Rs. molischianum*, an integral membrane protein of 16 polypeptides aggregating and binding to 24 bacteriochlorophyll-a (BChla) and 12 lycopenes.

Light Harvesting Complex

Photosynthetic organisms radically increase the efficiency and decrease the complexity of their energy gathering apparatus by surrounding the complex photosynthetic reaction centers with simple, pigment-rich protein aggregates known as light harvesting complexes or antenna complexes. The reaction center has a very rapid cycle time, on the order of 10^3 photons per second (van Grondelle and Sundstrom 1988), and cannot independently collect enough photons to saturate itself. With the inclusion of the antenna complexes, the bacterium can collect and channel to each reaction center much more light energy.

In most purple bacteria there are two basic types of light harvesting complexes: the light harvesting complex I, or LH–I, is found directly surrounding the photosynthetic reaction centers, while the light harvesting complex II, or LH–II, surrounds the LH–I–reaction center aggregates. LH–I absorbs at longer wavelengths than LH–II, typically with a strong absorption band between 870 and 1015 nm, and is found in all types of purple photosynthetic bacteria (Zuber, 1993). LH–II is found in some species (notable exceptions are *Rhodospirillum rubrum* and *Rps. viridis*) of bacteria, and typically has one absorption band between 820 and 860 nm, and another around 800 nm (Zuber, 1985). For *Rs. molischianum*, the LH–II complex displays two peaks at 800 and 850 nm and is often referred to as the B800–850 complex. Since the photosynthetic reaction center absorbs in the deep infrared (960 nm for the *Rps. viridis* reaction center) there is a clear energetic hierarchy in the light-harvesting system, with the LH–II complex absorbing light at the highest energy, surrounding LH–I which absorbs at a lower energy, which in turn surrounds the reaction center which absorbs at the lowest energy. This arrangement naturally channels energy from the outer regions of the antenna complex to the reaction center.

Understanding of the mechanism of the photosynthetic reaction center has been greatly enhanced by the determination of its three-dimensional structure (Deisenhofer and Michel, 1989). However, structural information about light-harvesting complexes is still limited to spectroscopic and biochemical characterization (Hawthornthwaite and Cogdell, 1991; Sundstrom and van Grondelle, 1991; Zuber and Brunisholz, 1991). The LH–II complex of *Rs. molischianum* has been crystallized and X-ray diffraction data have been collected up to 2.4 Å resolution (Koepke et al, 1996; Michel, 1991). To resolve a structure from measured diffraction intensities requires knowledge of phases which is unobtainable from a single diffraction experiment. Conventionally, the phase problem is solved by means of the multiple isomorphous replacement method. An alternative solution to the phase problem is to phase the structure by using a homologous structure in a procedure called molecular replacement (Lattman, 1985; Rossmann, 1972). In this method, a homologous probe structure is fit into the unit cell of the unknown structure and used to generate an initial phasing model for the unknown structure. At the time when this project was initiated, there existed no homologous structure to LH–II of *Rs. molischianum*. We attempted to predict the structure of *Rs. molischianum* and intended to use the predicted structure as a probe structure in the molecular replacement method to resolve the 2.4 Å X-ray diffraction data into an atomic structure. We report here the predictions that we have made with emphasis on structure prediction methods. At the end, the current prediction will be compared with the recently published structure for LH–II from *Rps. acidophila* by McDermott et al, who have successfully solved the phase problem for their structure by conventional means (Mcdermott et al, 1995).

All light-harvesting complexes display a remarkable similarity in the way they are constructed (Zuber, 1985; Zuber and Brunisholz, 1991). The basic structural unit is a heterodimer of two small polypeptides, commonly referred to as α and β apoproteins, both shorter than 60 amino acids, which non-covalently bind BChla and carotenoid molecules. These heterodimers aggregate into a large complex, functioning as light harvesting antennae. The size of the aggregate depends on the type of light-harvesting complex and varies from species to species, ranging from a putative hexamer for LH–II of *Rb. sphaeroides* (Boonstra et al, 1993) to a hexadecamer for LH–I of *Rs. rubrum* (Karrasch et al, 1995).

Various models have been proposed for the light-harvesting complex of purple bacteria (Olsen and Hunter, 1994; Zuber, 1986; Zuber and Brunisholz, 1991). The majority of these models are concerned with secondary structural features and the topology of the heterodimers. However, no atomic level modeling of the aggregated complex has been attempted before. Our goal is to build a model structure for LH–II of *Rs. molischianum* and to use it as a probe structure in the framework of the molecular replacement method. The ultimate correctness of the predicted structure can be tested by its ability to serve as a successful search model to resolve the X-ray diffraction data in terms of a consistent electron density profile to which an atomic structure can be configured.

Method

In practice, the task of prediction is divided into three stages: (1) predict the secondary structure of the α- and β-apoproteins from their amino acid sequences; (2) build the tertiary structures for the α- and β-apoproteins by comparative modeling; (3) fold the tertiary structures into an aggregated complex (quaternary structure) by means of molecular dynamics simulations and energy minimization under the constraints of experimental data and the predicted secondary structure features. Finally, the molecular replacement test was performed using the predicted structure as a probe structure to resolve the unknown crystal structure. A flowchart of the entire procedure is provided in Figure 1.

The molecular dynamics simulations and energy minimizations described in this chapter were carried out using the program X-PLOR (Brünger, 1992). All the simulation protocols were programmed with the versatile X-PLOR script language. An integration time step of 1 fs was chosen in the Verlet algorithm. The simulation of LH–II placed the protein in a vacuum. The parameters and charges used for the system were, respectively, the CHARMm all-atom parameter file `parallh22x.pro` and the CHARMm all-atom partial charge file

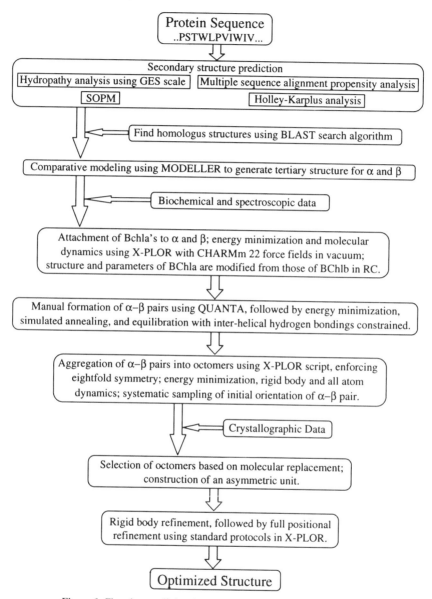

Figure 1. Flowchart outlining the entire structure prediction procedure.

topallh22x.pro (Brooks et al, 1983; A MacKerell et al, 1995) except for BChla. The partial charges and parameters for BChla were taken from those used in [Treutlein et al, 1988] for BChlb except for slight modifications to accommodate BChla. A cut-off distance of 12 Å for nonbonded interactions and a dielectric constant $\varepsilon = 1$ were employed.

Prediction of Secondary Structure

The LH–II complex of *Rs. molischianum* consists of two BChla-binding apoproteins α and β with the following sequences:

α: SNPKDDYKIWLVINPSTWLPVIWIVATVVAIAVHAAVLAAPGSNWIALGAAKSAAK
β: AERSLSGLTEEEAIAVHDQFKTTFSAFIILAAVAHVLVWVWKPWF

The smallest compositional unit of LH–II contains a pair of α and β apoproteins, three BChla and 1.5 lycopene molecules (Germeroth et al, 1993). It has
been determined by sedimentation equilibrium experiments that the native LH–
II complex is an octamer of such $\alpha\beta$ units (Kleinekofort et al, 1992). The space
group for the crystal is P42$_1$2 with cell dimensions of 92 × 92 × 209 Å.

Since the LH–II complex is an integral membrane protein, we performed
hydropathy analysis to identify the putative transmembrane segments (Kyte and
Doolittle, 1982; White, 1994). The transmembrane segments of apoproteins are
usually forced to adopt an α-helical conformation due to constraints of the
hydrophobic core of the membrane (Engelman et al, 1986). Hydropathy analysis
assumes that transmembrane segments are comprised mainly of hydrophobic
residues because of the low solubility of polar side chains in nonpolar lipid
bilayers. It takes about 20 amino acids (in an α helix) to span the hydrocarbon
regions of fluid bilayers that are typically 30 Å thick. Shown in Figure 2 is a
hydropathy plot for the α- and β-subunits based on the GES hydrophobicity
scale (Engelman et al, 1986) with a window size of 20 amino acids. The GES
scale is derived from the free energy cost for transferring amino acids from the
interior of a membrane to its water surroundings. On such a scale, a peak of

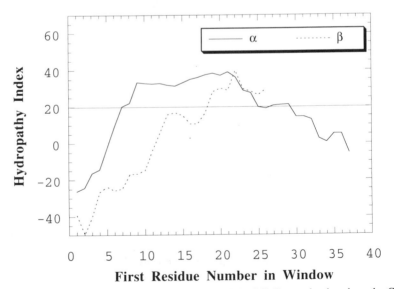

Figure 2. Hydropathy plot of α- and β-apoproteins of the LH–II complex based on the GES
hydrophobicity scale (see text) with a window size of 20 amino acids.

20 kcal/mol or higher identifies a transmembrane segment. Figure 2 clearly shows that a transmembrane segment exists for both the α- and β-apoproteins. The highest peak occurred at a first residue number of 21 and 22 for the α- and the β-apoprotein respectively. The transmembrane segments were thus identified as α-Val-21:α-Ala-40 and β-Thr-22:β-Trp-41. Hydropathy analyses with other two widely used hydrophobicity scales, i.e., Kyte and Doolittle (Kyte and Doolittle, 1982) and Eisenberg consensus (Eisenberg et al, 1982) scales, generated essentially the same hydrophobic core as the GES scale.

In addition to the hydropathy analysis, we have also carried out a multiple sequence alignment propensity analysis using the method of Persson and Argos (Persson and Argos, 1994) which combines two sets of propensity values (one for the middle, hydrophobic portion and one for the terminal region of the transmembrane span) to determine the transmembrane segments from multiply aligned amino acid sequences. A novel aspect of this method is the use of evolutionary information in the form of multiple sequence alignments as input in place of a single sequence. The method was shown to be more successful than predictions based on a single sequence alone. A total of 12 homologous sequences of LH–II and LH–I complexes have been aligned (see Figure 3) and analyzed. As shown in Figure 4, the transmembrane segment determined with this method spans from Trp-18 to Val-37 for the α-apoprotein and from Thr-22 to Trp-41 for the β-apoprotein. The hydropathy analysis identifies the region of the transmembrane segment. Multiple sequence alignment propensity analysis

α-apoprotein

Rs. molischianum B800-850	snpkddYKIWLVINPSTWLPVIWIVATVVAIAVHAAVLAAPGFNWIALGAAKsaak----
Rc. gelatinosus DSM 149 B870	-----mWRIWRLFDPMRAMVAQAVFLLGLAVLIHLMLLGTNKFNWLDGAKKApvasa---
Rs. rubrum B870	-----mWRIWQLFDPRQALVGLATFLFVLALLIHFILLSTERFNWLEGASTQpvqts---
Rb. capsulatus B870	--mskfYKIWLVFDPRRVFVAQGVFLFLLAVLIHLILLSTPAFNWLTVATAKhgyvaaaq
Rb. sphaeroides B870	--mskfYKIWMIFDPRRVFVAQGVFLFLLAVMIHLILLSTPSYNWLEISAAKynrvavae
Rps. marina B880	-----mWKVWLLFDPRRTLVALFTFLFVLALLIHFILLSTDRFNWMQGAPTApaqts---
Rp. viridis B1015	eyrtasWKLWLILDPRRVLTALFVYLTVIALLIHFGLLSTDRLNWWEFQRGLpkaa----
Rs. molischianum B800-850	snpkddYKIWLVINPSTWLPVIWIVATVVAIAVHAAVLAAPGFNWIALGAAKsaak----
Rp. viridis B1015	eyrtasWKLWLILDPRRVLTALFVYLTVIALLIHFGLLSTDRLNWWEFQRGLpkaa----
Rc. gelatinosus DSM 149 B800-850	---mnQGKVWRVVKPTVGVPVYLGAVAVTALILHGGLLAKTDWFGaywnggkkaaaa---
Rb. sphaeroides 2.4.1 B800-850	---mtNGKIWLVVKPTVGVPLFLSAAVIASVVIHAAVLTTTTWLPayyqgsaavaae---
Rb. capsulatus B800-850	---mnNAKIWTVVKPSTGIPLILGAVAVAALIVHAGLLTNTTWFAnywngnpmatvvava
Rps. acidophila Ac70.50 B800-850	---mnQGKIWTVVNPSVGLPLLLGSVTVIAILVHAAVLSHTTWFPaywqgglkkaa----
Rps. palustris 2.6.1 B800-850	---mnQGRIWTVVNPGVGLPLLLGSVTVIAILVHYAVLSNTTWFPkywngatvaapaaa-

β-apoprotein

Rs. molischianum B800-850	----AERSLSGLTEEEAIAVHDQFKTTFSAFIILAAVAHVLVWVWKPWf--------
Rc. gelatinosus DSM 149 B870	--aeRKGSISGLTDDEAQEFEKFWVQGFVGFTAVAVVAHFLVWVWRPWl--------
Rs. rubrum B870	--evKQESLSGITEGEAKEFEKIFTSSILVFFGVAAFAHLLVWIWRPWvpgpngys-
Rb. capsulatus B870	-adkNDLSFTGLTDEQAQELHAVYMSGLSAFIAVAVLAHLAVMIWRPWf-------
Rb. sphaeroides B870	-adkSDLGYTGLTDEQAQELHSVYMSGLWPFSAVAIVAHLAVYIWRPWf-------
Rps. marina B880	aeidRPVSLSGLTEGEAREFEGVFMTSFMVFIAVAIVAHILAWMWRPWipgpegia-
Rp. viridis B1015	--adLKPSLTGLTEEEAKEFEGIFVTSTVLYLATAVIVHYLVWTAKPWiapipkgwv
Rc. gelatinosus DSM 149 B800-850	ddaNKVWPSGLTTAEAEELQKGLVDGTRIFGVIAVLAHILAYAYTPWlh-------
Rb. sphaeroides 2.4.1 B800-850	-ddlNKVWPSGLTVAEAEEVEKQLILGTRVFGGMALIAHFLAAAATPWlg------
Rb. capsulatus B800-850	-mtdDKAGPSGLSLKEAEEIESYLIDGTRVFGAMALVAHILSAIATPWlg------
Rps. acidophila Ac70.50 B800-850	--adDVKGLTGLTAAESEELHKHVIDGTRVFFVIAIFAHVLAFAFSPWlh------
Rps. palustris 2.6.1 B800-850	addpNKVWPTGLTIAESEELHKHVIDGSRIFVAIAIVAHFLAYVYSPWlh------

Figure 3. Sequence alignment of the α and β-apoproteins of the LH–II complex of *Rs. molischianum* with α- and β-apoproteins of LH–II and LH–I complexes from other photosynthetic bacteria. Coding: Bold–Highly conserved; Shaded–Nearly conserved. Alignment was done using program MACAW (Multiple Alignment Construction and Analysis Workbench) (Schuler et al, 1991).

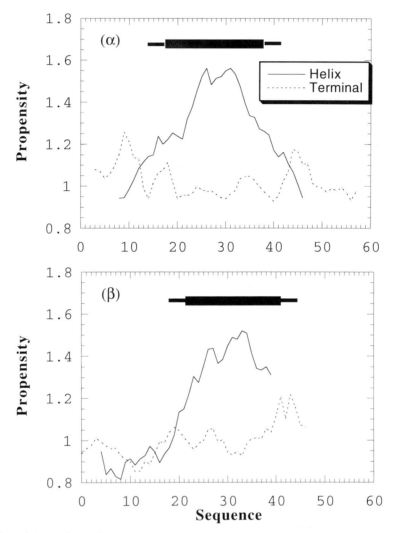

Figure 4. Propensity profile and assignment of 20 residue long transmembrane span (thick bar) and four residue long helical terminal (thin bar) for α- and β-apoproteins of the LH–II complex based on the multiple sequence alignment propensity analysis method of Persson and Argos. Solid line—transmembrane helix; dashed line—terminal region.

further pinpoints the most probable site of the 20 residue long transmembrane segment and the four residue long terminal sequence at both ends.

To confirm the above secondary structure prediction, we performed more sequence analyses with various secondary structure prediction methods including SOPM (self optimized prediction method) and Holley-Karplus analysis (Geourjon and Deleage, 1994; Holley and Karplus, 1989; 1991). SOPM takes structure classes into account and iteratively optimized prediction parameters to in-

Table I. SOPM Prediction[a]

α-apoprotein

```
SNPKDDYKIWLVINPSTWLPVIWIVATVVAIAVHAAVLAAPGSNWIALGAAKSAAK
CCCCCCCCEEEEECCCCCEEEHHHHHHHHHHHHHHHHHHHHHHCCCCHEEHHHHHHHHH
```

β-apoprotein

```
AERSLSGLTEEEAIAVHDQFKTTFSAFIILAAVAHVLVWVWKPWF
HHHHHCCCCHHHHHHHHHHHHHHHHHHHHHHHHHHHHHHHHHHHH
```

[a] Self optimized prediction method (SOPM) for protein secondary structure prediction (Geourjon and Deleage, 1994).

Table II. Holley-Karplus Prediction[a]

α-apoprotein

```
SNPKDDYKIWLVINPSTWLPVIWIVATVVAIAVHAAVLAAPGSNWIALGAAKSAAK
-------eEEEEe-------hhHHHHHHHHHHHHHhhhhh-----hhHHHHHHHHh--
```

$\beta - apoprotein$

```
AERSLSGLTEEEAIAVHDQFKTTFSAFIILAAVAHVLVWVWKPWF
-------hHHHHHHHHHhh-hhhh-hHHHHHHHhhhhHHH-----
```

[a] Neural network based informational approach for protein secondary structure prediction (Holley and Karplus, 1989; 1991).

Table III. Alignment of Homologous Sequences[a]

alpha		SNPKDDYKIWLVINPSTWLPVIWIVATVVAIAVHAAVLAAPGSNWIALGAAKSAAK
1R1E|E	97:106	---KDDYGEWRVV---
2RCR|M	196:225	---------LFYNPFHGLSIAFLYGSALLFAMHGATILA---------------
1ACB|E	22:26	------------------------------------AVPGS------------
1CPC|L	68:72	--------------------------------------APGGN-----------
1TYP|A	91:102	--NWKALIAAKNKA-
beta		AERSLSGLTEEEAIAVHDQFKTTFSAFIILAAVAHVLVWVWKPWF
1AAM	349:365	---SFSGLTKEQVLRLREEF------------------------
1GRA	381:388	------GLTEDEAI----------------------------
256B|A	85:97	----------KEAQAAAEQLKTT--------------------
2RCR|L	122:133	----------------------AFAILAYLTLVL--------
1EPS	22:36	------------------KTVSNRALLLAALAH----------
2BBQ|A	52:61	----------------------------LRSIIHELLW-----
2RCR|L	266:272	-------------------------------WVKLPWW

[a] 1R1E: Eco Ri endonuclease (E.C.3.1.21.4) complex with TCGCGAATTCGCG; 2RCR: Photosynthetic reaction center from *Rhodobacter sphaeroides*; 1ACB: Alpha-Chymotrypsin (E.C.3.4.21.1) complex with Eglin C; 1CPC: C-Phycocyanin; 1TYP: Trypanothione reductase (E.C.1.6.4.8); 1AAM: Aspartate aminotransferase (E.C.2.6.1.1) mutant; 1GRA: Glutathione reductase (E.C.1.6.4.2) (oxidized) complex; 256B: Cytochrome b562 (oxidized); 1EPS: 5-enol-pyruvyl-3-phosphate synthase (E.C.2.5.1.9); 2BBQ: Thymidylate synthase (E.C.2.1.1.45) complex.

crease prediction quality. The Holley-Karplus prediction is an information-based neural network approach. SOPM and Holley-Karplus predictions are listed in Tables I and II, respectively. The results are consistent with the assignment of transmembrane segments, for both the α- and β-apoprotein, derived above. Both analyses demonstrate that the transmembrane segments have a high tendency to form an α-helix.

The secondary structure assignments were further verified and improved by homology modeling. Although there existed no structure which was highly homologous to either α- or β-apoprotein as a whole, we have found structures in the PDB (Protein Data Bank) which are homologous to multiple fragments of α- and β-apoproteins. Table III lists some of the homologous fragments to α- and β-apoproteins resulting from a PDB BLAST search (Altschul et al, 1990). Detailed PDB BLAST search results are given in Table IV. A homology with 26% identity and 50% positive exists between a segment of the α apoprotein [α-Leu-11 to α-Ala-40] and the transmembrane helix D of the M

Table IV(A). PDB BLAST Search Results[a]

α-apoprotein

---

(1) pdb|1R1E|E Identities = 6/10 (60%), Positives = 7/10 (70%)
```
Query:      4 KDDYKIWLVI 13
               KDDY W   V+
Sbjct:     97 KDDYGEWRVV 106
```

(2) pdb|2RCR|M Identities = 8/30 (26%), Positives = 15/30 (50%)
```
Query:     11 LVINPSTWLPVIWIVATVVAIAVHAAVLAA 40
               L   NP    L + ++   +  + A+H A + A
Sbjct:    196 LFYNPFHGLSIAFLYGSALLFAMHGATILA 225
2nd str^b:     TTS HHHHHHHHHHHHHHHHHHHHHHHHHHHT
```

(3) pdb|1ACB|E Identities = 4/5 (80%), Positives = 4/5 (80%)
```
Query:     39 AAPGS 44
               A PGS
Sbjct:     22 AVPGS 26
```

(4) pdb|1CPC|L Identities = 4/5 (80%), Positives = 4/5 (80%)
```
Query:     40 APGSN 45
               APG N
Sbjct:     68 APGGN 72
```

(5) pdb|1TYP|A Identities = 8/12 (66%), Positives = 9/12 (75%)
```
Query:     44 NWIALGAAKSAA 55
               NW AL AAK+ A
Sbjct:     91 NWKALIAAKNKA 102
2nd str^b:    -HHHHHHHHHHH
```

---

[a] Basic local alignment search tool (BLAST) search (Altschul et al, 1990).
[b] Secondary structure determined by DSSP (Kabsch and Sander, 1983). H: 4-helix (α-helix); E: extended strand, participates in β-ladder; T: H-bonded turn; S: bend.

Table IV(B). PDB BLAST Search Results

β-apoprotein

---

(1) pdb|1AAM| Identities = 6/16 (37%), Positives = 12/16 (75%)
```
Query:        4 SLSGLTEEEAIAVHDQ 19
                S SGLT+E+ + + ++
Sbjct:      349 SFSGLTKEQVLRLREE 364
2nd str:        EEE---TTTTTTSSSS
```

(2) pdb|1GRA| Identities = 7/8 (87%), Positives = 8/8 (100%)
```
Query:        7 GLTEEEAI 14
                GLTE+EAI
Sbjct:      381 GLTEDEAI 388
2nd str:        E--HHHHH
```

(3) pdb|256B|A Identities = 7/13 (53%), Positives = 9/13 (69%)
```
Query:       11 EEAIAVHDQFKTT 23
                +EA A  +Q KTT
Sbjct:       85 KEAQAAAEQLKTT 97
2nd str:        HHHHHHHTHHHHH
```

(4) pdb|2RCR|L Identities = 7/12 (58%), Positives = 8/12 (66%)
```
Query:       26 AFIILAAVAHVL 37
                AF ILA + VL
Sbjct:      122 AFAILAYLTLVL 133
2nd str:        HHHHHHHHHHHT
```

(5) pdb|1EPS| Identities = 7/15 (46%), Positives = 11/15 (73%)
```
Query:       21 KTTFSAFIILAAVAH 35
                KT +  ++LAA+AH
Sbjct:       22 KTVSNRALLLAALAH 36
2nd str*:       HHHHHHHHHHHHHHH
```

(6) pdb|2BBQ|A Identities = 4/10 (40%), Positives = 7/10 (70%)
```
Query:       30 LAAVAHVLVW 39
                L ++ H L+W
Sbjct:       52 LRSIIHELLW 61
2nd str:        HHHHHHHHHH
```

(7) pdb|2RCR|L Identities = 4/7 (57%), Positives = 5/7 (71%)
```
Query:       39 WVWKPWF 45
                WV  PW+
Sbjct:      266 WVKLPWW 272
2nd str:        HHT-TTS
```

---

subunit of the photosynthetic reaction center of *Rb. sphaeroides* [M subunit, residues 196:225] (Michel et al, 1986). To establish the statistical significance of this alignment, we performed a statistical analysis with the BESTFIT program in GCG package (Devereux et al, 1984). Using a gap generating penalty of 3.0 and a gap extension penalty of 0.1, the BESTFIT program generates exactly the same alignment as shown in Table III with a quality of 19.2. The average quality for 100 randomized alignments in which the query sequence is randomly permuted (shuffled) is 12.7 with a standard deviation of 1.3. That gives rise to a Z-score of 5, which indicates a "possibly significant" alignment according to (Lipman and Pearson, 1985; Pearson, 1990). Perhaps a more convincing support for this alignment is the fact that structurally, both proteins exist as α-helical transmembrane segment, and functionally, both proteins contain bacteriochlorophyll-binding residues. Also, two short segments of β-apoprotein are highly homologous to two corresponding segments in the L subunit of the photosynthetic reaction center of *Rb. sphaeroides* [see Table III]. Reaction center L subunit sequence 122:133 AFAILAYLTLVL is located in the center of the transmembrane helix C (Michel et al, 1986), which corresponds well with our secondary structure assignment of the transmembrane segment for the β-apoprotein. Sequence WVKLPWW near the C-terminal of the reaction center L subunit corresponds well with sequence WVWKPWF of the β-apoprotein of LH–II.

Homology modeling can also be used to improve secondary structure assignment. In case that no clear-cut secondary structure assignment can be made, the secondary structural features of the homologous structure can be employed to establish the secondary structure identity of α- and β-apoproteins. Specifically, the N and C termini of the transmembrane helix for the α-apoprotein were set to Ser-16 and Ala-41 in analogy to the homologous transmembrane helix D of the reaction center M subunit and in consideration of the known fact that all residues in the NPS (residue 14:16) and PGSN (residue 41:44) segments have a high tendency to form reverse turn (Levitt, 1978). Similarly, the C terminus of the transmembrane helix for the β-apoprotein was set to Lys-42 in analogy to the homologous reaction center L subunit sequence 266:272 WVKLPWW and in consideration of the proline residue. The N terminus of the transmembrane helix for the β-apoprotein was determined to be Tyr-10 in analogy to the highly homologous glutathione reductase sequence GLTEDEAI. This assignment of the secondary structure at the N terminal is consistent with both SOPM and Holley-Karplus predictions.

Table V. Assignment of Secondary Structure and Topology[a]

	α-apoprotein
Topology	------------TTTTMMMMMMMMMMMMMMMMMMMMMMMTTTT----IIIIIIIIII-
Sequence	SNPKDDYKIWLVINPSTWLPVIWIVATVVAIAVHAAVLAAPGSNWIALGAAKSAAK
Secondary structure	CCCCCCCCEEEEECCHHHHHHHHHHHHHHHHHHHHHHHHHHHHHCCCCCHHHHHHHHHHC

	β-apoprotein
Topology	-------------TTTTMMMMMMMMMMMMMMMMMMMMMMTTTT---
Sequence	AERSLSGLTEEEAIAVHDQFKTTFSAFIILAAVAHVLVWVWKPWF
Secondary structure	CCCCCCCCCHHHHHHHHHHHHHHHHHHHHHHHHHHHHHHHHHHHCCC

[a] M—Transmembrane region; T—Terminal (interfacial) region; I—Interfacial helix.

The final secondary structure assignment for both the α- and β-apoproteins is listed in Table V. It is a consensus assignment derived from a combination of all the prediction methods used. It should be pointed out that in addition to a transmembrane helix, an interfacial helix of 10 residues [α-Ile-46 to α-Ala-55] has also been identified for the α-apoprotein at the C-terminal. This assignment is supported by the following observations: (1) residues in the sequence IALGAAKSAA have a high propensity to form an α-helix as evident from SOPM and Holley-Karplus analyses [see Tables I and II] and other propensity analyses which we have performed; (2) the homologous fragment KALIAAKNKA from trypanothione reductase, as listed in Table III, is an α-helix; and (3) as shown in Figure 5, the segment is highly amphiphilic and suitable for sitting at the interfacial region. The Trp and charged Lys residues face the lipid head group and all the hydrophobic residues face the interior of the membrane. This helical wheel representation of amphiphilic helix can be quantified in terms of hydrophobic moment (Eisenberg et al, 1982; Eisenberg, 1984; Eisenberg et al, 1984).

A reverse-turn segment for the α-subunit, PGSN, was assigned based on a propensity analysis with Levitt's scale (Levitt, 1978). Shown in Figure 6 is the propensity profile of the transmembrane helix and reverse-turn for the α-apoprotein. Solid circles indicate the transmembrane helix propensity (for a single sequence) based on the Persson and Argos scale (Persson and Argos, 1994); open circles represent the reverse-turn propensity according to Levitt (Levitt and 1978). All four residues show a reverse-turn propensity higher than one. In

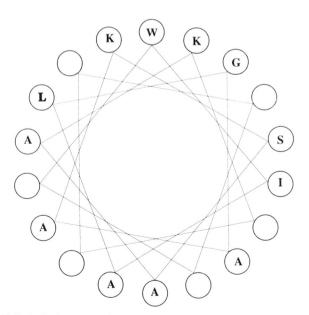

Figure 5. Helical wheel representation of the interfacial α-helix from α-Trp-45 to α-Lys-56.

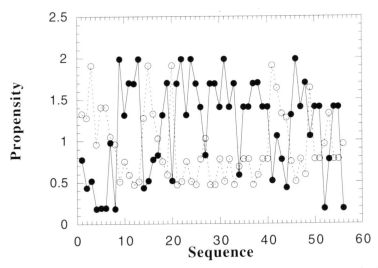

Figure 6. Propensity profile of transmembrane helix and reverse-turn for α-apoprotein. Solid circles—transmembrane helix propensity base on Persson and Argos scale (Persson and Argos, 1994); open circles—reverse-turn propensity by Levitt (Levitt, 1978).

summary, we have identified a [transmembrane helix–reverse-turn–interfacial helix] motif for the α-apoprotein. Such a motif has also been observed in other membrane proteins (Kühlbrandt et al, 1994; White, 1994).

Construction of the α and β Apoproteins

Based on the predicted secondary structure feature, tertiary structures for both the α- and β-apoproteins were built by means of comparative modeling using the program MODELLER (Sali and Blundell, 1993). In comparative modeling, the homologous structure is used as a template for the unknown structure. In our case, homologous fragment structures, as aligned in Table III, are employed in such a procedure. In MODELLER (Sali and Blundell, 1993), the three dimensional model of the unknown protein is obtained by satisfying spatial restraints in the form of probability density functions (pdfs) derived from the alignment of the unknown with one or more homologous structures. The pdfs restrain C_α–C_α distances, main-chain N–O distances, main-chain and side-chain dihedral angles. The optimization is carried out by the variable target method that applies the conjugated gradient algorithm to positions of all non-hydrogen atoms. In the present case, the transmembrane helical segments of the α-apoprotein and of the β-apoprotein, together with the interfacial helix of the α-apoprotein were built with the program MODELLER. To build the [transmembrane helix–reverse-turn–interfacial helix] motif for the α-apoprotein, the transmembrane helix and the interfacial helix were optimally linked with the reverse-turn fragment. In the normal practice of comparative modeling, the reverse-turn fragment is selected from a fragment library to generate the best fit between main secondary structure segments. Here, the reverse-turn fragment was built by superimposing two homologous fragments `AVPGA` and `APGGN` as listed in Table III. The rest of the terminal residues were also added using the corresponding homologous structures as templates. The α- and the β-apoproteins so constructed were each optimized by energy minimization with fixed protein backbone.

Placement of BChla's and Construction of $\alpha\beta$ Heterodimer

BChla is an integral part of the pigment-protein complex. Placement of BChla in the complex is essential to model building. Resonance Raman spectra demonstrated that all Mg^{2+} of all BChla molecules in the LH–II complex of *Rs. molischianum* are 5-coordinate (Germeroth et al, 1993). Multiple sequence alignment of light-harvesting complexes (see Figure 3) reveals three highly conserved His residues. Most likely, the histidines form the binding site for the three BChla's, a hypothesis supported by site-directed mutagenesis experiments on related systems (Bylina et al, 1988; Crielaard et al, 1994; Visschers et al, 1994; Zuber, 1986). As can be seen from sequence alignment (Figure 3), the motif α-Ala-30 ... α-His-34 ... α-Leu-38 is highly conserved (see Figure 7). It has been

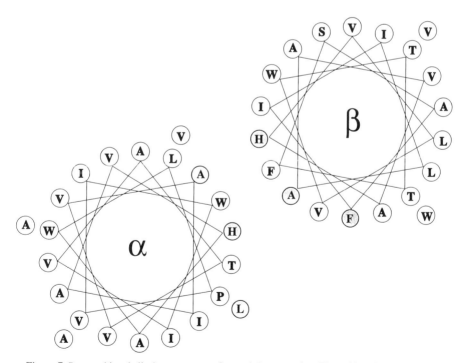

Figure 7. Proposed interhelical arrangement of α- and β-apoproteins. The residues in shaded circles are highly conserved.

reported that the corresponding Ala (-4 residue of the conserved His) in LH–I of *Rb. capsulatus* can only be substituted with small residues, Gly, Ser and Cys (Bylina et al, 1988) indicating the structural importance of "smallness" of the residue in the -4 position to His. A similar binding pocket β-Phe-27 ... β-Ala-31 ... β-His-35 is also highly conserved (see Figure 7) in the β-apoprotein. The conserved Phe residue in the -8 position to His is also significant (Zuber and Brunisholz, 1991). However, there exists no such conserved binding pocket for β-His-17, making the assignment of β-His-17 as the binding site for B800 BChla less certain.

Residue conservation can not only provide clues about the helix-pigment interactions, but also provide clues about helix-lipid interactions. The lipid facing residues tend to be more evolutionarily variable than internal residues (Donnelly et al, 1993; von Heijne and Manoil, 1990). The highly conserved residues, for the α-apoprotein (α-Ala-30 ... α-His-34 ... α-Leu-38) are four residues apart and thus aligned on the same side of the helical wheel, and so are the β-apoprotein (β-Phe-27 ... β-Ala-31 ... β-His-35). Taking into account the conserved residues in both the α- and the β-apoprotein, the interhelical arrangement, as shown in Figure 7, is the most probable one. The side of highly conserved residues, on both the α- and the β-apoprotein, faces inward and the other side is in contact with lipid bilayers. It has been observed that many transmembrane α helices are amphipathic, with opposing polar and nonpolar faces oriented along the long axis of the helix (Rees et al, 1989; Segrest et al, 1990; von Heijne and Manoil, 1990). Energetically, the apolar surfaces of transmembrane helices provide a good interface to the lipid bilayers, the polar side of inward-facing residues tends to be in the interhelical region to eliminate possible contacts between polar residue and lipid (Rees et al, 1989). In our case, transmembrane cores for both the α- and the β-subunit are highly hydrophobic as shown in the helical wheel representation of Figure 7. Except for the BChla binding histidine residues, the core of both transmembrane segments consists of nearly all nonpolar residues. On one hand, this potentially useful rule for arranging interhelical packing is not applicable to the current case. On the other hand, it may be a good indication of a lack of interhelical contacts in the transmembrane core for the LH–II complex.

As we build the heterodimer, the following well-known facts were also taken into consideration: (1) based on the observation of cleavage of part of the N-terminal domains of the α- and β-apoproteins of LH–I of *Rs. rubrum* on the cytoplasmic side of the membrane by partial hydrolysis with protelytic enzymes (Brunisholz et al, 1984), both α- and β-apoproteins should be oriented with their N-terminals towards the cytoplasm; (2) strong circular dichroism (CD) signals suggest exciton interactions between pairs of BChla molecules (Zuber, 1993 and references therein); (3) linear dichroism data indicate that the B850 Bchla's are approximately perpendicular to the membrane and the B800 BChla is parallel to the membrane (Kramer et al, 1984); (4) Fourier-transform Resonance Raman spectroscopy and site-directed mutagenesis (Fowler et al, 1994) of a related light-harvesting complex indicate that another highly conserved residue, α-Trp-

45, is hydrogen-bonded to the 2-acetyl group of BChla. All these observations impose constraints on the structure: (1) the $\alpha\beta$ heterodimer should be arranged interhelically with two B850 BChla binding histidine residues facing each other as shown in Figure 7; (2) the two B850 BChla's are paired and oriented perpendicular to the membrane plane. Therefore, the basic unit of the LH–II complex was configured with the B850 BChla pair sandwiched between two helices of the $\alpha\beta$ heterodimer. We will come back to the consequence of hydrogen bond constraint in the following section.

Construction of the Complete Octamer

The construction of the complete aggregated complex is based on the two stage model suggested in (Popot and Engelman, 1990) which assumes that the individual helices are formed prior to the formation of the helix bundle. We developed a protocol to aggregate the transmembrane helices into an octamer of eight $\alpha\beta$ heterodimers by means of molecular dynamics simulations and energy minimization under the constraints of experimental data (see Figure 1).

Our procedure consists of three essential steps: In a first step, the α- and β-apoproteins were constructed by comparative modeling based on information obtained through homology and secondary structure analyses as described above. The optimized tertiary structures for the α- and the β-apoprotein were preserved in the subsequent simulations by applying (1) for helical backbone harmonic restraints to the distances between the i-th carbonyl and the $(i + 4)$-th amide nitrogen and to the distances between the i-th and the $(i + 4)$-th C_α; by applying (2) in the turn region the harmonic restraints to the dihedral angles of the main chain. No restraint was applied to all coiled terminal residues. Two BChla's binding to β-His-17 and β-His-35 were manually attached to the β-apoprotein using the program QUANTA, and a third BChla was attached similarly to α-His-34. We employed the heavy atom coordinate of BChla from the crystal structure of BChlb in the photosynthetic reaction center of *Rps. viridis* (Deisenhofer et al, 1985) and added explicit hydrogens using X-PLOR's function *hbuild*. The binding conformation between BChlb and His in the crystal structure of *Rps. viridis* was employed for placement of the BChla's. The X-PLOR utility *patch* was used to build the ligand bond between magnesium and the nitrogen of His, and was followed by rigid body minimization between the apoprotein and the BChla's. Subsequently, energy minimization runs were performed with harmonic restraints, followed by three 1 ps molecular dynamics runs at consecutively increasing temperatures of 100, 200 and 300 K to equilibrate the system. The lycopenes were not included in the model structure.

In the second step, the complete octamer was constructed enforcing an eight-fold symmetry. A self rotational search (Brünger, 1992; McRee, 1993; Lattman, 1985) of the X-ray diffraction data indicated that, most likely, the LH–II complex possesses an eightfold symmetry. Thus, the task of constructing

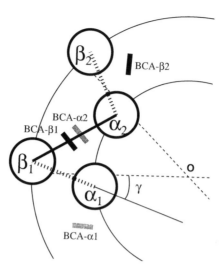

Figure 8. Definition of $\alpha\beta$ pair geometries. Heterodimer: $\alpha\beta$ dimer formed between β_1 and α_2, linked by a solid line; Neighboring pair: $\alpha\beta$ dimer formed between β_1 and α_1, linked by a dashed line; γ is defined as the angle between the vector linking two helices and the radial vector from the center of the octamer to the center of the neighboring pair.

the octamer was reduced to building a protomer for the eightfold octamer. There were two possible choices for the protomer: the $\alpha\beta$ heterodimer or a neighboring $\alpha\beta$ pair as shown in Figure 8. Since we intended to perform an intermediate step optimization for the $\alpha\beta$ dimer, the neighboring $\alpha\beta$ pair was chosen as the protomer. The rationale behind our choice was that in the $\alpha\beta$ heterodimer ($\alpha_2\beta_1$ in Figure 8), the two helices separated by the B850 BChla pair are too far apart to generate any significant interaction to be optimized. In contrast, in the neighboring $\alpha\beta$ pair ($\alpha_1\beta_1$ in Figure 8), inter-unit hydrogen bonds exist. As mentioned above, the α-Trp-45 is hydrogen-bonded to the 2-acetyl group of BChla. However, there is no direct information about which of the two B850 BChla's is involved in such a hydrogen bond (Fowler et al, 1994). Based on the optimized tertiary structure for both α and β subunits,we found that it is more feasible spatially to form an inter-unit hydrogen bond, i.e., a hydrogen bond between α-Trp-45 and 2-acetyl group of BChla binding to β-His-35. Since open hydrogen bonds are extremely unstable in a lipid environment, we also arranged to have β-Trp-41 hydrogen bonded to the 2-acetyl group of BChla binding to α-His-34. We used QUANTA to place the neighboring pair (β_1 and α_1 as shown in Figure 8) so that: (1) the $\alpha\beta$ heterodimers are arranged in a configuration as shown in Figure 7; (2) hydrogen bonds were established between α-Trp-45 and the 2-acetyl group of BChla attached to β-35 His, and between β-Trp-41 and the 2-acetyl group of BChla binding to α-His-34. We applied the "soft van der Waals" option in X-PLOR to minimize the energy content of neighboring $\alpha\beta$ pair, followed by rigid body minimization with two helices and three BChla's as five rigid bodies. Then, a simulated annealing was applied by heating the system to 2000 K and slowly cooling it down. This procedure was followed by a 100 ps dynamics run at 300 K, obtaining an equilibrated structure of the neighboring $\alpha\beta$ pair.

In a third step, the equilibrated neighboring $\alpha\beta$ pair with all three bound BChla's were combined into an octamer by means of long time molecular dynamics simulations and energy minimization with eightfold symmetry and all the restraints as described above enforced throughout the entire simulation. An iterative protocol consisting of multiple cycles was employed to optimize the octamer structure. Each cycle started with a 200 step rigid body minimization (with the entire protomer as a rigid body), followed by a 2.5 ps rigid body dynamics run at 600 K, then a 5 ps molecular dynamics run at 300 K, ending with a 200 step Powell minimization. The radius of the octamer was monitored to detect convergence. Initially, the octamer was constructed with an outer diameter as large as 100 Å to avoid any close inter-unit contact. During the first cycle, a large cutoff distance for nonbonded interactions was required (we chose 15 Å), since the initial inter-unit distances were large. The iterative process was terminated when the final radii at the end of two consecutive runs differed by less than 0.25 Å. At the end, the octamer was minimized again with the Powell algorithm until it converged to a minimum energy configuration or until a limit of 700 cycles was reached. It normally took about 50–100 ps of equilibration (10 to 20 iterations) before the radii converged. Test runs with longer times indicated that the radius of the octamer began to fluctuate around its average value after 50 ps. The underlying physical principle behind this simulation protocol is energy minimization perturbed by successive dynamic equilibration, which can be viewed as a dynamical analogy to simulated annealing.

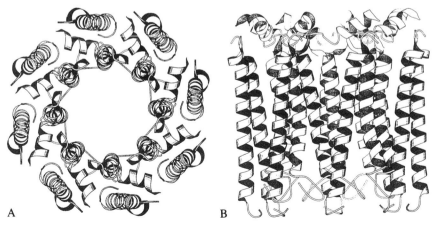

A   B

Figure 9. Schematic structure of the LH–II complex: (a) top view with C terminus pointing upward, showing a ring structure with 70 Å diameter, with α inside and β outside; (b) side view with C terminus on top. [produced with MOLSCRIPT (Kraulis, 1991)].

Protein folding is related to the problem of global minimization, the solution of which relies heavily on proper initial configurations. Starting from a heterodimer built within the constraints of all the biochemical and spectroscopic data dramatically reduces the phase space to be sampled, enhancing the chance of placing the initial structure in the basin of attraction of the global minimum. A systematic sampling of initial configurations has also been attempted by varying the relative orientations (angle γ in Figure 8) of the vector linking two helices and the radial vector from the center of the octamer to the center of the neighboring pair. The best predicted structure is determined by the molecular replacement procedure as described in the following session. The structure that gives rise to the lowest R value is deemed the most optimal one.

Shown in Figure 9 is one of the octamers we have optimized. The octamer forms a ring with an outer diameter of about 70 Å with α inside and β outside. This 70 Å outer diameter covers the side chains of the β-apoprotein. Typically, the radius of the eight β poproteins (measured from the center of the helices) varied between 27.8 and 32.8 Å; the radius of the eight α apoproteins varied between 15.1 and 21.0 Å. The Mg to Mg separation in the $\alpha\beta$ heterodimer was about 8.8 Å. The Mg to Mg separation between adjacent heterodimers was 16.0 Å. The inner helices were tilted against the membrane plane normal by about 6° and the outer helices were tilted by about 9°. The resulting BChla's are optimally oriented to capture light coming from all directions and are close enough for efficient energy transfer (see the following section on exciton prop-

Figure 10. Schematic drawing of the $\alpha\beta$ heterodimer, i.e., the monomer "unit" of LH–II. Shown are the predicted secondary structure and locations of three Bchla's binding to α-34 His, β-17 His and β-35 His. The C terminus is on top.

erties). As depicted in Figure 10, the B850 BChla pair are sandwiched between two helices of the $\alpha\beta$ heterodimer. Such an arrangement has recently been suggested for the LH–I complex of *Rs. rubrum* to interpret the ring–shaped 8.5 Å resolution projection map (Karrasch et al, 1995). It should be pointed out that our model shows the expected placement of the two B850 Bchla's perpendicular to the membrane, near the periplasmic side. However, the B800 BChla is not oriented parallel to the cytoplasmic side of the membrane as expected from LD data due to a lack of helix turn residues near β-His-17. As stated above, the placement of BChla at β-His-17 is itself problematic. The reason we placed it there is that we did not have any better clues at the time when the work reported here was carried out.

Future Work

It is worth mentioning that the optimized octamers are not selected based on the criterion of energy minimization. Although the vacuum environment is similar to a lipid bilayer in terms of hydrophobicity, the peptide-lipid interactions are not explicitly included in the conformational energy. Instead, the molecular replacement test is employed as the ultimate test of the correctness of the model. The predicted structures are presently employed as a search model in the framework of the molecular replacement method as implemented in the program X-PLOR (Brünger, 1992). In the molecular replacement method, a six-dimensional search is required to find the best match between observed and calculated diffraction data. In practice, the method is implemented as a three-dimensional rotational search followed by a three-dimensional translational search (Lattman, 1985; Rossmann, 1972). We also perform a Patterson correlation refinement preceding the translational search to filter the peaks of the rotational search (Brünger, 1990). The preliminary results have been encouraging as demonstrated by the model's ability to probe the position of molecules in the crystal. The rotational search orients the α helices of the octamers in a direction parallel to the c-axis of the crystal unit cell. The translational search further places the octamers in the site of the four-fold axis of the crystal. Currently, for our best model with 6-12 Å resolution data, the R value after rigid body refinement is 49.8% and decreases to 29.8% after positional refinement.

Work is in progress at two fronts to resolve the X-ray diffraction data for the LH–II of *Rs. molischianum*: (1) a more systematic sampling of initial octamer configurations is undertaken; (2) a rigid body simulated annealing protocol is being developed and will be implemented in both the Patterson correlation refinement and the rigid body refinement procedures (Brünger, 1990; 1992) in place of the rigid body minimizer to improve the convergence radius of the minimizer. It is our hope that the structure will be further refined to resolve the 2.4 Å diffraction data. If successful, this work will be the first using an *ab initio* predicted structure as a probe structure in the framework of molecular replacement to solve the phase problem in X-ray crystallography structure determination. The methodology developed may be useful for structure prediction of other integral membrane proteins.

Comparison with the Structure of LH–II of *Rps. acidophila*

This chapter is based on an article we recently submitted for publication in Protein Science (Hu et al, 1995).[1] As this chapter was written, the crystal structure of LH–II from *Rps. acidophila* determined by conventional multiple isomorphous replacement method was published by McDermott et al (Mcdermott et al, 1995). This permits us to compare our prediction to an observed structure, albeit not for the same protein. There exists significant sequence homology between the α-subunit from LH–II of *Rps. acidophila* strain 10050 and *Rs. molischianum*. For the β-subunit, the two sequences are homologous, but to a lesser extent. In Table VI, a comparison between the predicted secondary structure assignment for *Rs. molischianum* and the X-ray resolved secondary structure for *Rps. acidophila* (Mcdermott et al, 1995) is given.

The predicted secondary structure of LH–II of *Rs. molischianum* compares well with that of the *Rps. acidophila* crystal structure. In particular, the predicted [transmembrane helix–reverse-turn–interfacial helix] motif for the α-apoprotein was observed in the crystal structure.

For comparison, the overall structure of the aggregated LH–II complex of *Rps. acidophila* is briefly outlined below. For a complete description, interested readers are referred to the original publication (Mcdermott et al, 1995). The LH–II complex of *Rps. acidophila* is a ring-shaped aggregate of nine αβ heterodimers with nine-fold symmetry. "The transmembrane helices of nine α-apoprotein are packed side by side to form a hollow cylinder of radius 18 Å. The nine helical β-apoproteins are arranged radially with the α-apoproteins to form an outer cylinder of radius 34 Å. The α-apoprotein helices are parallel to the ninefold axis to within 2°, and the β-apoprotein helices are inclined by 15° to this axis" (Mcdermott et al, 1995). The two B850 BChla's are bonded to the two conserved histidine residues at the periplasmic side, sandwiched between the αβ heterodimer and oriented perpendicular to the membrane plane. The B800 BChla is oriented parallel to the membrane plane, and bonded to the formyl group of the f-Met residue at the N terminus of the α-apoprotein. Overall, the B850 BChla's form almost a symmetrical ring, with an Mg to Mg distance of 8.7 Å within the αβ heterodimer and 9.7 Å between the adjacent heterodimers. It should be pointed out that the aggregated LH–II complex of *Rs. molischianum* consists of eight αβ heterodimers while that of *Rps. acidophila* consists of nine heterodimers. The overall helical assembly of the LH–II complex of *Rs. molischianum* described in the previous section corresponds well with that of *Rps.*

---

[1] A modified version of this article has been published in *Protein Science*, 4:1670, 1995.

Table VI. Comparison of Secondary Structures

	α-apoprotein
Rs. molischianum	SNPKDDYKIWLVINPSTWLPVIWIVATVVAIAVHAAVLAAPGSNWIALGAAKSAAK
Prediction[a]	CCCCCCCCEEEEECCHHHHHHHHHHHHHHHHHHHHHHHHHHHCCCCCHHHHHHHHHHC
Rps. acidophila	---MNQGKIWTVVNPAIGIPALLGSVTVIAILVHLAILSHTTWFPAYWQGGVKKAA
X-ray[b]	---C333333333HHHHHHHHHHHHHHHHHHHHHHHHHHHCCCHHHHHHH??????

	β-apoprotein
Rs. molischianum	AERSLSGLTEEEAIAVHDQFKTTFSAFIILAAVAHVLVWVWKPWF-
Prediction	CCCCCCCCCCHHHHHHHHHHHHHHHHHHHHHHHHHHHHHHHHHHCCC-
Rps. acidophila	-----ATLTAEQSEELHKYVIDGTRVFLGLALVAHFLAFSATPWLH
X-ray	-----CCCCCChhhhHHHHHHHHHHHHHHHHHHHHHHHHHHHHHCCCC

[a] H—Helix; E—β-Sheet; C—Coil; 3—3 Turn Helix; ?—Unresolved.
[b] The secondary structures for LH–II of *Rps. acidophila* are extracted from (Mcdermott et al, 1995).

acidophila. The biggest discrepancy lies at the placement of BChla's. While our predicted structure place the B850 BChla pair unevenly, the X-ray structure shows that all B850 BChla's are arranged nearly symmetrically in a ring conformation. As expected, the B800 BChla is oriented parallel to the membrane plane in the crystal structure. A surprise is the binding of the B800 BChla to the formyl group of fMet residue at the N terminus of the α-apoprotein, which for the purpose has to dive into the interior of the membrane by as much as 9 Å. As stated before, our difficulties in placing the B800 BChla are attributed to a lack of pertinent biochemical information. It should also be born in mind that although in the majority of LH–II, even LH–I complexes, there exists a corresponding Met residue at the N terminus of the α-apoprotein, the N terminus of the α-apoprotein of *Rs. molischianum* is a Ser residue. As result, a slightly different conformation for *Rs. molischianum* is expected. The difference in the number of the constituting αβ dimers (eight vs. nine) is a clear indication of a difference as well. The extent of this difference will be clear once the structure of LH–II from *Rs. molischianum* has been determined by x-ray crystallography at 2.4 Å resolution.

Exciton Properties

The ring-like structure of the light harvesting system with its approximate eight-fold symmetry poses the question in how far this architecture of the protein is an optimal choice for its function, namely, to absorb sunlight and transfer the energy absorbed to the photosynthetic reaction center. We want to demonstrate here that the architecture, which realizes a cyclic arrangement of chlorophyls, is indeed optimal, yielding an excitonic structure which can serve very well for the energy transfer. For this purpose we idealize the arrangement of the chlorophyls to one in which a 2N-fold symmetry axes exists for the 2N chromophores. Adjacent chromophores are oriented to each other in close analogy to the arrangement of the special pair in the photosynthetic reaction center of *Rps. viridis* (Deisenhofer et al, 1985) for which the coupling energies are well known (see e.g., [Eccles et al, 1988]).

The proximity of the chlorophyls leads to a coupling of their excited states such that the stationary excited states are actually linear combinations of excited states of the individual chromophores, denoted by j where $j = 1, 2, \ldots 2N$. The relevant excited state of the chlorophyls is the so-called Q_y state. The chlorophyls in LH–II of *Rps. acidophila* and *Rs. molischianum* are positioned such that the transition dipole moments $\vec{D}_j = \langle\text{ground}|\vec{\mathcal{D}}|\text{Chl}_j^*\rangle$ are oriented in the plane of the membrane (Kramer et al, 1984; Zuber, 1993). Here Chl$_j$ denotes the Q_y excited state of the j-th chlorophyl and $\vec{\mathcal{D}}$ denotes the dipole moment operator. The transition dipole moments of neighboring cholophyls are approximately anti-parallel. If one assumes otherwise a perfect 2N symmetry, the transition dipole moments are given by the Cartesian vectors

$$\vec{D}_j = D_0 \begin{pmatrix} \cos\phi_j \\ \sin\phi_j \\ 0 \end{pmatrix}, \qquad \phi_j = \frac{\pi j(N+1)}{N}, \quad j = 1, 2, \ldots, 2N. \quad (1)$$

The system of 2N chlorophyl has 2N states in which a single chlorophyl is in the excited Q_y state. These states are formally written

$$|j\rangle = |\text{Chl}_1\,\text{Chl}_2 \cdots \text{Chl}_j^* \cdots \text{Chl}_{2N}\rangle, \quad j = 1, 2, \ldots, 2N. \quad (2)$$

The states $|j\rangle$ are coupled which each other through Coulomb interactions between the chlorophyls. At present, we neglect all couplings, except those between adjacent chlorophyls. From studies of the excitonic structure of the photosynthetic reaction center this coupling is well known. It holds for the Hamiltonian matrix in the adopted notation

$$\langle j|\hat{H}|j \pm 1\rangle = V_o \quad (3)$$

where V_o is positive and approximately equal to 500 cm^{-1} in case of the special pair of the photosynthetic reaction center (Eccles et al, 1988). The diagonal elements of the Hamiltonian are

$$\langle j|\hat{H}|j\rangle = E_0 \quad (4)$$

where E_o is the excitation energy of the Q_y states of the individual chlorophyls. All other matrix elements vanish in the stated approximation, i.e.,

$$\langle j|\hat{H}|k\rangle = 0 \quad \text{for } k \neq j, j \pm 1. \quad (5)$$

For the eigenstates of the Hamiltonian holds

$$\hat{H}|\widetilde{n}\rangle = \epsilon_n|\widetilde{n}\rangle \quad (6)$$

where

$$|\widetilde{n}\rangle = \frac{1}{\sqrt{2N}} \sum_{j=1}^{2N} e^{ijn\pi/N}|j\rangle \quad (7)$$

and

$$\epsilon_n = E_o + 2V_o \cos \frac{\pi n}{N}. \tag{8}$$

In these expressions n are integers in the range $n = -N + 1, -N + 2, \ldots, N$. The state of lowest energy, for the present case with $V_o > 0$, is

$$\epsilon = \epsilon_N = E_o - 2V_o. \tag{9}$$

The states $|\widetilde{n}\rangle$ represent superpositions of the excited states of the individual chlorophyls and are referred to as excitons. Some of these exciton states carry oscillator strength, i.e., they are capable of absorbing and emitting photons. This property is governed by the transition dipole moments

$$\langle \text{ground}|\vec{\mathfrak{D}}|\widetilde{n}\rangle = \frac{1}{\sqrt{2N}} \sum_{j=1}^{2N} e^{ijn\pi/N} \vec{D}_j \tag{10}$$

Using (1) one can derive

$$|\langle \text{ground}|\vec{\mathfrak{D}}|\widetilde{n}\rangle|^2 = ND_o^2(\delta_{n,-N+1} + \delta_{n,N-1}). \tag{11}$$

Hence, only two excitonic states carry oscillator strength. These states have identical energies

$$\epsilon_\pm = E_o - 2V_o \cos \frac{\pi}{N} \tag{12}$$

which is for large enough N approximately

$$\epsilon_\pm \approx \epsilon + 2V_o \left(\frac{\pi}{N}\right)^2. \tag{13}$$

We can summarize our findings stating that the coupling of the excited Q_y states in LH–II, in the case of a $2N$-fold symmetry axes for the arrangement of chlorophyls, leads to an exciton spectrum with lowest energy ϵ as given by (9) carrying zero oscillator strength, two degenerate excited states of approximate energy $\epsilon + 2V_o(\pi/N)^2$, carrying all the oscillator strength and $2N - 3$ further excited states of higher energy, with maximum energy $\epsilon + 2V_o$, none of which carries oscillator strength. Assuming a coupling strength $V_o = 400 \text{ cm}^{-1}$, i.e., 20 percent less than the value for the special pairs in photosynthetic reaction centers, one can estimate

$$\epsilon_\pm \approx \epsilon + 0.1\text{eV} \left(\frac{\pi}{N}\right)^2. \tag{14}$$

such that, for $N = 8$, holds $\epsilon_\pm \approx \epsilon + 0.015$ eV. This implies that the energy separation between the lowest energy ϵ and the energy $\epsilon_\pm$ of the optically allowed exciton states is less than the thermal energy at room temperature. As a result, one can expect that in thermal equilibrium electronic excitations of LH–II are strongly allowed and, hence, can transfer their energy efficiently towards the reaction center.

It is of interest to speculate how this optimal situation could be interrupted, resulting in less than optimal energy transfer as may be desirable in case of intense light. For this purpose the protein could move the chlorophyls pairwise together, such that the central Mg^{2+} ions lie as close as possible to each other, given a certain separation of the chlorophyl planes. In this case the exciton Hamiltonian would become (specifying only non-vanishing matrix elements)

$$\hat{H} = \begin{pmatrix} E_o & V_1 & & & & & V_2 \\ V_1 & E_o & V_2 & & & \cdots & \\ & V_2 & E_o & V_1 & & \cdots & \\ & & V_1 & E_o & V_2 & & \\ & & & V_2 & E_o & & \\ & \vdots & \vdots & & & \ddots & \\ V_2 & & & & & & E_o \end{pmatrix} \qquad (15)$$

In this case, the matrix elements V_1 would be large and negative and the matrix elements V_2 would be small and positive. The resulting exciton spectrum would consist of a narrow band of optically forbidden exciton states of energy $E_o + V_1 - V_2 < E < E_o + V_1 + V_2$, separated by a wide energy gap $\Delta \approx 4V_1$ which could easily exceed thermal energies from a band of optically allowed exciton states of energy $E_o - V_1 - V_2 < E < E_o - V_1 + V_2$. In this case the optically allowed states would not be thermally accessible. Energy transfer would involve optically forbidden exciton states and, hence, would be less efficient, resulting in internal conversion of the absorbed photon energy in the LH–II proteins.

ACKNOWLEDGMENTS

This work was supported by the Carver Charitable Trust and the National Institute of Health (NIH grant P41RR05969). We would like to thank Antony Crofts and Zan Schulten for many helpful discussions. James Schnitzer participated in this work during the early phase of the project.

REFERENCES

Allen J, Yeates T, Komiya H, Rees D (1987): Structure of the reaction center from *Rhodobacter sphaeroides* R-26: The protein subunits. *Proc Natl Acad Sci USA* 84:6162

Altschul SF, Gish W, Miller W, Myers EW, Lipman DJ (1990): Basic local alignment search tool. *J Mol Biol* 215:403–410

Argos P, Rao J, Hargrave P (1982): Structural prediction of membrane-bound proteins. *European Journal of Biochemistry* 128:565

Arkin I, Adams P, MacKenzie K, Lemmon M, Brünger A, Engelman D (1994): Structural organization of the pentameric transmembrane α-helices of phospholamban, a cardiac ion channel. *Embo Journal* 13:4757

Blundell TL, Sibanda BL, Sternberg MJ, Thornton JM (1987): Knowledge-based prediction of protein structures and the design of novel molecules. *Nature* 326:347

Boonstra AF, Visschers RW, Calkoen F, van Grondelle R, van Bruggen EF, Roekema EJ (1993): Structural characterization of the B800-850 and B875 light-harvesting antenna complexes from *Rhodobacter-Sphaeroides* by electron microscopy. *Biochimica et Biophysica Acta* 1142:181

Brooks BR, Bruccoleri RE, Olafson BD, States DJ, Swaminathan S, Karplus M (1983): CHARMm: a program for macromolecular energy, minimization, and dynamics calculations. *J Comp Chem* 4(2):187–217

Brünger AT (1990): Extension of molecular replacement: A new search strategy based on Patterson correlation refinement. *Acta Cryst* A46:46–57

Brünger AT (1992): *X-PLOR, Version 3.1, A System for X-ray Crystallography and NMR*. The Howard Hughes Medical Institute and Department of Molecular Biophysics and Biochemistry, Yale University

Brunisholz RA, Wiemken V, Suter F, Bachofen R, Zuber H (1984): The light-harvesting polypeptides of *Rhodospirillum rubrum*. II. localisation of the amino-terminal regions of the light-harvesting polypeptides B870-α and B870-β and the reaction-centre subunit L at the cytoplasmic side of the photosynthetic membrane of *Rhodospirillum rubrum* G-9+. *Hoppe-Seylers Zeitschrift fur Physiologische Chemie* 365:689

Busetta B (1986): Examination of folding patterns for predicting protein topologies. *Biochimica et Biophysica Acta* 870:327

Bylina E, Robles S, Youvan D (1988): Directed mutations affecting the putative bacteriochlorophyll-binding sites in the light-harvesting I antenna of *Rhodobacter capsulatus*. *Israel Journal of Chemistry* 28:73

Chou P, Fasman G (1978): Prediction of the secondary structure of proteins from their amino acid sequence. *Advances in Enzymology and Related Areas of Molecular Biology* 47:45

Cohen FE, Abarbanel RM, Kuntz ID, Fletterick RJ (1986): Turn prediction in proteins using a pattern-matching approach. *Biochemistry* 25:266

Cornette J, Cease K, Margalit H, Spouge J, Berzofsky J, DeLisi C (1987): Hydrophobicity scales and computational techniques for detecting amphipathic structures in proteins. *Journal of Molecular Biology* 195:659

Cowan SW, Schirmer T, Rummel G, Steiert M, Ghosh R, Pauptit RA, Jansonius, JN, Rosenbusch JP (1992): Crystal structures explain functional properties of 2 *E. coli* porins. *Nature* 358(6389):727–733

Cramer W, Engelman D, Heijne GV, Rees D (1992): Forces involved in the assembly and stabilization of membrane proteins. *Faseb Journal* 6:3397

Creighton T, ed. (1992): *Protein folding*. New York: WH Freeman

Crielaard W, Visschers R, Fowler G, van Grondelle R, Hellingwerf K, Hunter C (1994): Probing the B800 bacteriochlorophyll binding site of the accessory light-harvesting complex from *Rhodobacter sphaeroides* using site-directed mutants. I. Mutagenesis, effects on binding, function and electrochromic behaviour of its carotenoids. *Biochim Biophys Acta* 1183:473

Deisenhofer J, Michel H (1989): The photosynthetic reaction centre from the purple bacterium *Rhodopseudomonas viridis*. *EMBO J* 8:2149

Deisenhofer J, Epp O, Mikki K, Huber R, Michel H (1985): Structure of the protein subunits in the photosynthetic reaction center of *Rhodopseudomonas viridis* at 3 Å resolution. *Nature* 318:618–624

Devereux J, Haeberli P, Smithies O (1984): A comprehensive set of sequence analysis programs for the vax. *Nucleic Acids Research* 12:387

Donnelly D, Overington JP, Ruffle SV, Nugent JH, Blundell TL (1993): Modeling α-helical transmembrane domains: the calculation and use of substitution tables for lipid-facing residues. *Protein Science* 2:55

Eccles J, Honig B, Schulten K (1988): Spectroscopic determinants in the reaction center of *Rhodo-pseudomonas viridis. Biophys J* 53:137–144

Eisenberg D (1984): Three-dimensional structure of membrane and surface proteins. *Annual Review of Biochemistry* 53:595

Eisenberg D, Schwarz E, Komaromy M, Wall R (1984): Analysis of membrane and surface protein sequences with the hydrophobic moment plot. *Journal of Molecular Biology* 179:125

Eisenberg D, Weiss R, Terwilliger T, Wilcox W (1982): Hydrophobic moments and protein structure. *Faraday Symposia of the Chemical Society* 17:109

Engelman D (1982): An implication of the structure of bacteriorhodopsin: globular membrane proteins are stabilized by polar interactions. *Biophys J* 37:187

Engelman DM, Steitz TA, Goldman A (1986): Identifying nonpolar transbilayer helices in amino acid sequences of membrane proteins. *Ann Rev Biophys Biophys Chem* 15:321–353

Fasman G (1989a): Protein conformational prediction. *Trends in Biochemical Sciences* 14:295

Fasman G, ed. (1989b): *Prediction of protein structure and the principles of protein conformation.* New York: Plenum

Fowler G, Sockalingum G, Robert B, Hunter C (1994): Blue shifts in bacteriochlorophyll absorbance correlate with changed hydrogen bonding patterns in light-harvesting 2 mutants of *Rhodobacter Sphaeroides* with alterations at α-Tyr-44 and α-Tyr-45. *Biochemical Journal* 299:695

Garnier J, Osguthorpe D, Robson B (1978): Analysis of the accuracy and implications of simple methods for predicting the secondary structure of globular proteins. *Journal of Molecular Biology* 120:97

Geourjon C, Deleage G (1994): SOPM: a self optimised prediction method for protein secondary structure prediction. *Protein Engineering* 7:157

Germeroth L, Lottspeich F, Robert B, Michel H (1993): Unexpected similarities of the B800-850 light-harvesting complex from *Rhodospirillum molischianum* to the B870 light-harvesting complexes from other purple photosynthetic bacteria. *Biochemistry* 32:5615–5621

Hawthornthwaite AM, Cogdell RJ (1991): Bacteriochlorophyll binding proteins. In: *Chlorophylls,* (Scheer H, ed) pp. 493–528, Boca Raton: CRC Press

Henderson R, Baldwin JM, Ceska TA, Zemlin F, Beckmann E, Downing KH (1990): Model for the structure of Bacteriorhodopsin based on high-resolution electron cryomicroscopy. *J Mol Biol* 213:899–929

Holley LH, Karplus M (1991): Neural networks for protein structure prediction. *Methods in Enzymology* 202:204

Holley LH, Karplus M (1989): Protein secondary structure prediction with a neural network. *Proc Natl Acad Sci USA* 86:152–156

Hu X, Xu D, Hamer K, Schulten K, Koepke J, Michel H (1995): Predicting the structure of the light-harvesting complex II of *Rhodospirillum molischianum. Protein Science* 4:1670–1682

Jähnig F (1989): Structure prediction for membrane proteins. In: *Prediction of protein structure and the principles of protein conformation,* (Fasman G, ed) p. 707, New York: Plenum

Johnson M, Srinivasan N, Sowdhamini R, Blundell T (1994): Knowledge-based protein modeling. *Critical Reviews in Biochemistry and Molecular Biology* 29:1

Kabsch W, Sander C (1983): Dictionary of protein secondary structure: pattern recognition of hydrogen-bonded and geometrical features. *Bioploymers* 22:2577

Karrasch S, Bullough P, Ghosh R (1995): 8.5Å projection map of the light-harvesting complex I from *Rhodospirillum rubrum* reveals a ring composed of 16 subunits. *EMBO J* 14:631

Kleinekofortg W, Germeroth L, van der Broek J, Schubert D, Michel H (1992): The light-harvesting complex II (B800/850) from *Rhodospirillum molischianum* is an octamer. *Biochimica et Biophysica Acta* 1140:102–104

Koepke J, Hu X, Muenke C, Schulten K, Michel H (1996): The crystal structure of the light harvesting complex II (B800–850) from *Rhodospirillum molischianum. Structure* (submitted)

Kramer HJM, van Grondelle R, Hunter CN, Westerhuis WHJ, Amesz J (1984): Pigment organization of the B800-850 antenna complex of *Rhodopseudomonas sphaeroides. Biochim Biophys Acta* 765:156–165

Kraulis P (1991): MOLSCRIPT—a program to produce both detailed and schematic plots of protein structures. *J Appl Cryst* 24:946–950

Kühlbrandt W, Wang D-N, Fujiyoshi Y (1994): Atomic model of plant light-harvesting complex by electron crystallography. *Nature* 367:614

Kuhn L, Leigh J (1985): A statistical technique for predicting membrane protein structure. *Biochimica et Biophysica Acta* 828:351

Kyte J, Doolittle RF (1982): A simple method for displaying the hydropathic character of a protein. *J Mol Biol* 157:105

Lattman E (1985): Diffraction methods for biological macromolecules. Use of the rotation and translation functions. *Methods in Enzymology* 115:55

Levitt M (1978): Conformational preference of amino acids in globular proteins. *Biochemistry* 17:4277

Lipman D, Pearson W (1985): Rapid and sensitive protein similarity searches. *Science* 227:1435

Lohmann R, Schneider G, Behrens D, Wrede P (1994): A neural network model for the prediction of membrane-spanning amino acid sequences. *Protein Science* 3:1597

Mackerell A (1995): unpublished research

Mcdermott G, Prince S, Freer A, Hawthornthwalte-Lawless A, Paplz M, Cogdell R, Isaacs N (1995): Crystal structure of an integral membrane light-harvesting complex from photosynthetic bacteria. *Nature* 374:517

McRee D (1993): *Practical Protein Crystallography.* San Diego: Academic Press

Michel H (1991): General and practical aspects of membrane protein crystallization. In: *Crystallization of membrane proteins,* (Michel H, ed) p. 74, Boca Raton, Florida: CRC Press

Michel H, Weyer KA, Gruenberg H, Dunger I, Oesterhelt D, Lottspeich F (1986): The 'light' and 'medium' subunits of the photosynthetic reaction centre from *Rhodopseudomonas viridis:* Isolation of genes, nucleotide and amino acid sequence. *EMBO J.* 5:1149

Olsen JD, Hunter CN (1994): Protein structure modelling of the bacterial light-harvesting complex. *Photochem Photobiol* 60:521

Pearson W (1990): Rapid and sensitive sequence comparison with FASTP and FASTA. *Methods in Enzymology* 183:63

Persson B, Argos P (1994): Prediction of transmembrane segments in proteins utilising multiple sequence alignments. *Journal of Molecular Biology* 237:182

Popot J (1993): Integral membrane protein structure–transmembrane α-helices as autonomous folding domains. *Current Opinion In Structural Biology* 3:532

Popot J, de Vitry C (1990): On the microassembly of integral membrane proteins. *Annual Review of Biophysics and Biophysical Chemistry* 19:369

Popot J, Engelman D (1990): Membrane protein folding and oligomerization: the two-stage model. *Biochemistry* 29:4031

Popot J, de Vitry C, Atteia A (1994): Folding and assembly of integral membrane proteins: An introduction. In: *Membrane protein structure: experimental approaches,* (White, S., ed) p. 41, New York: Oxford University press

Presnell SR, Cohen BI, Cohen FE (1992): A segment-based approach to protein secondary structure prediction. *Biochemistry* 31:983

Rao JM, Argos P (1986): A conformational preference parameter to predict helices in integral membrane proteins. *Biochimica et Biophysica Acta* 869:197

Rees D, DeAntonio L, Eisenberg D (1989): Hydrophobic organization of membrane proteins. *Science* 245:510

Ring C, Cohen F (1993): Modeling protein structures: construction and their applications. *Faseb Journal* 7:783

Rooman M, Wodak S (1988): Identification of predictive sequence motifs limited by protein structure data base size. *Nature* 335:45

Rossmann M, ed (1972): *The Molecular Replacement Method.* New York: Gordon and Breach

Sali A, Blundell TL (1993): Comparative protein modelling by satisfaction of spatial restraints. *Journal of Molecular Biology* 234:779

Schuler G, Altschul S, Lipman D (1991): A workbench for multiple alignment construction and analysis. *Proteins: Structure, Function, and Genetics* 9:180

Segrest J, Loof HD, Dohlman J, Brouillette C, Anantharamaiah G (1990): Amphipathic helix motif: Classes and properties. *Proteins, Struct Funct Genet* 8:103

Sundstrom V, van Grondelle R (1991): Dynamics of excitation energy transfer in photosynthetic bacteria. In: *Chlorophylls,* (Scheer H, ed) pp. 627–704, Boca Raton: CRC Press

Treutlein H, Schulten K, Deisenhofer J, Michel H, Brünger A, Karplus M (1988): Molecular dynamics simulation of the primary processes in the photosynthetic reaction center of *Rhodopseudomonas viridis.* In: *The Photosynthetic Bacterial Reaction Center: Structure and Dynamics,* (Breton J, Verméglio A, eds) volume 149 of *NATO ASI Series A: Life Sciences* pp. 139–150. Plenum New York

Tuffery P, Etchebest C, Popot J, Lavery R (1994): Prediction of the positioning of the seven transmembrane α-helices of bacteriorhodopsin. A molecular simulation study. *Journal of Molecular Biology* 236:1105

van Grondelle R, Sundstrom V (1988): Excitation energy transfer in photosynthesis. In: *Photosynthetic Light–Harvesting Systems,* (Scheer H, ed) pp. 403–438, Berlin, New York: Walter de Gruyter and Co

Visschers RW, Crielaard W, Fowler GJ, Hunter CN, van Grondelle R (1994) Probing the B800 bacteriochlorophyll binding site of the accessory light-harvesting complex from *Rhodobacter sphaeroides* using site-directed mutants. II. A low temperature spectroscopy study of structural aspects of the pigment-protein conformation. *Biochim Biophys Acta* 1183:483

von Heijne G (1994a): Decoding the signals of membrane protein sequence. In: *Membrane protein structure: experimental approaches,* (White S, ed) p. 27, New York: Oxford University press von Heijne G (1994b): Membrane proteins: from sequence to structure. *Annual Review of Biophysics and Biomolecular Structure* 23:167

von Heijne G (1992): Membrane protein structure prediction—hydrophobicity analysis and the positive-inside rule. *Journal of Molecular Biology* 225:487

von Heijne G (1988): Transcending the impenetrable: how proteins come to terms with membranes. *Biochimica et Biophysica Acta* 947:307

von Heijne G, Manoil C (1990): Membrane proteins: from sequence to structure. *Protein Engineering* 4:109

Weiss M, Kreusch A, Nestel U, Welte W, Weckesser J, Schulz G (1991): The structure of porin from *Rhodobacter capsulatus* at 1.8 Å resolution. *FEBS Lett* 280:379

White SH (1994): Hydropathy plots and the prediction of membrane protein topology. In: *Membrane protein structure: experimental approaches,* (White SH, ed), New York: Oxford University press

Zuber H (1993): Structural features of photosynthetic light-harvesting systems. In: *The Photosynthetic Reaction Center,* (Deisenhofer J, Norris JR, eds) p. 43, San Diego: Academic Press

Zuber H (1986): Structure of light-harvesting antenna complexes of photosynthetic bacteria, cyanobacteria and red algae. *Trends Biochem Sci* 11:414

Zuber H (1985): Structure and function of light-harvesting complexes and their polypeptides. *Photochem Photobiol* 42:821

Zuber H, Brunisholz R (1991): Structure and function of antenna polypeptides and chlorophyll-protein complexes: Principles and variability. In: *Chlorophylls,* (Scheer H, ed) pp. 627–692, Boca Raton: CRC Press

Zvelebil MJ, Barton GJ, Taylor WR, Sternberg MJ (1987): Prediction of protein secondary structure and active sites using the alignment of homologous sequences. *J Mol Biol* 195:957

16

Monte Carlo Models of Spontaneous Insertion of Peptides into Lipid Membranes

JEFFREY SKOLNICK AND MARIUSZ MILIK

Introduction

The export of proteins from one compartment to another requires that they be transported across cellular membranes. In addition, membrane proteins must somehow insert into the lipid bilayer after they have been synthesized. Because of their biological importance, the problems of protein insertion into membranes as well as the mechanism(s) of transport across membranes have been extensively studied (Singer, 1990). Experimental data strongly suggest that in vivo most large membrane proteins do not spontaneously insert into membranes (Singer, 1990). Rather, a complex cellular machinery is used. While the basic elements of this machinery are known for many systems, the mechanism and the source of the translocation energy are still unclear (Gierash, 1989; Rapoport, 1992; 1991). On the other hand, when the translocation machinery is blocked or absent, many in vitro experiments provide evidence that some short proteins can insert into phospholipid vesicles (Maduke and Roise, 1993) or into membranes (Wolfe et al, 1985). These experimental results suggest that long and short proteins may translocate by different mechanisms in the cell (von Heijne, 1994) or that the translocation machinery is used to catalyze what is fundamentally a spontaneous insertion process (Jacobs and White, 1989); this may be particularly true for the case of long, amphiphilic sequences. Thus, a number of experiments suggest that spontaneous insertion can occur in membrane–protein systems and that spontaneous insertion can be explained on the basis of the thermodynamics of these systems. However, the detailed mechanism of such spontaneous insertion is not fully understood, nor is the panoply of interactions that drive the insertion process fully characterized. In this review, we describe recent Monte Carlo simulations that are designed to address these important questions.

Biological Membranes
K. Merz, Jr. and B. Roux, Editors
© Birkhäuser Boston 1996

Membrane–protein systems are inherently complicated multicomponent systems having a very large number of degrees of freedom. Therefore, they are very difficult to describe from either a thermodynamic or microscopic viewpoint. These systems exhibit a very complex balance between the enthalpy and entropy of the protein insertion process. The change in entropy associated with peptide insertion can be substantial. In addition to the structural changes of the lipid accompanying protein insertion, the free energy of the water will also change. One also has to deal with changes in hydrogen bond energy, as well as van der Waals and electrostatic interactions. Thus, if one wishes to examine the mechanism of insertion of a peptide from water into a lipid, in order to make the problem computationally tractable, simplified models must be studied. Their advantage is that they can span the time scale of the insertion process; on the other hand, a number of the fine atomic details may not be accounted for. Whether or not such details provide the dominant interactions will determine the validity of the simulations described in this chapter. However, comparison with the full atom simulations of Roux and coworkers (see the chapter by Roux and Woolf in the present volume) indicates that in a number of respects these simplified models have captured the essence of the physics.

The simulations presented here represent different levels of simplification. We begin with a very schematic lattice model of a membrane bound peptide chain, then turn to a more realistic off-lattice model exhibiting a more elaborate membrane structure. Finally, we describe models where the structure of the membrane is represented by set of schematic lipid molecules, but the possibility of conformational transitions in the peptide is ignored. Depending on the level of detail, different questions can be answered by these simulations. Overall, though, the objective of these simulations is to provide insights into the general mechanism of peptide insertion, to explore whether the orientation and conformation of membrane peptides can be predicted, and to examine some possible roles the membrane may play in the peptide insertion process.

Lattice Model of Peptide Insertion

Lattice models, while schematic and somewhat artificial, offer a number of advantages that are particularly important in the case of modeling of large and complicated biomolecular systems (Godzik et al, 1993). The possibility of precalculating the set of Monte Carlo moves that modify the chain configuration and the use of simplified interaction potentials can result in a substantial reduction in simulation time (sometimes by up to three orders of magnitude) in comparison with the analogous off-lattice model. Thus, certain problems only become computationally tractable when a lattice representation is used. However, the advantage of computational speed may be counterbalanced by the imposition of some artificial, geometrical constraints onto the modeled system. Obviously, some questions related to fine structural details cannot be addressed. Fortunately, by going to a sufficiently high resolution lattice, lattice simulations have proven to be very useful for modeling the basic thermodynamic and kinetic properties of systems such as globular proteins (Kolinski and Skolnick, 1994).

Description of the Model

The first lattice model used to investigate the mechanism of spontaneous insertion of peptides into lipid membranes is due to Milik and Skolnick (Milik and Skolnick, 1992). They describe the model peptide molecule as a sequence of diamond-lattice vectors, where three successive vectors represent one peptide residue (amino acid). Thus, the model schematically depicts the backbone NH, C^α and carbonyl groups. Here, the peptide chain can form three regular kinds of secondary structure: right and left handed helices (with four residues per turn) and an expanded zigzag state, which is a low resolution model of β structure. In order to simplify the calculation, the lipid is modeled as a structureless, effective medium which simply occupies a region of space, and the details associated with the existence of lipid molecules, which in reality form the lipid bilayer, are ignored. The bilayer only implicitly enters by its effect on the values of the hydrophobic and hydrogen bond potentials which depend on the z coordinate of the residue. Additional details follow below.

The Potential

The potential used in the lattice model contains four parts: a term, depicting how the peptide residues interact with each other and with the virtual lipid phase; a torsion potential for the peptide chain that represents the intrinsic propensity of residues to adopt a given type of secondary structure; a hydrogen bond potential; and an ordering potential which is designed to imitate steric interactions of the peptide chain with lipid molecules.

The interaction between peptide residues is represented by a mean-force Lennard-Jones (L-J) type potential (8–4), centered on the C^α carbons. The repulsive part of L-J potential is used when residues (e.g., hydrophils with hydrophobs) do not have any attractive region:

$$v_{LJ} = 4\varepsilon[(\sigma/r)^8 - (\sigma/r)^4]. \tag{1}$$

The parameters ε are different for different residue types. The σ parameter is constant and is defined such that for residues having an attractive region, V_{LJ} has a minimum at the distance equal to the length of one diamond lattice vector; i.e., $\sigma = \sqrt{3}/\sqrt[4]{2}$. The magnitude of the L-J interaction also depends on the z coordinates of both residues (depending on whether the residue is hydrophobic, hydrophilic, or lipophilic, pair interactions between residues are different in the water and in the hydrocarbon phase).

The energy of transport of a residue from the water into the hydrocarbon–membrane environment is represented by a one-body hydrophobic potential which depends on its type and its z coordinate (this mimics the effect of the

Table 1a. Values of Hydrophobic Potential
for Model Residues

Symbol of residue	E_{hph} [kT]
a	4.0
b	0.0
c	-1.5
d	-1.5
e	-4.5
f	-4.5

Table 1b. Values of ε Parameters of the Two-Body Interaction
for Pairs of Residues Inside of the Model Membrane

	a	b	c	d	e	f
a	0.0	0.2	1.0	0.5*	2.0*	2.0*
b		0.2	0.0	0.5*	0.5*	0.5*
c			1.5	0.5*	1.5	0.5*
d				0.5*	0.5*	0.5*
e					1.5	0.5*
f						0.5*

* Values with asterisks denote a repulsive interaction between the
residues.

virtual lipid membrane). Within the bilayer region, hydrophobic residues have a
lower energy, whereas hydrophilic residues have a higher energy. The converse
holds true in the aqueous region of space.

In order to represent the variety of natural amino acids, six types of peptide
residues are used, which differ by values of the L-J ε parameters, as well as
by the one-body hydrophobic potential values. The values of the parameters for
the different residue types used in these simulations are presented in Tables 1a
and 1b.

The hydrogen bond potential is independent of residue type and may be
depicted by:

$$E_{Hb} = \begin{cases} \varepsilon_{Hb}, \text{ when } r_{ij} = dist_1 \\ 0, \text{ when } dist_1 < r_{ij} < dist_2 \\ 0, \text{ when } r_{ij} > dist_2 \text{ and atoms are in the lipid phase} \\ \varepsilon_{HbW}, \text{ when } r_{ij} > dist_2, \text{ and atoms are in the water phase} \end{cases} \quad (2)$$

where: r_{ij} is the distance between the nitrogen or carbonyl atom of the residue
i and the appropriate atom of residue j; $dist_1$ is the length of the sum of two
diamond lattice vectors; $dist_2$ is the length of the sum of three diamond lattice

vectors in the trans conformation; $\epsilon_{Hb} = -1.5\ kT$ is the energy of intramolecular hydrogen bond interaction; and $\epsilon_{HbW} = -0.5\ kT$ is the energy of the hydrogen bond of a peptide with water.

Secondary structure propensities are controlled by a torsion potential. In the case of the diamond lattice model, a very simple potential has been used which discriminates against left-handed helices. Both right-handed helical and extended conformations are isoenergetic. Furthermore, in the bilayer region, the domination of helical structures is due to better saturation of hydrogen bonds.

The membrane phase, composed of lipid hydrocarbon chains, is ordered and anisotropic. One might expect that a fragment of the peptide chain which is buried in the membrane will locally disrupt this order and that the magnitude of the disruption will depend on the orientation of the fragment with respect to the bilayer. In the lattice model, this property of the membrane system has been addressed by introducing an orientational energy term of the form:

$$E_{ord} = \varepsilon_{ord} \sin^2(\theta) \tag{3}$$

where: ε_{ord} is a coefficient and θ is the angle between the end-to-end vector of a peptide fragment and the main lipid axis (which lies along the z coordinate). The value of the ε_{ord} parameter was small in comparison with the other potential parameters used in the present model and assumes values from 0.15 to 0.3 kT.

Results

The model has been used to simulate the spontaneous insertion process of two types of transmembrane structures whose sequences are given in Table 2. The first sequence consists of two identical hydrophobic, transmembrane helices connected by a hydrophilic linker. The second is more complicated and has been chosen to emulate the composition of the N-terminus of secreted proteins (having an uncleaved signal sequence). The composition of this peptide can be schematically depicted as: (charged fragment) − (hydrophobic fragment) − (hydrophilic linker) − (amphipathic fragment).

In all of the computational experiments presented here, the model peptide chain starts from a random conformation in the water phase. All the runs have been repeated 15 times, with different seeds for the random number generator. In the case of hydrophobic helices with a hydrophilic linker, the insertion process starts with adsorption of the chain on the lipid interface. This adsorbed,

Table 2. Sequences Used in the Lattice Model Simulations

Sequence number	N terminus	First transbilayer fragment	Linker	Second transbilayer fragment	C terminus
1	bb	ccddccddccddccddccdd	aaaaaa	ccddccddccddccddccdd	bb
2	aa	eeffeeffeeffeeffeeff	bbbbbb	ccddccddccddccddccdd	aa

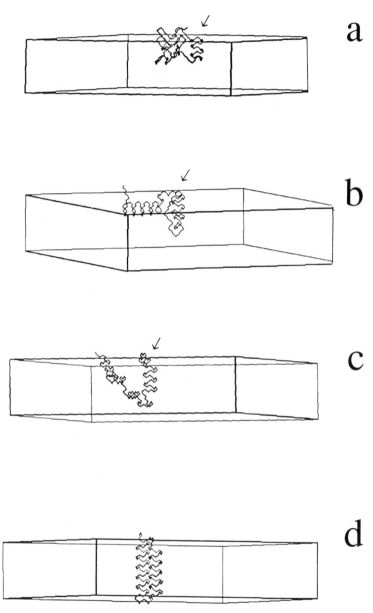

Figure 1. Snapshots of the insertion process of the diamond lattice chain into the model membrane. The arrows point to the *N*-terminus of the chain.

unordered structure represents a deep and broad local free energy minimum, and the chains spend about half of the simulation time in this minimum. The transition into the global free energy minimum for this sequence (a transmembrane hairpin) is associated with a large energetic barrier (about 20 kT) caused by the need to disrupt peptide-water hydrogen bonds. However, in all simulation runs, both chain ends eventually cross the membrane and result in the formation of a stable, transmembrane structure.

The second sequence has been prepared to illustrate Engelman and Steitz's helical hairpin hypothesis (Engelman and Steitz, 1981), which was later extended by Jacobs and White (1989). According to this hypothesis, fragments of proteins transported through the membrane first preassemble at the membrane interface, and then they are transported through the membrane as helical blocks. Our simulations, in which the sequence of the chain emulates an initial fragment of secreted protein, confirms this mechanism of transport. Figure 1 displays critical points of this process. First, the chain is adsorbed onto the membrane surface (Figure 1a). Then, the amphipathic fragment forms an adsorbed helix whose main axis is parallel to the interface, and the hydrophobic fragment forms a helical structure whose main axis is perpendicular to the interface. Both fragments are connected by a long, disordered linker (Figure 1b). Gradually, the transmembrane structure grows and pulls the amphipathic fragment through the membrane (Figure 1c), and finally, the antiparallel, transmembrane, helical structure forms (Figure 1d). The diamond lattice simulations indicate that the insertion process of peptide chains into the lipid membrane can be explained on a schematic level as arising from the effects of hydrophobic and hydrogen bond interactions.

Off-Lattice Model of Peptide Insertion

As mentioned above, lattice models may impose some artificial, geometrical constraints which could affect the polypeptide structure and dynamics. Therefore, an off-lattice extension of the model depicted above is a natural next step (Milik and Skolnick, 1993). Additionally, this offers the possibility of more realistically representing both the peptide chain and the model of the membrane.

Description of the Model

The protein is represented by a chain of balls with centers at the C^α carbon positions. The distance between adjacent balls equals the mean distance between neighboring C^α carbons in protein structures from the Brookhaven Protein Data Bank. The radii of these balls are taken from a statistical analysis of a database of protein structures done by Gregoret and Cohen (1990). The Monte Carlo Dynamics method is used to calculate the equilibrium properties of this model. The model system diffuses in conformational space by a set of discrete micro-modifications of the chain conformation. The probability of acceptance for a

newly generated state depends on the energies of old and new states, according
to the asymmetric Metropolis scheme (Binder, 1984). The set of elementary
chain micro-modifications consists of spike moves for the ends and internal
residues and long range sliding moves. On the level of secondary structure,
uniform (i.e., residue independent) helical propensities are assumed for all the
simulations and for all residues.

Interaction Scheme

Proceeding in a similar fashion to the lattice model described above, the hydrophobic effect exerted by the membrane is introduced as a z coordinate dependent effective potential which represents the different environments associated
with the water, the lipid, and the lipid-water interface. Thus, the hydrophobic
interaction energy for the residue of interest is not only a function of the amino
acid type but also of its z coordinate in the Monte Carlo box. In contrast to the
schematic lattice model described above, in this case the hydropathy scale is
derived from the experimental data of Jacobs and White (1987) for the interaction of tripeptides with model lipid bilayers and from Roseman's partition data
for model compounds in octanol/water systems (Roseman, 1988). Additionally,
the self-solvation effect is used to scale the thermodynamic data, according to
Roseman's proposition (Roseman, 1988).

    The interface of a lipid membrane contains mostly lipid head groups. It
has very specific thermodynamic properties, which are different from both the
hydrocarbon and water phases. Some experimental data indicate that certain
amino acids interact specifically with this region (Jacobs and White, 1987).
This is suggestive of the importance of the interfacial region in the first stage
of the insertion process, the adsorption of a peptide on the membrane interface.
Therefore, the interface region is treated more elaborately here than it has been
in the previous lattice model.

    Conceptually, the process of transport of a protein fragment from water
into the hydrocarbon phase can be divided into two main stages: adsorption
on the interface and transport from the interface into the hydrocarbon phase.
According to Jacobs and White (1989), the free energy of peptide adsorption
onto the membrane interface is represented by:

$$\Delta G_{w \to if} = \Delta G_{ifh} + \Delta G_{imm} \tag{4}$$

where ΔG_{ifh} is the free energy of partially burying the protein fragment in the
lipid phase (hydrophobicity), and ΔG_{imm} is the change of free energy due to
reduction of external degrees of freedom of the peptide chain.

    The burial energy, ΔG_{ifh}, is approximated in the present model as a linear
function of the total accessible surface area of a residue, according to equation:

$$\Delta G_{ifh} = f C_s A_s \tag{5}$$

where C_s is a solvation parameter, taken from the literature (Jacobs and White,
1989), that equals -22.0 cal/(Mol Å^2), and A_s is the total accessible surface

area of the peptide in an extended state (Lesser et al, 1987). f represents the fractional area of a residue that is buried in the adsorbed state; for the simulations described here, $f = 0.56$.

The change of free energy related to the reduction of the external degrees of freedom, ΔG_{imm}, is more difficult to estimate (Jahnig, 1983); it is omitted in the present model because the entropic part of the ΔG_{imm} is at least partially considered in the Monte Carlo scheme.

The free energy of transport of a protein fragment in an extended conformation from the interface into the hydrocarbon phase is defined analogously to the Jacobs and White definition (Jacobs and White, 1989):

$$\Delta G_{ext} = (1 - f)C_s A_{Gly} + \Delta G_{bb} + \Delta G_{sc} \tag{6}$$

where: A_{gly} denotes the accessible surface area of a glycine residue (used as a representation of backbone surface area). ΔG_{bb} is the free energy of transfer of non-hydrogen-bonded backbone polar groups. In the present model, it is assigned a value of 4.1 kcal/Mol (Milik and Skolnick, 1993). ΔG_{sc} is the free energy of side chain transfer from water into a hydrocarbon phase; it has been corrected for the self-solvation effect. The values for the naturally occurring amino acids are obtained from Roseman (Roseman, 1988).

Finally, the transport of a helical protein fragment from the membrane interface into the hydrocarbon phase is related to the following free energy change:

$$\Delta G_{hlx} = \Delta G_{ext} + \Delta G_{Hb} + \Delta G_{lost} \tag{7}$$

where ΔG_{Hb} is the free energy of hydrogen bond formation, set here to -3.8 kcal/Mol. ΔG_{lost} represents the decrease of side chain hydrophobicity due to the change of the accessible surface area of side chains during formation of a helical conformation; the values of this parameter are taken from Roseman (Roseman, 1988). Table 3 contains a compendium of the calculated values of ΔG_{ext} and ΔG_{hlx} for the naturally occurring amino acids.

Finally, all protein fragments (in either helical or extended conformations), when transported into the lipid membrane, disrupt the packing of lipid chains, and therefore they change the ordering of the hydrocarbon phase. As in the case of the lattice model (Equation 3), the change of free energy associated with this effect may be calculated from:

$$\Delta G_{lip} = C_{ord} \sin^2(\theta) \tag{8}$$

where: C_{ord} is a coefficient with values ranging from 0.05 to 0.15 kcal/Mol, and θ is the angle between the end-to-end vector of a polypeptide fragment comprised of four residues and the normal to the membrane surface (z axis).

Results

All simulations described below commence with a random chain conformation located somewhere in the aqueous phase. Most simulations have run for 5×10^6 Monte Carlo steps, where one Monte Carlo step is defined as the time in

Table 3. Values (in kcal/mol) of Free Energy of Transport
from the Interface to the Interior of the Lipid Membrane
for Amino Acids in Nonhelical (ΔG_{ext}) and Helical (ΔG_{hlx})
Conformations

Residue	ΔG_{ext}	ΔG_{hlx}
Ala	2.58	−0.82
Cys	1.25	−2.05
Asp	4.82	1.82
Glu	5.03	2.53
Phe	0.01	−3.09
Gly	3.25	−0.55
His	4.34	1.54
Ile	0.23	−3.17
Lys	5.71	2.31
Leu	0.23	−2.97
Met	1.58	−1.92
Asn	5.52	2.62
Pro	−1.10	−0.60
Gln	5.37	2.27
Arg	7.14	3.64
Ser	3.35	−0.05
Thr	2.83	−0.57
Val	1.07	−2.33
Trp	0.39	−2.81
Tyr	4.23	1.63

which the algorithm tried on average to move every residue of the model chain.
Information about the system is stored every 2500 Monte Carlo-steps into a
trajectory file. This file is then used for the analysis of the behavior of the
model system during the simulation.

Sequences of known, natural membrane bounded peptides and bacteriophage coat proteins have been used in the computational experiments described
here. The sequences are presented in the one-letter amino acid notation (Bechinger et al, 1991); bold fonts indicate hydrophobic amino acids.

MAGAININ2

Magainin2 is a member of a family of peptides found in frog skin. Magainins
have some antibacterial activity, and, according to NMR experiments, they are
predominantly helical in membranes. The sequence of magainin2 is: **GIGKFLH**
S**A**KK**FG**K**AFV**GE**IM**NS.

Figure 2a presents the final conformation of the chain. In the model, the
magainin2 chain is not transported into the hydrocarbon phase; rather, it forms
a highly stable, helical structure, which is adsorbed on the interface. This final
configuration is found in all runs on the magainin2 sequence.

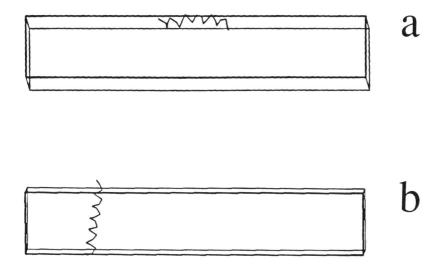

Figure 2. Predicted orientation and structure of (a) magainin2, and (b) M2δ peptides.

Figure 3a contains additional information about this simulation, which confirms that magainin2 adsorbs onto the membrane interface with one face of the helix formed by hydrophobic residues. The mean value of the z coordinates for the chain residues are presented in Figure 3a as a function of the residue number in the sequence. The bars on the figure denote the values of the mean absolute deviations of the data. Based on Figure 3a, we predict that the magainin2 chain forms a stable, helical structure adsorbed on the membrane interface. As indicated by the relatively small values of the mean deviations, vertical mobility is negligible.

M2δ PEPTIDE

The M2δ peptide is a peptide that forms transmembrane helical structures and has the sequence (Bechinger et al, 1991): **EKMSTAISVLLAQAVFLLLTSQR**. The hydrophobic residues form clusters in the middle of the M2δ sequence, thereby rendering its hydrophobic pattern similar to signal peptide sequences. On the other hand, in magainin2 the hydrophobic and hydrophilic residues are mixed together. The contrasting behavior of the two sequences lends support to the supposition that differences in hydrophobicity pattern, rather than overall sequence hydrophobicity, decide whether a given peptide forms a transmembrane structure.

Simulations on the M2δ peptide start from the same initial configurations and physico-chemical conditions as is the case for magainin 2. Figure 2b presents the final, transmembrane structure, which is stable during the last 80% of the

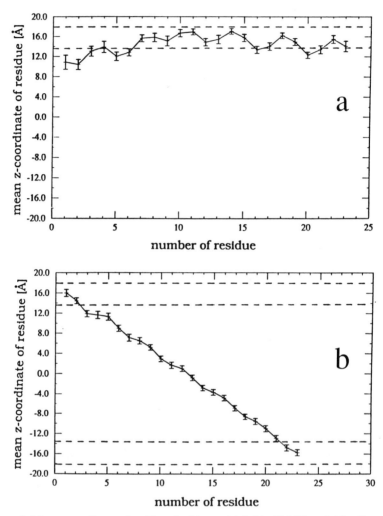

Figure 3. Mean z coordinates of residues for (a) magainin2 (a), (b) M2δ, and (c) $pf1$ and fd models.

simulation. Figure 3b displays the mean z coordinates of the M2δ residues. The result is very different from the magainin2 case. The residues of M2δ are uniformly distributed along the z coordinate, which lies across the membrane. The small mean deviation values indicate that the vertical motion of the residues is negligible after the transmembrane structure forms. Thus, M2δ forms a stable, transmembrane helix.

Taken together, these two examples demonstrate that on the basis of sequence information alone, this model can distinguish between transmembrane and surface adsorbed peptides.

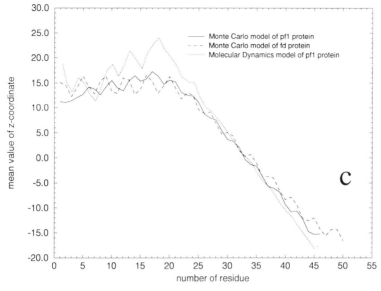

Figure 3. (Continued)

FILAMENTOUS BACTERIOPHAGES $Pf1$ AND fd COAT PROTEINS

Pf1 and fd represent two classes of filamentous bacteriophages identified on the basis of their X-ray diffraction patterns (Clark and Gray, 1989; Nakashima et al, 1975). Virions of these phages consist of a single stranded DNA ring coated by a protein layer. There are only two virion proteins, and the major protein (B) constitutes more than the 90% of the protein coat. All the major coat proteins for filamentous bacteriophages are relatively short (about 50 amino acids), with large percentages (close to 100%) of α-helical secondary structure. According to experimental data (Shon et al, 1991), in the membrane both proteins form structures containing transmembrane and interface adsorbed helices which are connected by a flexible linker. Thus, these sequences have been used in order to test the model on larger and more complicated protein structures. In our simulations, the behavior of both model pf1 and fd sequences are very similar; therefore, the process of insertion is presented here for the example of the fd coat protein. Figure 4 presents snapshots of the more important stages of the insertion process. Starting from a random conformation outside of the membrane, the peptide chain very rapidly adsorbs onto the surface of the membrane and forms a slightly distorted helical structure (Figure 4a). Then, the chain waits for a energy fluctuation that is sufficient to begin to transport the hydrophilic C-terminus of the protein across the membrane (Figure 4b). When the C-terminus crosses the membrane (Figure 4c), the final structure forms very quickly and remains stable during the remainder of the simulation (Figure 4d).

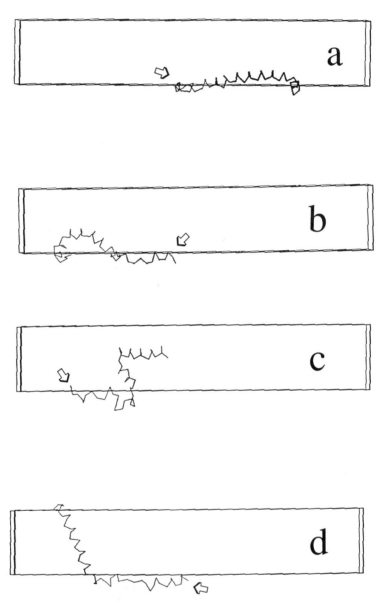

Figure 4. Snapshots from a typical simulation of the membrane insertion mechanism of the fd filamentous bacteriophage coat protein. The arrows point to the N-terminus of the chain.

In order to check the stability of the final state, a simulation ten times longer than usual ($5 \cdot 10^7$ MC steps) has been run. The simulation starts from the final, assembled conformation of the fd chain, with all parameters as in previous runs. In spite of the very fast lateral diffusion of the peptide in the membrane phase, both the orientation of the helices and the location of the turn are stable during the simulation. This confirms that the final structure obtained during the simulation is in a deep free energy minimum.

The sensitivity of the model to temperature has been tested by simulating the peptide-membrane system at a temperature of 270K, instead of the usual 305K. The behavior of the system at the lower temperature is analogous to that at higher temperature. Insertion of the fd coat proteins into the membrane phase is not substantially more difficult at lower temperature; thus, over the examined temperature range, the final structure does not depend on temperature.

One of the important features of our model is the possibility of obtaining information about the position in the sequence of the transbilayer regions, adsorbed fragments, and turns. This information is not assumed in our model, because uniform helix propensities are employed for the entire chain. However, it could be possible that the predicted protein structure is an artifact of the geometry of the model membrane. To eliminate this possibility, a set of runs with different values of the membrane thickness have been undertaken. The results of simulations on systems in which the thickness of the lipid phase ranged from 21 Å to 27 Å show that the final structure of the model membrane protein does not change. Thus, we are confident that the position of the predicted turn and the orientation of the helical fragments does not depend on the membrane geometry. Rather, the structures of the membrane proteins predicted by the present model depend strongly on the hydrophobicity pattern of the protein sequence and do not depend on the simulation parameters (at least over the range we investigated).

Figure 3c shows the average values of the z coordinates of the C^α carbons of pf1 and fd coat proteins, obtained using the present method. In spite of their very low sequential homology, the similarity of the hydrophobic patterns implies a very similar structure. Both protein structures are predicted to have an interfacial adsorbed helix, which is 20 residues long, and a transbilayer helix stretching from residues 22 to residue 42 in pf1 and to residue 43 in fd. In Figure 3c, we also superimpose the mean z coordinates obtained by Roux and Woolf for pf1 (see chapter by Roux and Woolf in this volume). Their results, which are obtained using a different model and method, (all atom models sampled with molecular dynamics), are very similar to ours. The positions of the transmembrane helix and linker are very close in the Monte Carlo and molecular dynamics (MD) simulations. This confluence of results provides additional evidence that the results obtained in the present set of simulations provide some insight into the nature of the process of insertion of peptide chains into lipid membranes.

Translocation of Polymer Chains across a Curved Membrane

Another application of the Monte Carlo method to protein–membrane systems is due to Baumgärtner and Skolnick (1994a,b). Because they are designed to investigate the role of the membrane in the insertion process, these models use a more elaborate representation of the membrane phase at the expense of a very simplified model of the protein chains, which lack any secondary structure. The cases of planar and curved membranes have been examined; the objective of this series of simulations is to determine whether membrane curvature exerts any effect on peptide translocation.

The Membrane Model

The membrane is comprised of an assembly of dumbbells tethered by one end to a penetrable interface. Because the dumbbells are tethered on both sides of the interface, they form a bilayer structure. The dynamics of the membrane is modeled by random moves of dumbbells in the monolayer (the length of each tether can vary from 0 to 0.7 model units) and by flip-flop moves, in which dumbbells can migrate from one layer to another. The last move is particularly important in the case of the model of a curved membrane, in which the planar densities of lipids are different on the cis, outside, and trans, inside, halves of the bilayer, and in which the density ratio is difficult to evaluate a priori. Every dumbbell is formed from two hard spheres of diameter $\sigma_L = 1.1$ (in the model units), connected by a rigid rod of length 1.32 (in the model units). The spheres represent the head and tail parts of lipids. In this realization, the bilayer is formed from 1000 dumbbells. The curved membrane is a sphere of radius 9.373. The equilibrium density of lipids in the cis half of the bilayer is 0.55, and in the trans half, it is 0.35. The difference in densities results in different average widths of the layers. The average width for the layers is 1.47 for the cis layer and 1.21 for the trans layer.

The Polymer Model

The polymer is modeled by a bead-spring chain, containing 20 hard spheres of diameter 0.69, connected by harmonic springs with a potential:

$$U(l) = K(l - l_0)^2 \tag{9}$$

with $K = 5$ and $l_0 = 0.7$. The average radius of such a chain is comparable with the width of one half of the model membrane.

The Potential

All the elements of the model system, lipid heads and tails and beads of the polymer chain, interact with a Morse potential of the form:

$$V(r) = \varepsilon[exp(-2\alpha(r - r_m)) - 2\exp(-\alpha(r - r_m))], \tag{10}$$

with parameters $\alpha = 4.0$ and $r_m = r_0 + \ln 2/\alpha$. The value of r_0 is set to be $(\sigma_L + \sigma_P)/2$ and $V(r_m) = -1.0$. The position of the polymer in the membrane phase is monitored by the mean penetration depth $\langle \Lambda \rangle$, defined as the average distance of the center of mass of the polymer from the central plane of membrane. Positive values of $\langle \Lambda \rangle$ correspond to the cis side and negative values to the trans side of the sphere.

Results

The mean penetration depth value is calculated for flat and curved membranes, as a function of adhesion strength, γ/kT. At high temperatures, for $\gamma/kT < 0.45$, the polymer remains outside of both flat and curved membranes. For lower temperatures $(0.45 < \gamma/kT < 0.8)$, the polymer in the curved membrane is spontaneously transported from the cis to the trans side of the membrane. This effect does not appear in the case of flat membrane, in which the polymer multiply crosses the bilayer. The total average interaction energy $\langle U/N \rangle$ between the polymer and the lipids, as a function of γ/kT, is very similar in the case of curved and flat membranes. Additionally, the average energy of a monomer as a function of its z coordinate is very similar in both cases. Therefore, the authors conclude that the process of transport is controlled by entropy and fluctuation effects. Indeed, if the average potential across the spherical bilayer is used in an effective medium model, the polymer spends roughly equal amounts of time on the trans and cis side of the curved bilayer. Furthermore, in the dumbbell model of the bilayer, the fluctuations of the orientational order parameter are two times larger on the trans than on the cis side. The polymer chain, after penetrating into the membrane, prefers to remain in the entropically more favorable, trans half of the bilayer. This portion of the membrane is of lower density; thus, the configurational entropy of the chain is larger. These simulations strongly indicate that there may be situations where explicit treatment of the lipid is extremely important, and thus, caution must be exercised in the use of effective medium membrane models. Certain physical properties (e.g., conformation) may be well described, but detailed insight into the transport process may require incorporation of the membrane molecules at some level of detail.

Model of Polymer Electrophoresis Across a Membrane

The model depicted above, with an additional electric field E, was used to simulate the process of electrophoresis of chain molecules across flat plane membranes. A short chain (length $N = 4$) of hard spheres, connected with harmonic bonds is used to model polymer chain. The electric field E is taken to be perpendicular to the flat membrane surface. The field effect on the dynamics of the polymer chain was simulated by modifying the dependence of the acceptance of a Monte Carlo move on the relative change of position during the move. The displacement from position r to r' is accepted, when $\exp[-E \cdot (r' - r)] > \eta$, where η was a pseudo-random number. The model of the membrane used in this simulation is analogous to the case of flat membrane depicted in the previous model and was formed by 1000 dumbbells, representing lipid molecules.

Results

As was the case in the previous simulations of Baumgärtner and Skolnick (1994b), the process of transportation was depicted by the position of center-of-mass of the polymer. The translocation time τ, the time needed for polymer to cross the membrane (from $\Lambda(0) > 0$ to $\Lambda(\tau) < 0$), was calculated as a function of field strength. According to authors, polymer electrophoresis across a fluctuating membrane can be related to a Kramer's problem, describing the average transition time for a Brownian particle escaping over a potential barrier. In the high field limit, the permeability, calculated as $\mu = \tau_0/\tau$ (where τ_0 is time of transition without the membrane present), is independent of the field. At low fields, the permeability exponentially decreases. In spite of its simplicity, the model gives some insights into the process of protein chain translocation across lipid membranes. Particularly interesting are experiments involving translocation of a chain across a frozen membrane; these emphasize the importance of fluctuation of the lipids in this process. Without lipid fluctuations, the polymer must randomly diffuse until it finds a hole through the entire membrane, a very unlikely process, whereas when the lipids fluctuate, the chain can burrow through the membrane.

Conclusions

There are a number of qualitative conclusions that emerge from the study of simplified models of peptide insertion into bilayers. First, for relatively small proteins, in accord with experimental evidence, peptide insertion in a bilayer can be spontaneous. Second, just as in globular proteins, the sequence determines the conformation of the model peptide. The location and orientation of a peptide with respect to a bilayer appears to be determined by the pattern of hydrophobic and hydrophilic residues. Thus, hydrophobic interactions play a very important part in the assembly process. Third, in addition to dictating the general features of the mechanism of spontaneous insertion, hydrogen bonds provide an additional, very important contribution to peptide stability in the bilayer. In the membrane interior, an isolated peptide can only hydrogen-bond to itself, whereas in the aqueous phase, a protein chain can readily exchange an internal hydrogen bond for one with solvent. Regardless of the absolute magnitude of a hydrogen bond, in the bilayer phase, the free energy cost of the absence of hydrogen bonds is very large. Thus, in agreement with the Engelman Steitz (1991) hypothesis of peptide insertion as modified by Jacobs and White (1989), in these simulations peptides first assemble helices on the membrane interface; then the almost fully assembled helices insert into the bilayer. Fourth, by building a model that incorporates the above features along with an experimentally determined transfer free energy scale, it is now possible to predict the orientation and conformation with respect to the bilayer of membrane associated peptides. It is especially encouraging that the results are in agreement with the more detailed models of Roux and Woolf in the present volume. This offers the prospect of simulating more complicated systems that are beyond the range of computer resources for more detailed models, and/or the possibility of simulating a larger number of membrane peptide systems than would be possible for models at full atomic resolution. However, the treatment of peptides that are devoid of tertiary interactions is actually the simpler part of the more general problem of predicting the conformation of membrane proteins. What is still lacking is an effective free energy scale that describes intraprotein interactions in a lipid. Because of the paucity of experimentally determined crystal structures, pair potentials of mean force describing interactions of protein residues in a bilayer are unavailable. This makes it very difficult to predict whether a leucine would prefer to interact with another hydrophobic residue in the protein core or whether it would prefer to interact with lipid molecules instead. While we described a schematic lattice model that uses a conjectured pair potential to obtain some qualitative insights, this is still a far cry from an interaction scheme that is applicable to real proteins.

At this juncture, it is clear that the membrane can be treated as a structureless effective medium to model some properties of transmembrane proteins. However, the complete neglect of the structure of the bilayer can miss important physical effects, in particular, those associated with the entropy of the bilayer. For example, for a structureless polymer diffusing across a curved membrane, only when the lipid molecules are explicitly present does almost irreversible transport from the outside to the inside of the sphere occur. Granted the radii of curvature required to produce these effects are high, but they are not unphysical. Membranes having high curvature are found at various intracellular membrane bounded compartments (Alberts et al, 1994); thus, the effect of quasi-irreversible transport may have some physical basis. On the other hand, these simulations ignore the possibility of conformational transitions of the polymer as it diffuses through the bilayer. Clearly, what is required is the marriage of models having the possibility of secondary structure formation, as in the off-lattice membrane peptide model described above, with models that include a more detailed treatment of the membrane. Again, the level of detail required to treat the membrane is uncertain. Initially, to reduce the requisite simulation time, dumbbells might be used, but subsequently, more detailed models of membranes will have to be investigated. Overall, though, the use of reasonably simplified models of both the peptide and the membrane appear to be a very promising direction for providing not only qualitative insights into the peptide insertion process, but also for providing structural models of membrane peptides at moderate resolution.

REFERENCES

Alberts B, Bray D, Lewis J, Raff M, Roberts K, Watson JD (1994): *Molecular Biology of the Cell*. New York: Garland

Baumgärtner A, Skolnick J (1994a): Spontaneous translocation of a polymer across a curved membrane. *Phys Rev Lett* 74:2142–45

Baumgärtner A, Skolnick J (1994b): Polymer electrophoresis across a model membrane. *J Phys Chem* 98:10655–58

Bechinger B, Kim Y, Chirlian LE, Gessel J, Neumann JM, Montal M, Tomich J, Zasloff M, Opella SJ (1991): Orientations of amphiphilic helical peptides in membrane bilayers determined by solid-sate NMR spectroscopy. *J Biomol NMR* 1:167–73

Binder K (1984): *Applications of the Monte Carlo Method in Statistical Physics*. Heidelberg: Springer-Verlag

Clark BA, Gray DM (1989): A CD determination of the α-helix contents of the coat proteins of four filamentous bacteriophages: fd, IKe, Pf1 and Pf3. *Biopolymers* 28:1861–73

Engelman DM, Steitz TA (1991): The spontaneous insertion of proteins into and across membranes: The helical hairpin hypothesis. *Cell* 23:411–22

Gierash LM (1989): Signal sequences. *Biochemistry* 28:923–30

Godzik A, Kolinski A, Skolnick J (1993): Lattice representations of globular proteins: How good are they? *J Comp Chem* 14:1194–1202

Gregoret LM, Cohen FE (1990): Novel method for the rapid evaluation of packing in protein structures. *J Mol Biol* 211:959–74

Jacobs RE, White SH (1989): The nature of the hydrophobic binding of small peptides at the bilayer interface: implications for the insertion of transbilayer helices. *Biochemistry* 28:3421–37

Jacobs RE, White SH (1987): Lipid bilayer perturbations induced by simple hydrophobic peptides. *Biochemistry* 26:6127–34

Jahnig F (1983): Thermodynamics and kinetics of protein incorporation into membranes. *Proc Natl Acad Sci USA* 80:3691–95

Kolinski A, Skolnick J (1994): Monte Carlo simulations of protein folding. Lattice model and interaction scheme. *Proteins* 18:338–52

Lesser GJ, Lee RH, Zehfus MH, Rose GD (1987): Hydrophobic interactions in proteins. *Protein Engin* 14:175–9

Maduke M, Roise D (1993): Import of mitochondrial presequence into protein free phospholipid vesicles. *Science* 260:364–367

Milik M, Skolnick J (1993): Insertion of peptide chains into lipid membranes: An off-lattice Monte Carlo Dynamics model. *Proteins* 15:10–25

Milik M, Skolnick J (1992): Spontaneous insertion of peptide chains into membranes: A Monte Carlo model. *Proc Natl Acad Sci USA* 89:9391–95

Nakashima Y, Wiseman RL, Konigsberg W, Marvin DA (1975): Primary structure of side chain interactions of Pf1 filamentous bacterial virus coat protein. *Nature* 253:68–71

Rapoport TA (1992): Transport of proteins across the endoplasmic reticulum membrane. *Science* 28:931–36

Rapoport TA (1991): Protein transport across the endoplasmic reticulum membrane: Facts, models, mysteries. *FASEB J* 5:2792–98

Roseman MA (1988): Hydrophilicity of polar amino acid side chains is markedly reduced by flanking peptide bonds. *J Mol Biol* 200:513–22

Shon J, Kim Y, Colnago LA, Opella SJ (1991): NMR studies of the structure and dynamics of membrane-bound bacteriophage Pf1 coat protein. *Science* 252:1303–5

Singer SJ (1990): The structure and insertion of integral proteins in membranes. *Annu Rev Cell Biol* 6:247–96

von Heijne G (1994): Sec-independent protein insertion into the inner *E. coli* membrane. A phenomenon in search of an explanation. *FEBS Lett* 346:69–72

Wolfe PB, Rice M, Wickner W (1985): Effects of two sec genes on protein assembly into the plasma membrane of *Escherichia coli*. *J Biol Chem* 260:1836–41

17

Molecular Dynamics of Pf1 Coat Protein in a Phospholipid Bilayer

BENOÎT ROUX AND THOMAS B. WOOLF

Introduction

The anchoring, stabilization, and function of membrane proteins is of central importance for understanding numerous fundamental biological processes occurring at the surface of the cell. In recent years, extensive efforts have been devoted to develop such powerful tools as X-ray crystallography (Deisenhofer and Michel, 1989; Cowan et al, 1992; Weiss and Schulz, 1992; Picot et al, 1994), electron microscopy (Henderson et al, 1990) and nuclear magnetic resonance (Cross and Opella, 1994) to determine the three-dimensional structure of membrane proteins. Despite this progress, many of the factors responsible for the function of biomembranes are still poorly understood. This is due, in large part, to the extreme difficulties in applying experimental methods to obtain detailed information about the phospholipid bilayer environment and its influence on the structure, dynamics, and function of membrane proteins. In a simplified view, the membrane-solution interface is often pictured as a relatively sharp demarcation between the hydrophilic and hydrophobic regions (Edholm and Jahnig, 1988; Milik and Skolnick, 1993). Its dominant effect is usually represented as that of a thermodynamic driving force partitioning the amino acids according to their solubility (Eisenberg et al, 1982; Engelman et al, 1986; Wesson and Eisenberg, 1992); hydrophobic amino acids are more likely to be found within the hydrocarbon core of the membrane, whereas charged and polar amino acids are more likely to be found in the bulk solvent. It is clear that such an approximation ignores the true complexity of the membrane, a partly ordered and partly disordered liquid-crystaline bilayer constituted of phospholipid molecules (Gennis, 1989). An understanding of biomembranes thus requires a better characterization of lipid-protein interactions at the molecular level.

Biological Membranes
K. Merz, Jr. and B. Roux, Editors
© Birkhäuser Boston 1996

At the present time, even qualitative information gained by performing detailed molecular dynamics simulations of protein-membrane complexes can be valuable since only scarce information is available from experiments about the structure and dynamics of these systems. The molecular dynamics method consists in solving Newton's classical equations of motion on a computer to calculate the trajectory of all the atoms as a function of time (Brooks et al, 1988; see also the chapter by Pastor and Feller in this volume). The calculated classical trajectory (though an approximation to the real world) provides detailed information about the system which may not be easily accessed experimentally. However, there are many difficulties in performing simulations of a protein embedded in a solvated phospholipid bilayer, and despite the growing research in the field of pure lipid bilayers (Pastor, 1994), there have been very few simulations of membrane proteins with all atomic detail (Edholm and Johansson, 1987; Edholm et al, 1995; Huang and Loew, 1995; Stouch, 1993; Xing and Scott, 1992; 1989). A major problem is the construction of a realistic initial configuration of the solvated protein-membrane system. Traditionally, initial configurations of a protein in bulk solvent are constructed using a box of pure solvent that has been equilibrated independently (Brooks et al, 1988). The protein (usually the X-ray structure) is overlaid on the solvent box, and the overlapping solvent molecules are then deleted. This simple approach does not work in the case of membrane proteins due to the large size of the phospholipid molecules and the significant spatial extent of the lipid acyl chains. In practice, it is almost impossible to remove a small number of phospholipid molecules from an equilibrated pure bilayer to create a cavity of the appropriate dimensions for inserting the protein. To avoid this problem, a novel construction protocol has been introduced for assembling the initial configurations of the protein-membrane complex from preequilibrated and prehydrated phospholipid molecules (Woolf and Roux, 1996; 1993). The approach has been used to construct a large number of initial configurations for the gramicidin channel embedded in a dimyristoyl phosphatidycholine (DMPC) bilayer and extensive comparison with experimental results has been performed (Woolf and Roux, 1996; 1994b).

The goal of this chapter is to provide an introduction to the state-of-the-art techniques and methods that are involved in setting up a computer simulation of a membrane protein embedded in a solvated phospholipid bilayer. To illustrate the method, the step-by-step procedure for constructing, equilibrating, and performing a molecular dynamics simulation of the coat protein of filamentous bacteriophage Pf1 in a dipalmitoyl phosphatidylcholine (DPPC) membrane is described in detail. The Pf1 coat protein represents an attractive model system for the study of membrane proteins. This 46-residue protein can exist both as an integral membrane protein, spanning the host cell membrane, and as a structural part of the viral coat, surrounding the virus DNA. Assembly of a new virus particle begins at the membrane of an infected cell, where the coat protein undergoes a significant structural transformation to aggregate and form a symmetrical outer shell around the extruding viral DNA (Azpiazu et al, 1993a,b). Analysis of X-ray and neutron diffraction from fibers indicates that the conformation of Pf1 in the viral coat is a long α-helix (Liu and Day, 1994; Nambudripad et al, 1991a,b). Most information about the structure of the membrane-bound form of Pf1 has been obtained from solid state NMR with oriented samples and from high resolution two-dimensional NMR in dodecylphosphocholine micelles (Schiksnis et al, 1987; Shon et al, 1991). These studies show that the membrane-bound conformation of Pf1 has a long transmembrane hydrophobic helix and a shorter amphipathic helix lying parallel to the membrane surface; a short loop connects the two helices and the remaining residues at the N-terminus and the C-terminus, located on the opposite side of the membrane, are mobile. Because the membrane-bound form of Pf1 possesses many of the key structural elements typical of larger intrinsic membrane proteins, theoretical investigations of this system can help us to better understand the assembly process and gain more insight into the important factors playing a role in the stability of complex protein-membrane macromolecular assemblies.

The Pf1 coat protein has been the object of two theoretical studies (Milik and Skolnick, 1993; Tobias et al, 1993). A molecular dynamics simulation of the protein in vacuum was performed by Tobias et al (1993) using a detailed atomic model. Starting from a completely uniform α-helical conformation, it has been observed that the structural fluctuations are dominated by a mobile internal loop from residues 14 to 18 connecting two relatively rigid helices. However, the amphipathic helical segment formed by residues 1 to 13 adopts an incorrect orientation in vacuum; the polar and nonpolar faces are perpendicular to the plane containing the two helices whereas it is expected that the two faces should be in the plane with the nonpolar face buried in the membrane hydrocarbon and the polar face sticking into the solvent above the membrane surface. It was concluded that simulations in an environment reproducing the dominant features of the membrane bilayer are needed to obtained a correct protein conformation (Tobias et al, 1993). Milik and Skolnick (1993) have performed Monte Carlo simulations, in which the influence of the membrane was approximated by a mean-field potential, to study the membrane insertion process of Pf1 (see also the chapter by Skolnick and Milik in this volume). An extended representation of the polypeptide chain has been used in which each residue was modeled as a single effective sphere and specific energy terms controlled the stability of the α-helical conformation of the backbone. A membrane-bound configuration in qualitative agreement with the results from solid state NMR has been obtained (Shon et al, 1991); the C-terminus forms a transmembrane helix, and the N-terminus forms a helix that is adsorbed on the bilayer surface (Milik and Skolnick, 1993). The present work reports on the first molecular dynamics simulation of Pf1 coat protein embedded in an explicit phospholipid bilayer.

In the next section, the potential function and the microscopic system of the Pf1 phage coat protein in a DPPC bilayer are described. This section is then followed by a detailed description of the general protocol used to set up the simulation of the solvated protein-membrane complex. The resulting system is equilibrated for 150 ps, and a simulation of 200 ps is performed. Preliminary results about the average structural properties and lipid-protein interactions obtained from this trajectory are presented and analyzed. Due to its increasing relevance to membrane proteins (Cross and Opella, 1994; see also the chapter by Ketchem et al in this volume), the calculation of solid state NMR properties such as ^{15}N chemical shifts and deuterium quadrupolar splittings (DQS) is briefly discussed. The chapter ends with a summary of important aspects of molecular dynamics simulations of membrane protein systems.

Methodology

Microscopic Model and Computational Details

The simulation system represents a model for oriented multilayer samples such as used in solid state NMR experiments (Shon et al, 1991). The microscopic model is composed of the Pf1 coat protein, 33 DPPC molecules (15 and 18 in the upper and lower halves of the bilayer, respectively), and 1308 water molecules, for a total of 8879 atoms. This number of waters corresponds to a hydration level of 45% weight water. The membrane-bound three-dimensional conformation of Pf1 is modeled using all available information from NMR experiments (see next section). Hexagonal periodic boundary conditions are applied in the XY direction, with a center-to-center distance of 39.4 Å between neighboring cells to mimic an infinite planar bilayer. The center of the bilayer membrane is located at $Z = 0$. A translation distance of 69.0 Å is used along the Z-axis to simulate a periodic multilayer system. The all-atom PARAM 22 force field of CHARMM (Mackerell et al, 1992), which includes lipid molecules (see the chapter by Schlenkeich et al in this volume) and the TIP3P water potential (Jorgensen et al, 1983), is used for the calculations. For the dynamics, the nonbonded list is truncated at 11 Å using a group-based cutoff. The van der Waals and electrostatic interactions are smoothly switched over a distance of 3.0 Å. The trajectory is calculated in the microcanonical ensemble with constant energy and volume. The temperature is set to 330 K, above the gel-liquid crystal phase transition of DPPC (Gennis, 1989). The length of all bonds involving hydrogen atoms is kept fixed with the SHAKE algorithm (Ryckaert et al, 1977). The equations of motion are integrated with a time-step of 2 fs. The ABNR algorithm is used for all minimizations (Brooks et al, 1983).

Modeling the Pf1 Structure

Although the Pf1 coat protein has been extensively studied experimentally (Liu and Day, 1994; Nambudripad et al, 1991a,b; Schiksnis et al, 1987; Shon et al, 1991), there is not enough information at the present time to completely determine the three-dimensional structure of the membrane-bound form. Moreover, the exact location of the protein relative to the bilayer surface is not known. To obtain a complete three-dimensional structure for Pf1, it is thus necessary to supplement the available experimental data with further analysis and molecular modeling. Based on NMR investigations (Shon et al, 1991), it is known that residues 1 to 5 are mobile, 6 to 13 form an amphipathic α-helix oriented parallel to the membrane-solution interface, 14 to 18 form a mobile connecting loop, 19 to 42 form a long membrane-spanning hydrophobic α-helix, and 43 to 46 are mobile. Residues 26 to 39, in the middle of the long transmembrane helix, are highly hydrophobic and are probably buried inside the hydrocarbon core of the bilayer. The boundaries of the hydrocarbon region of a DPPC bilayer are around ± 13 Å (Gennis, 1989). On the basis of the α-helical conformation of residues 19 to 42, the distance between the C^{α} of Tyr25 and Tyr40 is around 22 Å, which suggests that they are located around ± 11 Å from the center of the bilayer. Such a location for the side chain of Tyr25 and Tyr40 is consistent with their amphiphilic character and is in general accord with the high propensity of residues with aromatic side chains to be found near the membrane-solution interface (Landolt-Marticorena et al, 1993). The structure and orientation of the amphipathic N-terminus segment is more difficult to determine. As shown in Figure 1, construction of an amphipathic helical wheel (Schiffer and Edmunson, 1967) demonstrates that the pattern of polar and nonpolar side chains of residues 6 to 13 can form an amphipathic α-helix with one hydrophilic face, composed of residues Ser6, Glu9, Ser10, and Thr13, and one hydrophobic face, composed of residues Ala7, Val8, Ala11, and Ile12. However, residues 1 to 5 cannot be part of the amphipathic 6–13 helix even though it is also composed of

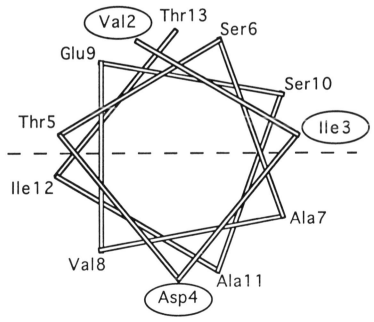

Figure 1. Illustration of the structural constraints imposed by the sequence of residues from Val2 to Thr13 using an amphipathic helical wheel (Schiffer and Edmunson, 1967).

amino acids with polar and nonpolar side chains (e.g., Asp4 would be located on the same side as Val8). It may be expected that residues 1 to 5 adopt a three-dimensional conformation (or multiple conformations) in which the polar side chains are exposed to the bulk, whereas the nonpolar residues are exposed to the hydrocarbon region.

To help in determining a plausible backbone conformation and orientation for the polar and nonpolar side chains, a configurational search has been performed using molecular dynamics simulations and energy minimizations in which the influence of the bilayer is represented with an empirical mean-field potential. Incorporating the influence of the environment, even on the basis of a very approximate mean-field potential, is essential for obtaining a configuration that satisfies the constraints imposed by the pattern of polar and nonpolar amino acids (Tobias et al, 1993). The dominant influence of the environment can be incorporated in a mean-field potential based the solubility of the amino acids (Edholm and Jahnig, 1988; Milik and Skolnick, 1993; see also the chapter by Skolnick and Milik in this volume). Assuming that the membrane normal is oriented along the Z-axis, a simple membrane mean-field is constructed as a sum over the protein atoms:

$$w(z) = \sum_i w_i \times \begin{cases} e^{-(|z|-z_0)^2/\Delta z^2} & \text{if } |z| \geq z_0 \\ 1 & \text{otherwise} \end{cases}, \tag{1}$$

where w_i is an empirical free energy associated with the transfer of the ith atom from the bulk to the membrane, and z_0 and Δz are the position and the width of the interface, taken to be 14.5 and 5.0 Å, respectively. Milik and Skolnick (1993) have used a very similar mean-field potential to study the membrane insertion process of Pf1. In their potential, the membrane-solution interface is located at 15 Å and the width is 4.5 Å.

Given the lack of information about the membrane-solution interface and the approximations involved in the simple mean-field potential Equation (1), no attempt has been made to adjust the values of the constants w_i in accord with the experimental free energy of transfer of individual amino acids. The mean field has been applied only to the non-hydrogen atoms of the side chains. A positive value of 1 kcal/mol has been attributed to the polar side chains, favoring the configurations in which they are located in the region corresponding to the bulk solution. Similarly, a negative value of -1 kcal/mol has been attributed to the nonpolar side chains, favoring the configurations in which they are located inside the region corresponding to the hydrophobic core. It has been observed during molecular dynamics performed in vacuum that strong electrostatic interactions of charged side chains with the peptide backbone (Asp14, Asp18, and Lys20 in particular) are responsible for inducing a conformation that is inconsistent with the data (Tobias et al, 1993). To avoid this spurious effect, the partial charges of the polar side chains are scaled by $1/\sqrt{80}$ to yield the normal charge-charge Coulomb interactions, $q_1 q_2/\epsilon r_{12}$, in accord with the dielectric constant of bulk water. The backbone charges are not scaled because they are involved in α-helical hydrogen bonds and are expected to be shielded from the solvent in the membrane-bound conformation.

The structure is initially constructed as a long α-helix oriented parallel to the Z-axis. Due to lack of information about their conformations, the side chains are constructed from the most probable backbone-dependent rotamers observed in α-helices of 132 protein taken from the Brookhaven Protein Database (Dunbrack and Karplus, 1993). The backbone dihedrals of Gly15 are modified manually using computer graphics to bring the amphipathic N-terminus helix at the membrane surface. Several cycles of molecular dynamics simulation at 300K and energy minimization are then performed to relax the structure. Harmonic restraints are applied between the carbonyl oxygens and the amide hydrogens of the backbone during the simulation to maintain the hydrogen bonds of the 6–13 and 19–42 helices. The resulting structure is shown in Figure 2. Despite

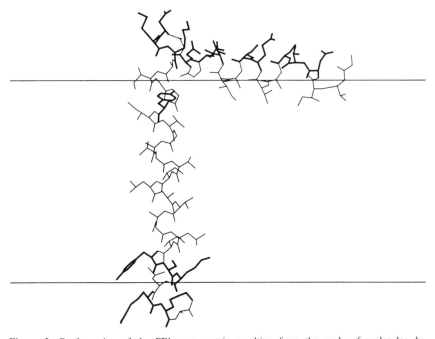

Figure 2. Conformation of the PF1 coat protein resulting from the cycle of molecular dynamics and energy minimization in the presence of the mean-field given by Equation (1). The hydrophilic residues are highlighted in bold. The membrane-bulk interfaces of the mean-field, located at ±14.5 Å, are indicated with two horizontal lines. The sequence of the 46 amino acid Pf1 protein is (helices are indicated by brackets): Gly1-Val2-Ile3-Asp4-Thr5- [Ser6-Ala7-Val8-Glu9-Ser10-Ala11-Ile12-Thr13]-Asp14-Gly15-Gln16-Gly17-Asp18-[Met19-Lys20-Ala21-Ile22-Gly23-Gly24-Tyr25-Ile26-Val27-Gly28-Ala29-Leu30-Val31-Ile32-Leu33-Ala34-Val35-Ala36-Gly37-Leu38-Ile39-Tyr40-Ser41-Met42]-Leu43-Arg44-Lys45-Ala46.

the difference in methodologies, the membrane-bound configuration is qualitatively similar to that obtained in the previous study of Milik and Skolnick
(1993). The polar residues (indicated with bold lines) are located outside the
membrane region, the nonpolar residues are inside, and the two Tyr side chains
are near the interface. The main difference concerns the segment 1 to 5, which
is in an α-helical conformation in the two previous studies, whereas it adopts
a coil-like conformation in the present model structure. This appears to be necessary to shield the nonpolar residues Val2 and Ile3 from the bulk solvent (see
Figure 2). The long hydrophobic helix is oriented along the Z-axis, parallel
to the bilayer normal. In the study of Milik and Skolnick (1993), the helix is
tilted by approximately 30 degrees relative to the membrane normal although
the thickness of the hydrophobic region (30 Å), should have been sufficient to
cover the full length of the helix. The ^{15}N backbone chemical shifts calculated
on the basis of the membrane-bound three dimensional structure are shown in
Figure 3. The values for Glu9, Tyr25, and Tyr40 from the model structure are in

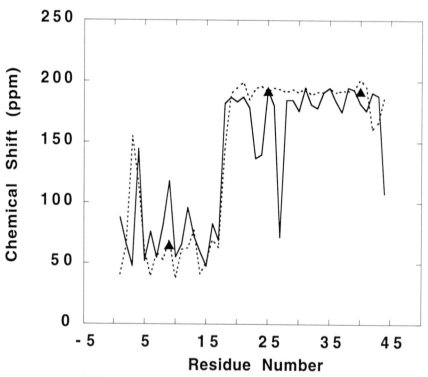

Figure 3. Backbone ^{15}N chemical shift for the 46 residues of the protein. The values calculated
on the basis of Equation (2) from the initial Pf1 structure (dashed line) and the average from the
molecular dynamics trajectory (solid line) are shown. The experimental values for ^{15}N labeled Pf1
at Glu9, Tyr25 and Tyr40 are indicated by triangles (Shon et al, 1991). The values were extracted
from Figure 1 of Cross and Opella, (1994).

good agreement with the available experimental data (Shon et al, 1991). In the absence of more experimental data, the present structure represents a reasonable model for the membrane-bound conformation.

Determination of the Cross-Sectional Area

The total cross-sectional area for the simulation of the protein-membrane system must be carefully determined because it has an important influence on the state of the bilayer (Chiu et al, 1995; Heller et al, 1993; Woolf and Roux, 1996). The membrane-bound conformation of Pf1, shown in Figure 2, has a marked asymmetrical shape due to the amphipathic segment lying parallel to the membrane surface (residues 1 to 13) and the transmembrane helix (residues 19 to 42). The cross-sectional area of the protein as a function of the position along the Z-axis is shown on Figure 4. It can be seen that the cross-section varies from 180 Å to a maximum of about 370 Å along the bilayer normal. Since the cross-section area of one DPPC in the liquid-crystaline L_α phase is approximately 64 Å^2 (Gennis, 1989; Nagle, 1993), the difference between the upper and lower parts of the protein corresponds roughly to the cross-section of three DPPC molecules.

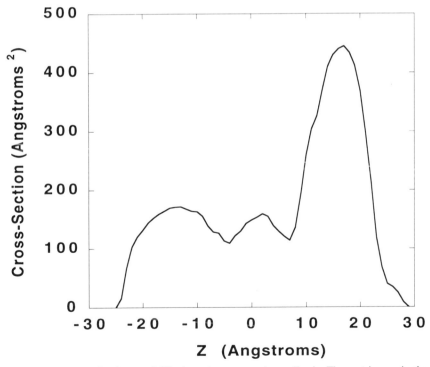

Figure 4. Cross-sectional area of Pf1 along the transmembrane Z-axis. The protein was in the conformation shown in Figure 2 and a probe radius of 1.3 Å was used.

It is necessary to account for the such a large difference in the cross-section of the protein on each side of the membrane in determining the appropriate number of lipids to include in the upper and lower halves of the bilayer in the microscopic model. To accommodate the large difference in cross-section, the simulation system is constructed with 15 and 18 DPPC molecules in the lower and upper halves of the bilayer, respectively. The total cross-sectional area of the entire DPPC:Pf1 system is 1340 Å^2. This corresponds to a center-to-center distance of 39.4 Å between neighboring hexagons in the periodic XY plane. Although the present simulations are carried out at constant volume, it should be emphasized that a reasonable estimate of the cross-section of the protein and the asymmetric number of lipids in the the upper and lower halves of the bilayer are also required with constant pressure algorithms in which the dimension of the simulation system is variable (Chiu et al, 1995; Pastor, 1995).

Construction of a Starting Configuration

It is desirable to construct an initial configuration of the protein-membrane complex that is as close as possible to that of an equilibrated system. The construction method used here has been developed previously by Woolf and Roux for the simulation of the gramicidin A channel in a DMPC bilayer (Woolf and Roux, 1996; 1993). This approach represents an extension of the work of Pastor and coworkers to investigate pure lipid bilayers (Pastor et al, 1991; Venable et al, 1993; see also the chapter by Pastor and Feller in this volume). The general strategy for creating a reasonable starting configuration for the DPPC:Pf1 system consists in randomly selecting 33 lipids from a preequilibrated and prehydrated set, placing them around Pf1, and finally reducing the number of core-core overlaps between heavy atoms through systematic rotations (around the Z-axis) and translations (in the XY plane) of the DPPC and Pf1. To provide the initial XY positions for each lipid, the full DPPC molecules are first represented as single effective particles corresponding to the average DPPC cross-sectional area. The packing of the effective lipid particles around the protein is determined from a molecular dynamics simulation in which the large effective particles are constrained along $Z = \pm 18$ Å and moving in the XY plane with the hexagonal periodic boundary conditions. Large Lennard-Jones spheres of 4.8 Å radius are used to model the effective DPPC. The resulting XYZ positions of the large spheres are utilized for the initial XYZ placement of the phosphate of each DPPC. The initial configuration of the large effective lipid particles around the upper and lower parts of Pf1 after dynamics and minimization are shown in Figure 5. The large effective spheres are packed around the hydrophobic helix according to a distorted triangular lattice (in the lower half), whereas the packing around the amphipathic helix is more complicated (in the upper half). It would have been difficult to model such a packing arrangement by manually adjusting the position of the effective particles.

The preequilibrated conformers for each DPPC molecule were taken randomly from a set of 2000 that was previously generated from Monte Carlo simulations of an isolated DPPC molecule in the presence of a mean-field (De Loof

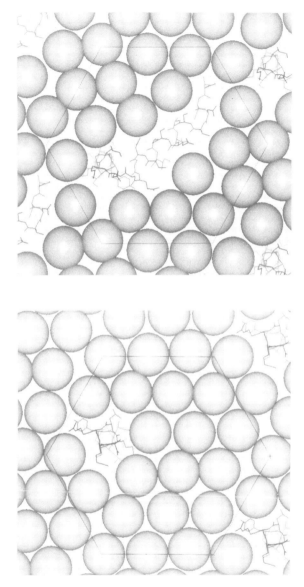

Figure 5. Representation of the large effective DPPC-like spheres modeled as Lennard Jones particles surrounding the upper (top) and lower (bottom) part of the protein in the XY plane.

et al, 1991; Pastor et al, 1991; Venable et al, 1993). The parameters of the mean-field were empirically adjusted to reproduce such experimental observables as deuterium quadrupolar splitting (DQS) order parameters and ^{13}C NMR relaxation times of the acyl chains. The conformers generated by the mean-field Monte-Carlo simulations, in so far as they agree with the available experimental

Figure 6. The 33 (15 at the top and 18 at the bottom) preequilibrated and prehydrated DPPC molecules chosen randomly from the library of 2000 for the initial construction of the system.

data, are representative of the phospholipid molecules found in a bilayer membrane in thermal equilibrium with its normal axis in the Z direction. The 33 DPPC molecules chosen for assembling the initial configuration of the DPPC:Pf1 system are shown in Figure 6. To provide the primary hydration for the polar headgroup, approximately twenty water molecules are constructed around both the phosphate and the choline group based on a molecular dynamics simulation of o-phosphorylcholine (o-PC) in bulk solution (Woolf and Roux, 1994a).

Because the initial configuration is first assembled with totally uncorrelated phospholipid conformers, there is a large number of unrealistic hard core overlaps. To improve the initial configuration, a global search, using systematic rigid body rotations and translations of the lipids (with their primary waters) has been performed to reduce the number of such bad contacts. Systematic rotations have been performed consisting of rigid body rotations around the Z-axis in 10 degree increments for each individual lipid. Systematic translations in the XY plane have been performed consisting of 0.25 Å steps around a square of side 2 Å centered on the initial placement of each lipid. A bad contact is defined as a distance of less than 2.6 Å between two non-hydrogen atoms of the system. After the global search based on systematic rotations and translations, the remaining contacts have been removed by energy minization. For each cycle, the protein is fixed and the Lennard-Jones radii of the lipids are gradually increased.

To obtain a microscopic system with 45% weight water, the remaining bulk solvent is constructed using a box of 3624 waters equilibrated using the same XY hexagonal periodic boundary conditions. The water box is translated along the Z-axis and its position is adjusted to give the correct total number of waters for the system. All waters that projected into the hydrocarbon interior of the bilayer, between, ± 14 Å in the Z direction are deleted. After the overlay, energy minimization is performed for 200 steps. Dynamics for equilibration consist of three different stages. In the initial stages, Langevin dynamics is used to apply a uniform temperature of 330 K throughout the system, the Pf1 backbone is fixed, and planar harmonic restraints with a force constant of 1.0 kcal/mol-Å^2 are applied to the DPPC glycerol C_2 atoms that deviate by more than 1.0 Å from reference Z-values of ± 16.0 Å. The harmonic restraints are gradually decreased, so that by the end of 50 ps, the full system is completely free. In the following 25 ps, the velocities are scaled every 0.5 ps to adjust the temperature. No velocity scaling is applied during the last 75 ps of equilibration and the temperature remains stable.

Calculation of Solid State NMR Properties

Because solid state NMR spectroscopy is a method of increasing relevance for studying membrane proteins (Cross and Opella, 1994; see also the chapter by Ketchem et al in this volume), it is of interest to describe how observed properties can be computed from a molecular dynamics trajectory. The time scale of solid state NMR is much slower than that of rapid molecular motions (Cross and Opella, 1985; Seelig and Seelig, 1980). For this reason, observed properties such as the chemical shift and the deuterium quadrupolar splitting (DQS) correspond to a time-average over rapidly fluctuating quantities. To obtain the observed properties, an average over instantaneous values must be performed. For example, the chemical shift for a specifically ^{15}N labeled site is given by a time-average over a large number of configurations of the projection of the instantaneous second-rank shielding tensor (Cross and Opella, 1985):

$$\sigma_{obs} = \left\langle \hat{Z} \cdot \left[\sum_{i=1}^{3} \hat{\mathbf{e}}_i(t) \, \sigma_{ii} \, \hat{\mathbf{e}}_i(t) \right] \cdot \hat{Z} \right\rangle \tag{2}$$

where $\sigma_{ii}(t)$ and $\hat{\mathbf{e}}_i(t)$ are, respectively, the instantaneous magnitude and direction of the principal tensor components and $\hat{Z}$ is a unit vector in the direction of the bilayer normal (Ketchem et al, 1993). Similarly, the DQS order parameter, S_{CD}, for a specifically deuterated site is (Seelig and Seelig, 1980):

$$S_{CD} = \left\langle \frac{3\cos^2(\theta(t)) - 1}{2} \right\rangle \tag{3}$$

where the brackets represent a time-average over a large number of configurations for which $\theta(t)$ is the instantaneous angle between the director of the C–D bond and the bilayer normal.

The magnitude and orientation of the three components of the ^{15}N backbone chemical shift tensor of polypeptides have been determined experimentally from powder spectra (Cross and Opella, 1985; Teng and Cross, 1989). Typically, the largest component of the tensor, σ_{33}, has a magnitude of 201 ppm and an orientation approximately parallel to the N–H bond in the amide plane. The component σ_{22} (perpendicular to the amide plane) has a magnitude of 55 ppm and the component σ_{11} (in the amide plane) a magnitude of 28 ppm (Teng and Cross, 1989). To compute the instantaneous chemical shift, the tensor components are built in the local molecular frame of a backbone site on the basis of all the atomic configurations taken from the trajectory. The principal axis of the third component (201 ppm) is constructed in the H-N-C plane, with an angle of 105 degrees relative to the N–C bond. The principal axis of the second component (55 ppm) is constructed perpendicular to the H-N-C plane and that of the first component (28 ppm) is obtained from a cross product of the second and third principal axes (i.e., $\hat{\mathbf{e}}_1 = \hat{\mathbf{e}}_2 \times \hat{\mathbf{e}}_3$). This approximation ignores the rapid fluctuations of the tensor component magnitudes caused by variations in the local backbone geometry (Woolf et al, 1995).

Results and Discussion

Protein Structure and Dynamics

The average position of all protein atoms along the Z-axis is shown on Figure 7. The amphipathic helix lies approximately parallel to the membrane surface, around 15–20 Å, although the residues near the N-terminus are lower by almost 5 Å. The residues 1–3 are located around 10–15 Å, whereas the residues 10–20 are closer to 20 Å. The side chain of Ile3 is the lowest part of the amphipathic segment. The long hydrophobic helix runs parallel to the Z-axis from +20 to −20 Å. The ^{15}N backbone chemical shifts calculated from

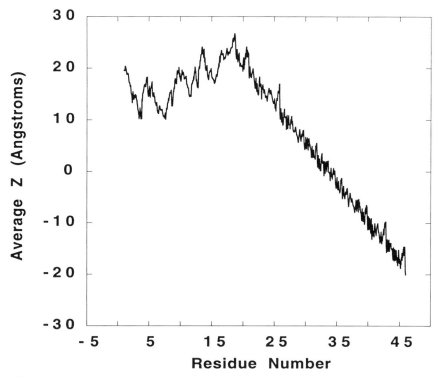

Figure 7. Average position of all the Pf1 atoms along the axis parallel to the bilayer normal. The rms fluctuations were on the order of 2 to 3 Å.

the 200 ps trajectory are shown on Figure 3. The values are more scattered than those obtained from the initial configuration, indicating the presence of significant conformational fluctuations. The average value for Tyr25 and Tyr40 are around 190 ppm, in qualitative agreement with experimental data (Shon et al, 1991). In contrast, the average value for Glu9 is close to 100 ppm, whereas the observed experimental value is around 65 ppm (Shon et al, 1991). A possible explanation for the discrepancy may be the existence of structural fluctuations in the amphipathic helix occurring slowly on the NMR time-scale. The rms fluctuations in the chemical shift of Glu9 calculated from the trajectory are on the order of 40 ppm. Such structural fluctuations could give rise to the larger width (approximately 40 ppm) observed in the experimental ^{15}N spectra of Glu9 relative to the narrow peaks observed for Tyr25 and Tyr40 (see Figure 2 of Shon et al, 1991). The configuration of the C^α backbone trace is shown on Figure 8 at five different times during the 200 ps trajectory. The backbone flexibility is dominated by a mobile internal loop from residues 14 to 18 connecting the amphipathic helix to the long hydrophobic helix. The fluctuations of the two helices have the character of a hinge bending motion (Brooks et al, 1988). Similar fluctuations were observed in previous molecular dynamics of Pf1 performed

Figure 8. Superposition of five configurations of the carbon α trace taken from the trajectory. The configurations are separated by 40 ps.

in vacuum (Tobias et al, 1993), and in mean-field Monte Carlo simulations of a simplified model (Milik and Skolnick, 1993; see also the chapter by Skolnick and Milik in this volume). These results suggest that the helix near the N-terminus may be undergoing some slow motions relative to the membrane surface. Nevertheless, the present trajectory is too short for an extensive configurational sampling of this process.

The occurrence of dihedral transitions has been monitored to examine the dynamics of the side chains. During the trajectory, isomerizations of the dihedral angle χ_1 have been observed for Val2, Ile3, Leu30, and of the angle χ_2 for Ile26 and Ile39. Similar fluctuations of the non-polar side chains of the gramicidin channel have been observed during a trajectory in a DMPC bilayer (Woolf and Roux, 1996; Woolf et al, 1994). Solid-state deuterium NMR spectra of CD_3-labeled side chains have indicated the presence of local motions for Val2, Thr5, Leu43, and Ala46 (Shon et al, 1991). The DQS order parameters S_{CD}, calculated on the basis of Equation (3), are around 0.06 for the methyl groups of Thr5 (C^γ), Leu43, (C^δ), and Ala46 (C^β), in qualitative agreement with the experimental data. This is consistent with the dihedral time-series which indicate that the methyl group of Ala46 and Thr4 are mobile and undergo multiple rotations. The isoleucines (Ile3, Ile26, and Ile39) appear to be surprisingly mobile with isomerizations of the dihedral angles χ_1 and χ_2. These large side chains are in direct contact with the hydrocarbon of the lipids (see the average density profile below). However, the calculated order parameters for Val2 (C^γ) and Ile3 (C^δ) are 0.64 and -0.33, much larger than the values observed experimentally. Structural fluctuations occurring on a time-scale that is longer than the present trajectory (e.g., side chain dihedral transitions and collective hinge bending motions) may be responsible for the discrepancy.

Average Density Profile and Lipid Structure

The density profile of the main components along the bilayer normal provides information on the organization of the DPPC:Pf1 system. The density profile for the heavy atoms of Pf1, the water, the hydrocarbon chains, the ester oxygens of the glycerol region of the DPPC lipids, and the phosphate and the nitrogen of the PC headgroup are shown on Figure 9. The protein occupies the region ±25 Å from the bilayer center. The lower protein density, between -25 and 10 Å corresponds to the hydrophobic helix. The higher protein density, between 10 and 25 Å, corresponds to the amphipathic helix. From 25 Å to 10 Å, the water density varies from a value of $0.0334/\text{Å}^3$, corresponding to the bulk solvent density, to zero in the hydrocarbon core region.

The general features of the lipid density profile are similar to those observed for pure DPPC bilayers (Engelman and Lewis, 1983; Merz and Damodaran, 1994; Stouch, 1993; Venable et al, 1993; Zaccai et al, 1979; see also the chapter by White and Wiener in this volume). The distribution of the polar headgroups is spread over 15 Å. The P and N atoms of the PC headgroups span the region from 15 to 25 Å and the carbonyl oxygens of the ester group of the lipids, the region from 10 Å to 20 Å. The P and N atoms of the headgroup are located within the same region, reflecting the fact that the headgroup dipoles are approximately aligned parallel to the membrane plane on average. The density of the hydrocarbon chains is reduced near the center of the bilayer system, in accord with both experiment and other molecular dynamics simulations, However, the distribution of the hydrocarbon chains is asymmetric, and the position of

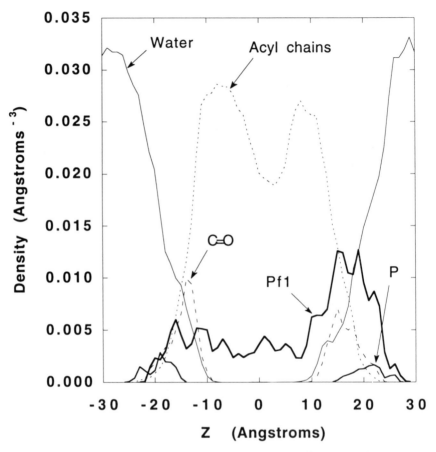

Figure 9. The density profile of the main components for the DPPC:Pf1 system along the Z-axis. The density profile of the heavy atoms of Pf1, water and the hydrocarbon chains, the ester oxygens of the glycerol region and the headgroup phosphate and nitrogen of the DPPC lipid molecules are shown.

the density minimum is shifted toward the amphipathic helix in contrast with the case of pure bilayers. This is due both to the different number of DPPC molecules on each side of the bilayer (18 and 15) and the increased disorder of the hydrocarbon chains on the side of the amphipathic helix (see below). It might be expected that the configuration of the lipids is affected due to the constraint of filling the empty space under the amphipathic helix with hydrocarbon chains. A few configurations of the hydrocarbon chains in contact with the protein are shown in Figure 10. The lipid chains in the neighborhood of the long hydrophobic helix appear to be more ordered, whereas those in contact with the amphipathic helix are more disordered. The histogram of the individual carbon positions along the hydrocarbon chains is shown in Figure 11. It is clear that the chains have a very different distribution on each side of the bilayer. The

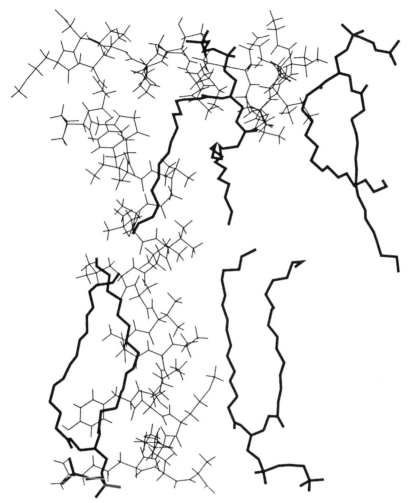

Figure 10. Typical configurations of a few lipids surrounding Pf1. The snapshot was taken 100 ps after the beginning of the simulation.

histograms are more spread out on the side of the amphipathic helix and more sharply peaked on the side of the hydrophobic helix.

Order parameters extracted from DQS can be used to characterize the acyl chain order of the phospholipid molecules (Seelig and Seelig, 1980). The DQS order parameters calculated from the simulation on the basis of Equation (3) are shown on Figure 12. The acyl chains are slightly more disordered on the side of the amphipathic helix relative to the other side. In contrast to the histogram of the carbon position, only a very slight difference between the two sides of the bilayer is observed. Based on the analysis of the chain distribution, the large differences in the distribution of the acyl chains indicate that

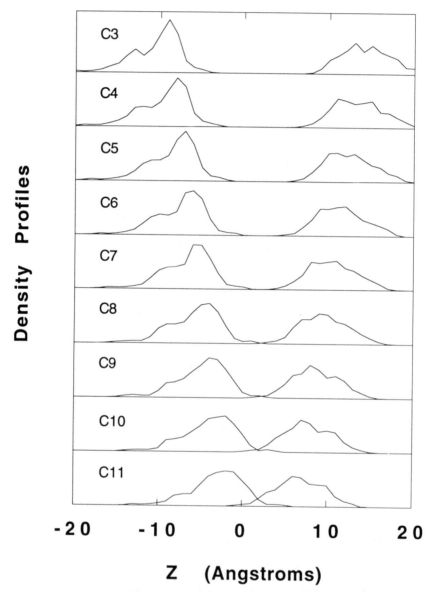

Figure 11. Density profile of the acyl chain carbons of the DPPC molecules along the Z-axis. The density profile of each carbon was obtained by averaging over the Sn-1 and Sn-2 acyl chains.

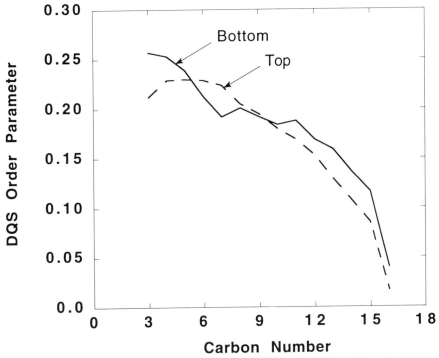

Figure 12. Deuterium quadrupolar splitting (DQS) order parameters of the acyl chains of the DPPC molecules. The order parameter of each carbon was calculated by averaging over the Sn-1 and Sn-2 chains on the basis of Equation (3). All calculated values were less than zero and were multiplied by minus one. The rms fluctuations were on the order of 0.3 to 0.4.

the DQS is not very sensitive to the change in structure of the hydrocarbon region in the present case. This could provide an explanation of why the presence of proteins in lipid bilayer have generally not resulted in a dramatically large change of the experimentally observed order parameters of the acyl chains (Davis, 1983).

Solvation of the Side Chains

The density profiles shown in Figure 9 indicate that the protein overlaps with regions corresponding to very different environments. In particular, the atoms of the amphipathic helix occupy a very broad region between 10 and 25 Å over which the surrounding varies from pure hydrocarbon to bulk water, whereas the long hydrophobic helix is almost entirely buried in the hydrocarbon core and is not exposed to water. Considerations of the average density profiles along the membrane normal does not provide a complete picture of the protein solvation. To examine how the individual polar and nonpolar side chains are solvated, the radial distribution functions of all the side chains with the main components of

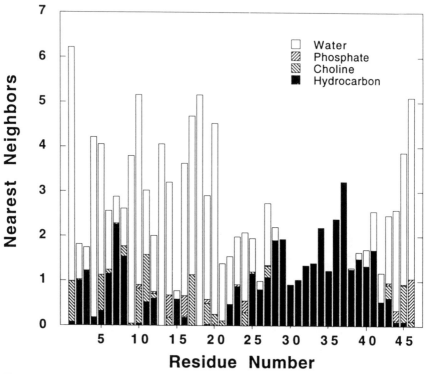

Figure 13. Contribution from the main components of the membrane system (water oxygens, acyl chains carbons, phosphate and choline polar headgroups) to the solvation of the side chains of Pf1. The average solvation number were calculated by counting the number of nearest neighbors within a distance of 4 Å around each side chain. The number of neighbors were averaged by normalizing with respect to the total number of atoms in the side chains.

the membrane system have been calculated. The average number of water, acyl chain carbons, and headgroups surrounding the side chains are shown in Figure 13. It is observed that the polar side chains are exposed to water whereas the nonpolar side chains are exposed to hydrocarbon. Thus, the solvation requirements imposed by the amphipathicity of the protein are satisfied on average, even though the membrane-solution interface is very broad. For some nonpolar residues, the average environment is more complex. For example, the Ala11 side chain is equally exposed to water, hydrocarbon, and the choline moiety of the headgroup.

The radial distribution function of particular side chain atoms of residues 2 to 13 with water are shown in Figure 14. It is clear that only the polar side chains are exposed to water. Residues 6 to 13 form an amphipathic helix with one face exposed to solvent and one face buried in the hydrocarbon. Residues 2 to 5 are in a disordered, coil-like conformation. Such a conformation is necessary to satisfy the constraints imposed by the pattern of polar and nonpolar

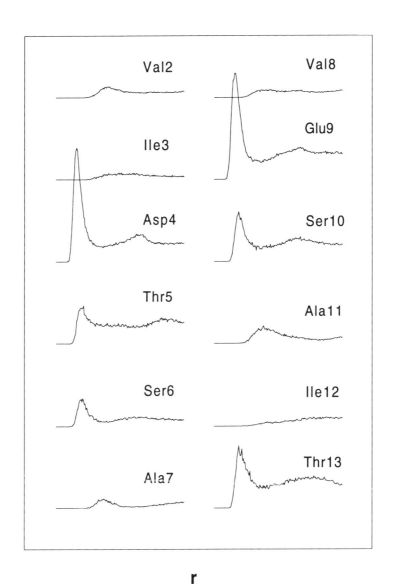

Figure 14. Radial distribution function between the water oxygen and particular side chain atoms for the amphipathic segment: Val2(C^γ)-Ile3(C^δ)-Asp4(O^δ)-Thr5(O^γ)–[Ser6(O^γ)-Ala7(C^β)- Val8(C^γ)-Glu9(O^η)-Ser10(O^γ)-Ala11(C^β)-Ile12(C^δ)-Thr13(O^γ)]. The residues 6–13 are in a helical conformation. The residues 2–5 adopt a coil-like conformation to bury the nonpolar Val2 and Ile3 in hydrocarbon.

residues. As illustrated in Figure 1 using a helical wheel (Schiffer and Edmunson, 1967), residues 2 to 5 cannot be part of an ideal amphipathic helix with the nonpolar residues all located on the same side. Residues 2 to 5 adopt a coil-like conformation during the molecular dynamics simulations performed in the presence of the membrane mean-field Equation (1) at the modeling stage. However, despite the change in secondary structure occurring at residue 6, it can be seen from Figure 14 that the nonpolar side chains Val2 and Ile3 are buried in the hydrocarbon core, and the polar residues Asp4 and Thr5 are exposed to water.

Generally, the charged side chains retain a first hydration shell and do not associate directly with the polar headgroup of the lipid molecules (but see below). This suggests that the interaction of charged residues with the polar heads is indirect and may have a long range electrostatic character. By using peptide analogs of amphipathic helices, it has been shown that the localization of the positively charged residues near the membrane-solution interface and the negatively charged residues at the center of the polar face is important for lipid affinity (Epand et al, 1987; Segrest et al, 1990). An obvious interpretation of this result is that positive residues localized near the membrane-solution interface can better interact with the negatively charged phosphate group, whereas negative residues localized at the center of the polar face of an amphipathic helix better interact with the positively charged choline. As observed in Figure 13, a number of contacts of the side chains of Pf1 with the headgroups are present. There is a slightly larger number of contacts between the protein and the choline groups than with the phosphate group. Analysis shows that the choline group are involved in nonpolar interactions rather than electrostatic interactions. In particular, the choline is associated with Gly1, Thr5, Ala11, Gly17, and Lys45. The lysine side chain is relatively nonpolar even though the $N^\zeta H_3^+$ end group carries a positive charge. No direct association of the choline group with a negatively charged side chain is observed. Similarly, no direct association of the phosphate groups with a positively charged side chain is observed. In contrast, a stable associated pair formed by the COO^- of Asp14 and the phosphate PO_4^- of DPPC is found. A detail of the hydrated complex is shown in Figure 15. The stability of the like-charge pair is maintained by the presence of three bridging water molecules, making hydrogen bonds simultaneously with the two negatively charged groups. Similar hydration structures have also been observed in simulations of chloride ion pairs in water (Dang and Pettitt, 1990). Based on potential of mean force calculations, it is believed that such negative ion pairs are thermodynamically stable in water. This result, although not statistically significant, suggests that direct association of Asp and Glu with the negative phosphate group of the lipids is possible. In conclusion, the present results warn against simplified electrostatic arguments in describing the interactions involving the lipid headgroups with the protein side chains.

The presence of amino acids with aromatic side chains such as tryptophan, tyrosine, and phenylalanine near the interfacial region is a recurrent feature of several membrane proteins; e.g., see the bacterial photosynthetic reaction center

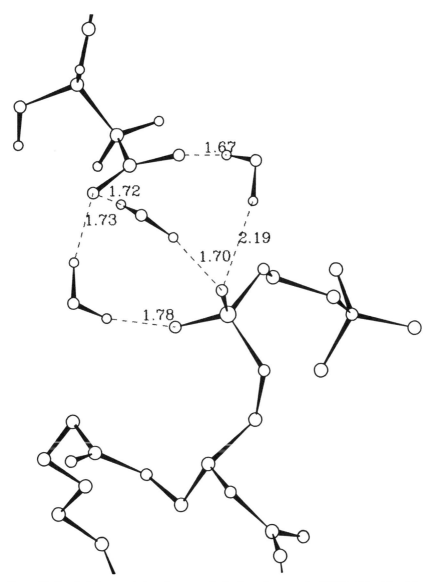

Figure 15. Detail of the association of two negatively charged groups, the COO⁻ of Asp14 and the phosphate of DPPC. The three bridging water molecules are shown (distances are indicated in Å).

(Deisenhofer and Michel, 1989), porins (Cowan et al, 1992; Weiss and Schulz, 1992), Pf1 and fd coat proteins (McDonnell et al, 1993; Shon et al, 1991), prostaglandin H synthase (Picot et al, 1994), gramicidin A (Ketchem et al, 1993; Woolf and Roux, 1996; 1994b), the peptide fusion of hemmagglutinin (White, 1992), and a large series of α-helical human type I membrane proteins (Landolt-Marticorena et al, 1993). The interfacial region is particularly favorable to accommodate the hydrophobic aromatic ring and the hydrogen bonding hydroxyl group of the tyrosine side chain. In the membrane-bound conformation of Pf1, both Tyr25 and Tyr40 are located near the boundary of the hydrocarbon core of the membrane. During the trajectory, it has been observed that Tyr25 makes a hydrogen bond with water molecules while Tyr40 makes a hydrogen bond with the ester carbonyl of the glycerol backbone of a DPPC molecule (shown in Figure 16). It is possible that the presence of tyrosines near the interface provides further stability to anchor the membrane-bound form of the coat protein. It is also possible that the tyrosine side chains may play a functional role. It has been suggested that the aromatic side chains are important for tight and cooperative binding of the 144-residue gene 5 protein of bacteriophage Pf1,

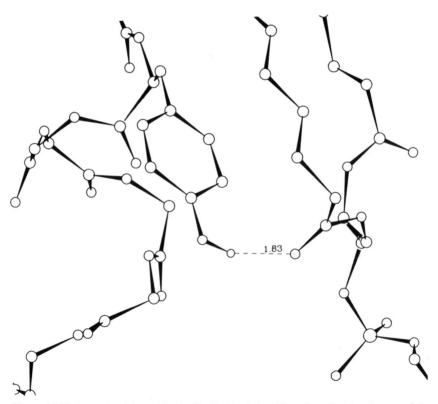

Figure 16. Hydrogen bond formed by the Tyr40 side chain with a glycerol carbonyl group of the lipid observed during the simulations (the distance is indicated in Å).

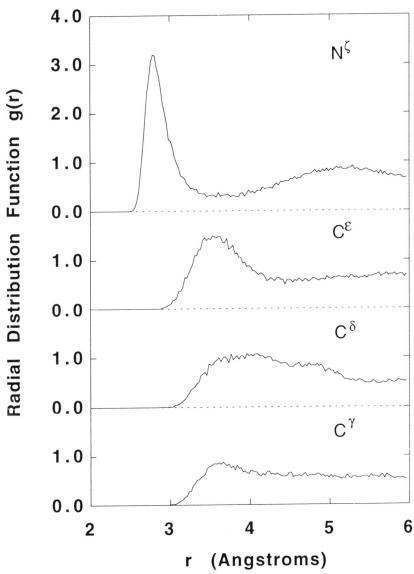

Figure 17. Water oxygen side chain radial pair distribution function for the side chain atoms of Lys20.

presumably through base stacking, to single-stranded DNA during replication of the phage (Plyte and Kneale, 1991). Such interactions are also proposed based on an analysis of fiber diffraction in the case of the Pf1 coat protein (Liu and Day, 1994).

A suggestion, termed the "snorkel model," has been made that the long and flexible lysine side chains are favorably located near the membrane-solution interface by extending from the hydrocarbon core to the membrane surface in amphipathic helices (Segrest et al, 1990). To examine the solvation of Lys20, the radial distribution functions of all the non-hydrogen side chain atoms with water are calculated. The results are shown in Figure 17. It is observed that the charged end group is well solvated, and the remaining part of the side chain is progressively shielded from the water (i.e., $C^\gamma < C^\delta < C^\epsilon$). This can also be seen in a snapshot of the water configuration near Lys20, shown in Figure 18.

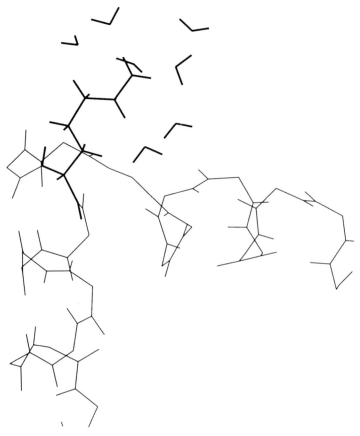

Figure 18. Detail of the hydration of Lys20. Mostly the N^ζ is well-hydrated whereas the non-polar remaining part is shielded from the solvent.

The side chain extends almost vertically, the average position of the C^{α} and N^{ζ} being around 20 and 23 Å, respectively (see also Figure 7). Thus, the present results support the general idea of the snorkel model (Segrest et al, 1990).

Conclusion and Summary

The structure and dynamics of the Pf1 coat protein embedded in a fully hydrated DPPC bilayer has been examined using molecular dynamics simulations. The initial configuration is constructed from preequilibrated and prehydrated phospholipid molecules. The approach has been previously developed for studying the gramicidin A channel in a DMPC bilayer (Woolf and Roux, 1996; 1994b; 1993; Woolf et al, 1994). Due to the marked asymmetry of the Pf1 coat protein (see Figure 4), the present application is more challenging than that of the gramicidin channel which is a relatively simple transmembrane β-helical structure. Since no particular problems have been encountered, the results show that the method is general and can be used to assemble initial configurations of protein–membrane systems in the case of proteins of arbitrary shape and size. The main advantage of the present approach is that it avoids the long simulation times that are usually needed to equilibrate such systems.

During the trajectory the protein exhibits significant structural fluctuations, e.g., a connecting loop causes a hinge bending motion between the two helices. Such fluctuations occur on a slow time scale and are incompletely averaged in the present trajectory. The average density profiles of the main components of the system show that the membrane-solution interface region is much broader than is traditionally pictured. There is an ordering of the lipid acyl chains in direct contact with the long hydrophobic helix but a disordering of the chains in contact with the amphipathic helix. The structural constraints imposed by the amphipathicity are satisfied on average, even though the membrane-solution interface is very broad. Polar side chains are exposed to water on average whereas the nonpolar side chains are usually exposed only to the hydrocarbon. Interactions of the protein side chains with the lipid headgroups do not appear to be governed by simple electrostatics. The choline moiety associates with nonpolar side chains. One associated pair of two negatively charged groups stabilized by bridging water molecules, the COO^{-} of Asp14 and the phosphate PO_4^{-} of a DPPC, has been observed.

The initial protein conformation has been constructed using all available structural information obtained from experiments. Despite the numerous experimental studies of the Pf1 coat protein, (Liu and Day, 1994; Nambudripad et al, 1991a,b; Schiksnis et al, 1987; Shon et al, 1991), there is not enough information to determine the three-dimensional structure of the membrane-bound form. To complete the missing information, a simulation has been performed in the presence of a mean-field potential to mimic the membrane environment and obtain a starting model of the membrane-bound conformation of Pf1. It is clear that the method used to model the protein structure is not unique. In particular, the membrane mean-field potential Equation (1), can be refined in accord with current hydropathy scales (Eisenberg et al, 1982; Milik and Skolnick, 1993; Turner and Weiner, 1993; Wesson and Eisenberg, 1992). In addition, the conformation of the side chains, based of a library of rotamers (Dunbrack and Karplus, 1993), could have been modeled using considerations based on sequence homology with proteins of known structures. For example, the helix formed by residues 6 to 13 might be stabilized further by modeling the conformation of Pf1 Ser6-Ala7-Val8-Glu9 as a helix Capping Box (Harper and Rose, 1993), based on its homology with the sequence Ser12-Ala13-Ala15-Glu16 found at the beginning of helix A (residues 13–28) in hemoglobin (Honzatko et al, 1985). Nevertheless, it is likely that more experimental data are needed to improve the structural model of Pf1. This highlights the importance of experimental structural information for meaningful theoretical studies of biological systems with molecular dynamics simulations.

The insight generated by the present molecular dynamics simulation has begun to provide a sound basis for viewing protein-lipid interactions at the molecular level. This suggests that molecular dynamics simulations will continue to be a powerful tool in exploring the microscopic details of lipid-protein interactions. Further work is in progress on other membrane protein systems.

ACKNOWLEDGMENTS

B. Roux is supported by grants from the Medical Research Council of Canada and from the Fonds pour la Recherche en Santé du Québec. We are greatful for the Service Informatique of the Université de Montréal for their support of this work.

REFERENCES

Azpiazu I, Gomez-Fernandez JC, Chapman D (1993a): Biophysical studies of the pf1 coat protein in the filamentous phage, in detergent micelles, and in a membrane environment. *Biochem* 32:10720–10726

Azpiazu I, Haris PI, Chapman D (1993b): Conformation of the pf1 coat protein in the phage and in a lipid membrane. *Biochem Soc Trans* 21:82S

Brooks BR, Bruccoleri RE, Olafson BD, States DJ, Swaminathan S, Karplus M (1983): CHARMM: A program for macromolecular energy minimization and dynamics calculations. *J Comput Chem* 4:187–217

Brooks CL, Karplus M, Pettitt BM (1988): Proteins. A theoretical perspective of dynamics, structure and thermodynamics. In: *Advances in Chemical Physics,* Vol. 71. Prigogine I, Rice SA, eds. New York: John Wiley

Chiu SW, Clarck M, Subramaniam S, Scott HL, Jakobsson E (1995): Incorporation of surface tension into molecular dynamics simulation of an interface: A fluid phase lipid bilayer membrane. to appear *Biophys J*

Cowan SW, Schirmer T, Rummel G, Steiert M, Gosh R, Pauptit RA, Jansonius JN (1992): Crystal structures explain functional properties of two *E. coli* porins. *Nature* 358:727–733

Cross TA, Opella SJ (1994): Solid-state NMR structural studies of peptides and proteins in membranes. *Curr Op Struc Biol* 4:574–581

Cross TA, Opella SJ (1985): Protein structure by solid state nuclear magnetic resonance. residues 40 to 45 of bacteriophage fd coat protein. *J Mol Biol* 182:367–381

Dang LX, Pettitt BM (1990): A theoretical study of like ion pairs in solution. *J Phys Chem* 94:4303–4308

Davis JH (1983): The description of membrane lipid conformation, order and dynamics by 2H-NMR. *Biochim Biophys Acta* 737:117–171

Deisenhofer J, Michel H (1989): The photosynthetic reaction center from the purple bacterium rhodopseudomonas viridis. *Science* 245:1463–1473

De Loof HD, Harvey SC, Segrest JP, Pastor RW (1991): Mean field stochastic boundary molecular dynamics simulation of a phospholipid in a membrane. *Biochemistry* 30:2099–2113

Dunbrack RL Jr, Karplus M (1993): Backbone-dependent rotamer library for proteins. Application to side-chain prediction. *J Mol Biol* 230:543–574

Edholm O, Jahnig F (1988): The structure of a membrane-spanning polypeptide studied by molecular dynamics. *Biophys Chem* 30:279–292

Edholm O, Johansson J (1987): Lipid bilayer polypeptide interactions studied by molecular dynamics simulation. *Eur Biophys J* 14:203–209

Edholm O, Berger O, Jahnig F (1995): Structure and fluctuations of bacteriorhodopsin in the purple membrane: a molecular dynamics study. *J Mol Biol* 250:94–111

Eisenberg D, Weiss RM, Terwilliger TC (1982): The helical hydrophobic moment: A measure of the amphiphilicity of a helix. *Nature* 299:371–374

Engelman DM, Lewis BA (1983): Lipid bilayer thickness varies linearly with acyl chain length in fluid phosphatidylcholine vesicles. *J Mol Biol* 166:211–217

Engelman DM, Steitz TA, Goldman A (1986): Identifying nonpolar transbilayer helices in amino acid sequences of membrane proteins. *Annu Rev Biophys Biophys Chem* 15:321–353

Epand RM, Gawish A, Iqbal M, Gupta KB, Chen CH, Segrest JP, Anantharamaiah GM (1987): Studies of synthetic peptide analogs of the amphipathic helix. Effect of charge distribution, hydrophobicity, and secondary structure on lipid association and lecithin:cholesterol acyltransferase activation. *J Biol Chem* 262:9389–9396

Gennis RB (1989): *Biomembranes Molecular Structure and Function*. New York: Springer-Verlag

Harper ET, Rose GD (1993): Helix stop signal in proteins and peptides: The capping box. *Biochemistry* 32:7605–7609

Heller H, Schaefer M, Schulten K (1993): Molecular dynamics simulation of a bilayer of 200 lipids in the gel and in the liquid-crystal phases. *J Phys Chem* 97:8343–8360

Henderson R, Baldwin JM, Ceska TA, Zemlin F, Beckmann E, Downing KH (1990): Model for the structure of bacteriorhodopsin based on high-resolution electron cryomicroscopy. *J Mol Biol* 213:899–929

Honzatko RB, Hendrickson WA, Love WE (1985): Refinement of a molecular model for lamprey hemoglobin from *Petromyzon marinus*. *J Mol Biol* 184:147–164

Huang P, Loew GH (1995): Interaction of an amphiphilic peptide with a phospholipid bilayer surface by molecular dynamics simulation study. *J Biomol Struct & Dyn* 12:937–955

Jorgensen WL, Impey RW, Chandrasekhar J, Madura JD, Klein ML (1983): Comparison of simple potential functions for simulating liquid water. *J Chem Phys* 79:926–935

Ketchem RR, Hu W, Cross TA (1993): High-resolution conformation of gramicidin A in lipid bilayer by solid-state NMR. *Science* 261:1457–1460

Landolt-Marticorena C, Williams KA, Deber CM, Reithmeier RAF (1993): Non-random distribution of amino acids in the transmembrane segments of human type *i* single span membrane proteins. *J Mol Biol* 229:602–608

Liu DJ, Day LA (1994): Pf1 virus structure: helical coat protein and dna with paraxial phosphates. *Science* 265:671–674

Mackerell AD Jr, Bashford D, Bellot M, Dunbrack RL, Field MJ, Fischer S, Gao J, Guo H, Joseph D, Ha S, Kuchnir L, Kuczera K, Lau FTK, Mattos C, Michnick S, Nguyen DT, Ngo T, Prodhom B, Roux B, Schlenkrich B, Smith J, Stote R, Straub J, Wiorkiewicz-Kuczera J, Karplus M (1992): Self-consistent parametrization of biomolecules for molecular modeling and condensed phase simulations. *Biophys J* 61:A143

McDonnell PA, Shon K, Kim Y, Opella SJ (1993): fd coat protein structure in membrane environments. *J Mol Biol* 233:447-463

Merz KM, Damodaran KV (1994): A comparison of DMPC- and DLPE-based lipid bilayers. *Biophys J* 66:1076–87

Milik M, Skolnick J (1993): Insertion of peptide chains into lipid membranes: An off-lattice Monte Carlo dynamics model. *Prot Struct Funct Gen* 15:10–25

Nagle JF (1993): Area/lipid of bilayers from NMR. *Biophys J* 64:1476–1481

Nambudripad R, Stark W, Makowski M (1991a): Neutron diffraction studies of the structure of filamentous bacteriophage pf1. Demonstration that the coat protein consists of a pair of alpha-helices with an intervening, non-helical surface loop. *J Mol Biol* 220:359–379

Nambudripad R, Stark W, Opella SJ, Makowski M (1991b): Membrane-mediated assembly of filamentous bacteriophage pf1 coat protein. *Science* 252:1305–1308

Pastor RW (1995): unpublished

Pastor RW, (1994): Molecular dynamics and Monte Carlo simulations of lipid bilayers. *Curr Op Struct Biol* 4:486–492

Pastor RW, Venable RM, Karplus M (1991): Model for the structure of the lipid bilayer. *Proc Natl Acad Sci USA* 88:892–896

Picot D, Loll PJ, Garavito RM (1994): The X-ray structure of the membrane protein prostaglandin H_2 Synthase-1. *Nature* 367:243–249

Plyte SE, Kneale GG (1991): The role of tyrosine residues in the dna-binding site of the pf1 gene 5 protein. *Protein Engin* 4:553–560

Ryckaert JP, Ciccotti G, Berendsen HJC (1977): Numerical integration of the cartesian equation of motions of a system with constraints: Molecular dynamics of *n*-alkanes. *J Comp Chem* 23:327–341

Schiffer M, Edmunson AB (1967): Use of helical wheels to represent the structures of protein and identify segments with helical potential. *Biophys J* 7:121–135

Schiksnis RA, Bogusky MJ, Tsang P, Opella SJ (1987): Structure and dynamics of the pf1 filamentous bacteriophage coat protein in micelles. *Biochemistry* 26:1373–1381

Seelig J, Seelig A (1980): Lipid conformation in model membranes and biological membranes. *Quat Rev Biophys* 13:19–61

Segrest JP, De Loof H, Dohlman JG, Brouillette CG, Anantharamaiah GM (1990): Amphipathic helix motif: Classes and properties. *Prot Struct Funct Gen* 8:103–117

Shon K, Kim Y, Colnago LA, Opella SJ (1991): NMR studies of the structure and dynamics of membrane-bound bacteriophage pf1 coat protein. *Science* 252:1303–1305

Stouch TR (1993): Lipid membrane structure and dynamics studied by all-atom molecular dynamics simulations of hydrated phospholipid bilayers. *Mol Sim* 10:335–362

Teng Q, Cross TA (1989): The in situ determination of the 15N chemical-shift tensor orientation in a polypeptide. *J Mag Res* 85:439–447

Tobias DJ, Klein ML, Opella SJ (1993): Molecular dynamics of pf1 coat protein. *Biophys J* 64:670–675

Turner RJ, Weiner JH (1993): Evaluation of transmembrane helix prediction methods using the recently defined nmr structures of the coat proteins from bacteriophages m13 and pf1. *Biochim Biophys Acta* 1202:161–168

Venable RM, Zhang Y, Hardy BJ, Pastor RW (1993): Molecular dynamics simulations of a lipid bilayer and of hexadecane: An investigation of membrane fluidity. *Science* 262:223–6

Weiss MS, Schulz GE (1992): Structure of porin refined at 1.8 Å resolution. *J Mol Biol* 227:493–509

Wesson L, Eisenberg D (1992): Atomic solvation parameters applied to molecular dynamics of proteins in solution. *Protein Sci* 1:227–35

White J (1992): Membrane fusion. *Nature* 258:917–924

Woolf TB, Roux B (1996): Structure, energetics and dynamics of lipid-protein interactions: A molecular dynamics study of the gramicidin A channel in DMPC bilayer. *Prot Struct Funct Gen* 24:92–114

Woolf TB, Roux B (1994a): The conformational flexibility of *o*-phosphorylcholine and *o*-phosphorylethanolamine: a molecular dynamics study of solvation effects. *J Am Chem Soc* 116:5916–5926

Woolf TB, Roux B (1994b): Molecular dynamics simulation of the gramicidin channel in a phospholipid bilayer. *Proc Natl Acad Sci USA* 91:11631–11635

Woolf TB, Roux B (1993): Molecular dynamics simulations of proteins in lipid membranes: the first steps. *Biophys J* A354

Woolf TB, Desharnais J, Roux B (1994): Structure and dynamics of the side chains of gramicidin in a DMPC bilayer. In: *NATO ASI Series: Computational Approaches in Supramolecular Chemistry,* Vol. 426, Wipff G, ed. Dordrecht, The Netherlands: L Kluwer Academic Publishers

Woolf TB, Malkin VG, Malkina OL, Salahub DR, Roux B (1995): The backbone ^{15}N chemical shift tensor of the gramicidin channel: A molecular dynamics and density functional study. *Chem Phys Lett* 239:186–194

Xing J, Scott HL (1992): Monte Carlo studies of a model for lipid-gramicidin A bilayers. *Biochim Biophys Acta* 1106:227–232

Xing J, Scott HL (1989): Monte Carlo studies of lipid chains and gramicidin A in a model membrane. *Biochem Biophys Res Commun* 165:1–6

Zaccai G, Buldt G, Seelig A, Seelig J (1979): Neutron diffraction studies on phosphatidylcholine model membranes. II. Chain conformation and segmental disorder. *J Mol Biol* 134:693–706

Index

Ab initio, chapter 2, 34–35, 107, 326
acetic acid, 53
adamantane, 259, 271
alamethicin, 281, 283, 286, 288, 289, 292, 450
algorithm, 311, 325, 507, 564–567
  for constant pressure MD, 5–7
  for constant temperature MD, 7
allosteric transition, 393, 395
amphipathic helix, 559, 568, 576, 578
  peptides, 281, chapter 9
anesthetics, 474
annexins, 382–389
antibiotic peptides, 281
antimicrobial peptides, 281–282
aromatic residues, 386–387, 580

bacteriorhodopsin, 233–235
Band 3 protein, 442
bending rigidity, 176, 294
benzene, 261, 264–265, 267, 269
binding enthalpy, 421
binding isotherms, 407, 409, 415
blood coagulation, 389
boundary conditions NPT, 109
  periodic, 107–109, 112
  stochastics, 108
Brownian MD, 17, 26

calcium, 356, 382, 385, 394–395
calcium channel blocker, 268
calcium-myristoyl switch, 393
chain melting, 418, 443, 420–421

channels, 289
channels; *see* Gramicidin A
chemical shift anisotropy, 233, 300, 304, 308, 562, 568,
cholesterol (interaction with lipids), 92–94
cholesterol, 100, 93–94, 207–209, 227, 468, 468, 477
choline, 44–48
circular dichroism (CD), 282, 284–286
clotting factors, 389
compressibility, 294, 451–454
compressional wave propagation, 455–456
confidence interval, 11–12, 25
conformational chain disorder, 140, 159–171, 330, 334, 339, 572–575
conformational change, 371–377, 385, 393, 395
constant pressure MD, 5–7
constant temperature MD, 7
cross-sectional area (of Pf1), 563
cross-sectional area, 205–207, 445, 478
crystal calculations, 36, 60–67
crystallographic R-factor, 136
crystallography, chapter 12
curvature of membrane, 550
cutoff, 118–119, 507, 521
cytochrome c, 414, 416, 442, 446–447, 488

Debye length, 434
Debye-Hückel, 412
Debye-Waller factor, 133

deformations of bilayer membranes, 225
density profile, 134–139, 263, 331–332,
     341–343, 571–572, 574
detailed balance, 86
deuterium ^{2}H NMR, 180
deuterium order parameter, 91, 94,
     140–141, 186, 191, 202–207, 256,
     272, 335, 340, 568, 573–575
deuterium quadrupolar splitting, 186
diffraction neutron, 129–132, 135
diffraction x-ray, 129–132, 135
diffusion (lateral diffusion of lipids and
     proteins in membranes), 479–480
   in a potential, 23–25
   of small solutes in membranes, chapter
     8, 256
   rotational, 177, 216–217
   translational, 18–21, 178
dipolar coupling, 181
dipolar splitting ^{1}H–^{15}N, ^{13}C–^{15}N, 300,
     304, 308, 319
displacement correlation function, 21
DLPC, 220, 288, 289
DLPE, 60–67
DMPC, 60–67, 89, 111, 140, 221, 227,
     228, 261, 301, 328
DMPC-DSPC mixture, 468
DMPS, 430, 442
domains, 421, 437, 426
   growth, 151
   of phospholipids, 148–159
   size of, 154, 150
DOPC, 129–130, 137–140, 417
DOPE, 337
DOPG, 415, 417
DPhPC, 286, 290, 292
DPPC, 68, 164, 168, 171, 223, 229, 230,
     442, 566
DPPC-cholesterol, 468
DPPC-Pf1 coat protein, 572
drug, 474, 477
drug design, 255

EFG domains, 390
electrophoresis across membranes, 551
electrostatic, 579
electrostatic partial atomic charges, 33–34,
     40–43, 80–81, 106–107, 328–329
electrostatic (truncation of), 108

equation of state, 408
ethanolamine, 32, 48–50
ethyl acetate, 54, 57–60
ethyl trimethylamonium (ETMA), 43–44
Ewald sums, 118–119

fd virus coat protein, 546–548
first order transition, 432
fluid state, 418, 440
force field, 8, 12, 106–107, 259
   parallh22, 506
   PARAM22 parameters, chapter 2,
     75–81, 310, 558
   parameters Lennard-Jones, 33–34,
     40–43, 54–56, 79
   parameters partial atomic charges,
     33–34, 40–43, 80–81, 106–107,
     328–329
   water model SPC/E, 107
   water model TIP3, 558
Fourier transform Infra Red (FTIR),
     chapter 6
free energy, 411–414, 436
free energy (of membrane deformation),
     294–296
free energy potential, 294–296, 542–544
free volume, 260, 262, 266, 272
fusion inhibiting peptides, 337
fusion peptide, chapter 11

G-proteins coupled receptor, 483
gauche conformers, 166–171, 204
gel state, 418, 440
Gla domain, 389–390
glycophorin, 537
Gouy-Chapman theory, 412
gramicidin A channel, chapter 10, 69,
     171, 327,
   sequence of, 300
   structure of, chapter 10, 300

head group dynamics, 346
   interactions with side chain, 579
heat capacity, 423, 425, 430, 439, 441, 452
helical wheel, 283, 515, 517, 559
helix airpin hypothesis, 541
helix propensity, 510–514, 516
hexagonal phases, 469
hexagonal phases propensity, 491

homology modeling, 513–514
hydrocarbon chains, 90
hydrogen bonding, 270, 305, 317, 579, 580, 582
hydrophobic interactions, 544
hydrophobic mismatch, 475, 490
hydrophobic moments, 515

immiscibility, 475
Infra Red (IR), chapter 6
initial conditions, 4, 7–8, 564–567
insertion (of peptide), 536, 549
integral protein, 435, 437, 490
interactions hydrophobic, 544
  lipid-cholesterol, 92–94
  lipid-protein, 92–94
  of head group with side chains, 579
  peptide-membrane, 141, 295
interfaces, 113, 139, 272, 560
interfacial binding, chapter 12, 381, 389
ion pairs, 579
irreducible tensor representation, 182, 239–245
isomerization, 259–260
isomerization rates, 9–12
isothermal compressibility, 478

jumps, 258, 264–266

kinks, 93, 170, 204, 476

Langmuir isotherm, 407, 408
lateral pressure, 478
lattice models, 95–101, 436, 422–429, 536
Lennard-Jones parameters, 33–34, 40–43, 54–56, 79
light harvesting complex, chapter 15
lipases, 377–381
lipid phases, 7
  gel, 153, 163–164
  hexagonal, 469
  hexagonal phase propensity, 491
  inverted hexagonal H_{II}, 167–171
  liquid-crystal L_α, 163–164, 167–171, 198–208, 296
  microphase separation, 148–159
  ripple, 96–101
  transition, 90, 96, 164
liquid crystallography, 128, chapter 5

LPPC, 60–67
lysis, 281–282

M2δ, 327, 545, 546
magainin 2, 281, 327, 544
Markov chain, 86
mean field free energy, 294–296
mean field potential, 16, 95–101, 537–539, 542–544, 550, 560
melittin, 281, 283, 327
membrane curvature, 550
membrane dipole potential, 112–117, 273–274
methyl acetate, 50–54
methyl propionate, 54, 57–60
micelle SDS, 68
microphase separation, 148–159
molecular frame, 181, 188
molecular replacement, 504
molecular vibrations (wagging, scissors, and rocking modes), 146–147, 148–155, 157, 159
Monte Carlo, 35, chapter 3, 118–119, 436, 422–429, chapter 16, 536
  configurational bias Monte Carlo (CBMC), 101–102
  dynamics, 540, 548
  force biased, 101–102
  importance sampling, 86
  kinetic, 87–88
  Metropolis, 86, 93
  smart methods, 101–102
myelin basic protein, 430

n-Alkane mixtures, 151–153
n-Heptane-4-d_2, 161
Na-K-ATPase, 491
neutron diffraction and scattering, 129–132, 135, 482
nifedipine, 257, 268, 269
NMR, chapter 7
  ^{13}C NMR, 222, 229
  chemical shift anisotropy, 233, 300, 304, 308
  deuterium order parameter, 140–141, 186, 191, 202–207, 256, 272
  deuterium quadrupolar splitting, 186, 300, 304, 308
  dipolar coupling, 181

NMR, (*continued*)
  dipolar splitting $^1H-^{15}N$, $^{13}C-^{15}N$, 300, 304, 308, 319
  ^2H NMR, 180
  irreducible tensor representation, 182, 239–245
  lineshape, 196, 200
  ^{31}P NMR, 209
  relaxation, 209–224
  Solid State NMR, chapter 10
nonbonded interaction parameters, *see* force field

order parameters, 91, 94, 140–141, 186, 191, 202–207, 256, 272, 335, 340, *see also* NMR deuterium order parameters
oriented circular dichroism (OCD), 284–286
oriented samples, 284–294, 301

partition coefficient, 256
penalty function, 307–308
peptide insertion, 296, 536, 549
peripheral membrane protein, chapter 12, 395, 406, 410, 435, 446, 484
permeability, 451, 454, 479–480
permeation, 257
Pf1 virus coat protein, chapter 17, 546–547
phase; *see* lipid phase
phase diagram, 286
phase equilibria, 466
phase transition, 90, 96, 164, 467
phospholipase A2, 448, 484–488
phospholipases, 356–371
phospholipids, 32, 60–67
  DLPC, 220, 288, 289
  DLPE, 60–67
  DMPC, 60–67, 89, 111, 140, 227, 221, 228, 261, 301, 328
  DMPS, 430, 442
  DOPC, 129–130, 137–140, 417
  DOPE, 337
  DOPG, 415, 417
  DphPC, 286, 290, 292
  DPPC, 68, 168, 164, 171, 223, 229, 230, 442, 566
  LPPC, 60–67
  POPC, 68
  POPE, 169–170

phosphosynthetic reaction center, 504
Poisson process, 87
POPC, 68
POPE, 169–170
pores, 289
pressure, 110
program CHARMM, chapter 2, 8, 33–34, 310, 316, 506, 558
  AMBER, 328–329
  BLAST, 512
  BOSS, 35
  CHARMM PARAM22, 310, 558
  DSSP, 512
  Gaussian, 34–35
  MOLVIB, 34
  SOPM, 510–511, 515
  TORC, 310
  XPLOR, 506
protein folding, 561, chapter 15
protrusion, 489

quadrupolar coupling, 181
quadrupolar deuterium splitting, 300, 304, 308

radial distribution function, 333, 345
recoverin, 392
reorientational correlation function, 13–18
Rhodopsin, 229–231
Rhodospirillum molischianum, chapter 15
ripple phases, 96–101
rotational isomerization, 90–92
rotational relaxation, 13–18

SDS, 68
secondary structure prediction, 514
sequence analysis, 508–509
simulated annealing, 305–307, 520–521
snorkel model, 582
solid state NMR, 562; *see also* nuclear magnetic resonance, NMR
solvation of head group, 566
  of side chains, 576
solvent radial distribution function, 577, 581
specific heat, 477
spectral density, 214, 346
spectroscopy CD; *see* circular dichroism, CD

FTIR; *see* Fourier Transform infra red, FTIR
IR; *see* infra red, IR
NMR; *see* nuclear magnetic resonance, NMR
OCD; *see* oriented circular dichroism, OCD
spin Hamiltonian, 180
statistical error, 11–12, 25
stochastic processes jump model, 23
   Poisson process, 11–13, 23–25, 87
structural constraints, 300
   determination, 132, 302
   fluctuations, 336, 570
   refinement, 134, chapter 10
structure prediction, 508, 514, 521, 561
supramolecular assembly, 492, 573
surface pressure, 407
surface tension, 5, 110–112
swelling, 481

tetramethylamonium (TMA), 37–43
thermal disorder, 133
Thermodynamic ensembles;
   canonical, 85
   NγT, 110
   NpγH, 6
   NPAH, 6

NVE, 109
NVT, 109
thermodynamics of surfaces, 5
thickness, 294
transfer free energy, 544
transition temperature, 482
transmembrane helix, 508, 510, 569
transmembrane protein domains (TMD), 483
tryptophan, 386–387
tyrosin residues, 580

undulation (wave propagation), 455–456
undulations, 224, 481

vibrational frequencies, 38, 52, 53, 146–147
virial expression, 110

water model SPC/E, 107
   TIP3P, 558
water orientation (near interface), 273
wave propagation, 455–456
Wigner matrix, 182, 186, 187, 214
wobbling in a cone, 13–18

x-ray diffraction, 129–132, 135

Related Birkhäuser Titles

The Protein Folding Problem and Tertiary Structure Prediction
K.M. Merz, Jr. and S.M. Le Grand, Editors
ISBN 0-8176-3693-5
1994, 582 pp.

Biochemistry of Cell Membranes: A Compendium of Selected Topics
S. Papa and J.M. Tager, Editors
ISBN 3-7643-5056-3
1995, 376 pp. (Molecular and Cell Biology Updates series)

Progress in Membrane Biotechnology
J.C. Von Gomez-Fernandez, D. Chapman, and L. Packer, Editors
ISBN 3-7643-2666-2
1991, 340 pp. (Advances in Life Sciences series)